Engineering Professionalism
and Ethics

Engineering Professionalism and Ethics

Edited by

JAMES H. SCHAUB, Ph.D., P.E.

University of Florida

and

KARL PAVLOVIC, Ph.D.

University of Florida

With

M. D. MORRIS, P.E.

ROBERT E. KRIEGER PUBLISHING COMPANY
MALABAR, FLORIDA

Original Edition 1983
Repring Edition 1986

Printed and Published by
ROBERT E. KRIEGER PUBLISHING COMPANY, INC.
KRIEGER DRIVE
MALABAR, FLORIDA 32950

Copyright © 1983 by John Wiley & Sons, Inc.
Reprint by Arrangement

All rights reserved. No part of this book may be reproduced in any form or by any electronic or mechanical means including information storage and retrieval systems without permission in writing from the publisher.
No liability is assumed with respect to the use of the information contained herein.

Printed in the United States of America

Library of Congress Cataloging-in-Publication Data

Engineering professionalism and ethics.

    Reprint. Originally published: New York : Wiley, 1983.
    Includes index.
    1. Engineers.   2. Engineering.   3. Engineering Ethics.
I. Schaub, James Hamilton, 1925-   II. Pavlovic, Karl.
III. Morris, M. D. (Morton Dan)
TA157.E623 1986   174'.962   86-2956
ISBN 0-89874-954-9

10 9 8 7 6 5 4 3

Preface

Future historians may choose to call the twentieth century, at least in the West, the "Century of the Professions." This will reflect the effect upon Western society of the culmination of a slowly developing experiment in the direct and indirect organization of life, work, and social relations through professional affiliation. If so, the recorders of the latter half of the century undoubtedly will describe it as a time of debate over the significance of the results of the experiment. Although the appearance of professions and of a society largely organized around professions may be a historically new phenomenon, neither experiments in social organization nor the debates that inevitably ensue are new. Both the experiments and the debates are expressions of a very old, perennial, and apparently fundamental Western conflict over the nature and extent of constraints upon the behavior of individuals and groups. This conflict has existed and grown throughout our history, not only between different individuals but, increasingly, within the individual.

The engineering profession will be seen as particularly important within the historical context because, at the same time that the experiment in the professions has been developing, Western society has been engaged in another experiment. This is the creation of a society that finds its material and, to a large extent, its spiritual base in technology. The outcome of the debate in the engineering profession, thus, will have a double significance. It concerns not only the shape of our society but also the continuation of Western society as it currently exists.

Every book represents an evaluative position—a bias that may or may not be shared by its readers—this book no less for its being an anthology of the observations and positions of people who often agree upon little among themselves. Our position is that it is better to know the range of differing opinions than not; better to know what is happening around you than not; better to know what you are doing than not. One thing that is occurring and that affects every member of this society is the debate and conflict over the social impact of the professionalization of engineering and the associated ethical issues. Every member of society contributes to that debate and conflict. We present this anthology, therefore, as representative of the current state of the debate and conflict and as an introduction to these questions for engineers and would-be engineers. It is presented also for members of the public in general whose interest may not be as great as we believe it should be.

Other components of our position will help to explain the form and substance of the book. We believe the debate should be represented so far as possible as it has been and continues. Some of the writings may be judged by professional scholars as lacking the requisite amount of sophistication. To the extent that this judgment is not itself simply a professional prejudice, we can only respond that so too, then does the debate. The parties to the debate have their own, partially self-interested positions. The articles that appear here demonstrate this. Since the debate is over interpretation and appropriation of the past to our present and future situation, we see it as essentially historical. The current issues of professionalism and ethics in engineering are what they are because of the history of technology and developments in engineering, in general, and of the growth and development of the concepts of professionalization within engineering in particular.

We wish to point out to all of our readers that the use of the impersonal pronoun "he" by the editors and the authors whose works have been included in this anthology is intended to refer to both sexes. This explanation is necessary to acknowledge the contribution and participation of both men and women in the engineering profession without introducing changes in the contributors' text.

We would like to acknowledge a great debt of gratitude to the "Humanities Perspectives on the Professions" program offered through the Center for Studies in the Humanities at the University of Florida with the sponsorship and partial funding of the National Endowment for the Humanities. This program made it possible for us to engage in the activities that led ultimately to this book. In addition, we received both inspiration and encouragement from our participation in a variety of seminars, short courses, and workshops sponsored by several professional societies, the National Science Foundation and the National Endowment for the Humanities, and from the pioneering academic work at Rensselaer Polytechnic Institute and the Illinois Institute of Technology.

This anthology could not have been compiled without the cooperation, interest, and support of the numerous authors whose work we have drawn upon and the many periodicals and presses that have generously permitted us to reprint from their publications. Their contributions are gratefully recognized.

We are particularly grateful to Professor John Fielder of Villanova who read our original manuscript with care and perspicuity. His observations and criticism resulted in a much improved manuscript.

Special thanks are due Peggy Schaub, Kristina Pavlovic, Wendy Hofer, Sarah Byers, and Joanne Stevens for their secretarial support and patience in the preparation of the manuscript.

Gainesville
November 1982

JAMES H. SCHAUB
KARL PAVLOVIC

__________Acknowledgments__________

The authors gratefully acknowledge the permissions of the publishers and/or authors to reprint the following material:

CHAPTER 1   THE GROWTH AND DEVELOPMENT OF THE PROFESSION

E. RUBINSTEIN, "IEEE and the Founder Societies," © 1976 by the Institute of Electrical and Electronics Engineers, Inc. Reprinted with permission from IEEE SPECTRUM, Vol. 13, No. 5, May 1976, Pages 76–84.

E. RUBINSTEIN, "IEEE and the 'Founders'—II," © 1976 by the Institute of Electrical and Electronics Engineers, Inc. Reprinted with permission from IEEE SPECTRUM, Vol. 13, No. 6, June 1976, pages 67–72.

W. H. WISELY, "The Influence of Engineering Societies on Professionalism and Ethics." Reprinted with permission from *Ethics, Professionalism and Maintaining Competence*, A.S.C.E. Specialty Conference, March 1977, Pages 51–62. © 1977— American Society of Civil Engineers.

E. T. LAYTON, JR., "The Engineer and Business." Reprinted from *The Revolt of the Engineers* with permission of the author and the former Press of Case Western Reserve University. Copyright © 1971 by The Press of Case Western Reserve University.

CHAPTER 2   ENGINEERS AND THE PUBLIC

J. BARZUN, "The Professions Under Siege." Copyright © 1978 by *Harper's Magazine*. All rights reserved. Reprinted from the October 1978 issue by special permission.

S. C. FLORMAN, "Moral Blueprints." Copyright © 1978 by *Harper's Magazine*. All rights reserved. Reprinted from the October 1978 issue by special permission.

F. COLLINS, "The Special Responsibility of Engineers." Reprinted, with permission, from Volume 196, Article 10 of the *Annals of the New York Academy of Sciences*, Pages 448–450. Copyright 1973, by The New York Academy of Sciences. All rights reserved.

T. L. TURNICK, "Public versus Client Interests—An Ethical Dilemma for the Engineer." Reprinted with permission from *Engineering Issues*, January 1975, Pages 61–75. © 1975—American Society of Civil Engineers.

R. L. Schott, "The Professions and Government: Engineering as a Case in Point." Reprinted with permission from *Public Administration Review*, March/April 1978. © 1978 by The American Society for Public Administration, 1225 Connecticut Avenue, N.W., Washington, D.C. All rights reserved.

M. Tribus, "The Engineer and Public Policy-Making." © 1978 by the Institute of Electrical and Electronics Engineers, Inc. Reprinted, with permission of the author and publisher, from IEEE Spectrum, Vol. 15, No. 4, April 1978, Pages 48–51.

M. Manheim, "Values and Professional Practice." Reprinted from *Values and the Public Works Professional* (© Curators of the University of Missouri), the proceedings of a workshop conducted by Drs. D. L. Babcock and C. A. Smith for the American Public Works Association, with funding from the National Science Foundation and the National Endowment for the Humanities.

"The Consultants' Competitive Negotiation Act," State of Florida and "Suggested Key Steps in Administering the Consultants' Competitive Negotiation Act," reprinted with permission of Florida Institute of Consulting Engineers from *Directory of Consulting Engineers*, 1978–79 edition.

CHAPTER 3   OBLIGATIONS TO EMPLOYERS AND PEERS

J. E. Ullmann, "The Responsibility of Engineers to Their Employers." Reprinted, with permission, from Volume 196, Article 10 of the *Annals of the New York Academy of Sciences*, pages 417–428. Copyright 1973, by The New York Academy of Sciences. All rights reserved.

B. F. Gordon and I. C. Ross, "Professionals and the Corporation." Reprinted with permission from *Research Management*, Vol. V, No. 6, 1962, pages 493–505.

D. Christiansen, "Conflict of Interest." © 1975 by the Institute of Electrical and Electronics Engineers, Inc. Reprinted, with permission, from IEEE Spectrum, Vol. 12, No. 6, June 1975, Page 33.

"Conflicts of Interest Pose No Big Problem." Reprinted, with permission of the copyright owner, The American Chemical Society, from *Chemical and Engineering News*, October 1961.

W. J. Haga, "Can an Engineer join a Union and Still Be Professional?" Reprinted with permission from *Astronautics and Aeronautics*, December 1975, Pages 31–42.

CHAPTER 4   THE ETHICAL DILEMMA

J. B. Stockdale, "The World of Epictetus." Copyright © 1978 by The Atlantic Monthly Company, Boston, Mass. Reprinted with permission from *The Atlantic*, April 1978.

H. Jonas, "Technology and Responsibility: Reflections on the New Tasks of Ethics." Reprinted by permission of the author and The New School for Social Research from *Social Research*, Vol. 40, No. 1, Spring 1973, Pages 31–54.

and *Science, Technology, & Human Values,* copyright © 1974 by The President and Fellows of Harvard College.

Anonymous, "I Gave Up Ethics—to Eat." Reprinted from *Consulting Engineer,* ® 1975, Volume 45, Number 6, Pages 39–41. © Copyright by Technical Publishing, A Division of Dun-Donnelley Publishing Corp., A Company of the Dun & Bradstreet Corporation.

B. J. Lewis, "The Story Behind the Recent National Scandals Involving Engineers." Reprinted with permission from *Engineering Issues,* April 1977, Pages 91–98. © 1977—American Society of Civil Engineers.

C. Barnett, "Three Leave GE . . . " Reprinted with permission of the author from *New Engineer,* May 1976, Pages 34, 41–42.

H. Popper and R. V. Hughson, "How Would *You* Apply Engineering Ethics to Environmental Problems?" Reprinted by special permission from *Chemical Engineering,* November 2, 1970. Copyright © 1970, by McGraw-Hill, Inc., New York, N.Y. 10020.

H. Popper and R. V. Hughson, "Environmental-Ethics Panel Offers Views and Guidelines." Reprinted by special permission from *Chemical Engineering,* March 8, 1971. Copyright © 1971, by McGraw-Hill, Inc., New York, N.Y. 10020.

CHAPTER 5   PROFESSIONALISM: IN WHOSE INTEREST?

R. Nader, "The Engineer's Professional Role: Universities, Corporations, and Professional Societies." Reprinted with permission of the American Society for Engineering Education from *Engineering Education,* February 1967, Pages 450–454.

M. Goland, "Can Professionalism Be Attained within the Corporate Structure?" Reprinted with permission of the author and *Journal,* Florida Engineering Society, March 1974.

R. Baum, "The Limits of Professional Responsibility." Reprinted from *Values and the Public Works Professional* (© Curators of the University of Missouri), the proceedings of a workshop conducted by Drs. D. L. Babcock and C. A. Smith for the American Public Works Association, with funding from the National Science Foundation and the National Endowment for the Humanities.

R. L. Whitelaw, "The Professional Status of the American Engineer: A Bill of Rights." Reprinted with permission of *Professional Engineer* from August 1974.

M. D. Bayles, "Against Professional Autonomy." Reprinted with permission of the author and the *National Forum* from *National Forum* Summer 1978, Pages 23–26. Copyright © Honor Society of Phi Kappa Phi 1978.

M. F. Lunch, "Do the Professions Have a Future?" Reprinted with permission of the author and the Michigan Association of the Professions from *Legislative Monitor,* April 27, 1979 and May 11, 1979.

M. F. Lunch, "Addendum to 'Do the Professions Have a Future?' By permission of the author.

E. M. YODER, "PE Guest Comment." Reprinted with permission of *The Washington Star* and *Professional Engineer* from *Professional Engineer,* June 1979.

M. L. COGAN, "The Problem of Defining a Profession." Reprinted from Volume No. 297 of *The Annals of The American Academy of Political and Social Science.* Copyright 1955, by the American Academy of Political and Social Sciences. All rights reserved.

R. B. REICH, "Professional Licensing and Public Policy." Reprinted with permission of the author from a paper given to the Regents Conference on the Professions, The University of the State of New York, November 1979.

C. NELSON AND S. PETERSON, "Ethical Decisions for Engineers: Systematic Avoidance and the Need for Confrontation." Reprinted with permission from *Civil Engineering Education,* Volume 1, Pages 628–634. © 1979—American Society of Civil Engineers.

CHAPTER 6   WHISTLE BLOWING

R. NADER, "An Anatomy of Whistle Blowing." From *Whistle Blowing,* edited by Ralph Nader, Peter Petkas, and Kate Blackwell. Copyright © 1972 by Ralph Nader. Reprinted with permission of Viking Penguin, Inc.

D. CHRISTIANSEN, "Fact vs. Feeling." © 1976 by the Institute of Electrical and Electronics Engineers, Inc. Reprinted, with permission, from IEEE Spectrum, Vol. 13, No. 3, March 1976, Page 35.

K. VANDIVIER, "Engineers, Ethics and Economics." Reprinted with permission of American Society of Civil Engineers from *Proceedings,* Conference on Engineering Ethics, May 1975, Pages 20–24.

R. M. ANDERSON, JR., *et al.,* "The Bay Area Rapid Transit (BART) Incident." Reprinted with permission of Bentley College and University Press of America from *Proceedings* of the Second National Conference on Business Ethics, 1979. Copyright © by Bentley College, Waltham, Mass., 1979.

"Engineering Ethics: The Amicus Curiae Brief of the Institute of Electrical and Electronics Engineers in the BART Case." Reprinted with permission of The Institute of Electrical and Electronics Engineers, Inc., from *IEEE Technology and Society* (formerly IEEE CSIT Newsletter) from Issue No. 12 December 1975.

F. SAWYIER, "The Case of the DC-10 and Discussion." Reprinted by permission of the author from a study developed for the Center for the Study of Ethics in the Professions, Illinois Institute of Technology, December 1976.

V. WEIL, "The Browns Ferry Case." Reprinted by permission of the author from a study developed for the Center for the Study of Ethics in the Professions, Illinois Institute of Technology, January 1977.

S. BOK, "Whistleblowing & Professional Responsibilities." Reprinted with permission from *Professional Engineer,* June 1979.

CHAPTER 7   CODES OF ETHICS AND ENFORCEMENT

D. MEAD, "Why a Code of Conduct?" Reprinted with permission from *Standards of Professional Relations and Conduct,* Manual of Engineering Practice, No. 21,

Pages 10–13, 29–30, published by the American Society of Civil Engineers, 1941.

"ECPD: CODE OF ETHICS OF ENGINEERS." Reprinted with permission of the Engineers' Council for Professional Development, Inc., as approved by the Board of Directors, October 1974.

"NSPE: CODE OF ETHICS FOR ENGINEERS." Reprinted with permission of the National Society of Professional Engineers from NSPE Publication No. 1102, revised July 22, 1978.

"IEEE: CODE OF ETHICS FOR ENGINEERS." © 1975 by the Institute of Electrical and Electronics Engineers, Inc. Reprinted, with permission, from IEEE Spectrum, February 1975, Vol. 12, No. 2, Page 65.

"American Public Works Association: Code of Ethics." Reprinted with permission from the *Directory*, American Public Works Association.

A. G. OLDENQUIST AND E. E. SLOWTER, "Proposed: A Single Code of Ethics for All Engineers." Reprinted with permission from *Professional Engineer*, May 1979.

A. C. LAUCHLAN, "When in Rome Do as the Romans." Reprinted with permission from *Civil Engineering*, January 1979, Page 93. 1979—American Society of Civil Engineers.

J. McMINN, "Ethics Spun from Fairy Tales." Reprinted with permission from *Engineering Issues*, April 1977, Pages 165–181. © 1977—American Society of Civil Engineers.

J. A. ROMANO, "The Fable of the Consulting Engineer." Reprinted from *Consulting Engineer*, ® 1979, Volume 53, Number 5, pages 127–132, © Copyright by Technical Publishing, A Division of Dun-Donnelley Publishing Corp., A Company of the Dun & Bradstreet Corporation.

D. H. PLETTA, "Ethical Standards for the Engineering Profession: Where Is the Clout?" Reprinted with permission from *Professional Engineer*, July 1975, Pages 40–42.

F. BAGLEY, "Ethics, Unethical Engineers and ASME." Reprinted with permission from *Mechanical Engineering*, July 1977, Pages 42–45. Copyright © 1977 by the American Society of Mechanical Engineers.

"Professional Responsibility and the Dispatching of Police Cars—A Case Study" and "MCC Report in the Matter of Virginia Edgerton." Reprinted with permission of the Institute of Electrical and Electronics Engineers, Inc., from *IEEE Technology and Society*, No. 22, 1979.

CHAPTER 8   PROFESSIONAL REGISTRATION AND MAINTENANCE OF COMPETENCE

M. FINE, "Registration Viewed as Bond to Practice of Learned Professions." Reprinted with permission from *Professional Engineer*, December 1979.

M. F. LUNCH, "Legislators Play Dealers' Choice with P.E. Registration Laws." Reprinted from *Consulting Engineer*, ® 1975, Volume 45, Number 3, Pages 45–48. © Copyright by Technical Publishing, A Division of Dun-Donnelley Publishing Corp., A Company of the Dun & Bradstreet Corporation.

M. J. KOLHOFF, "For the Industry Exemption . . ." Reprinted with permission from *Professional Engineer*, March 1976.

G. J. KETTLER, "Against the Industry Exemption . . ." Reprinted with permission from *Professional Engineer*, March, 1976.

J. D. CONSTANCE, "Getting Your P.E. Why? How?" Reprinted from *New Engineer*, October 1976, with permission of the author.

A. C. INGERSOLL, "Qualifications for Continued Practice." Reprinted with permission from *Ethics, Professionalism and Maintaining Competence*, A.S.C.E. Specialty Conference, March 1977, Pages 235–247. © 1977—American Society of Civil Engineers.

"Continued Professional Development in Florida: PE's Set Up Criteria for Certificate." Reprinted with permission from *Professional Engineer*, December 1978.

"Continuing Professional Competence—Summary of a Position Paper by the National Council of Engineering Examiners," August 1979. Reprinted with permission of National Council of Engineering Examiners.

Contents

5 PROFESSIONALISM: IN WHOSE INTEREST?   **273**

6 WHISTLE BLOWING   **339**

7 CODES OF ETHICS AND ENFORCEMENT   **417**

8   PROFESSIONAL REGISTRATION AND MAINTENANCE OF COMPETENCE   507

The Growth and Development of the Profession

I hold each man a debtor to his profession; from the which as men of course do to seek to receive countenance and profit, so ought they of duty to endeavor themselves, by way of amends, to be a help and ornament thereunto.

Francis Bacon (1561–1626)
Maxims of the Law

The architects [engineers] who have aimed at acquiring manual skill without scholarship have never been able to reach a position of authority to correspond to their pains, while those who relied only upon theories and scholarship were obviously hunting the shadow, not the substance.

Vitruvius, Book I
Ten Books on Architecture

Engineering as the skilled working of materials into structures and machines has been practiced for thousands of years. These skills were used to satisfy human propensities ranging from improved material existence through the expression of spiritual aspiration to the efficient elimination of rival men and societies. As a skill it consisted mainly of accumulated trial and error knowledge applied by artisans. Although the invention of geometry as a systematic body of theoretical knowledge was early taken up and employed by this skill knowledge, it was not until the scientific revolution in the eighteenth and nineteenth centuries that the engineer in the modern sense emerged. The engineer as the professional, scientifically trained applier of theoretical knowledge of the structure and dynamics of materials to the needs of individuals and societies is a thoroughly modern phenomenon. The engineer is seen today as the mediator between science and society.

The industrial revolution that followed the scientific revolution worked far-reaching changes upon the shape and internal structure of Western society. The capital accumulations of the commercial nations of northern and southern Europe that created an urban industrial society waged evermore efficient economic and military warfare upon each other and elevated the material standard of living to incredible heights for relatively large segments of Western society. It would have been extremely unusual if the engineers, who were in a sense the hands through which this was accomplished, did not come to view themselves as a distinctly useful group deserving a privileged status in society and possessing a unique responsibility toward society. With this increasing importance of engineering in the mideighteenth century, the tendency to form organizations of people with common professional and technical interests developed. In England, the Smeatonian Society (1771) and later The Institution of Civil Engineers (1818) marked the first beginnings of professional engineering societies.

In the United States, the first half of the nineteenth century was one of expansion and a companion growth of engineering. As early as 1824, the Franklin Institute offered its facilities to engineers, but the earliest formal attempt to organize an engineering society was an unsuccessful one by civil engineers in 1839. However, by 1848, the Boston Society of Civil Engineers was formed and has been a continuing organization to this time. The civil engineer of this era was any engineer other than those in military service.

In 1852, the American Society of Civil Engineers and Architects was organized in New York City and has remained in continuous existence since then, although the architects formed their own society some five years later. The purpose of the organization is well defined by the constitution adopted at its organization:

> The professional improvement of its members, the encouragement of social inter-course among men of practical science, the advancement of engineering and its several branches, and of architecture, and the establishment of a central point of reference and union among its members.
>
> Among the means to be employed for attaining these ends shall be the periodical meetings for the reading of professional papers, and the discussion of scientific subjects, . . .
>
> Civil, geological, mining and mechanical engineers, architects and other persons who, by profession, are interested in the advancement of science shall be eligible as members.

From this beginning, numerous other engineering societies were organized by discipline. By 1908, the five founder societies of ASCE, AIME, ASME, AIEE (predecessor of IEEE), and AIChE were established and housed in close proximity to each other in New York City. Once the practice of engineering acquired a coherent and at least nominally national internal organization, the major outlines of modern engineering were in place. Two out of the three defining characteristics of contemporary engineering, formal schooling in basic and applied science and professional organization as a distinct sector of society, were fixed and then worked out in detail by a series of events only within the last 100 years.

The late nineteenth century and early portion of the twentieth century in America was a period of exuberant growth and enthusiastic hope for engineering as a profession and its potential contributions to the improvement of society. The number of engineers multiplied dramatically and with this the membership in the professional societies. A strong symbiotic relationship with industrial corporations evolved. Technical schools and programs proliferated, a distinctive engineering curriculum evolved, and formal education became the standard route for entry into the profession.

The middle part of this century followed with an era of calm expansion and studied innocence. America became a thoroughly technological society with engineers increasingly occupying crucial positions and exercising crucial functions within the private and public sectors of American society. The majority of engineers practicing their profession today grew up and were trained as engineers during this middle period. For these engineers the struggle for the acceptance of engineering as profession that occurred during the nineteenth century was no more than the dying rumble of a past war. Only a very few of these practitioners foresaw the current problems confronting engineering. Many of these problems are rooted in the profession's view of what it ought to accomplish and how it should be organized and in society's expectations of the profession—views and expectations formed in the nineteenth century and matured during the early part of this century.

A general questioning of science and technology began in the late sixties and continues today. With this questioning the innocence of engineering as a profession has come to an end. Now everything is both debatable and debated—the existing goals, means, and organization of the engineering profession have all been called into question by the public, the government, and by members of the profession itself.

The history of the last 100 years of engineering has contributed more to the present form and scope of the crises, dilemmas, and problems facing engineering today than the millenia that preceded that period. It is precisely because the practice of engineering became organized as a profession that the current debates and conflicts center around the issues of professionalism and that engineering finds itself caught in the general questioning of professionalism that exists today.

The attempt to unify engineers into a coherent group on the basis of professional commonality of technical interests and practice when technical interests and forms of practice are so diverse virtually assured that fragmentation into subgroups would be a persistent problem. The claims to professionalism on the basis of service and responsibility to society that were advanced alongside the natural alliances formed with business virtually guaranteed engineers would periodically be caught in the conflicts between the public and business. Professional societies would periodically find themselves in the dilemma of being called upon to police their members in the interests of the public to the detriment of their employers or vice versa. Because engineering adopted formal education as the route for entry into the profession, its fate is now intertwined with the problems and crises that formal education is undergoing today.

For these reasons the readings in this introductory chapter concentrate on the period of the professionalization of engineering practice and, in particular, the time since the turn of the century. The emphasis here is upon the role of the professional societies, the relation of engineering and business, and the development of engineering education.

The opening two articles in this chapter trace the history of the development of engineering societies in the United States. Written by E. Rubinstein, they provide an excellent summary of the conflicts and problems faced by the early societies in their development. One of the serious problems faced by the growing engineering profession was, and continues to be, the proliferation of organizations with each representing a segment of the profession.

There has never been a single voice to speak for engineers though numerous attempts have been made to organize for this purpose. In January 1980, another effort to establish an "umbrella" society for all engineers resulted in the formation of the American Association of Engineering Societies (AAES). There are indications this organization may succeed. By late 1980, it was a federation representing over fifty professional engineering societies with more than 1 million members. The larger and most prestigious engineering societies have affiliated with AAES. AAES is organized with four councils to provide guidance and unified action in the areas of education, public affairs, international cooperation, and engineering affairs. The latter council is concerned particularly with the practice of engineering in the public interest.

In "The Influence of Engineering Societies on Professionalism and Ethics," William H. Wisely discusses the role of the engineering societies in fostering professionalism in general and their specific roles in the development of codes of ethics within the broader concepts of professionalism. The development of professionalism and its ethics component necessitates some discussion of the proliferation of engineering societies and current problems with acceptance of codes of ethics. Wisely concludes that the engineering ethic might well be stated in one brief sentence: "Service of the public interest with integrity and honor."

E. T. Layton's book *The Revolt of the Engineers* has become an accepted standard for the history of engineering development in this country. The selection drawn from his book for use in this chapter highlights the internal conflict faced by engineers between loyalty to the employer and clients and the autonomy implicit in the concepts of professionalism.

In the final selection in this chapter, James H. Schaub and K. R. Pavlovic describe the development of engineering education from the traditional master–apprentice relation to the current highly organized formal curricula found in colleges and universities. In this respect it should be noted that, historically, formal college-level education in engineering is still in a formative stage. The roots of the changing patterns of engineering education lie not only in the increase in scientific knowledge and new technology available to engineering, but also in the professionalization of engineering and in engineering's heavy involvement with corporate industrialization in the twentieth century. Further, the growing public demand that engineers assume more and more social responsibility for the results of their work complicates the format that engineering education should follow.

1.1

IEEE and the Founder Societies

ELLIS RUBINSTEIN

- The American Society of Civil Engineers (ASCE) publishes an officially sanctioned document (updated annually) entitled "Recommended Salary Ranges." This document provides employers and employees with a guide to fair salary practice for civil engineers based on professional grades and yearly changes in the average salary level of civil engineers entering the job market. Further, some ASCE sections check with local employers, and if it is determined that an employer is paying below recommended levels, members attempt to assist the employer by discussing with him available methods and time frames by which to improve his employees' lot.
- The American Institute of Mining, Metallurgical, and Petroleum Engineers (AIME) spends approximately $100 000 annually on self-generated non-technical programs. This includes the institute's total expenditure in the area of education.
- The American Society of Mechanical Engineers (ASME) allocates more than one third of its entire expense budget—more than $3 million—to a growing codes and standards operation.
- The American Institute of Chemical Engineers (AIChE) spends nearly one of every ten dollars on education activities. Consequently, AIChE is budgeting more than $8 per member on education, whereas its closest competitor in this regard is the ASCE, which spends less than $5 per member on education.

These statements serve two purposes: They introduce the major U.S.-based engineering societies that, along with IEEE, have come to be known as the Founder Societies, and they demonstrate that, although headquartered in the same building, these societies have very different attitudes toward what services their members want. In fact, the five Founder Societies, from their earliest days, have been very different in their approach to member programs, and the purpose of this article is to delve into both their origins and their present activities so that the IEEE member may better evaluate the services the Institute of Electrical and Electronics Engineers provides.

CLUES TO THE PRESENT IN THE PAST

Each of the five Founders has, to some degree, been stamped by conditions existing at its very birth. Common to all of the societies at their inceptions were two founding principles: that practitioners of a common discipline should have available to them a current flow of information on the state of the art, and that they should have a common ground for meeting—be it for business or social purposes. These two objectives of the professional society have remained the only continuously shared characteristics of the Founder Societies over the decades, and they account for the fact that, today, each of the five societies spends by far the largest amount of its funds (see Table I) on technical meetings and technical publications. In fact, it could be said that until these two objectives could be met, no professional engineering society could prove enduring.

The birth of the first enduring U.S. engineering society occurred on November 5, 1852. On that day, 12 men gathered in the office of Alfred W. Craven, chief engineer of the Croton Aqueduct Department in New York City, to found what was to become the ASCE. Precedents for the meeting abounded. Societies of medical practitioners existed; societies of lawyers existed; but in the field of engineering in the U.S., there had been more failures than successes. In Europe, it was different. The very first viable engineering society had been founded in 1818—the British Institution of Civil Engineers. Modeled after this group, civil engineering societies sprang up in Holland in 1843, in Belgium and Germany in 1847, and in France in 1848. In each case, a prime reason for the society's success was that the members were located relatively near headquarters and could attend meetings and correspond easily. National civil engineering societies had failed in the U.S. during the 1830s largely because of the difficulties created by vast distances. Consequently, the first long-lasting society to be formed in the U.S. was a regional society—the Boston Society of Civil Engineers, formed in 1848. Three years later, however, U.S. civil engineers were ready to try again for a national organization.

In 1851, Craven and his 11 New York colleagues drafted a letter that said in part:

> Civil, geological, mining and mechanical engineers, architects and other persons who, by profession, are interested in the advancement of science, shall be eligible as members.
>
> It is anticipated that the union of the three branches of civil and mechanical engineering and architecture will be attended by the happiest results, not with the view to the fusion of the three professions in one, but as in our country, from necessity, a member of one profession is liable at times to be called upon to practice to a greater or lesser extent in the others, and as the line between them cannot be

I. Vital Statistics, as Supplied by Each of the Founder Societies. Only Gross Differences Should Be Noted Because Accounting Practices Vary Somewhat from Society to Society. A Comparison of Recent Dues Histories as Affected by Inflation Can be Found in the March *Spectrum*, p. 65, along with a Discussion of Institute Financing Problems.

| | | | | | Activities (Budgeted Dollars per Member, $) | | | | | | | |
Society	Member-ship	Staff Size	Expense Budget (Millions of Dollars)	Expense Budget (Dollars/ Members)	Re-gional[1]	Tech-nical	Standards	Publica-tions	Educa-tional	Profes-sional	Inter-society	Corpora-tive and Admini-strative[2]
IEEE	180 000	275	15.6[3]	86.67	5.56	37.22[4]	3.33	40.00[4]	2.22	6.72[5]	2.22	16.67
ASME	73 170	198	8.9	121.63	7.12	25.40[4]	43.46	33.55[4]	1.34	1.87	1.22	24.56
ASCE	71 800	115	5.7	79.39	4.88	19.19[4]	0.56	39.44[4]	4.84	5.88	1.82	17.70
AIME	56 152	79	4.2	74.80	1.78	14.25	—	39.18	1.78	—	1.78	16.03
AIChE	36 322	70	3.1	85.35	2.75	17.90[4]	—	49.56[4]	8.26	1.38	1.65	16.52

[1]Figures do not include what local entities spend out of their own funds.
[2]Figures include: accounting, data processing, committee and board expenses, member correspondence and development.
[3]Does not include approximately $4 million in so-called interdepartmental transfers.
[4]Includes technical publication costs—therefore, a portion of these monies have been "double accounted" under both Technical and Publication Activities.
[5]Represents allocation for U.S. members only.

drawn with precision, it behooves each, if possible, to be grounded in the practice of the others; and the bonds of union established by membership in the same Society, seeking the same end, and by the same means, will, it is hoped, do much to quiet the unworthy jealousies which have tended to diminish the usefulness of distinct societies formed heretofore by the several professions for their individual benefit. . . .

Clear from these words is the intention of the 12 New Yorkers to create a mechanism for the exchange of knowledge. It is this impulse that would lead to the implementation of programs with which IEEE members (as well as members of the other Founder Societies) are today familiar: meetings were called from the beginning to discuss technical matters; headquarters offices were rented (1867) to accommodate meetings; a library was begun (1867); the first technical publication, called the *Transactions of the ASCE*, was initiated (1873), followed shortly thereafter (1876) by the *ASCE Proceedings*; standards were issued (1875); and, finally, technical divisions were created (1922). With the exception of standards, these activities are what the five Founders have in common today.

But what of their differences? Obviously, the 12 founders of the ASCE envisioned a single professional society encompassing all engineers. So the first differences appear in the very splintering of that ideal. And, in fact, the causes of this splintering are clues to the very particular natures of each of today's societies.

The AIME was the second national engineering society to be formed. While the founders of the ASCE purported to represent mining engineers as well as civil engineers, in reality ASCE quickly became an extremely elitist organization. It established both high membership qualifications and a difficult mechanism for acceptance of new members. In practice, New Yorkers controlled the society for many years and, in one respect, this made sense: Regional squabbles had destroyed an earlier attempt to form a national society. As a direct consequence, however, the majority of engineers, even in civil engineering, never were represented during the early years of the ASCE, and the stage was therefore set for new societies.

The first of these, the American Institute of Mining Engineers as it was then called, was formed in 1871 at a meeting in Wilkes-Barre, Pa. In reaction to ASCE policy, AIME's membership requirements were extremely low. One didn't have to be a long-practicing engineer to be accepted. The consequence was that, from the beginning, AIME was flooded by members who were businessmen rather than technologists. The opposite was the case in ASCE.

It is this distinction between the formative principles of the two societies that leads us to a broadstroke description of the differences in today's engineering societies. Although all the Founders engage in similar technical activities, they differ drastically in their degrees of involvement in nontechnical arenas, and dictating their degrees of involvement over the years in nontechnical activities was, to a great extent, their relative degree of industry (as opposed to professional) orientation. Eugene Zwoyer, ASCE Executive Director, told *Spectrum*, "We've always felt there's no separation between those things termed professional and those termed technical." By this, he means that the ASCE, to a greater or lesser degree during its 124-year history, has always engaged in all activities seen as in the professional interest of the civil engineer. This has included, over the years, programs designed to benefit the ASCE member financially as well as technically.

As early as 1872, for example, a letter was circulated to the membership offering employment services through the auspices of the society secretary—the part-time, volunteer chief administrative officer of the society. In 1875, a proposal was put before the membership to provide life insurance through "the Civil Engineers' Insurance League." And in 1881, ASCE authorized an appeal to the U.S. Congress to assist nonmilitary civil engineers in the procurement of public works jobs—then being given mainly to the Army's corps of engineers.

This appeal to Congress raises a second category of nontechnical activities—those in the area of public affairs. Early in the history of the ASCE, society members determined that the public image of civil engineers was as important to the individual member's professional life as was his strictly economic concerns. In the 1870s, U.S. bridges were failing at a rate of 25 annually. Dams had failed tragically as well. These events spurred the ASCE to create *ad hoc* committees for the express purpose of reporting on means of averting such accidents. When, in 1875, a third committee report—this time on New York City's rapid transit facilities—was adopted in a furor of controversy, the directors of ASCE decided to appoint a special committee on engineering subjects in the public domain.

Other not-strictly-technical activities that were engendered by the previously mentioned bridge and dam failures included turn-of-the-century efforts in the areas of registration, licensing, and ethics. By 1910, a decade of committee work produced the first "model law" in licensing (passed in New York State). And in 1914, ASCE adopted a code of ethics.

Finally, ASCE was involved in the educational arena almost before any of the other Founder Societies came into existence. In 1874, a Cornell University professor—a member of ASCE—first proposed that the society determine engineering curricula guidelines. Two years later, ASCE and AIME formed a Joint Committee on Technical Education, which met at the Franklin Institute in Philadelphia, Pa., during the Centennial Exhibition. This is the first intersociety activity on record.

From this selection of early ASCE efforts, it should be clear that the society had, almost from its inception, shown a commitment to nontechnical activities. And one major reason for this commitment was that business interests, which over the last century or so have often proved sensitive to nontechnical programs, were excluded from the ASCE. They were excluded in the beginning by high membership requirements. Later, their influence was diluted by a unique aspect of ASCE: most of its members (see Table II) are not employees of industry.

To say this, however, is in no way to pretend that the ASCE was always a "liberal" society: In 1888, for example, the society decided specifically not to allocate any funds for educational activities—a policy that wasn't to change until 1910. Furthermore, the leadership of the ASCE in the early 1900s had to be "dragged kicking and screaming" into its licensing and ethics activities—only in the face of a young member revolt did the leadership agree to such actions. Also, it should not be assumed that the ASCE *funded* nontechnical activities from the beginning. As Table III indicates, for more than half a century, publications, meetings, and overhead were the only categories of expenditure. But despite these caveats ASCE was far more professionally conscious than was the AIME.

AIME Executive Director Joe B. Alford openly states that, more than any of the societies, AIME is "industry oriented." There is nothing new about this. From the

II. Categories of Engineering Employment in Percent of Total Membership[1]

Area of Practice	ASCE	AIME	ASME	IEEE	AIChE	AIA	ABA	AMA
Private practice (principals and employees)	29	9	10	10	4[2]	68	76	63
Government (Federal, state, and local)	40	4	9	9	32			
Industry (including construction)	23	76	73	68	24	32	24	37
All other	8	11	8	13	37			

[1]All figures, with the exception of AIChE statistics, come from W. H. Wisely, *The American Civil Engineer*, New York, ASCE, 1974.
[2]Self-employed only.
[3]Educators only.

beginning, AIME admitted industry leaders, whatever their engineering qualifications. Rossiter W. Raymond, a highly trained and respected mining engineer, was chosen as the first secretary of the AIME, and he used this position literally to rule the institute for 40 years. In an early speech, he proudly stated that AIME's membership included "common miners, laborers, mine foremen, and people that cannot spell." He went on to boast that 20 percent of the most distinguished "captains of industry" in the U.S. were listed on the AIME membership roster. It is the convincing theory of Edwin T. Layton, Jr., author of *The Revolt of the Engineers* that this combination effectively paralyzed AIME during its first 50 years—that is, in regard to professional activities. As editor of the AIME journal, Raymond refused to print discussions of professional topics such as the social responsibilities of engineers. He called ethics concerns "hysteria," and in the early 1900s he bucked the tide of engineer activity by censoring any consideration of conservation. In short, according to science and technology historian Layton, Raymond nearly destroyed his institute. Said a leading technical journal in 1908, "As an engineering society [AIME] is a failure."

INHERITANCE: MODERN-DAY POLICY

Table I shows that these broad differences in philosophy continue to this day. Whereas ASCE has developed a relatively ambitious package of nontechnical member programs, AIME has barely scratched the surface. In fact, in an interview with *Spectrum*, AIME Executive Director Alford pointed out that, alone among the five Founder Societies, AIME voted against the adoption of employment guidelines. The institute has further refused to create a liaison with the National Society of Professional Engineers (NSPE). But, unlike in 1900, AIME's recent conservatism is member-generated, rather than the policy of the chief executive officer.

Explains Mr. Alford, "Most AIME members are employees of, and accustomed to, high-risk entrepreneurial industries. For example, only one of ten wells drilled in the U.S. finds any oil or gas. Consequently, people gravitating to such high-risk enterprises tend to be rugged individualists, not likely to look to their society for nontechnical support."

Again an important caveat must be made, however. AIME has not always been a conservative society, just as ASCE has not always been a relatively liberal one. In 1908, H. M. Chance led a revolt against AIME secretary Raymond and the board of directors. At a meeting held at the Engineers' Club in New York City, 115 charter members of the Mining and Metallurgical Society of America (MMSA) agreed that their new organization should:

1. Represent professional mining engineers on matters of ethics and public welfare.

2. Develop strict membership requirements similar to those of the ASCE, thereby excluding business interests.

3. Create a strong regional organization to prevent the society from being controled by a single clique.

The MMSA, as it was to be known, failed quickly but it triggered a powerful reform movement within the AIME. In 1911, AIME secretary Raymond resigned. An effort by AIME's board of directors at the Annual Meeting of 1912 to solidify industry control of the society by creating a "corporate class" of wealthy members who would pay higher dues and constitute a conservative elite within the society was crushed by an insurgent group armed with proxies. The board, in fact, was thrown out and, during the next several years, the reformists beat back conservative counterattacks, largely thanks to their discovery and revelation of serious financial mismanagement by the disenfranchised. Furthermore, the insurgents succeeded in raising the membership qualifications of the institute—a process that has continued to the present so that, today, AIME boasts standards similar to those of IEEE. But the most significant victory of the reformists came in 1915 in the

III.   History of ASCE Expenditures*

Year	General Services, $	Publications, $	Technical, $	Professional, $	Total, $
1853	115	—	—	—	115
1875	6 216	3 112	—	—	9 328
1900	27 476	17 863	—	—	45 339
1925	169 203	55 265	10 498	7 841	242 807
1950	373 942	313 188	20 987	31 624	739 736
1973	1 867 027	2 009 890	303 363	258 316	4 438 596

*Source: W. H. Wisely, *The American Civil Engineer*, New York, ASCE, 1974.

form of a constitutional amendment that permitted AIME, for the first time, to take positions on public policy issues. It was this act that paved the way for AIME to collaborate in the great progressive engineering movement of the early 1920s— something that will be discussed shortly.

It must be noted that, like most rebellions, AIME's sudden liberalization wasn't to last more than a decade. There are probably two reasons for this: In the first place, AIME has grown over the years by expanding its technical base. The first move in this direction occurred in 1918 when the institute absorbed the American Institute of Metals and changed its name (though not its abbreviation) to the American Institute of Mining and Metallurgical Engineers. In 1922, a Petroleum Division was created within AIME; six years later, there was an Iron and Steel Division; and so on. By 1974, the society had become the American Institute of Mining, Metallurgical, and Petroleum Engineers and had a unique structure. There were four constituent societies: the Society of Mining Engineers, the Metallurgical Society, the Society of Petroleum Engineers, and the Iron and Steel Society. Of these four, only two remain in New York City today; one is now in Salt Lake City, Utah, and one in Dallas, Tex. The effect of this accordion-like history may be twofold. On the one hand, with the exception of the more "discipline-oriented" metallurgists, each constituent group has tended to be a conservative industry-oriented group. This pattern is borne out by recent votes on public issue proposals. On the other hand, the more recent decentralization of the AIME is both an indication of the conflicting nontechnical interests held by the membership and a serious handicap to the kind of unified action necessary for public policy adoption.

In addition, there is a second reason for AIME's return to conservatism. The progressive engineering movement of the post-World War I period that has so far only been alluded to in passing created a powerful industry backlash that effectively curtailed not only AIME's adventure in the nontechnical arena but also those of ASME and IEEE—at least until 1970. And that brings up the question of the comparative natures of the three remaining Founder Societies—the ASME, the IEEE, and the AIChE. . . .

COMPROMISERS OF INDUSTRY AND PROFESSIONALISM

Whereas the ASCE has, throughout most of its century and one quarter, been concerned with both the technical and nontechnical aspects of engineering professionalism, and whereas AIME, as we have seen, only briefly flirted with the nontechnical side, ASME and IEEE have generally taken a kind of middle ground—but one that is characterized by sudden swings of the pendulum. Today, both societies are heavily committed to nontechnical goals. Table I shows that only ASCE rivals IEEE in dollars spent per member on "professional" activities. As for ASME, while its dollar-per-member figure in this category is moderate, its objectives are ambitious and it may be that, barring a retrenchment on those ambitions, the figure will rise rapidly in the next several years. But it is important to remember that, in both cases, neither ASME nor IEEE was allocating any funds to speak of for self-generated professional programs during the 40 years prior to

1970. What caused these turnabouts and what were the factors that determined from their inceptions the natures of ASME and IEEE?

ASME was the third national engineering society to be founded—the year was 1880. IEEE is, of course, the product of a 1963 merger of two societies: the American Institute of Electrical Engineers (AIEE), founded in 1884, and the Institute of Radio Engineers (IRE), founded as recently as 1912. IEEE, therefore, must be discussed in two parts as its character is an amalgam of rather different traditions.

According to historian Layton, both ASME and AIEE can be seen as attempts to find a compromise between ASCE's bias toward professionalism and AIME's bias toward industry—but they compromised in somewhat different ways. Rather like AIME, ASME permitted a relatively open membership of which well over 50 percent were owners and managers of business. Nevertheless, Layton documents an early power clique of what he calls "creative professionals" who controlled the society until 1904 on the principle that they were the producers of new knowledge whereas the majority business faction represented mere users of new knowledge. AIEE, on the other hand, is described by Layton as initially a high-standard, closed-membership society rather like ASCE, except that the powers behind the throne were "businessmen."

The original call for the organizational meeting that was to produce the AIEE was issued by an inventor and patent-law reformer (and a chemist by education) named Nathaniel S. Keith—hardly an industry type. However, within months Keith resigned as AIEE's first secretary, ceding his potential power to Joseph P. Davis, vice president of a telephone company. At the same time, Norvin Green was AIEE's first president. Green was a medical doctor, a legislator, and a businessman, but not a professional engineer. Most notably, Green was the president of Western Union. And Green was just the first in a line. In 1894, for example, then AIEE president Edwin J. Houston called for the institute to "be the acknowledged center of the industry" through the coming decade. With expectations such as this, it seemed unlikely that AIEE would involve itself either in public controversy or in the ethical or economic concerns of the employee-engineer. One notable institute objective voiced in Keith's original call to meeting of the AIEE had been to "protect" the engineer against "unfavorable legislation." Needless to say, that objective had been quickly forgotten.

ASME, however, quite soon became involved in legislative (if technical) issues. Frederick Hutton, president of the society in 1902, later wrote a history of ASME's first 35 years in which he pointed out the studied balance of the first society officers: Two represented mechanical engineering academia, two came from the ranks of the machine tool designers and builders, a third pair were mechanical engineers employed in mining, one was a metallurgist, another a chemical engineer, two were builders of engines, one was a railway man, one an arms manufacturer, and one was an Army engineer while another came from the Navy.

Perhaps it was this range of interests that encouraged a diversity of society activities. In 1881, ASME's members voted against compulsory metrication in hopes of influencing Congress. In 1885, the first standard was issued on steam boilers, no doubt partially in reaction to accidents that were occuring at the time (the society, while allowing the publication of this standard, refused to endorse it officially, fearing subsequent suits and/or the implications in terms of commercial temptations of such a precedent). In 1889, international contacts were solidified through a mass voyage of members to England. And probably before 1900, the society attempted to influence Congress in such areas as construction material testing and Patent Office reform.

Then, around the turn of the century, in both ASME and AIEE reactions set in against previous policies—ASME moved toward the "right," while AIEE moved toward the "left." In fact, despite the early professional aspirations of the ASCE and ASME, society chronicler Layton credits AIEE as being a forerunner of the great progressive movements of the 1920s.

Why would this have been so, considering AIEE's early business bias? Layton credits the very nature of electrical engineering at the century's end. Writes Layton:

> The professionalization of engineering was everywhere associated with the shift from a craft to a scientific base for the underlying technology, but in probably no case was this transition so rapid as in the electrical field... It lent itself to treatment by the highly sophisticated methods of mathematical physics. The coming of alternating-current power and the growing complexity of communication systems created problems that the older inventor types were ill-equipped to solve. The AIEE flourished with the arrival of a new type of scientist-engineer who could master these technological difficulties.
>
> Charles P. Steinmetz and Michael I. Pupin were leading examples of the new breed of electrical engineer. Pupin, Steinmetz, and a few kindred spirits set the pace for AIEE.... They did much to make the AIEE a dynamic and exciting society in the 1890s.

This excitement gradually carried over into the nontechnical affairs of the institute. The first such instance occurred in 1893 when a group of insurgents, using "Tammany methods" according to a commercial engineering publication at the time, defeated the board candidate for president. This signaled the gradual disintegration of the business clique that had held power for a decade. The board of directors next began to select as the official candidate whoever received the greatest number of member nominations. As a direct consequence, a minority could "elect" a president and by the 1900s very young men began to head up the institute. In 1901, Steinmetz, himself, was chosen president. The following year there were no less than 77 nominations and Charles Scott emerged as the winner. Scott was, at the time, 33 years old and already chief electrician at Westinghouse. Scott followed in Steinmetz's footsteps by pushing for engineering unity. He proposed (to no avail) a "national engineering congress" that would wield greater power than any individual society could. As a first step, he almost single-handedly convinced Andrew Carnegie to fund the construction of what was to become the eventual home of the Founder Societies.

But AIEE's nontechnical ventures were not limited to Scott's efforts. As early as 1905, the institute had authorized a committee on forest preservation. By 1911, this committee had become responsible for AIEE's total involvement in public policy—that is, in conservation legislation, reform of the Patent Office, and authorization of local sections to become involved in civic affairs, as well as in licensing laws, in standards, and, perhaps most remarkably, in the movement to regulate the electric utilities.

Later, as we shall see, it was largely the utilities that crushed ASME notions of progressive action, and perhaps in AIEE as well utility interests among the more powerful members triggered a backlash. The record isn't clear on this point, but what is clear is that a backlash began about 1912 and turned AIEE public policy activities along an industry-oriented path. AIEE standards activities suddenly paralleled those of the electric industry. When it came to conservation, AIEE refused to cooperate with existing conservation agencies. Under great pressure from the hierarchy, the progressive-minded local Sections traded their autonomy in public affairs for greater representation in headquarters deliberations. And in a remarkable and bitter battle that actually reached the New York State Supreme Court, business leaders in the electrical industry emerged victorious, having gained the right to full membership status no matter the extent of their professional qualifications. For nearly 60 years, AIEE was to wax conservative except as its members' professional interests were represented in the progressive intersociety efforts of the early 1920s.

And what of ASME during these first years of the twentieth century? As has been hinted, while AIEE was dabbling in reform, ASME was largely serving its industry interests. In 1904, the increasing membership of the society forced a constitutional amendment that opened ASME's doors to businessmen. The apex of this movement is exemplified by the following events as related by Layton:

> How these private interests operated was illustrated in 1914 when the ASME drew up a standard code for steam boilers. The preliminary draft, drawn up by a committee headed by John A. Stevens, a distinguished engineer, aroused violent protests from members who were owners and executives of affected companies and who objected to rules interfering with their businesses. Led by Henry Hess, a member of the society's governing council, they demanded a reconstruction of the committee and the replacement of Stevens. A compromise was reached whereby a special advisory committee as created on which all of the major trade associations involved were represented.... After the code was amended to suit them, the trade associations banded together to lobby for the enactment of the boiler code by the states.

This, apparently, was what it took to convince the ASME directors that publishing standards was a safe endeavor. Without the trade association manuevering, the unique $3 million ASME "business" might not exist today.

This first stage of ASME conservatism was not to last long, however. Soon after World War I, it was within the ASME that the great progessive movement of the early 1920s was to bloom.

BRIEF BLOSSOM OF ENGINEERING PROGRESSIVISM

It should be particularly instructive to us, today, to study the powerful reformist movements in engineering during the early years of the otherwise Roaring Twenties. This is because a half century later engineers seem to have come full circle in many ways. Only within the last ten years have the ASME and IEEE created Washington, D.C., offices. Whereas the ASCE prefers to see itself as merely increasing its nontechnical activities along a natural continuum (ASCE had a Washington office from 1941 to 1955, only to reopen the office in 1972), ASME readily admits to having undertaken a conscious alteration in its professional course. Says ASME Executive Director Rogers B. Finch, speaking of the roots of a resurgence of his society's interest in the nontechnical: "We had some astute leaders in the society who felt that a close look needed to be taken at some issues on the horizon." The consequence of this was a far-reaching 1970 Goals Conference that produced a wide spectrum of nontechnical objectives currently being pursued by the society (it is estimated that about 7 percent of the society's membership was involved, in one way or another, with the formulation of the new policies). As for IEEE, the addition of nontechnical programs was seen to have been even more drastic: the Institute's very constitution was amended, and the Executive Director and the Institute's lawyers agreed voluntarily to give up the "C3" tax status (originally estimated to cost $250 000 per year but, in practice, costing about one tenth the original estimate) still enjoyed to this day by the other Founder Societies—all designed to permit (and dramatize, perhaps) IEEE's new involvement in professional activities through the United States Activities Committee (now a Board).

AIChE, as well, has entered the professional arena, if a bit more modestly. In 1976, the "chemicals," with a membership of only a little over 36 000, have budgeted $50 000 for this purpose. And it should be noted (Table I) that these funds are over and above a whopping $300 000 budgeted for education activities. Even AIME, according to Executive Director Alford, has instituted a program for professional involvement of the society—administered by the Government–Energy–Minerals Committee, this program provides AIME members with facts and figures with which they can present their personal viewpoints to key legislators.

In the light of this movement of the Founder Societies in the 1970s, it seems worthwhile to consider the accomplishments and failures of a similar engineering movement in the 1920s.

No study of the progressivism of engineering in the 1920s is better than Layton's *The Revolt of the Engineers*. Layton credits the primary impetus of this movement to events (and men) occurring in (and belonging to) two of the Founder Societies: ASCE and ASME. As has been previously mentioned, the elitist membership requirements of the ASCE and its consequent conservatism around the turn of the century angered many young civil engineers. In 1914, a group of civils in municipal jobs for the city of Chicago, Ill., formed a short-lived group that quickly splintered into two factions: one pro-union; the other anti-union. The latter group decided to approach several prominent civil engineers known to be dissatisfied with the ASCE to interest them in the formation of the American Association of Engineers (AAE), which occurred without the older engineers' direct support in 1916. By 1918, however, these older reformists in the ASCE ranks had recognized that the AAE could be a viable entity. A year after adding their talents to those of the young "rebels," AAE successfully pushed through licensing bills in the state legislatures of New York and Virginia. During the same year, the AAE took a first step toward collective bargaining by drawing up salary schedules for engineers employed by the U.S. railroads. These schedules were then presented to a Federal mediating agency with the result being substantial wage increases for some AAE members. The effect of these successes was seen in the association's membership figures: from 2300 engineers in January 1919, the membership soared to over 20 000 by September 1920.

The greatest AAE success of all, however, was its influence in altering the course of the Founder Societies. In ASME, as in ASCE, there had been a revolt brewing as early as 1910 when a past president named Frederick W. Taylor in concert with like-minded mechanical engineers attempted to convince the ASME leadership to endorse a study by lawyer Louis D. Brandeis concluding that a proposed increase in rail rates should not be granted by the Federal regulatory body because of glaring inefficiencies in rail management. These were brilliant men. Six years later, Brandeis was to be appointed to the U.S. Supreme Court. Taylor, among many accomplishments, was the originator of a movement that swept industry and academia alike: scientific management. For those unacquainted with the term, there is no time here to dwell on the theory of scientific management other than to say that through scientific management, Taylor believed that nearly all the significant areas of society—much less business—could be "engineered" to maximum efficiency. At the heart of Taylor's belief was the implication that engineers should not merely improve their social status but rather "rule" society—that society should be governed not by politicians but by "technocrats." Clearly, Taylor was unlikely to be satisfied with the policies of his own or the other Founder Societies.

And at first, the ASME group in power was equally unsympathetic to Taylor. In 1910, it refused to endorse Brandeis's study for fear of alienating the railroads. In 1912, it rejected Taylor's own principles. But between 1914 and 1919, a follower of Taylor, Morris L. Cooke succeeded where Taylor had failed. Writes Edwin Layton of Cooke:

> He had a gift for politics and intrigue. He grasped the key fact that the hard core of resistance to scientific management came from the public utilities and railroads acting together as a sort of monopoly interest within the engineering profession. By shifting emphasis from the virtues of scientific management to the vices of the utilities, he was able to broaden the base of his appeal and link the efficiency crusade to the national progressive movement.

In practical terms, this meant that in 1914, Cooke forced the inclusion of a session on municipal engineering in the ASME annual meeting program—in doing this he was taking on the utilities. The following year, Cooke discovered and publicized the fact that utility minorities controlled not only ASME, but ASCE and AIEE as well. Perhaps Cooke's signal achievement, however, was to urge with success the creation in 1917 of the Engineering Council. This body was constituted by the four Founder Societies of that time. Even AIEE and AIME went along with the plan for one basic reason. The Founders had been shaken to the core by the formation, essentially by young civil engineers, of the previously mentioned American Association of Engineers. The Engineering Council lasted only two and one half years and was pretty much a failure (it did little more than adopt paler versions of AAE programs, including licensing, salary schedules, and an employment bureau); nevertheless, the Council represented the first major step of the Founder Societies toward unified professional and political action.

The second major step was the decisive one. In 1920, following major reappraisals of internal objectives by ASCE and ASME (the model for the previously mentioned 1970 ASME Goals Conference), the governing bodies of the four Founder Societies met to form the American Engineering Council. The first elected president of this body was none other than Herbert Hoover, a progressive member of the AIME. Hoover believed, much like Frederick Taylor, that engineers were "through the nature of their training, used to precise and efficient thought; through the nature of their calling, standing midway in the conflicts between capital and labor; and above all, being in their collective sense independent of any economic or political interest, they comprise a force in the community absolutely unique in the solution of many national problems."

This philosophy led Hoover to press his American Engineering Council (AEC) to undertake two major studies: The first was to investigate waste in industry; the second was to consider the ramifications of the 12-hour work shift, then the norm in industry. In both cases, the projects were staffed largely by "Taylorites" and the results of the studies were, to say the least, controversial. The AEC Committee on the Elimination of Waste concluded its 1921 report by assessing blame for inefficiency. Not surprisingly, industry felt attacked and it marshaled its forces in the Founder Societies to prevent the report from being published. Hoover, now Secretary of Commerce in the Harding administration, personally interceded, effecting a compromise: The report would be officially no more than a committee document while Hoover, himself, would write and sign an introduction to the report. Meanwhile, Hoover assigned his staff in the Commerce Department to carry out some of the report's recommendations.

As for the study of the 12-hour day, the AEC committee in charge concluded in 1922 that a three-shift, 8-hour day would be far more efficient, if not more humane. Once again, Hoover drafted an introduction, this time signed by President Harding, himself. But the high-level sanctioning of the report did not prevent an even greater storm of protest on the part of industry (particularly, the steel interests) than had greeted the report on waste. The latter had brought the AIME directors to the point of withdrawing their institute from the AEC. Only Hoover's special relationship with AIME had averted this. After the second report, nothing could stop them. They pulled out in 1924 joining ASCE, which for various reasons had never followed through after the Council's organizing conference by becoming a member. With only two Founder Societies left, the AEC's era of progressivism was over—so much so that ASCE's now conservative-minded board finally decided to sign its society up in the AEC in 1929. By then the Council had become a business ally, refusing to involve itself in the development of standards, fighting rural electrification, and lobbying against bills designed to outlaw wooden cars on railroads. By 1931 and 1932, the Council actually took stands against legislation written to limit the pollution of streams. The age of engineering society commitment to the public interest was over. By now, even the reformist American Association of Engineers had collapsed under the powerful competition of the Founder Societies' AEC. It would be 40 years before the Founders again heard the call for professionalism and the pity of it was that with Herbert Hoover as President of the United States, engineering could have gone a long way toward achieving one of its most persistent dreams: public respect on a par with the medical and legal professions.

THE "FLEDGLING" SOCIETIES—AIChE AND IRE

This discussion has so far barely mentioned the American Institute of Chemical Engineers and IEEE's second ancestral society, the Institute of Radio Engineers. The reason is that both developed comparatively late: AIChE was founded in Philadelphia, Pa., in 1908; IRE was founded in New York City in 1912. Furthermore, the AIChE didn't officially become a Founder Society until 1958 when it bought into the new (and present-day) home of the Founder Societies, the United Engineering Building at 345 East 47 Street in New York. And IRE only became part of the story on the Founder Societies by virtue of its 1962 merger with the AIEE. Nevertheless, both societies have a role in this discussion. They have their own unique histories; their present-day characteristics have much to do with conditions surrounding their formation (IEEE cannot be properly understood without reviewing IRE's origins and its remarkable growth prior to its merger with AIEE).

1.2

IEEE and the "Founders"—II

ELLIS RUBINSTEIN

THE BIRTH OF THE "CHEMICALS"

AIChE was founded in Philadelphia, Pa., in 1908, and was not officially designated as a Founder Society until 1958 when it bought into the new (and current) home of the Founder Societies, the United Engineering Center at 345 West 47 Street in New York City.

More than anything else, two circumstances seem to have defined the AIChE during its 68-year existence. First, although the term "chemical engineer" appears in a dictionary as early as 1839, arguments persisted well into the 1920s regarding the legitimacy of chemical engineering as a branch of engineering rather than as a mere offshoot of chemistry. Second, the AIChE to this day is overshadowed by the giant American Chemical Society, to which, according to AIChE executive officer F. J. Van Antwerpen, no less than 25 percent of the institute's membership also belongs.

The principal effect of these handicaps has been to have forced the institute, from its earliest days, to allocate much of its energy and financial resources to defining and propagating its constituency. It wasn't until 1888 that any U.S. university offered a specific course in chemical engineering (the Massachusetts Institute of Technology was the first), and it wasn't until 1892 that a university established a chemical engineering department (this time, the University of Pennsylvania). No wonder, then, that when Richard K. Meade, editor of the five-year-old journal, *Chemical Engineer*, called what was to be the organizational meeting that developed the concept of a chemical engineering society, his primary task was to justify the profession. Mechanical engineers, he pointed out to those assembled, were not the best qualified to build chemical plants. Neither were chemists. And as long as "electro-chemists" were leaders in both the American Chemical Society (ACS) and the American Institute of Mining Engineers, why not start an American Institute of Chemical Engineers?

It has been this battle for recognition that has made AIChE perhaps the leader among the Founder Societies in educational programs. A Committee on Chemical Engineering Education was appointed in the very first year of AIChE's existence. In 1922, the institute embarked on an ambitious study of chemical engineering curricula as offered at 78 U.S. colleges. In 1925, AIChE became the first engineering society to take on the responsibility of accrediting university engineering departments. This act predated the formation of the Engineers' Council for Professional Development, which is responsible for accreditation today. In light of this tradition, it should come as no surprise that, in 1976, the AIChE will spend nearly twice the dollars per member allocated by any of the other societies for education programs.

A secondary effect of AIChE's tenuous position over the years (particularly in relation to the rival ACS), and of its need to expend many of its dollars for education, has been its relative lack of an otherwise nontechnical tradition. As early as 1912, the AIChE Council appointed three delegates to provide inputs to a Washington, D.C., conference on patent problems. In 1919, the institute sent four resolutions to President Harding and the U.S. Congress (though, significantly, in support of chemical industry concerns at the very height of the progressive movement). And in 1921, it was active on both a state and Federal level, attempting to influence, for example, a reorganization of Federal bureaus into a Department of Public Works as well as the activities of the licensing board of New York State. But from 1930 on, the AIChE waxed conservative as evidenced by a 1953 Council resolution "that the enhancement of the status of the engineer is best promoted by his reliance upon his personal professional growth and accomplishment, and that his status as a professional man is endangered by reliance upon group efforts to act on his behalf in reaching short-range economic objectives..."

Today, according to chief executive officer Van Antwerpen, AIChE is wrestling with professional questions that are dividing the membership. A member vote was split down the middle over the question of establishing a Washington, D.C., office as ASCE, ASME, and IEEE have done. The AIChE Council subsequently decided in favor of institute representation in Washington. A second difficult question has been how the institute should react to member layoffs. Mr. Van Antwerpen says that many chemical engineers believe the institute should "white-list" companies that follow the Joint Employment Guidelines endorsed by most of the Founders, if not "black-list" companies that resort to mass layoffs. A surprising 25 percent of the AIChE members have declared themselves in favor of unions—a statistic that hints at the volatile makeup of the AIChE. So far, the institute distributes a booklet entitled "Professional Standards" as well as copies of the Joint Employment Guidelines; it meets with representatives of management to explain and promote the Guidelines; it provides an employment service; and it intercedes "on the *QT*," as Mr. Van Antwerpen puts it, with chemical engineering management when complaints regarding layoffs are received at headquarters.

THE RADIO ENGINEERS ORGANIZE

As for the IRE, which became a Founder Society only by virtue of its merger with AIEE in 1963, its character at the time of the merger was no less defined by the conditions surrounding its birth than were the characters of the other engineering societies. In 1907, the Society of Wireless Telegraph Engineers (SWTE) was formed in Boston, Mass., as an outgrowth of local seminars held by engineers employed by a single company: the Stone Wireless Telegraph Co. According to Edwin Layton, author of *The Revolt of the Engineers* (Cleveland, Ohio: Case Western Reserve University Press, 1971), the aim of the SWTE was to achieve closer relations to industry. AIEE, it must be recalled, was at the time a very progressive group led by Dr. Steinmetz, and Layton points to the fact that SWTE's initial membership requirements restricted acceptance to employees of Stone Wireless—hardly an indication of a nonbusiness orientation.

In 1908, Robert H. Marriott circulated a proposal that called for the formation of the Wireless Institute, a society that was to be patterned after AIEE but was to be restricted to engineers accomplished in wireless. Marriott's organization was centered in New York. When, by 1911, the Stone Wireless Telegraph Co. had been dissolved and many of SWTE's members had moved to Brooklyn, N.Y., as a result of the relocation of a second Boston wireless firm, Marriott's Wireless Institute also found itself struggling. The consequence was that, in 1912, Marriott and Alfred N. Goldsmith (see *Spectrum*, August 1974, pp. 32–35), representing the Wireless Institute, and John V. L. Hogan, an active SWTE member, agreed to call a joint meeting of members of both societies at Columbia University. At that meeting the IRE was born with a combined membership of 46 and international ambitions as witnessed by the decision to omit "American" from its name.

Two important factors contributed to both the success of IRE and its highly professional nature. First, radio engineering in 1912 was, as historian Layton puts it, "highly esoteric," thereby tending to exclude business interests in favor of technical engineers. Second, it will be remembered from Part I of this article that 1912 was the very year in which a "palace revolt" was accomplished by business interests in the AIEE. Suddenly, industrialists without professional qualifications could be admitted to full membership. What could have been more fortuitous for the fledgling IRE! Disgruntled AIEE members flocked to the IRE, and while this may have solidified the former's new industry faction, it emphasized the IRE's new professionalism.

If these events characterized IRE from the beginning, they seem not to have thrust IRE into the forefront of engineering progressivism in the early 1920s. The most probable reason for this is simply that it took all of the young institute's efforts to get itself organized during those early years. Not until 1924 could the institute even afford rented headquarters offices. But this is not to say that it was idle. In 1925, its first of many international sections was formed, and in 1927, when the Founder Societies were shunning Government involvement and taking on a decidedly pro-business stance, IRE was providing the newly established Federal Radio Commission with presumably unbiased technical advice. However, in the absence of pressure toward engineering involvement in nontechnical matters over the next three decades, IRE remained a strictly technically oriented society. By the time of the merger, though, IRE had substantially outgrown the AIEE. With nearly 100 000 members—more than twice that of AIEE—IRE dominated AIEE and gave IEEE what is today its relatively nonindustry, professional orientation.

Had it not been for the merger, AIEE might never have funded a major nontechnical effort. In fact, discussions with key individuals involved in the merger make one wonder whether AIEE could have survived the economically difficult 1970s by itself. It is certainly true that the leadership of both the AIEE and IRE were well aware in the early 1960s that the redundancies created by the continued separation of the electrical and electronics engineers into two camps made little sense. Apparently, the student members of the two societies were among the first to realize this, and the merger of numerous Student Branches gave impetus to the movement to marry the two societies as a whole. It is also on record that the case for a merger was brought to the two memberships in terms of the economic benefits that would accrue from the elimination of separate (and sometimes competitive) publications, and technical and professional staff functions. The vast majority of AIEE's members were willing to give up some of their independence in return for such economic benefits—a contention established by the outcome of the AIEE member vote in favor of the merger, which matched to within a percentage point that of the IRE members.

But what of AIEE's leaders who risked losing much of their power in the inevitable dilution of a merger? Without question, many of these men were aware of the potential benefits just mentioned. In addition, though, it has been said that AIEE, at the time of the merger, was mismanaged and badly overextended— perhaps a consequence of trying to compete with the much larger and more affluent IRE. If this is true, then these men were in the best position of all to realize the necessity of a merger: Altruism aside, perhaps they saw that only in a merger did they have any real chance for power.

As for the IRE, as has been said, it was ambitious and well-heeled. The opportunity of increasing its membership by 40 000 or more in one step was a tempting one. The argument must have been made that the gain in prestige that would result from being two and one half times the size of the next largest engineering society could not be ignored—and all this at no sacrifice to IRE's traditional technical independence while, simultaneously, benefiting from economies of scale. The result was that on January 1, 1963, after extended maneuvering, a great deal of hard work by their respective leaders, and highly favorable member votes, the Institute of Radio Engineers and the American Institute of Electrical Engineers legally became the IEEE.

FACTS OF ESPECIAL INTEREST TO IEEE MEMBERS

Several unrelated points regarding the present-day concerns of the Founder Societies may be of particular interest to *Spectrum*'s readers. One of these relates specifically to institute financing. The topic is dealt with in some detail in the March [1976] issue of *Spectrum*, pp. 64–69, but in the course of researching this article (for which interviews were granted by the chief executive officers of each of the Founder Societies) several new facts came to light. Of the three Founders (ASCE, ASME, and IEEE) that have heavily committed themselves to professional involvement in one form or another, all have faced unprecedented financial

Some Seminal Dates

1782—John Smeaton,English road and canal builder, signs himself as a "civil engineer" in presenting expert testimony to the courts.

1787—Law establishing U.S. Northwestern Territory allocates Federal funds for civil works construction.

1794—Reconstruction, in peacetime, of U.S. Army Corps of Engineers.

1802—U.S. Military Academy at West Point, N.Y., created by act of Congress to function as an arm of the Corps of Engineers; first U.S. graduate in engineering.

1818—British Institution of Civil Engineers is founded—first national engineering society.

1821—Congress directs Corps of Engineers to survey major roads and canals employing both "engineer officers" and "civil engineers."

1824—Franklin Institute, Philadelphia, Pa., established—first professional home of many engineers in U.S.

1825—Erie Canal project, training ground for seven future ASCE presidents, is completed.

1835—First civil engineering degree conferred by Rensselaer Polytechnic Institute.

1848—Boston Society of Civil Engineers, first permanent U.S. engineering organization, is founded.

1852—ASCE is founded in New York City.

1855—ASCE goes into 12-year eclipse.

1867—ASCE is revived.

1871—AIME is founded in Wilkes-Barre, Pa.

1880—ASME is founded in New York, N.Y.; Association of Engineering Societies, first major intersociety effort, is formed.

1884—AIEE is founded in New York, N.Y.

1908—AIChE is founded in Philadelphia, Pa.

1912—IRE is founded in New York, N.Y.

1963—IEEE is formed through merger of the AIEE and IRE.

---

requirements that have only been exacerbated by the high rate of inflation during the last several years. Each, however, has reacted somewhat differently:

IEEE, as *Spectrum* readers know, initiated a special U.S. member dues assessment of $5 in 1973 to cover the total costs of professional activities—all of which were U.S.-oriented. This assessment was raised in 1976 to $10. While the level of dues charged the U.S. members was, in both cases, determined solely by the Board of Directors without membership vote—this according to the IEEE constitution, which vests sole discretion for all dues changes in the Board of Directors—it must, nevertheless, be noted that the members agreed by vote, at the time of the constitutional amendment that resulted in the creation of the U.S. Activities Board, to fund, through increased dues, such an IEEE involvement. Today, a call has been issued by IEEE's Long Island Section Professional Activities Committee to amend the constitution so that members must vote on every dues increase. How does this compare with policies of other societies?

Neither the American Medical Association, the American Bar Association, the American Institute of Architects, nor any of the Founder Societies, with the single exception of the ASME, provides for a member vote on changes in the magnitude of annual dues. According to most executive officers of these societies, the reason is simply that such a procedure would create havoc in financial planning, especially in times of rapid economic fluctuation. The member's veto, they say, is his ability to drop his membership if he becomes dissatisfied. This, in the collective, would be a lightning-like signal to the Board of Directors.

Why then does ASME maintain the opposite policy? It most likely will not as of July of this [1976] year. ASME's board hasn't raised its dues for more than a decade. thanks to its unique standards income, but even ASME is prey to inflation and its members' desires for new programs. In 1972, ASME's membership turned down a proposed increase. Consequently, with dues hikes likely during the next several years, ASME's board has determined that, on the one hand, it cannot rely forever on a single source of income—standards—nor can it plan for the society's future if it cannot count on determining a proper dues level.

And as for ASCE, the third of the five Founders to be committed to costly professional activities, IEEE members will be interested to know that the ASCE dues assessment is expected to reach $70 before 1980.

An unrelated but fascinating activity of the ASCE will be of interest to the reader. Based on scandals resulting from payoffs made by Maryland civil engineering firms to former U.S. Vice President Spiro Agnew, the society's Board of Direction, upon the recommendation of the Committee on Professional Conduct, expelled three civil engineers from the society. In 1975, the increasingly tough committee considered 54 cases, of which four were dropped for lack of evidence, eight resulted in member suspensions for periods of one to five years, three resulted in admonishments, and one was closed when the individual involved resigned without prejudice. The rest of the cases were pending action in 1976.

ASCE Executive Director Eugene Zwoyer told *Spectrum* that the society is committed to the licensing of *all* engineers and that it has taken an active role in a variety of local and Federal legislative actions. On the local level, an ASCE technical committee is filing an *amicus curiae* brief in the case of the state of New Jersey vs. the city of Philadelphia, Pa., regarding solid waste disposal. Federally, the society is involved in environmental and transportation issues.

Rogers Finch, Executive Director of ASME, describes *his* society as more a compromiser than a leader. Nevertheless, ASME's 1970 Goals Conference (mentioned in Part I) produced a wide range of professional objectives now being implemented. Almost as interesting as these objectives, themselves, which won't be listed here, is the procedure adopted for that conference. Some 90 ASME members, constituting a cross section of the membership, produced a list of goals that were then discussed by 4000 members in the 140 Sections over a one-year period before a final report was written. This provides an interesting contrast to IEEE's recently completed, but not yet published, long-range planning report, produced in relative isolation by a committee appointed by the Board of Directors and currently being reviewed by Section and Group and Society officers before final action expected to be taken by the Board in September [1976] (see *Spectrum*'s interview with IEEE President Dillard, March [1976], p. 73).

Thumbnail chronology of major intersociety activities

On paper, at least, U.S. engineers of all disciplines were united under a single banner in 1852. That banner was the American Society of Civil Engineers (ASCE) founded in the same year in New York City. Destined to be the first enduring national engineering society in the United States, the ASCE, at its inception, opened its membership to all engineers and, indeed, its early membership included all types of "civilian" (non-military) engineers. This early semblance of unity was short-lived: In 1871, with the ASCE still a young and relatively weak organization, a group of mining engineers founded the American Institute of Mining Engineers (AIME) and a splintering process began that, over the next century, was to produce literally hundreds of national, state, and local engineering groups.

But the dream of engineering unity has been a hardy one over the years, as can be see from the [chronology] below. As early as 1875, the two existing national engineering societies, ASCE and AIME, were cooperating on projects of mutual interest. In one case, they were preparing for the 1876 Philadelphia Centennial, and in another, they formed a Joint Committee on Technical Education. Then, in 1880, four local societies, at the invitation of the Cleveland Engineers Club, met in Chicago, Ill., to form a confederation (to be called the Association of Engineering Societies) primarily for the purpose of publishing a joint technical journal that would widen the circulation of papers presented before local meetings. This confederation was soon overshadowed by the growing power of the national societies, but it lasted for 35 years, thereby becoming the first major intersociety effort.

Brief sketches of the principal subsequent intersociety efforts in the United States follow:

1904—the United Engineering Society (UES) was formed by three national societies, AIME, the American Society of Mechanical Engineers (ASME), and the American Institute of Electrical Engineers (AIEE). The initial purpose of the UES (today called the United Engineering Trustees—UET) was to administer the expenditure of about $1 million donated by Andrew Carnegie to build a common home for the engineering societies. The UES thus became the first forum of the national societies and, as such, expanded its activities in several directions. It spawned the Engineering Foundation in 1914 for the furtherance of research in science and engineering, for the advancement of the profession of engineering, and for the benefit of mankind. And in 1917, an Engineering Council was created as part of the UES. This Council took on the responsibility for providing a link between the Federal government and the Founders. In addition, it developed classification and salary schedules, and it created the Engineering Societies Service Bureau—an employment service later taken over by the American Engineering Council (see below) and eventually to become an independent organization renamed the Engineering Societies Placement Service. The three Founder Societies that created the UES became four when ASCE voted to move into the Engineering Societies Building in 1915. The American Institute of Chemical Engineers (AIChE) became the fifth and final member of what by then was the UET in 1958 when it agreed to contribute funds toward the construction of today's United Engineering Center at 345 East 47 Street in New York City.

1916—the American Association of Engineers was formed by young engineers who felt the established societies to be too conservative. Favoring licensing, the development of. salary scales, employment assistance, etc., the AAE grew from 2300 members in January 1919 to 20 000 members in September 1920. Eventually, however, this organization faded away due partly to competition from the American Engineering Council and partly to its own lack of initiative.

1920—the Federated American Engineering Societies (the brainchild of the four Founders) was created to challenge the growing power of the AAE by replacing the Engineering Council on a grand scale (some 71 engineering organizations attended the initial organizing conference). To manage this body, the American Engineering Council was formed, and under Herbert Hoover, the Council's first president, the AEC produced two remarkable studies described in *Spectrum*'s May issue, p. 84, before problems weakened, and eventually destroyed, the Council.

1932—the Engineers' Council for Professional Development was formed for the purpose of promoting and advancing engineering education. Its major functions throughout the years have been the accreditation of engineering curricula and the administration of a program of guidance for precollege students.

1934—the National Society of Professional Engineers (NSPE) was established to promote the concept of engineering registration. Now technical in character, NSPE promotes economic, social, and legislative aspects of the engineering profession.

1944—the Engineers Joint Council (known as the Joint Conference Committee in 1941–44) was formed by the presidents and executive secretaries of the five Founder Societies. It was later expanded to include over 40 national and regional engineering societies. In the mid-1960s, disagreement regarding its role and program caused some of its major society members to withdraw. The last decade has been one of reassessment— EJC has reconsidered its organization and programs; the societies have discovered their growing need for a strong focal point for the engineering profession. As a result, many societies that had withdrawn have rejoined EJC. The critical holdout at this point is IEEE, the only remaining Founder Society that has not yet reaffiliated with EJC. IEEE is currently debating reentry. Its eventual decision will have an important impact on the Council's future.

1964—the National Academy of Engineering (NAE) was formed as a result of a study begun early in 1960 by a special EJC committee established to explore the possibility of forming an engineering counterpart to the National Academy of Sciences. Although the first three presidents of NAE were past presidents of EJC, there is still ambiguity regarding the relationship between NAE and the community of engineering societies from which NAE sprang. The general objective of NAE is to advise the various branches of the Federal government in matters of national importance pertaining to engineering.

1975—the Association for Cooperation in Engineering (ACE) was formed as a result of recommendations developed by an EJC Task Force. It was to provide a means through which the top officers of engineering societies might act as direct representatives. The relatively simple structure of ACE (it requires no dues for example) makes it possible for societies not currently members of EJC to participate. IEEE presently provides the secretariat for ACE.

Carl Frey
Executive Director
Engineers Joint Council

1.3

The Influence of Engineering Societies on Professionalism and Ethics

WILLIAM H. WISELY[1]

"Professionalism," as interpreted in this paper, is based on the classic definition of a profession stated by Roscoe Pound as "the pursuit of a learned art in the spirit of public service." "Ethics" are construed here as the philosophies or principles which circumscribe such objectives. A professional code of ethics, then, is a statement of principles by which the practitioner may calibrate his personal attitude and conduct to the model approved by his peers.

The manifestation of professionalism and ethical consciousness on the part of the engineer occurs through his interaction with many people—his client or employer, his colleagues, his associates from allied professions, and with the public. Thus, the advancement and inculcation of the service motive and sense of honor that signifies professionalism is logically an exercise in group dynamics. The engineering societies are ideally constituted to serve this purpose. Indeed, an organization purporting to be a "professional society" is obligated to assume leadership in the enhancement of altruism and integrity as qualities of its membership.

There are at least five basic functions that can be provided by the engineering society in fostering professionalism:

1. To serve as a sounding board for the enunciation of professional philosophy and values.
2. To serve as a medium for formalizing standards (codes) of professional practice and ethical propriety.
3. To serve as a medium for education of the practicing engineer, the engineering student, and the public with regard to the manner in which the profession can best serve the public interest.

---

[1]Hon. M. ASCE; Executive Director Emeritus, ASCE.

4. To serve as a medium for the enforcement of professional standards.
5. As a by-product of the administration of standards, to provide feedback as to the problems and deficiencies that must be overcome by the practitioner.

How well have the many engineering societies actually fulfilled this role up to this time? How has the engineering profession fared in comparison to the professions of medicine, law and architecture?

BACKGROUND OF ETHICAL STANDARDS
IN NON-ENGINEERING PROFESSIONS

About 2500 B.C. in ancient Babylon the Code of Hammurabi enunciated in considerable detail the manner in which practitioners in what we now call medicine, architecture and engineering had to conduct themselves. This was actually a set of laws, however, with severe penalties for malpractice, innocent or otherwise.

In a way, King Hammurabi's code was a statement of ethical standards, although dictated outside the ranks of the practitioners to whom they applied.

Medicine

The Oath of Hippocrates, a brief and pragmatic set of ethical principles, was conceived in the fifth century B.C. and is even today an effective conduct guideline for the physician. It was Christianized about 1100–1200 A.D. by elimination of references to the pagan gods of ancient Greece. The oath demands a commitment to the rights of the patient but imposes no sanctions or penalties on the physician.

A British physician, Thomas Percival, published in 1803 his "Code of Medical Ethics." When the American Medical Association held its first meeting, in 1847, it acted to establish minimum requirements for education and training, and to adopt a set of Principles of Ethics. The latter drew heavily upon the Percival code.

The original AMA code prevailed generally for more than a century although major revisions were made in 1903, 1912 and 1947. In 1955 a move to distinguish between medical ethics and "matters of etiquette" resulted two years later in adoption of the present "Principles" comprising a preamble and ten comprehensive sections.

The AMA judicial council interprets ethical matters, and may impose sanctions against its members for ethical violations. Prime responsibility for disciplinary action, however, is left with local and state medical societies, each of which has its own code of practice.

Law

A series of lectures by Judge George Sharswood, published in 1854 under the title "Professional Ethics," were the source of the ethical standards adopted 54 years later by the American Bar Association. In 1887 the Alabama Bar Association enacted a code of ethics borrowed largely from the Sharswood lectures, and the 32

"Canons of Professional Ethics" adopted by the ABA in 1908 were in turn based upon the Alabama code.

By 1928, despite several amendments, an overall revision of the Canons was proposed. This move, followed by similar ones in 1933 and 1937, were all unsuccessful. It was not until 1970 that the present ABA "Code of Professional Responsibility" became effective, after five years of intensive committee activity.

A separate "Code of Judicial Conduct" was adopted by ABA in 1972, as an extensive revision of the "Canons of Judicial Ethics" originally introduced almost 50 years earlier. This code is applicable only to judges.

The ABA Committee on Ethics and Professional Responsibility interprets questions relating to both codes. Disciplinary action, however, is assumed entirely by the state bar associations, and is based upon the ethical canons and oath of admission to the bar which is in effect in each state. Judiciary discipline, of course, is administered by the courts.

Architecture

Almost from the time of its organization in 1857 the American Institute of Architects was plagued with professional practice problems incident to the prevailing custom of selecting architects through design competitions. As early as 1870 the Institute established a "Schedule of Terms" regulating the conduct of such competitions.

In 1909 AIA adopted a "Circular of Advice Relative to Principles of Professional Practice and the Canons of Ethics," a somewhat wordy but nevertheless effective general code. Although the design competition method of architect selection became less popular through the years, an updated version of that guideline was implemented in 1948.

The original Canons of Ethics of 1909 have gradually evolved into the AIA Standards of Professional Practice of today. A noteworthy change in enforcement procedure was made in 1955, when Regional Judiciary Committees were created to investigate complaints of professional misconduct before referral to the National Judiciary Board for appropriate action. Previously this function had been served only by a national committee of the AIA Board.

Ethical problem areas presently of concern to the AIA are advertising, expanded practice, foreign practice and free sketches.

EVOLUTION OF ETHICAL STANDARDS IN ENGINEERING

After the American Society of Civil Engineers was formed in 1852, such leaders as John B. Jervis and James P. Kirkwood referred to the high character and integrity demanded of the civil engineer in serving the interests of others that were committed to him. Yet, when it was proposed in 1877 that the Society adopt a declaration affirming the right of the engineer to exercise engineering judgement without lay interference, the Board of Direction concluded:

> That it is inexpedient for this Society to instruct its members as to their duties in private professional matters.

In 1893, the Cincinnati Association of ASCE members, led by Samuel Whinery, petitioned the Society to enact a code of ethics, but the Board declined to put the proposal before the membership. An effort to revive the issue in ASCE was also unsuccessful.

The breakthrough came in 1906, when Dr. Schuyler S. Wheeler presented his presidential address entitled "Engineering Honor" before the American Institute of Electrical Engineers. Immediate action was inspired in the Institute toward drafting a code of ethics. The draft that was forthcoming in 1907 was delayed first by parliamentary procedures and then by lapse of interest.

In the meantime, the American Institute of Consulting Engineers noted that The Institution of Civil Engineers in England had acted on February 22, 1910 to become the first engineering society to adopt professional conduct standards. Moving independently, AICE formulated a code of ethics with five articles derived from the six-point British code, plus seven new articles. On June 23, 1911, this became the first engineering code of ethics to be adopted in America.

The issue was revived in AIEE in 1911, and its "Principles of Professional Conduct" were adopted in March, 1912. Thus, the first engineering ethical standards to be drafted in America became the second to be actually adopted. In 1963, AIEE merged with the Institute of Radio Engineers to form the Institute of Electrical and Electronic Engineers (IEEE). A membership mandate about a decade later demanded greater emphasis on professional matters, one result of which was the adoption on December 4, 1974, of a new IEEE Code of Ethics. It was the product of "the combined efforts of a number of volunteer committees and boards within IEEE," and appears to be original in its composition.

The American Institute of Chemical Engineers was founded in 1908. Formulation of ethical standards followed only four years later, with adoption of a code in 1912.

The American Society of Mechanical Engineers then made the first attempt to accomplish a goal that has yet to be achieved—the development of a code of ethics that would be acceptable to engineers in all fields. The AIEE "Principles" were modified somewhat to that purpose, but the outcome was adopted only by ASME in June, 1914.

At this point, ASCE returned to the scene when Percival M. Churchill urged the Board of Direction to adopt the code compiled by ASME. The Board responded in 1913 by creating a task committee to study the need for ethical standards and to draft a short code if one was considered desirable. A few months later the Board received favorably a draft code comprised of six of the twelve articles in the 1911 Code of Ethics of the American Institute of Consulting Engineers.° With a few minor amendments this "short" code was adopted on September 2, 1914, by a membership vote of 1,997 in favor from 2,162 ballots cast.

The ASCE Code of Ethics is not significant for either its originality of context or its priority of adoption. Enactment of this code was, however, accompanied by a conscientious, continuing enforcement policy, the implementation of which was to give ASCE stature in this area unmatched anywhere else in the engineering world!

°Interestingly, only one of the articles used by ASCE related to the five that had been drawn by AICE from the code of the British Institution of Civil Engineers in 1911.

In September, 1976, ASCE replaced its much amended original code with a modified version of the then current "Code of Ethics of Engineers" of Engineers' Council for Professional Development, as noted later in this paper.

The National Society of Professional Engineers was not founded until 1934, but it has engendered considerable emphasis upon professionalism and ethics in engineering. A code of ethics was proposed in 1935. Although it was not adopted, NSPE was one of the first societies to endorse (1946) the Canons of Ethics formulated by Engineers' Council for Professional Development. In 1952 NSPE adopted 15 Rules of Professional Conduct, presumably to supplement the ECPD Canons, and in 1964 replaced this combination with a new NSPE Code of Ethics, which is now in effect.

NSPE leaves the burden of code enforcement to its state societies, but has acted at the national level in a few cases. Its Board of Ethical Review serves a commendable education function in its continuing program of published case history analyses.

The Quest for a Unified Engineering Code of Ethics

As AIEE provided leadership in formulating ethical standards, so must ASME be credited as the early prime mover in seeking to bring about acceptance of a single code by all engineering societies. After the abortive 1913 effort, ASME brought about formation of a Joint Committee on Proposed Universal Code of Ethics in 1921, with ASME, ASCE, AIME, AIEE, and the American Society of Heating and Ventilating Engineers (ASHRAE) participating. A commendable "Code of Ethics for Engineers" encompassing ten articles was submitted to the societies in 1922, but was not acceptable to all.

Simultaneously, the American Association of Engineers, founded in 1915 to advance the ethical and economic concerns of engineers, was engaged in formulation of its "Principles of Conduct," issued in 1923. The influence of AAE in the profession began to decline soon after this, however, so its ethical functions were not particularly significant.

Despite its lack of success in 1922, the Joint Committee on Proposed Universal Code of Ethics continued under the leadership of Prof. A. G. Christie of ASME in its endeavor to bring about agreement of the major societies on ethical standards. The joint committee persisted as the Independent Committee, adding AICE and SPEE to its list of sponsors. It became a functionary of the American Engineering Council shortly before the demise of that ill-fated unity federation in 1940, and finally rose again in 1941 under the auspices of the Engineers' Council for Professional Development. The new committee included representatives from ASME, ASCE, AIEE, AIChE, AIME, SPEE, the National Council of State Boards of Engineering Examiners (NCSBEE) and the Engineering Institute of Canada (EIC). It was chaired by Dugald C. Jackson, who had been involved in the drafting of the AIEE code in 1911.

This body lost no time in producing a brief, well-worded statement entitled "The Faith of The Engineer," which was adopted by ECPD in 1943. Although this declaration incorporates remarkably well the elements of professionalism and social responsibility that should inspire the engineer, it has been practically ignored by the engineering societies other than ASME. It has been perpetuated in publication each year in the Annual Report of ECPD, but it is unlikely that as many as one percent of all engineers are aware of its existence.

The same reactionary inertia on the part of the ECPD societies greeted the early drafts of a proposed set of canons of ethics. ASCE and ASME were especially unresponsive at this time. The crusade finally came to fruition in 1947, when all eight of the constitutent societies either adopted the Canons or "assented" to them. ASCE did not actually approve the Canons until 1950, and even then retained its own Code of Ethics, as did AIEE.

The 1947 ECPD Canons of Ethics contained enough similarities in format and language to establish clearly the parentage of the 1911 AIEE code. By 1955 it had been accepted to some degree by 82 national, state and local engineering organizations. This is probably the greatest progress to be made ever before or since toward the realization of a single set of ethical standards for all engineers.

The first up-dating of the ECPD Canons came in 1963, sixteen years after the original. This comprised three "Fundamental Principles of Professional Engineering Ethics," with some 21 articles circumscribing relationships with the public, with employers and clients, and with engineers. By this time there were ten constituent societies of ECPD, and acceptance of the revised Canons was by no means unanimous. AIChE not only adopted the revision, but also made it part of its bylaws, replacing its original 1912 code. By 1968, six of the ECPD societies had adopted the complete revision, two societies (ASCE and NCEE) had adopted the "Fundamental Principles" only, and two societies (IEEE and AIAA) had taken no action at all.

The most promising effort toward a unified code of ethics was led by George S. Rawlins in 1968–74. Through strong liaison with ASCE and NSPE the ECPD Ethics Committee produced a new Code of Ethics of Engineers, which was approved by the ECPD Board on October 1, 1974. This code is constituted of three levels:

The Fundamental Principles (4 elements)
The Fundamental Canons (7 articles)
Suggested Guidelines (56 elements)

It is intended that this format will enable any engineering society to accept at least one or, better, at least the first two parts.

As of 1976, ECPD was comprised of 14 participating societies. Eight of them had endorsed the Fundamental Principles, six the Fundamental Canons, and only one the Guidelines. ASCE, in a major deviation from its past attitude, acted in 1976 to replace its traditional code with a modified version of the new ECPD code. The modifications included one article of the Fundamental Canons and six items of the Guidelines.

Another significant move was made by the American Consulting Engineers Council in 1974 when it adopted the new ECPD Fundamental Canons and Guidelines as its official Code of Ethics.

A blow against the unified code was struck by IEEE, however, when it enacted in December 1974, a new code that was completely independent of the ECPD model.

And thus the quest continues!

Chronological Summary

The following chronological summary of ethical developments in the engineering profession provides a compact overview:

1893 and 1902. Proposals of ethical standards in ASCE rejected.

June, 1907. Code of ethics drafted in AIEE but not adopted.

February, 1910. Institution of Civil Engineers (Gr. Br.) adopts first engineering standards of professional conduct.

May, 1911. AICE adopts first American code of ethics, based on ICE standards.

March, 1912. AIEE adopts Code of Principles of Professional Conduct.

December, 1912. AIChE adopts code of ethics.

June, 1914. ASME adopts code based on AIEE "Principles."

September, 1914. ASCE adopts code comprising parts of AICE code of ethics.

April, 1920. ASME leads in formation of Joint Committee on Code of Ethics; 1922 draft Code of Ethics for Engineers fails to find general acceptance.

May, 1923. AAE adopts Principles of Professional Conduct intended for all engineers; acceptance limited.

January, 1941. ECPD undertakes development of ethical standards for all engineering.

1943 to 1950. ECPD Canons of Ethics for Engineers in process of acceptance by ECPD societies.

June, 1952. NSPE adopts 15 Rules of Ethical Conduct to supplement ECPD "Canons"; both replaced in 1957 by enactment of NSPE Rules of Professional Conduct.

1957. ASCE and AIA jointly produce Interprofessional Principles of Practice for Architects and Engineers.

1963. ECPD enacts major revision to its Canons of Ethics.

1974. ECPD approves new Code of Ethics of Engineers, replacing the former Canons of Ethics.

1974. IEEE enacts a new code independent of ECPD model.

1974. ACEC adopts a new Code of Ethics following ECPD Fundamental Canons and Guidelines.

1976. ASCE adopts modification of ECPD code to replace its traditional Code of Ethics.

Review of Professionalism and Ethics Activities

Only the American Medical Association made the adoption of ethical standards a first order of business at the time of its organization. In law such action came 30 years after the founding of ABA; in architecture the time lapse was 52 years; in engineering 59 years passed between the founding of the first national engineering society (ASCE) and adoption of the first engineering code of ethics by the American Institute of Consulting Engineers. It seems proper to conclude here that these callings did not actually acquire stature as true professions in America until they assumed ethical responsibilities in formal fashion. If so, medicine could be dated as a profession in 1847, law in 1908, architecture in 1909 and engineering in 1911 or later.

In medicine, law and architecture the existence of but one major professional society for each "learned art" has greatly simplified the establishment, administration, and enforcement of ethical standards, as well as related education functions within and outside those professions. In each case the standards are under constant review, and are amended as necessary to meet changing conditions. Standing committees or similar bodies are maintained to provide official interpretation of standards as needed, and to monitor current problems. Procedures for handling ethical violations have been developed through many years of experience, and enforcement has been diligently pursued regardless of cost and occasional legal counteraction.

In engineering, unfortunately, the pattern is not so neatly drawn. With more than 150 national organizations serving the many specialized areas of engineering practice in one way or another, only a handful manifest any real concern for the professional motivation and ethics of their members. If such concern is a basic attribute of a profession, then it is also a requirement in a "professional" society. It would seem that relatively few organizations in engineering can claim that distinction.

Even in the five founder societies the spectrum is broad. There can be no question about the Civils, who were first reluctant but who rose to worldwide leadership by 1955 as the foremost proponent of engineering ethics. The Mechanicals and Chemicals have adopted the new ECPD Fundamental Principles and Canons, and the Civils have adopted a modified version. The Electricals have apparently rejected the ECPD model in adopting a different code. The Miners are staying with their endorsement of the former ECPD Canons, having as yet taken no action on the 1974 revision. Except for the Civils, little visibility has been given to enforcement and educational efforts as related to ethical matters, although AIChE has a current booklet entitled "Professional Standards."

Outside of the founders, NSPE is a driving force, keeping its standards apace of the times, publishing case histories, encouraging its "associate" societies to emphasize ethical activities, and urging its state societies to pursue vigorous enforcement programs. Because it is left to the state societies enforcement lacks consistency and uniformity of emphasis across the nation.

The American Consulting Engineers Council, as a federation of state and local organizations having consulting engineering firms as their members, is limited in its ethical functions to the maintenance of standards and to education services. Imposition of sanctions against member firms because of the indiscretions of individuals is not feasible.

A few other engineering societies have subscribed to the ECPD code. Profession-wide, the emphasis given in engineering to professional motivation and conduct is spotty at best, and is essentially non-existent in far too many areas.

Ethics vs. Etiquette

It is noteworthy that those engaged with the framing of ethical standards in all four of the professions of medicine, law, architecture and engineering at one time or another were confronted with a common problem. This was the decision as to whether or not a distinction should be made between ethics as moral values and details of professional "etiquette" and, if so, how to make such a distinction. Some of the older engineering codes—particularly those of AICE and ASCE—were especially slanted to business development operations of consulting engineers, such as competitive bidding, advertising, unfair competition, attempt to supplant and malicious conduct.

The United States Department of Justice, in the late 1960's, highlighted this situation in dramatic fashion. On the premise that the professions were using their codes of ethics as justification for self-serving procedures, the DOJ cited several professional organizations for violation of the Sherman Anti-Trust Act. The first action was taken against the American Institute of Certified Public Accountants, with the charge that its rule prohibiting the selection of professional services by competitive bidding was restraining trade and interfering with interstate commerce. Similar actions against ASCE and AIA followed in 1971. Consent decree settlements were made by all three organizations, on the reasoning that it was better to give up the ethical standard while retaining the right to advise and educate users of professional services that competitive bidding as a means of selection was not in the public interest. Later, DOJ brought suit against NSPE on the same charges, and that suit is being contested. If not successfully defended, present activities to discourage competitive bidding are likely to be limited even more.

More recently, the Federal Trade Commission has joined the DOJ in the instigation of suits against AMA and ABA, charging that their ethical standards prohibiting advertising by physicians and lawyers are in restraint of trade. In 1975 the U.S. Supreme Court ruled that a mandatory fee schedule established by a state bar association was illegal. And in 1976 the DOJ has again attacked ASCE on the grounds that its prohibition of one engineer from attempting to supplant another who had been selected for an engagement was in violation of the 1972 consent decree.

It is quite clear from these developments that professional codes of ethical conduct must eventually be purged of any rules that interfere with the availability of professional services under the condition of the marketplace. It would seem that the exclusion of rules of business etiquette from codes of ethics is a matter of law rather than choice. Let the buyer beware!

Under these circumstances, it becomes more important than ever that the engineer be dedicated to a strong idealism that will insure his constant respect for the public welfare.

It will be noted that the campaign to emasculate professional ethical standards is not the result of a national policy dictated either by the executive or legislative arms of government. Instead, the movement has arisen from federal bureaucratic decisions in the General Accounting Office, the Department of Justice and the Federal Trade Commission. Let the public also beware!

THE ENGINEERING ETHIC

Because of its heterogeneity, engineering sorely needs a common engineering ethic, as the first step toward the unity that has so long been sought but never attained. The sharing of a moral foundation—a binding sense of idealistic motivation and purpose—would provide the single cornerstone upon which to build other blocks of commonality and cooperation.

Simply stated, the engineering ethic° is:

Service of the public interest with integrity and honor.

Clothed in inspirational language and set forth in an effective format, this ethic should be brought by every resource available into the heart and mind of every engineer, regardless of his technical specialty or mode of practice. It should be made as inherent to the art of engineering as is basic calculus although, hopefully, it would be remembered longer!

The engineering ethic as stated here is not specific, but it is nevertheless amenable to enforcement. It is readily possible for a body of peers to judge whether the action of a colleague is dishonest, dishonorable, or in violation of the public trust. Ethical norms need not be subject to the limits of law; they encompass a greater breadth of human conduct than can be defined by law.

The etiquette of professional practice, insofar as it may be dictated after the federal bureaucracy has its way, should be set forth separately, yet always within the concept of the engineering ethic. This is already the format of the 1974 ECPD Code of Ethics of Engineers. The ECPD etiquette rules, called "Guidelines," are classified into relationships with the public, with the client or employer and with other engineers. In this form they have not found universal acceptance by the engineering societies.

Does it not seem logical that classification of these guidelines according to the major modes of engineering practice might make them more applicable to the operations of the engineer at work? The special rules that apply to the engineer in industrial practice, private practice, public practice and education might be so arranged, if in this form they might have more meaning and have more influence upon the actions of the individual.

Engineers' Council for Professional Development is obligated by its very name to enhance professionalism, and it is finally beginning to receive support in its efforts to further ethical standards. Some of the societies (such as ASCE) which were once a part of the problem are now trying to aid in the solution. There is still, however, a frustrating lack of interest in ethics and professionalism, both within and outside the family of ECPD societies.

It is now broadly recognized that the inculcation of professional motivation and ethical consciousness is a part of the education process. It is thus more than ever appropriate that ECPD assume leadership and be supported as the medium for promulgation and establishment of the engineering ethics. The present ECPD Fundamental Principles and Fundamental Canons are already such a declaration, and the Guidelines are already a sound set of working rules of professional practice. They should not be allowed to wither on the vine.

Our objective must be to make the engineering ethic a part of the conscience of every student and practitioner of engineering. If the participating societies of ECPD will sincerely ally themselves to achieve this objective—laying aside their past disinterest, provincialism and chauvinism—engineering will acquire at last the professional maturity that it should have attained long ago.

°This might more appropriately be termed the "professional" ethic, as it is equally applicable to any profession.

1.4

The Engineer and Business

EDWIN T. LAYTON, JR.

The engineer is both a scientist and a businessman. Engineering is a scientific profession, yet the test of the engineer's work lies not in the laboratory, but in the marketplace. The claims of science and business have pulled the engineer, at times, in opposing directions. Indeed, one outside observer, Thorstein Veblen, assumed that an irrepressible conflict between science and business would thrust the engineer into the role of social revolutionary.[1]

While nothing like a soviet of engineers has appeared, the tensions between science and business have been among the most important forces shaping the engineer's role on the job, in his professional relations, and in the community at large. Veblen assumed that science and business made mutually exclusive demands on the engineer; in fact, however, they often complement one another. Both, for example, may benefit from technological progress. Nor is the existence of tensions necessarily detrimental to engineering work: attempting to resolve them may account for some of the engineer's drive and creativity.

Despite his mordant irony, Veblen was, in one sense, an optimist; he assumed that the tensions between business and science were resolvable, if only through a cataclysmic destruction of the former by the latter. In this he missed the essence of the engineer's dilemma which is, at base, bureaucracy, not capitalism. The engineer's problem has centered on a conflict between professional independence and bureaucratic loyalty, rather than between workmanlike and predatory instincts. Engineers are unlikely to become revolutionaries because such a role would violate the elitist premises of professionalism and because revolution would not eliminate the underlying source of difficulty. The engineer would still be a bureaucrat. Tensions with business have been dominant because in the American context economic development has been carried out principally through the agencies of private capitalism. But engineers in government have experienced quite analogous conflicts, if anything more severe than those of privately employed engineers; and it can be argued that the market system, in providing a final test acceptable to both the engineer and his employer, has served to buffer discord between the two. Perhaps the engineer's problem ultimately is marginality; he is expected to be both a scientist and a businessman, but he is neither. A social revolution would merely alter the terms of his marginality without ending it.

The engineer's relation to bureaucracy is not new; he is the original organization man. The scientifically trained, professional engineer has characteristically appeared on the technical scene at the point of transition from small to large organizations. Economically, it makes little sense for small enterprises to employ engineers; the gains are not worth the costs. Large corporations, on the other hand, can more readily support engineers and research establishments, since they represent a small percentage of their total costs. Large corporations can get substantial net returns from rather small percentage gains in efficiency. Where large investments are at stake, the engineer can serve a useful function in eliminating guesswork and minimizing risks. Technically, large works are more likely to involve complexities than are small ones; and the larger the project, the more likely it is that such difficulties will transcend the capabilities of artisans and businessmen.

In the eighteenth and through much of the nineteenth century, America developed quite diverse and advanced technologies without requiring a corps of scientifically trained experts. The Mississippi River steamboat, for example, was developed in an era of relatively small, highly competitive enterprise by the cut-and-try methods of the practical mechanic, rather than by the rational analysis of scientists. It was large-scale organizations, such as the navy and private corporations, that supported experts who could approach the steam engine through the laws of thermodynamics. Similarly, it was not the small ironmasters who first called on the aid of science, but the corporate giants of the post-Civil War era. From the start, engineers have been associated with large-scale enterprises.[2]

There were two stages in the emergence of the engineering profession in America in the nineteenth century. The first demand for engineers came from the construction of large public works, such as canals and railroads, particularly in the period 1816 to 1850. The organizations that undertook these works were pioneers in technology and were also among the largest enterprises in America, representing aggregations of capital that were huge for their day. The civil engineering profession was called into being to meet the technical needs of these organizations.

In 1816, the engineering profession scarcely existed in America. It has been estimated that there were only about thirty engineers or quasi-engineers then available; but by 1850, when the census first took note of this new profession, there were 2,000 civil engineers. Canal and railroad construction generated not only the demand for engineers but, in large measure, the supply as well. From an early stage, organizations employing engineers found it convenient to group their technical staffs into a hierarchy of chief engineer, resident engineers, assistant engineers, and the like. Within this bureaucratic context, regular patterns of recruitment and training emerged on the job, and early engineering projects like the Erie Canal and the Baltimore and Ohio Railroad became famous as training schools for engineers.[3]

The rising demand for engineers by industry began the second stage in the emergence of the engineering profession. The golden age for the application of science to American industry came from 1880 to 1920, a period which also witnessed the rise of large industrial corporations. In these forty years, the engineering profession increased by almost 2,000 percent, from 7,000 to 136,000 members. The civil engineer was overshadowed by the new technical specialists who emerged to meet the needs of industry: by the mining, metallurgical, mechanical, electrical, and chemical engineers. The astonishing growth of engineering continued, though at a less rapid rate, after 1920. In 1930, there were 226,000 engineers when the depression put a brake on expansion; by 1940, the number was but little higher—260,000. Postwar prosperity increased the size of the profession past the half-million mark in 1950, and to over 800,000 by 1960. Engineering is by far the largest of the new professions called forth by the industrial revolution.[4]

The rise of the engineering profession was accompanied by a scientific revolution in technology. The change was not a sudden one; in most cases engineering built upon and extended traditional techniques. But professionalism was associated with a slow incorporation of scientific methods and theory into technology and the accumulation of an esoteric body of technical knowledge. Professionalism was a means of preserving, transmitting, and increasing this knowledge. The transition from traditional rule-of-thumb methods to scientific rationality constitutes a change as momentous in its long-term implications as the industrial revolution itself.

The development of engineering education constitutes a sensitive indicator of the shift from art to science in technology. The early civil engineers were educated by self-study and on-the-job training. Only a minority received a college degree. By 1870, there were twenty-one engineering colleges, but only 866 degrees had been conferred. College education became increasingly common after 1870. By 1896, there were 110 engineering colleges. The number of students increased rapidly, from 1,000 in 1890 to 10,000 in 1900. Only with the twentieth century, however, did the college diploma become the normal means of admission to engineering practice.[5]

College training was a sign of a greater emphasis on science. But even here the change was evolutionary. The engineering curriculum of the latter nineteenth century placed as much or more emphasis on craft skills as upon scientific training. In 1875, Alexander L. Holley, a prominent mechanical engineer, argued that all the great engineering triumphs of his day owed more to art than to science. The aim of college education, Holley maintained, was to train "not *men* of good general education, but *artisans* of good general education," for as Holley emphasized, "the art must precede the science."[6] Despite an increasing emphasis on science, engineering educators down to 1920 seriously debated whether engineering students ought to learn the calculus. Some of them thought such courses were "cultural" embellishments to the curriculum. Only since the end of the Second World War has the balance in engineering education shifted unequivocally toward science.[7]

The origin of engineers carries with it built-in tensions between the bureaucratic loyalty demanded by employers and the independence implicit in professionalism. It is important to note, however, that these tensions are not, as Veblen thought, the outgrowth of a clash between the rationality of science and the irrationality of capitalism. The scientific knowledge possessed by the engineer is highly rational, but his professionalism derives from the mere possession of esoteric knowledge, not its specific content. Incomprehensibility to laymen, rather than rationality, is the foundation of professionalism. In essence, the professional values adopted by American engineers are the same as those of other professions. They may be summarized under the headings of autonomy, colleagual control of professional work, and social responsibility.[8]

Perhaps the most invariant demand by all professions is for autonomy. The classic argument is that outsiders are unable to judge or control professional work, since it involves esoteric knowledge they do not understand. Autonomy operates on at least two distinct levels: it applies to engineers in their corporate sense as an organized profession and to the individual engineer in relation to his employer. In both cases, conflicts have appeared between business demands and the ideal of professional independence. Businessmen usually concede that engineering societies should be free of external control, but in practice business domination is not uncommon. Employers have been unwilling to grant autonomy to their employees, even in principle. They have assumed that the engineer, like any other employee, should take orders. Some engineers, however, have maintained that the engineer, like the doctor, should prescribe the course to be followed and that the very essence of professionalism lies in not taking orders from an employer.[9] The employer, of course, has the power to reward and punish. But the value of the engineer generally hinges on his being a professional in the sense of being both a master of an ever-growing body of knowledge and a creative contributor to that knowledge. Such men are the ones most likely to be inspired by professional ideals. As a result, the role of the engineer represents a patchwork of compromises between professional ideals and business demands.

The argument for colleagual control of professional work is closely related to that for autonomy. Since professional work cannot be understood fully by outsiders, the person in charge of such work should be a member of the profession. In this manner, doctors have insisted that the heads of hospitals and medical schools should be members of the medical profession. In the same vein, engineers have maintained that engineers should be in charge of engineering work.[10] In practice, engineering departments are usually headed by engineers. But in the case of engineering, this principle can be extended much further. Engineering is intimately related to fundamental choices of policy made by the organizations employing engineers. This can lead to the assertion that engineers ought to be placed in command of the large organizations, public and private, which direct engineering. This is tantamount to saying that society should be ruled by engineers. A more representative manifestation of the ideal of colleagual control has been the repeated demand by the profession for reform of governmental public-works policy. Engineers have advocated the creation of a cabinet-level department of engineering to be headed by a civilian engineer.[11]

Although the arguments for autonomy and colleagual control are fundamentally similar, one of the basic dilemmas of modern engineers has been that these two goals are not completely compatible. Organizations like the federal government or a modern corporation have other ends in view than the best and most efficient engineering. Doctors are in a more fortunate position, since it may be assumed that the professional aim—health—is identical to the ends of the large organizations employing doctors, such as hospitals and medical schools. But unlike medicine, engineering serves purposes ulterior to itself. Colleagual control of engineering implies, in the extreme, a change in basic social values, to make those of engineering supreme. Such aspirations open up the possibility that outside organizations might reciprocally seek to control engineering, as something potentially dangerous to their purposes.

Professional freedom implies social responsibility. The professional man has a special responsibility to see that his knowledge is used for the benefit of the community. Social responsibility points in two directions: inwardly, at self-policing to prevent abuses by colleagues, and outwardly, to the making of public policy. In either case, it is with social responsibility that professionalism comes most clearly into conflict with bureaucracy. This may be seen in two possible meanings of the term "responsibility." On the one hand, there is the bureaucratic sense implied in the phrase "responsible public official." This denotes responsibility in executing policies, but not necessarily in formulating them. On the other hand, the term "professional responsibility" entails an independent determination of policy by the professional man, based on esoteric knowledge and guided by a sense of public duty. An assertion of professional responsibility, therefore, may signify a rejection of bureaucratic authority. In this manner, the scientists' crusade against the May-Johnson Bill for postwar control of atomic energy was both an assertion of a professional responsibility and a rebellion against General Groves and the formal hierarchy of the Manhattan Project.

For engineers, the most overt element of professionalism has been an obsessive concern for social status. Although the income and the power of engineers would of course be enhanced by professionalism, these ends have been given second place, at least verbally. Professionalism has induced engineers to seek greater deference, in particular, to gain the same social recognition accorded to the traditional learned professions, law and medicine. Spokesmen for the engineering profession have, in fact, frequently made status the fundamental aim, and other professional values means to this end. Thus, engineers have argued that in order to gain more status their profession should show a greater sense of social responsibility.[12]

Although engineers emphasize the importance of status, it is not clear that this distinguishes their goals from those of other interest groups. Engineers differ from nonprofessional groups chiefly in that they are more likely to rationalize their ambitions in terms of protecting the public. Following the cue of the older professions, engineers have secured the enactment of licensing laws, and they have endeavored to raise standards in education and practice. Such measures enable professions to limit competition and alter supply-and-demand relationships in their favor. Engineers have hoped to achieve in this way many of the same goals sought by labor unions through such devices as the closed shop and strike. Professional autonomy and control of the profession's work by colleagues, if fully realized, would lead to control of the conditions of work, an end pursued equally, if by other means, by labor unions. As an interest-group strategy, professionalism offers several advantages. It is "dignified," since the professional abjures "selfish" behavior, at least verbally, and gains group advantage on the pretext of protecting the public. Professionals emphasize the intimacy of the personal relationship with clients, rather than the cash nexus between the buyer and seller, employer and employee.

Professionalism for engineers is not exclusively a matter of esoteric knowledge. Engineers do not seek autonomy simply because they are professionals; to some extent they have adopted professionalism as a way of gaining autonomy. Professionalism was, in part, a reaction against organization and bureaucracy. It was a way to prevent engineers from becoming mere cogs in a vast industrial machine. Thus, in 1939, Vannevar Bush made an eloquent plea for a professional spirit in engineering. Without this spirit, Bush thought,

> we may as well resign ourselves to a general absorption as controlled employees, and to the disappearance of our independence. We may as well conclude that we are merely one more group of the population . . . forced in this direction and that by the conflict between the great forces of a civilized community, with no higher ideals than to serve as directed.[13]

Professionalism carries overtones of elitism that grate against the egalitarian assumptions of American democracy. Professionalism stresses hierarchy and the importance of the expert; its emphasis is on the creative few, rather than on the many. But professionalism, not itself democratic, may serve democratic ends. It is one means of preserving the ideal of the autonomous individual, without which democracy could scarcely exist. Democracy requires not only freedom, but an informed public opinion. One of the problems of the modern age is that many issues of public policy involve technical matters. Independent and informed judgments of these questions are badly needed by the public. Professionalism, because of its stress on social responsibility, offers one way of meeting this need by establishing a legitimate role for private judgment by engineers, protected and encouraged by an organized profession.

Whether the emphasis be on esoteric knowledge, public service, or selfish interest, professionalism requires that engineers identify themselves with their profession. Engineers must think of themselves as engineers before they can constitute a profession. There are several factors that link engineers together as a group and encourage self-consciousness. As with all professions, the fundamental tie is a common body of knowledge. Self-interest is another powerful cohesive force. Both find a natural focus in the professional society, which brings engineers together and gives them a sense of corporate identity. These factors have helped to produce a steady push toward professionalism among engineers.

Professionalism, however, has met powerful resistance from business. Business has been reluctant to grant independence to employees. The claims of professionals rest fundamentally on esoteric knowledge; to reduce the importance of this knowledge is to weaken the engineers' aspirations for autonomy. To some extent this may take the form of a depreciation of "theory" and an emphasis on practicality. But if businessmen tend to depreciate esoteric knowledge, they cannot wish it away. Experts have been of increasing, not diminishing importance. A more pervasive and effective argument is the priority of business needs over technical considerations.[14] This contention is not without force. The transition from art to science in engineering has been slow and partial. Even where technical knowledge has achieved a high state of perfection, its importance is limited by the exigencies of business. Engineers work in complex organizations. Engineering is only one factor among many that contribute to their success.

In the long run, the most effective check on professionalism by business has been a career line that carries most engineers eventually into management. The promise of a lucrative career in business does much to ensure the loyalty of the engineering staff. Conversely, it undermines engineers' identification with their profession. Social mobility carries with it an alternative set of values associated with the businessman's ideology of individualism. These values compete with, and to some degree conflict with, those of professionalism. Thus, professionals stress the importance of expert knowledge, but businessmen stress the role of personal characteristics, such as loyalty, drive, initiative, and hard work. Professions value lifetime dedication. But business makes engineering a phase in a successful career rather than a career in itself. Insofar as business treats engineering merely as a stepping stone to management, it represents a denial of much that professions stand for.

It is possible to distinguish several stages in the typical engineering career. The most important source for understanding the engineer's background is a study conducted by the Society for the Promotion of Engineering Education in 1924, as part of a large survey of engineering education. It was based on a questionnaire administered to 20 percent of all engineering freshmen admitted in the fall of that year, 4,079 students selected from thirty-two engineering schools so as to constitute a representative cross section. A very large proportion of the parents of these students were members of the old middle class: 42.5 percent were owners or proprietors of businesses. Of the remainder, more than one-quarter were members of the new middle class: 28.2 percent were employed in executive or supervisory positions and 5.6 percent were engineers or teachers. Another 13 percent of the total were skilled workers, but only 2.7 percent were unskilled workmen and 3.5 percent were clerks.[15]

The engineering students surveyed by the 1924 study were drawn from the poorer and less well-educated segments of the middle class. Almost all of the parents who were owners or proprietors were engaged in small mercantile enterprises or farming. Only 13 percent of the fathers had a college degree. Forty percent had graduated from high school, and the same number had a grammar school education or less. Sixteen percent had started but not finished high school. That the parents were not especially well-off is further suggested by the fact that 90 percent of the freshmen had to work a year before starting college.[16]

The parents of these students were overwhelmingly of Anglo-Saxon or northwestern European stock. Of the students, 96.2 percent were native-born, as were 73.6 percent of their parents and 60.7 percent of their grandparents. Of the grandparents not born in the United States or Canada, two out of three were from northwestern European countries. Only 10 percent of all the grandparents were from Latin, Slavic, or other countries.

More than three out of five of the students' families lived in small towns, villages, or on farms. Only 38.7 percent came from cities with a population of over 25,000.[17] A recent study of similar scope indicates that the proportion of engineers drawn from families of blue-collar workers has significantly increased. Presumably this has affected the ethnic background of the engineers concerned. But the large city continues to be underrepresented in the profession.[18]

The selectivity in recruitment of engineering students from middle-class, small-town, and old-stock families goes far to account for the profession's strong commitment to traditional individualism. One engineer, in tracing his own faith in rugged individualism to the self-reliance of his father, suggested that his own experience was typical of his generation.[19] Another engineer, commenting on the frontier individualism of his home community, noted that "no one growing up in such an environment could escape being influenced by it."[20] A frequent theme in discussions of success in engineering is the importance of an "inner drive," or an "inner urge," or more simply of initiative.[21] It is perhaps no accident that the term "rugged individualism" was given currency by an engineer, Herbert Hoover. But predisposition to individualism also provides the foundations for the development of a commitment to business.

Engineering education is susceptible to business influence in a number of ways. Businessmen serve as trustees of colleges, on alumni boards, and on committees of technical societies concerned with technical education.[22] Engineering educators suffer from the same divided loyalties as other engineers, and some of them have been important spokesmen for a business point of view. Many businessmen, and some engineering educators, have assumed that the buyer, business, has the right to determine both the technical skills to be taught and the ideas to be implanted in students' minds. Business demands on this score have ranged from training in "sound economics" to concern for international responsibilities.[23] In 1932, a group of engineering educators at Ohio State University became interested in training engineering students in social responsibilities; but before undertaking anything definite, they prudently sampled the opinions of business leaders on major issues of the day. One of the educators noted that the "industrial system" was not then working well. He reflected:

> What would happen if these socially awakened young men should come to the conclusion that its very fundamentals are wrong. . . . Manifestly such young reformers might find a cold reception in industry. . . . It is therefore well to ascertain what leading executives think about such matters.[24]

Despite strong business influence, however, the feeling of professional independence is probably stronger in engineering schools than elsewhere. Educators tend to think of themselves as independent practitioners. Shielded by traditions of academic freedom and in contact with other professions, engineering educators have been in a position to assert a professional commitment. It is not surprising that educators have been prominent in movements to uphold and advance professionalism. Perhaps the educators' most important impact has been on their students whom they inculcate with professional ideals.

The next stage in the engineer's development is that of professional engineering proper. The transition is quite marked geographically and environmentally, since most engineers leave their home states on completion of college—59 percent, as against an average of only 38 percent for all other college graduates.[25] The young graduate does not qualify immediately as a full-fledged professional. Four years at college are barely sufficient to lay the foundations upon which the young man must continue to build for a lengthy period thereafter. This process is not simply one of absorbing more knowledge; managerial and personal skills come into play also. The final judge of the young man's gradual advance is as much or more his employer as his professional colleagues. Major engineering societies recognize four stages in the engineer's development, which they embody in their grades of membership: from student to junior, to associate, and finally to full member. But most such societies place as one of their requirements for higher grades of membership that the candidate be in "responsible charge" of engineering work, a test as much of his bureaucratic status and business success as of his engineering knowledge. Success in business overlaps and may eclipse success in the engineer's profession.

Most engineers work in industrial bureaucracies, which are capable of exerting a considerable amount of pressure on the individual. The effectiveness of this influence is heightened when the individual seeks not only to keep his job, but to rise in the hierarchy. The price of success, in at least some cases, is total and undivided loyalty. An engineering educator made this point in 1935 when he asked a group of young engineers:

> Are you truly loyal to your employer's interests?
> Do you try to advance these on every opportunity?
> Do you have a fighting spirit for the reputation of 'our company' and 'our products'?
> The merging of your employer's interests into your own is one of the surest signs of real progress in business life.[26]

Conformist pressures on the job are not limited to company loyalty; they extend to all sorts of social and political norms. As one engineering society president, J. F. Coleman, noted, the employed engineer cannot be as open in expressing his opinion on socio-economic subjects as one who is a free agent.[27] One young engineer complained, in 1942, that on the job "everything possible was done to make alert youth conform to routine. Ideas were discouraged, social and political interests frowned upon."[28] Pressure appears to be especially great against heterodox ideas. A young mechanical engineer wrote, in 1941, that many engineers favored joining unions, but that they were "under the terror of jeopardizing their connections by coming out into the open with their opinions."[29] Another engineer wrote to a leading engineering journal, in 1940, "to my mind for individual engineers themselves to have published in *Mechanical Engineering* their own views of what they would like is very likely to bring about some action which may put us all on the spot."[30] A distinguished civil engineer, Daniel W. Mead, in 1929 advised young engineers to avoid unorthodox dress, since "conspicuous dress usually shows unfortunate idiosyncrasies which need to be eliminated."[31]

Employers have occasionally urged engineers to show more social responsibility. But by this they generally have in mind the defense of business, rather than independent action by the profession. Engineering support for business is often taken for granted. As one mining engineer told a group of younger engineers in 1931:

> We as employees owe the company our best efforts not only while at work but at all times, promoting always the company's interests by representing them favorably. This we can best do by active, harmonious participation in the community social, civil, and spiritual activity.[32]

He thought the engineer could best serve his employer by being a "booster imbued with the spirit of boosting at all times."[33] H. B. Gear, an engineer who became the vice-president of the Commonwealth Edison Company of Chicago, in

1942 maintained that engineering best served society as an "implement of management."[34] It is not clear whether Gear had in mind the sort of social service provided by Samuel Insull, the founder of Commonwealth Edison, but in any case Gear's idea of the preeminence of business considerations left little room for an assertion of social responsibility by an independent profession.

The coercive element with bureaucracies should not be exaggerated, however. Pressure internalized within the individual are far more effective. One of the most acute observers of the profession, William E. Wickenden, noted that engineers seeking social premises look to "the presuppositions of the man higher up who sets the engineer at his problem and reserves final judgment on his recommendations."[35] An ardent advocate of professional values, Arthur E. Morgan, sadly noted that "the engineer tends to reflect, somewhat uncritically, the social attitude of his employer."[36] He implored engineers to abandon their "moral servitude" and develop an independent philosophy.[37]

The final phase of the engineer's development, not achieved by all, is the transition from professional engineering to business management. A survey of the six largest American engineering societies, in 1946, revealed that over one-third of their membership was engaged in management.[38] A study of 5,000 engineering graduates by the Society for the Promotion of Engineering Education, in the 1920s, suggested that, on the basis of their sample, over 60 percent of all engineers made this shift eventually. There was a secondary drift from technical to nontechnical management and administration. Half had made this change by forty years of age, and the proportion increased to three-quarters after fifty-five.[39] No doubt monetary rewards have played an important role in encouraging engineers to move into management. Several studies of engineers have stressed the large premiums paid for managerial work as against purely technical engineering.[40] One result of this trend has been that engineers constitute a large and increasing segment of the upper echelons of business. Mabel Newcomer found that the proportion of engineers in the managerial elite of presidents and board chairmen grew from one in eight in 1900 to almost one in five in 1950.[41]

It is not uncommon for engineers who have risen into managerial positions to think of themselves as businessmen. John Mills, a discerning observer of his profession, noted that as engineers climb into supervisory positions, they tend to be blinded by visions of future promotions and to lose sight of colleagues lower down.[42] Although many such men no longer think of themselves as engineers, they may retain professional-society membership for business reasons. But professional commitment is clearly diminished. Others sever their ties with engineering completely. The 1960 census, which relied on the self-identifications of respondents for occupations, counted 541,000 engineers and scientists in manufacturing, the bulk of whom were engineers. A follow-up survey by the National Science Foundation discovered an additional 73,000 technically qualified persons who had been missed because they had identified themselves with nontechnical occupations, presumably managerial.[43]

A number of engineers have looked to the major professional associations to offset the power of business. Arthur E. Morgan, the first head of the Tennessee Valley Authority, was deeply concerned by the fact that the engineer employed in a bureaucracy "tends to be not a free agent, but a technical implement of other men's purposes." But he pessimistically concluded that "the lone individual is largely helpless."[44] Morgan, however, hoped that engineering societies might enable their members collectively to gain the freedom they lacked as individuals. He was aware this would not be easy; he pointed out:

> For engineering associations to be a democratic force in the public interest, they must be more than the summing up of the selfish interests of their members, and more than the reflections of the views of their several employers. . . . Otherwise an engineering organization may be only a more powerful means of regimenting the members along lines of policy which their employers happen to be following.[45]

The professional ideal expressed by Morgan was difficult to attain because business influence had penetrated the very citadel of professionalism, the engineering society. The most important agents of business influence within the organized profession are engineers who have moved into management; they constitute a large and powerful minority of senior and successful men. As the editor of *Mining and Metallurgy* observed: "naturally such men are highly influential in establishing the policies of engineering societies."[46]

Engineers who have risen into top management present a dilemma for engineering societies. Although it has been generally conceded that engineers engaged in technical management are still engineers, this is far from clear in the case of those who have gone up the corporate hierarchy to positions involving general management and administration. Some engineers have argued that such men are no more practicing their profession than a lawyer in a similar position is practicing his.[47] This argument implies that such men should be excluded from the higher grades of membership of engineering societies and, hence, from effective control of their profession. Other engineers have taken a contrary position. They have maintained that the engineer in top management has fulfilled the engineer's cherished ideal of success and that even those who have lost any active interest in technical matters should be encouraged to join engineering societies and participate in their government. In 1953, Frederick S. Blackall, Jr., the president of the American Society of Mechanical Engineers, maintained that

> if ASME . . . is one of the most important factors in moulding the character and competence of the technical staff, then too, most certainly the men of top management should make it their business to have a voice in its affairs. Here will be found a splendid sounding board for their views and platforms for their leadership.[48]

However, the fact that engineers in management tend to lose their identification as engineers presents a further difficulty. In 1927, for example, Blackall's society elected Charles M. Schwab as its president; yet in his *Who's Who* autobiography, Schwab listed himself as a "capitalist," down to 1934, and as a "steel manufacturer" thereafter; he did not mention his presidency of the American Society of Mechanical Engineers or even his membership in that organization.[49]

To some extent business influence is checked by the power of majority rule. But engineering societies are not perfectly functioning democracies, and there are some serious limitations to the control that rank-and-file members can exercise. Some societies long denied younger members the right to vote; even without formal restrictions on voting, the idea of a hierarchy of professional excellence implicit in membership grades tends to give power to the senior full members and to deny significant influence to junior members. In certain specific instances, employers have canvassed their technical staffs to line-up votes on particular issues. But more commonly, widespread ignorance and apathy give inordinate power to comparatively small minorities. Most members live in remote parts of the country and are little interested in society affairs. Nominations are controlled by committees whose actions are usually shrouded in secrecy; the election presents a mere formality.[50]

It is considered bad form to publicize the inner workings of engineering societies. Even in those rare cases where elections have been contested, both parties have usually taken precautions to keep the real issues secret. In 1914, Morris L. Cooke led a revolt against excessive dominance of the American Society of Mechanical Engineers by the utilities. His election to the society's governing board was opposed by a rival candidate put up by the utility interests. Not only did no mention of this appear in the ASME's official publications, but it was deliberately obscured in the informal circulars distributed by both sides. The opposition based its position on a technicality in the manner of Cooke's nomination. Cooke, though a foremost advocate of publicity in engineering-society affairs, also avoided the real issue for tactical reasons. In writing to a close friend concerning his circulars, Cooke noted that "the utility matter is kept in the background and can be used with telling effect, but we do not draw on ourselves the charge of washing the society's soiled linen in the public."[51]

Given the secrecy surrounding engineering-society affairs and the indifference of most members, the machinery of society government offers many opportunities for small but active groups to exercise control. Three committees have been particularly important as sources of business influence. They are those concerned with nominations, new members, and publications. By controlling nominations, a group can make sure that engineers sympathetic with its interests are in strategic positions in the society's government. Control of membership, in the long run, is the most important power of all, since this committee does much to determine the future composition of the society. Control of membership committees can also be employed to keep out of a particular society individuals not acceptable to particular business interests. Edward W. Bemis, who often took the public side in utilities-valuation cases, was advised not to apply for membership in the American Society of Mechanical Engineers because of the power of the utilities in the committees concerned with passing on new membership applications.[52]

Perhaps the most influential committees of engineering societies are those concerned with publications. Such committees can control the publication of technical papers, censor heretical opinions, or silence proposals that might embarrass particular business interests. The electric utilities have been especially active and successful in this respect. Their power in the American Institute of Electrical Engineers led to a prohibition of papers dealing with costs and rate making. According to Governor Pinchot of Pennsylvania, this practice seriously hampered the efforts of public agencies to regulate utilities.[53] Other societies, lacking such a prohibition, have been prevented from publishing papers opposed by utilities interests. In May of 1938, for example, Gregory M. Dexter submitted a paper to the American Society of Mechanical Engineers that dealt with electric rates of a subsidiary of Consolidated Edison in Scarsdale, New York. A long series of delays and rewritings ensued. Dexter claimed he would be asked to write it one way, and then he would find the "rules" had been changed and he would have to write it another way. Not until January, 1942, was the paper finally rejected. On the committee that ruled the paper unacceptable were several engineers connected in various ways with utilities interests, among whom was an advisory member who held the position of assistant engineer for Consolidated Edison.[54]

Virtually all of the major engineering societies in America have had more or less celebrated cases of censorship, involving a broad spectrum of business interests. Within the American Institute of Mining Engineers, for example, the representatives of the coal and oil industries have been accused of censorship. In 1916, W. H. Shockley claimed that a paper of his was suppressed at the instance of the anthracite section because it presented government statistics showing the wages of coal miners were inadequate.[55] In 1922, another member, Robert B. Brinsmade, complained that a paper of his was censored because of certain comments about oil policy. He sarcastically suggested that his letters would enable members to avoid opinions "which have been put on the prohibited index by the august Committee of Publications."[56]

This sort of censorship is made more effective by the codes of ethics adopted by some societies. These codes may prohibit criticism of fellow engineers and the discussion of engineering subjects in the general press. The codes of ethics of the American Institute of Electrical Engineers and the American Society of Mechanical Engineers both provided that "technical discussions and criticisms of engineering subjects should not be conducted in the public press, but before engineering societies, or in the technical press."[57] These codes of ethics are seldom enforced, so the sanction is perhaps limited. But it is not trivial, either. In order to circumvent the restriction on publication outside the society, Frederick W. Taylor had one of his works privately printed and gave a free copy to each member of the American Society of Mechanical Engineers prior to general circulation.[58] This expedient, however, was available only to those with considerable means.

Censorship, though not absolute, is a significant deterrent to free publication by engineers. In 1932, it led to the expulsion of Bernhard F. Jakobsen and James H. Payne from the American Society of Civil Engineers. Payne and Jakobsen, in 1930, had exposed certain malpractices in connection with the construction of a dam for Los Angeles County. They published their findings in the pages of a local newspaper, and they were critical of the chief engineer, among others. The two engineers who made the exposure were then expelled from the American Society of Civil Engineers for unprofessional conduct. It was later established that the contractor had bribed the chairman of the Board of Supervisors, who was sent to jail. The contractor was forced to return over $700,000 to the county. Despite a plea from the Los Angeles County Board of Supervisors to reconsider, the ASCE refused to reinstate Jakobsen and Payne.[59]

In many matters, such as censorship, engineers who hold managerial positions have been the principal agents of business influence in engineering societies. But they are by no means the only source of business power within the profession. Even if such men were excluded from the higher grades of membership—a favorite remedy for some reformers—business influence would remain substantial.

Virtally all American engineering societies are financially dependent on business. This is clearest in those smaller societies that have company members; here the subsidy is direct.[60] Other engineering societies are nominally supported solely by the dues of members. But in fact, they receive indirect financial support from business. A substantial number of members have their dues paid by their employers. A survey in 1947 revealed that, in the sample studied, some 30 percent of the employers regularly paid the dues of some of their employees in certain societies.[61] In 1940, a spokesman for the membership committee of the American Institute of Mining Engineers complained that qualified men were not joining the institute because they were waiting for their employers to pay their dues; he found it necessary to remind them that membership in the institute was a personal matter.[62]

A second form of indirect business subsidy is that of employers paying traveling expenses and allowing time off for employees to attend engineering-society meetings, especially when presenting a paper, serving on a committee, or participating in a discussion. The official journal of the American Society of Mechanical Engineers, *Mechanical Engineering*, commented editorially that engineering societies must accept such aid, since without it they "would have to shut up shop, as a majority of the work of engineering societies is done by its members who are encouraged by their employers to do this work."[63]

Business subsidies pose a delicate problem for engineers. Some societies have openly admitted the practice and defended it. *Mechanical Engineering* urged employers to look on engineering-society activities of an employee as an "assignment" and suggested that the employer "send him to a meeting, or encourage his participation in some other form of society activity, with the feeling that he is representing his company's as well as his own interests."[64] Other engineers, concerned with professional independence, have argued that such practices are harmful. One engineer maintained that subsidies from business made their recipients unfit to represent the engineering profession, and he urged support of the National Society of Professional Engineers since it was "for the benefit of the engineering profession primarily and not for the advancement of corporation interests."[65] A further problem here is that "pure" professional organizations, such as the NSPE, do not publish technical papers. It would be financially impossible for them to do so, even if employers would permit publication of papers by their employees by an organization outside their influence.

Although there are significant areas of conflict between business and professionalism, the dimensions of the clash should not be exaggerated. Neither could exist under present circumstances without the other. Modern business needs highly esoteric technical knowledge, and only professionals can supply it. Technologists need organizations in order to apply their knowledge; unlike science, technology cannot exist for its own sake. The problem has been to find suitable mechanisms of balance and accommodation. One of the basic problems of American engineers is that the balance has tended to shift too far in the direction of business, and accommodation has taken place largely on terms laid down by employers. The professional independence of engineers has been drastically curtailed. The losers are not just engineers. The public would benefit greatly from the unbiased evaluations of technical matters that an independent profession could provide. American business too might profit in the long run from the presence of a loyal opposition.

Notes

1. Thorstein Veblen, *The Engineers and the Price System* (New York, 1933), 70–76.

2. Louis C. Hunter, *Steamboats on Western Waters* (Cambridge, 1949), 175–180, 307–309; William F. Durand, *Robert Henry Thurston* (New York, 1929), 201, *passim*; W. Paul Strassman, *Risk and Technological Innovation: American Manufacturing Methods During the 19th Century* (Ithaca, 1956), 28–46.

3. Daniel Hovey Calhoun, *The American Civil Engineer, Origins and Conflict* (Cambridge, 1960), 22, 27–29, 48–50.

4. U.S. Bureau of Census, *Sixteenth Census, Population, Comparative Occupational Statistics for the United States, 1870–1940* (Washington, 1943), p. 111, table 8; Jay M. Gould, *Technical Elite* (New York, 1966), 172. Here and elsewhere I have rounded off figures for total number of engineers to the nearest thousand.

5. Society for the Promotion of Engineering Education, *Report of the Investigation of Engineering Education, 1923–1929*, 2 vols. (Pittsburgh, 1934), I, 541–547 (hereafter cited as SPEE, *Report of Investigation*), and David L. Fiske, "Are the Professions Overcrowded?" *Civil Engineering*, IV (June, 1934), p. 16, table I.

6. Alexander L. Holley, "The Inadequate Union of Engineering Science and Art," *Trans AIME*, IV (1875–1876), 191–192, 201.

7. C. R. Mann, "Report of the Joint Committee on Engineering Education," *Engineering Education*, IX (September, 1918), 19–26; H. D. Gaylord, "The Relation of Mathematical Training to the Engineering Profession," *Engineering Education*, VII (October, 1916), 54–55; Frank McKibben, "The Colleges and the War," *Engineering Education*, IX (May, 1919), 363; and George F. Swain, "The Liberal Element in Engineering Education," *Engineering Education*, IX (December, 1918), 97–107.

8. The literature on professions is vast, but I have been particularly influenced by the writings of Everett C. Hughes. See Everett C. Hughes, "Professions," in Kenneth S. Lynn, ed., *The Professions in America* (Boston, 1965), 1–14, and Everett C. Hughes, *Men and Their Work* (Glencoe, Illinois, 1958). For a discussion of professionalism by an engineer, see William E. Wickenden, "Toward the Making of a Profession," *Electrical Engineering*, LIII, pt. 2 (August, 1934), 1146–48.

9. Frederick H. Newell, "A Practical Plan of Engineering Cooperation," *Journal of the Cleveland Engineering Society*, IX (March, 1917), 311.

10. For an example, see Hunter McDonald, "Address at the Annual Convention," *Trans ASCE*, LXXVII (December, 1914), 1755.

11. For examples, see Clemens Herschel, "Address at the Annual Convention," *Trans ASCE*, LXXX (1916), 1307–14, and "News of the Federated American Engineering Societies," *Mechanical Engineering*, XLIII (June, 1921), 421–422.

12. C. O. Mailloux, "The Evolution of the Institute and of its Members," *Trans AIEE*, XXXIII, pt. 1 (January–June, 1914), 827–834.

13. Vannevar Bush, "The Professional Spirit in Engineering," *Mechanical Engineering*, LXI (March, 1939), 198.

14. For one engineering educator's reaction to demands for "practicality," see Abraham Press, "Education Versus Engineering," *Engineering Education*, VIII (September, 1917), 28–30. For an example of the stress on business needs, see H. B. Gear, "The Engineer as an Implement of Management," *Electrical Engineering*, LXI (August, 1942), 426–427.

15. SPEE, *Report of Investigation*, I, 161–164, 188.

16. *Ibid.*, 164–166, 189.

17. *Ibid.*, 162–163, 188–189.

18. Robert Perrucci, William K. Le Bold, and Warren E. Howland, "The Engineer in Industry and Government," *Journal of Engineering Education*, LVI (March, 1966), 239–240.

19. A. W. Robertson, "Industry's New Responsibilities," *Electrical Engineering*, L (September, 1931), 719.

20. Frank B. Jewett, "An Exciting and Pleasant Forty Years," *Electrical Engineering*, LVIII (April, 1939), 161.

21. William E. Wickenden, "The Young Engineer Facing Tomorrow," *Mechanical Engineering*, LXI (May, 1939), 347, 348, and John C. Parker, "Responsibilities in the AIEE," *Electrical Engineering*, LVIII (August, 1939), 331–332.

22. "Industry's Influence," *Mechanical Engineering*, LXVII (August, 1945), 499–500.

23. Thorndike Saville, "Engineering Education in a Changing World," *Journal of Engineering Education*, XLI (September, 1950), 5. See also L. W. W. Morrow, "Industry Demands and Engineering Education," *Electrical Engineering*, LIII, pt. 1 (April, 1934), 518–522.

24. A. Norman, "Industrial Fundamentals," *Journal of Engineering Education*, XXII (March, 1932), 537.

25. Ernest Havemann and Patricia Salter West, *They Went to College* (New York, 1952), 235.

26. A. G. Christie, "Engineers' Business Contacts," *Mechanical Engineering*, LVII (February, 1935), 88.

27. J. F. Coleman, "Reflections on the Status of the Engineer," *Trans ASCE*, XCIV (1930), 1346.

28. Walter J. Gray, "Present-Day Responsibilities of the Engineer" (letter to the editors), *Civil Engineering*, XII (December, 1942), 687.

29. Andrew A. Bato, "Unionization of Engineers" (letter to the editors), *Mechanical Engineering*, LXIII (June, 1941), 476.

30. James M. Sherilla (letter to the editors), *Mechanical Engineering*, LXII (May, 1940), 412.

31. Daniel W. Mead, "The Engineer and His Education," in Dugald C. Jackson and W. Paul Jones, eds., *The Profession of Engineering* (New York, 1929), 28.

32. Henry Coleman, "Loyalty," *Mining and Metallurgy*, XII (August, 1931), 374–375.

33. *Ibid.*

34. H. B. Gear, "Engineering as an Implement of Management," *Electrical Engineering*, LXI (August, 1942), 426–427.

35. William E. Wickenden, "The Social Sciences and Engineering Education," *Mechanical Engineering*, LX (February, 1938), 149.

36. Arthur E. Morgan, "The Faith of the Engineer," *Civil Engineering*, XII (August, 1942), 421.

37. Arthur E. Morgan, "Engineer's Share in Democracy," *Civil Engineering*, IX (November, 1939), 638.

38. Andrew Fraser, *The Engineering Profession in Transition* (New York, 1947), p. 45, table 3.5g. The societies were the American Society of Civil Engineers, the American Institute of Mining Engineers, the American Society of Mechanical Engineers, the American Institute of Electrical Engineers, the American Institute of Chemical Engineers, and the National Society of Professional Engineers.

39. SPEE, *Report of Investigation*, I, 231–232. See also William E. Wickenden and Eliot Dunlap Smith, "Engineers, Managers, and Engineering Education," *Journal of Engineering Education*, XXII (June, 1932), 846–847.

40. Fraser, *Engineering Profession in Transition*, p. 28, table 1.9a. See also "1930 Earnings of Mechanical Engineers," *Mechanical Engineering*, LIII (September, 1931), 655.

41. Mabel Newcomer, *The Big Business Executive* (New York, 1955), 90. In absolute terms, this increase has been still larger since the group of top executives grew from 284 in 1900 to 319 in 1925, and to 868 in 1950. Jay M. Gould has recently extended Newcomer's figures to 1964. He found that the proportion of engineers and scientists had increased from about one in five in 1950 to about one in three in 1964 (see his *Technical Elite*, 82–84).

42. John Mills, *The Engineer in Society* (New York, 1946), 116, 138–139.

43. Gould, *Technical Elite*, 130.

44. Arthur E. Morgan, "Engineer's Share in Democracy," *Civil Engineering*, IX (November, 1939), 638.

45. *Ibid.*

46. A. B. Parsons, "Superorganizing Professional Engineers," *Mining and Metallurgy*, XXIV (September, 1943), 392.

47. Robert E. Doherty, "The Engineering Profession Tomorrow," *Journal of Engineering Education*, XXXV (September, 1944), 9.

48. Frederick S. Blackall, Jr. "ASME's Importance to Management," *Mechanical Engineering*, LXXV (September, 1953), 752.

49. *Who's Who in America*, 1930–1931, pp. 194–195.

50. Morris L. Cooke, "On the Organization of an Engineering Society," *Mechanical Engineering*, XLIII (May, 1921), 323, 325. See also Morris L. Cooke, *Professional Ethics and Social Change* (New York, 1946), 10.

51. Morris L. Cooke to Frederick W. Taylor, October 30, 1914, file "Cooke, July–December, 1914," Frederick William Taylor Collection, Stevens Institute of Technology, Hoboken, New Jersey.

52. Charles W. Baker to Edward W. Bemis, September 21, 1916, and Morris L. Cooke to Bemis, September 27, 1916, file "ASME Engineering Ethics–1916," box 168, Papers of Morris L. Cooke, Franklin D. Roosevelt Library, Hyde Park, New York.

53. "President Lee Answers Governor Pinchot," *Electrical Engineering*, L (March, 1931), 215.

54. Gregory M. Dexter, "An Appeal to Members" (letter to the editors), *Mechanical Engineering*, LXIV (October, 1942), 757.

55. W. H. Shockley, "The American Institute of Mining Engineers as Censor—A Protest," *Mining and Scientific Press*, CXIII (October 21, 1916), 589–590.

56. Robert B. Brinsmade, "Freedom of Discussion in the Institute" (letter to the editors), *Engineering and Mining Journal-Press*, CXIV (December 16, 1922), 1063.

57. "Code of Principles of Professional Conduct of the American Institute of Electrical Engineers," *Trans AIEE*, XXXI, pt. 2 (June–December, 1912), 2229, and "Report of Committee on Code of Ethics," *Trans ASME*, XXXVI (1914), 26. The ASME dropped this provision from its code of ethics in 1922.

58. Frank B. Copley, *Frederick W. Taylor*, 2 vols. (New York, 1923), II, 281.

59. Bernhard F. Jakobsen, *Ethics and the American Society of Civil Engineers* (Los Angeles, 1955), 1–13, and "A.S.C.E. Asked to Reinstate Two Los Angeles Engineers," *Engineering News-Record*, CXVII (October 8, 1936), 525.

60. Engineers Joint Council, *Directory of Engineering Societies and Related Organizations* (New York, 1963), 10–49. About a third of the national societies listed have company members; the amounts they provide are indicated in the individual societies' annual financial statements, which are usually published in their transactions.

61. Engineers Joint Council Subcommittee on Survey of Employer Practices, "Employer Practices Regarding Engineering Graduates," *Mechanical Engineering*, LXIX (April, 1947), 307.

62. Ernest K. Parks, "Membership in the A.I.M.E. is Purely Personal" (letter to the editors), *Mining and Metallurgy*, XXI (August, 1940), 387.

63. "Afraid to Ask?" *Mechanical Engineering*, LXX (January, 1948), 3.

64. *Ibid.*

65. Harry E. Harris, "Unification of Engineering Societies" (letter to the editors), *Mechanical Engineering*, LX (February, 1938), 177.

1.5

The Development of Engineering Education in America

JAMES H. SCHAUB, P. E.
KARL RICHARD PAVLOVIC

All of the generally accepted definitions of the noun "professional" make reference to the use of a particular body of knowledge obtained by specialized education. Further, it has been said that the hallmark of any profession is an acceptance by the members of that profession of the responsibility for the education and training of those individuals preparing to enter into its practice. Historically engineering education has two roots: the apprentice training of neophytes by skilled practitioners and the formal university training in theoretical science that developed following the scientific revolution. The training of the engineers of Rome and the Middle Ages proceeded in the master-apprentice fashion and this time tested method continues into modern times through the practice aspects of cooperative education as well as industrial training programs.

A more formalized program of instruction was organized with the establishment of École des Ponts et Chaussées in 1747 by the great French bridge and road engineer, Perronet. Once started, formal engineering education spread rapidly with the Bergakademie in Germany (1766), École Polytechnique (1794), and the program of study in mechanical philosophy initiated at University College, London (1827).

Engineering education in the USA closely paralleled the development of the nation. Traditionally, the beginning of engineering education is credited to George Washington's call for volunteers to form a Corps of Engineers which he issued at Valley Forge in 1778. West Point, established in 1794 and reorganized in 1802 on the model of École Polytechnique, provided the training for many of America's early engineers. An interesting sidelight is reported by Mavis (1) in noting that the first student, Hinton Jones, at the first state university, the University of North Carolina, practiced as a civil engineer following his graduation in 1798.

Although West Point graduated men trained in engineering fundamentals, the honor of being the first designated engineering degree program is disputed by Norwich University which graduated two students with Master of Civil Engineering degrees in 1837 and Rensselaer Polytechnic Institute where a civil engineering degree was first awarded under an officially designated academic program in 1835. Engineering schools expanded from these beginnings until the Civil War when there were approximately a dozen engineering degree granting programs actively educating prospective engineers, generally in the civil, mechanical and mining disciplines.

In 1862, the Morrill Land Grant Act was passed by Congress after an initial defeat in 1857. This act provided federal public lands to be allocated to the various states to found and support colleges for "particularly such branches of learning as are related to agriculture and the mechanic arts." The colleges formed under the Morrill Act were modelled on the traditional liberal art colleges and offered four year degree programs. This put engineering in the unique position of having programs of this length available at a time when the companion professions of law and medicine required relatively little formal education and were relying almost exclusively on the master-apprentice relationship for training.

The value of an academic preparation for engineering practice was the subject of much discussion during the last half of the 19th century. Engineers differed sharply with some favoring practical training and others tending to favor a broad education in scientific principles to precede any practice experience. (2)

U.S. engineering drew heavily on European experience at this time. Many of the developing university programs followed generally the French tradition of emphasis on theoretical education while numerous eminent practitioners remained committed to the apprentice type training common in Great Britain. The relative merits of these two approaches were the subject of intense controversy within the profession from 1870 to 1900. A landmark discussion was held in conjunction with the Centennial Exposition of 1876. Engineers from practice and education agreed

> . . . that education to prepare real engineers must be broad—stressing essential underlying principles and a considerable range of so-called cultural studies. (1)

Following the 1876 meeting, the concept of a theoretical academic training gained more approval from the profession although a strong vocational strain continued through the early years of the 20th century. The parallel development of new materials (concrete, steel, etc.) and new technologies (electrical devices, automobiles, airplanes, etc.) did much to emphasize the need for the successful engineer to be well prepared in the theoretical aspects of the profession.

The practice element of engineering was continued in engineering education through the extensive use of shop type courses within the academic program. These reflected the order of development in the traditional areas of engineering, where skilled practice had preceded scientific understanding, and these areas of study emphasized the shop approach. In contrast, science preceded practice in electrical and chemical engineering and thus the curricula in these fields emphasized science to a greater extent. The growth of laboratory work throughout engineering education in all disciplines became a characteristic of the U.S. approach.

The spur of industrialization and the demand for engineers of all disciplines led to a rapid growth in engineering degree programs following the incentive generated by the Morrill Act. The formation of professional engineering societies during the last half of the 19th century generated still more interest in formal education for engineering practice. Unfortunately, the rapid growth of interest in engineering led to a proliferation of academic programs to satisfy highly specialized needs. Engineering education has yet to recover from this splintering. There is no commonality of education for today's engineers that can be compared to that in medicine, law, or other professions.

From the perspective of professionalization and ethics this formative period saw another significant development leave a lasting mark on engineering education. In the early decades of the 20th century increasing industrialization and the burgeoning industrial demand for engineers led to close collaboration between engineering schools and programs and the corporate industrial community. The ties and reciprocal relationships that evolved at this time continue to be a source of both benefit and disadvantage to engineering. Concerning professional unity, they mean that engineering education does not present to engineering students a clear and unequivocal example of a ranking of obligations and duties to different groups. Engineering education is rather the first example to the student of the ambiguous and confusing mass of conflicting claims upon loyalty, duty, responsibility, etc. that the practicing engineer encounters. Engineering education does not say with a single voice to the prospective engineering student "Your first duty is to your profession," or " . . . to your employer," or " . . . to yourself," or " . . . to society." It says rather all of these in different ways and at the same time. Since the 1920s and 1930s engineering education has been faced constantly, as a consequence, with the need to resolve this ambiguity.

The circumstances of the time made the corporate-education collaboration an obvious arrangement. First, engineering faculty and administrators were seeking to consolidate the education of future engineers within their curricula. In this they were proposing to assume the bulk of the required training in the traditional areas of engineering—areas that were already beginning to feel a tension between the consulting-type relations of the engineer and his client and wage-type relations between the engineer and his employer. At the same time they were also proposing to design and supply the training for new areas in engineering—areas overwhelmingly characterized by corporate industrial employment. It was not likely that engineering educators would cease consulting practitioners in the traditional areas concerning their needs. Nor was it likely that they would not begin to consult employers and practitioners in the new employment areas.

Second, increasingly the major portion of engineering employment and thus curricular growth was in the corporate industrial area. Finally, engineering education as preparation both for corporate employment or private practice (where it is a matter of business competitiveness) and for public service (where it was a matter of the best possible standard of living and quality of life for society) presupposed striving to keep engineering students at the leading edge of developments in their field. This is an expensive objective that requires keeping faculty involved in research as well as teaching and requires constant replacement of equipment that in other respects is still adequate and functional. Again, it was not likely that engineering educators would forego appealing to the immediate consumers of their products for the financial support all this required.

Out of this developing relation came the Cooperative Course at the University of Cincinnati and the Personnel and Placement System developed at Purdue University—both of which were widely copied throughout the country.

The Cooperative Course was begun at the University of Cincinnati in 1907

> ... the course was a six-year program of instruction in mechanical, electrical, and chemical engineering, and was carried on in cooperation with a number of Cincinnati's electrical and machinery companies. Students alternated week by week between the industrial shops and the college during the school year and were required to work full-time in the industries during the four-month "vacation." In total, they spent four years in the industries and two years in the college classroom. (3)

The program did not expand rapidly but by 1919 there were seventy-five companies participating and by 1929 twenty schools besides Cincinnati had adopted such a program.

In 1920 Andrey A. Potter became the dean of the engineering school at Purdue University and began to develop an elaborate personnel and placement system.

> Designed in cooperation with the Indiana Manufacturers' Association, it involved the accumulation of extensive information about students and alumni: school grades, intelligence test scores, aptitude-test scores, career aspirations, teacher evaluations, practical experience, hobbies, employer references and "character profiles." Good traits of character were the focus of Potter's system, and those students who rated low and were thus deemed "deficient" in such areas as loyalty, efficiency, and adaptability were counseled about ways of improving their standing. (3)

This system, like the Cooperative Course, was replicated at other institutions and even became the basis of the Westinghouse Corporation personnel system.

Since World War II similar developments have been observed in the basic sciences as the industrial corporate employment of graduates from the basic sciences has grown. Engineering was simply the first. Thus, engineering and engineering education can be said to be at the beginning of their "mature phase" of dealing with the complications and ramifications of this kind of collaboration with corporate industry. The other academic disciplines are for the most part only beginning to enter an "immature phase" in this respect.

Between 1907 and 1968, the various professional engineering societies acting in collaboration with the Society for the Promotion of Engineering Education and its successor, the American Society for Engineering Education, sponsored a number of studies of engineering education. (See the Mann Report—1918 (4), the Wickenden Report of 1923–29 (5), the Hammond Report—1940 (6), the Grinter Report—1955 (7), and the Goals of Engineering Education Report—1968 (8).) All of these studies contributed to an increase in the science component of engineering education and urged the inclusion of additional social science and humanities studies to broaden the scope of engineers' basic education.

During the past decade much of the concern with engineering education has been directed toward what has come to be known as the "professional school" concept. This concept is based upon a belief that current engineering first degree programs, typically of four years duration or the same length as they were with the establishment of the Morrill Act over 100 years ago, are inadequate for today's needs. Though content has been changed since 1862, the length of the degree program is the same even though engineering has seen tremendous, almost geometric, increases in the fundamental sciences and in the applied technologies that comprise engineering. Even more critical, perhaps, is the increased demand society is making upon engineers that has necessitated knowledge in areas far afield from the conventional technology of engineering practice. The need for education beyond the basic preparation in engineering is further emphasized by comparison to what has happened in the education for legal and medical practitioners. In both cases, the minimum education requirements established by these professions include a broad general education bachelor's degree followed by a three or four year professional education period.

> Engineering now stands nearly alone among the recognized professions by requiring only an undergraduate education program for entry. (9)

Other factors that have influenced the advocates of the professional school of engineering are a desire for more college control of admissions, faculty standards, budget, and similar administrative functions.

There is no unanimity among engineering educators or practitioners on the "professional school" question. It is a subject of continuing discussion and several new or experimental programs (Texas A & M University, University of Florida, University of Detroit, for example) are being closely watched.

Today all academic programs are suffering what has been termed "A Crisis in Engineering Education." The crisis label reflects the lack of adequate funding for an expensive academic program that more and more requires a period of education well beyond the traditional four-year program. Faculty are not available in the numbers needed because of attractive industrial opportunities. Space and equipment have deteriorated and become obsolete. New technology has placed additional demands on resources for equipment. And, what may be the most critical, relatively few U.S. educated students are continuing their education through the advanced training that is required for engineering faculty. The question becomes, What will we do for teachers ten years hence? The graduate school program today simply does not have the prospective teachers in the educational pipeline.

Many groups from professional engineering societies to the National Academy of Engineering and from private practitioners to industrial giants are considering "The Crisis in Engineering Education" and seeking means to overcome its effects. In many ways, the situation today is parallel to that of a century ago with engineering education facing critical discussions of curriculum, funding, needs created by new techniques, and a growing demand for engineering graduates capable of using the new tools effectively to serve society.

"Education is, like success, a journey, not a destination" (10) and the engineering profession, educators and practitioners, must be prepared to continually re-evaluate both the route and the method by which the journey is conducted. There is no one best way to proceed. Engineering education demands continuing discussion and attention from all segments of society. Any successful solutions will require strong and positive leadership from the profession if the professional societies are to fulfill their obligation to educate in the best possible way those who will be the next generation of practitioners.

References

1. F. T. Mavis, "History of Engineering Education," *Journal of Engineering Education*, December 1952, pp. 214–221.

2. M. A. Calvert, *The Mechanical Engineer in America, 1830–1910*, Johns Hopkins Press, Baltimore, 1967.

3. D. F. Noble, *America by Design: Science, Technology and the Rise of Corporate Capitalism*, Oxford University Press, Oxford, 1977.

4. C. R. Mann, "A Study of Engineering Education," *Bulletin No. 11*. Carnegie Foundation for the Advancement of Teaching, 1918.

5. W. E. Wickendon, Report of the Investigation of Engineering Education, 1923–29, *Proceedings*, Society for the Promotion of Engineering Education, Vol. I, 1930, Vol. II, 1934.

6. H. P. Hammond, "Aims and Scope of Engineering Curricula," *Journal of Engineering Education*, March 1940, pp. 556–566.

7. L. E. Grinter, "Report on Evaluation of Engineering Education (1952–1955)," *Journal of Engineering Education*, September 1955, pp. 25–63.

8. "Final Report: Goals of Engineering Education," Committee chaired by E. A. Walker, American Society for Engineering Education, January 1968.

9. M. R. Lohmann, "Professional Schools and Programs," *Civil Engineering Education*, Vol. II, American Society of Civil Engineers, 1974, p. 128.

10. W. R. Park, "The Educated Engineer: More Wise Words from Wise Guys," *Consulting Engineer*, November 1978, p. 24.

Selected Additional Readings

Calvert, M. A., *The Mechanical Engineer in America, 1830–1910*, The Johns Hopkins Press, Baltimore, 1967.

Canjar, L., "The Professional Engineering School", a paper presented to the School of Chemical Engineering, Oklahoma State University, as part of the Phillips Petroleum Company Lecture Series in Chemical Engineering, 1972.

Cross, H., *Engineers and Ivory Towers*, edited by R. C. Goodpasture, McGraw-Hill Book Company, Inc., New York, 1952.

Dalton, G., Thompson, P., and Wilson, I., "An E.E. for All Seasons", *IEEE Spectrum*, December 1976, pp. 42–46.

Finch, J. K., *Trends in Engineering Education*, Columbia University Press, New York, 1948.

Grayson, O. P., "A Brief History of Engineering Education in the United States", *Engineering Education*, December 1977, pp. 246–264.

Grinter, L. E., "Report on Evaluation of Engineering Education (1952–1955)," *Journal of Engineering Education*, September 1955, pp. 25–63.

Hammond, H. P., " Aims and Scope of Engineering Curricula," *Journal of Engineering Education*, March 1940, pp. 555–566.

Layton, E. T., *The Revolt of the Engineers*, Press of Case Western Reserve University, Cleveland, 1971.

Lohmann, M. R., "Professional Schools and Programs," *Civil Engineering Education*, Vol. II, American Society of Civil Engineers, New York, 1974, pp. 125–137.

Mann, C. R., "A Study of Engineering Education," *Bulletin No. 11*, Carnegie Foundation for the Advancement of Teaching, 1918.

Mavis, F. T., "History of Engineering Education," *Journal of Engineering Education*, December 1952, pp. 214–221.

National Society of Professional Engineers, Board of Directors, "Policy on Professional Schools of Engineering," July 1971.

National Society of Professional Engineers, Board of Directors, "Guidelines for Recognition of Professional Schools of Engineering," January 1980.

Noble, D. F., *America By Design: Science, Technology and the Rise of Corporate Capitalism*, Oxford University Press, Oxford, 1977.

Park, W. R., "The Educated Engineer: More Wise Words from Wise Guys," *Consulting Engineer*, November 1978, pp. 24–28.

Rogers, G. L. and Schaub, J. H., "The Professional School—A Sleeping Giant," *Civil Engineering Education*, Vol. I, Part 2, New York, 1974, pp. 752–763.

Schaub, J. H., "Professional Schools and Traditional Schools: Engineering Education in Transition," *Engineering Issues*, American Society of Civil Engineers, April 1977, pp. 105–110.

Walker, E. A., Committee chaired by, "Final Report: Goals of Engineering Education," American Society for Engineering Education, January 1968.

Wandmacher, C., "Education for the Practice of Civil Engineering: An Historical Perspective—How We Seem to Have Arrived and Where We Are!" *Civil Engineering Education*, Vol. I, American Society of Civil Engineers, New York, March 1979, pp. 33–49.

Wickenden, W. E., "Report of the Investigation of Engineering Education, 1923–29," Society for the Promotion of Engineering Education, Vol. I, 1930, Vol. II, 1934.

Wisely, W. H., *The American Civil Engineer, 1852–1974*, American Society of Civil Engineers, New York, 1974.

2

Engineers and the Public

A true engineer—considers his duties as a trust.

John Jervis, 1876

Concern for man himself and his fate must always be the chief interest of any technical endeavor.

Albert Einstein

It is well for a man to respect his vocation whatever it is and to think himself bound to uphold it and to claim for it the respect it deserves.

Charles Dickens

Professionalism is a state of mind and a way of life.

James F. Shivler, Jr.
Past President
National Society of
  Professional Engineers

Does the individual engineer have any special obligations to the public—obligations that other individuals do not have? Do engineers collectively have a special obligation to the public? More generally, what is exactly the relation of engineering in general to society?

Engineering has become a profession, and because professions, engineering included, have historically based their status and claims to a distinct status within society on "service to the public," there is the presumption today that engineers do have special obligations to the public. Discussion of this question takes place within the more general criticism of all professions today.

The question thus develops in two parts: first, considering the engineer as a member of a profession and thus not acting alone in relations to the rest of society and, second, considering the engineer as an individual with certain capabilities acting as an individual, irrespective of whether a member of a profession.

In "The Professions under Siege," Jacques Barzun analyzes the relation between professions and society and the individual's situation within that relation. According to Barzun the problem lies with the professions having been taken for granted too long by the public and by professionals themselves. Under such circumstances few concern themselves with underlying assumptions and long-term consequences of behavior. The professionals simply assume that their activity will benefit society, and the public simply assumes that they will receive benefit rather than harm. When harm, rather than good, results from this two-sided inattentiveness, the professions, rightly or wrongly, are blamed. Barzun calls for a reappraisal of the professions and of the expectations of them by the public and, more importantly, by the professions themselves. He argues that, to preserve whatever is good in the present shape and form of the professions, the professions will have to learn to effectively police and critically evaluate themselves.

Samuel Florman in "Moral Blueprints" takes a position somewhat different from that of Barzun. Florman argues that so far as engineering and science are concerned, the current crisis of the professions has little to do with ethics and the ethical or unethical behavior of engineers and scientists as regards the public. He argues rather that the public and society would be better served if engineers and scientists were not expected to always hold paramount the health, safety, and welfare of the public. For Florman it is not reasonable to consider engineers as members of a profession having a special obligation to the public.

Looking now at the second aspect of the question, the position is increasingly advanced that the very nature of engineering practice places certain obligations on the practicing engineer. In "The Special Responsibility of Engineers," Frank Collins maintains that engineers, because of their knowledge and experience, are in the best position to predict the consequences and effects of the technology that they design and troubleshoot. As a consequence, he argues, engineers have the responsibility to participate in shaping the policy that puts that technology to use. Terry Turnick in "Public versus Client Interests" discusses the irony of the twin effects of advances in technology and science: (1) society becomes increasingly interdependent through its reliance upon technology, while (2) the specialization required to produce that technology makes each individual increasingly incapable of viewing the whole in its interdependencies. Turnick suggests that engineers caught in the dilemmas generated by this situation have no simple way out. He believes, however, that the severity of the dilemmas can be mitigated.

Both Richard Schott in "The Professions and Government" and Myron Tribus in "The Engineer and Public Policy-Making" analyze engineers, the engineering profession, and the nature of engineering education. They discuss those characteristics of the profession that work against engineers displaying a greater awareness of the social impact of technology and assuming a role of greater participation in public policymaking. For example, engineers are often thought to be oriented more toward things than people, to have a low tolerance for ambiguity, to place a high value on orderliness and efficiency, to be committed to truthfulness and the belief that truth will out in the end, and are viewed as team players. Schott ends on a pessimistic note, while Tribus sees individual engineers as being able to overcome these shortcomings.

Manheim believes that engineers have restricted their participation in broader discussions of technology because they are reluctant to deal with values—

particularly the values of the groups affected by engineering work. He suggests concrete ways in which values may be brought into the day-to-day activity of public works engineering.

Recently the government and private groups have argued that the traditional prohibition against competitive bidding for engineering services results in needless additional cost to the public. The courts have ruled that the professional societies may not formulate or enforce such prohibitions (see Lunch articles in Chapter 5). This has opened up the danger inherent in competitive bidding that a decrease in quality will result when only price is considered. Public officials and engineers involved in public works engineering have created the concept of "competitive negotiation" in an attempt to find a middle ground here and to protect the public's interest in quality engineering without needless expense. The final pieces in this chapter are an example of a state act that defines and describes the process of competitive negotiation for engineering services and a state engineering society's interpretation of the procedures set out in the act.

The Professions under Siege

JACQUES BARZUN

Something new has happened when the heads of two of the three branches of our government publicly attack two of the leading professions. The President has called down the lawyers and doctors in turn; the Chief Justice has twice criticized the men of law. But the feelings behind these acts of censure are not new; they have agitated the public and the press for a decade or more, and it is evident to all that the learned professions are not the splendid companies, held in awe and respect, that they once were.

The doctors, formerly worshiped as omniscient Good Samaritans, are now seen as profiteers, often of doubtful competence. Lawyers have never been popular, but they did seem the defenders of private and civil rights in time of need. Now they are thought neglectful and extortionate, when not actually dishonest. The poor academics, who won sudden prestige in the war years because they knew so much about so much that was useful, lost it all by their fecklessness in the troubles of 1965–68. As for the scientists, demigods since Darwin, they became objects of suspicion after Hiroshima. Their work was amoral; they dabbled in either treason or warmongering, and in any case their view of the universe was probably at the root of the modern malaise. Latest on the carpet, the austere, unfathomable accountant is being shown up as a master of misrepresentation, a cordon bleu at cooking the books. With his fall, the idea of "the professional man" is near to being swallowed up in contempt. For if the engineers seem to escape, it is because the public never sees them.

This comprehensive resentment against the visible professions does not spring wholly from experience, although facts at second hand or in print are not lacking. Part of the animus comes from the general unrest and impatience with authority in the Western world, coupled with the belief that anything long established is probably corrupt.

But to judge a profession rightly is not easy at a distance, and indignation on some immediate ground usually lacks perspective. For example, the medical profession has been under especially heavy fire; clearly, one reason is its formerly undisputed preeminence: The disillusion is proportionate. Until lately, doctors in this country were revered as in an earlier day only the clergy could hope to be. The physician simply displaced the clergyman when Western man shifted his allegiance from religion to science and technology and when medicine could show that it, too, was scientific. Such facts tell us something about the status of any profession; it goes up and down like the stock market, in response to things other than net worth.

VULNERABLE INSTITUTIONS

More than a century ago, Oliver Wendell Holmes—the great Holmes, father of the Justice, and a remarkably original physician—noticed this effect of culture on medical practices themselves: "The truth is that medicine, professedly founded on observation, is as sensitive to outside influences, political, religious, philosophical, imaginative, as is the barometer to the changes of atmospheric density." This conditioning is ignored, usually, both by those inside a profession and by those outside.

A profession is an institution, and as such it cuts a figure in public that may or may not match the prevailing habits and merits of the practitioners. The insiders genuinely believe in that figure; they live by it in more ways than one, and they can hardly help thinking of The Profession as going on forever in the same glorious way, altering itself only as it improves performance by new skill. Then, suddenly, it comes to grief through the outsiders' dissatisfaction, expressed in various unpleasant ways. Physicians have seen their brethren buffeted by malpractice suits, charged and convicted of fraud in handling federal and local health moneys; accused of mismanaging hospitals and clinics both medically and financially. With the mistrust has come the belief that doctors care only for money and make too much; that they cover up one another's homicidal mistakes; and that their lobbying as a closed corporation is holding back the advent of a much improved national health.

The parallel grievances about lawyers include, besides exorbitant fees, total disregard of the interests of society, mutual protection through the bar associations' connivance at error and fraud, and calculated deception of the public through the use of purposely mystifying language.

As for teachers in school and college, the grumbling is still diffuse, in part because the work they do or bungle is not observable case by case and is not paid for directly. Academic incompetence, indifference, misdirection of effort may be known or suspected, as they widely are today, but clear-cut instances of malpractice are hard to prove, and so far the few lawsuits against schools generally have failed.

The rights and wrongs of the charges will continue to be argued—as they should be—but what is of greater moment to the nation is the present state of disaffection from what used to be considered not only the best brains in the country, but also the most dedicated and self-sacrificing. It is true that young men and women in large numbers still want to join the ancient professions; they continue to hope for one of the too few places of training. But if, as one hears, the competition sometimes goes on by corrupt means, it would seem that little idealism enters into the search for admission. Or it may be that between the first resolve and the diploma the training itself fails to sustain the true professional faith.

At any rate, even those who are close to the training ground express disenchantment with their own kind. A recent writer to the *New York Times* who is an affirmative-action officer at an Eastern university asserts flatly that we are "so controlled by the professionals, so passive to our own interests," that danger threatens. And she explains the "rising popular discontent with the so-called truths of all disciplines and the so-called standard accepted practices of all professions" by the "limited vision of those who control these areas of human thought and endeavor."

Long ago, Bernard Shaw made a somewhat terser charge. "Every profession," he said, "is a conspiracy against the laity." The epigram should not be dismissed as a joke or a needless exaggeration. The only overemphasis is in the word *conspiracy*, which implies a secret purpose to overreach the public. Yet it is that very imputation of making the most of closely held secrets that becomes the common man's idea of a profession when it begins to lose the public's faith and regard. And there lies the danger. For the obvious next idea that occurs to the aroused critics is to demand strong supervision from outside. The affirmative-action officer wants the professions and disciplines to represent all groups in the community—that is, become democratic associations "controlling all aspects of a particular area of human concern." In various forms, this same idea is being urged as the remedy for the unsatisfactory state of affairs. Meanwhile leagues and individuals publish guidelines for outwitting or defeating professional malice, even in dentists.

To go in pain or anxiety and ask for expert help with a bludgeon up one's sleeve has something so topsy-turvy about it that the scene would be high comedy if issues of life and well-being were not involved. Better relations between layman and professional than we now seem to enjoy can hardly be brought about in mutual suspicion and hostility. Nor can there be improvement through collective bargaining. The essence of those relations is individual, from which it follows that some clearer idea of what a profession is must once again become common property among insiders and outsiders both.

According to Dr. Abraham Flexner, the famous critic and reformer of medical education fifty years ago, to be medically trained implies "the possession of certain portions of many sciences arranged and organized with a distinct practical purpose in view. That is what makes it a 'profession'." The key words here are: "a distinct practical purpose in view," for which "special training is required." Since the laity, by definition, has no such purposes and lacks special training, a profession is necessarily a monopoly. In modern societies this monopoly is made legal by a license to practice; but the professions have always managed to form a guild, a trade union, claiming the exclusive right to practice the art. From the tribal medicine man to the priest-physicians of the days of Hippocrates and to those now certified by the National Boards, no secret has been made of this exclusion, this separation of the profession from the rest of the people. Rather, it is a source of pride to the professionals; and they justify the monopoly by calling it essential to the safety of the public. But between monopoly and conspiracy the line of demarcation is hard to fix and easy to step over.

The upshot is that a profession is by nature a vulnerable institution. It makes claims; it demands unique privileges; and it has to perform. But "it" of course does not exist as a single entity; it is a dozen or a few hundred or many thousands of individuals, who differ as widely as all other human beings, yet who, as professionals, are expected to act in a standard manner and to be invariably successful in their art. At this point one might conclude that a profession was not merely vulnerable but naturally unstable, a scheme beyond human strength to live up to.

Nonetheless, professionals of all kinds have existed for thousands of years, and it is this apparent continuity that gives them the illusion of immortality. What every professional should bear in mind is the distinction between a profession and a function. The function may well be eternal; but the profession, which is the cluster of practices and relationships arising from the function at a given time and place, can be destroyed—or can destroy itself—very rapidly. The priest-physician is gone, like the priest-astrologer—two plausible combinations of roles. The town crier has disappeared. The broadcaster of news and advertising carries out the same function in a totally different form. The coachman has vanished and given place to the bus or engine driver and the airplane pilot. The miller is now a set of steel rollers and other machinery close to automatic; yet we still get our flour, and so much more white and pure that no living pest wants to touch it.

Even the professions that look ancient and continuous have radically changed, and not solely in the materials of their respective arts. The schoolteacher or university professor differs from Plato as well as from Abelard, and even more perhaps from the academic trained as a minister, who until recently lectured on six different subjects, all with the aid of the ancient classics. The modern tax and corporation lawyer does not work in the manner of Demosthenes or Cicero, nor do physicians dress or behave or talk Latin like those faithfully represented in the comedies of Molière.

These changes, both external and intrinsic, are linked to the influences that Oliver Wendell Holmes listed, and it would seem reasonable that in a time of public suspicion and outcry, the many professions would examine themselves in the light of these influences—political, social, religious, and philosophical. This should be done not just through the associations' committees, which are likely to deal in whitewash, but individually and silently, which is the only way to reawaken a sense of fact narcotized by habit.

To do so with honesty and lucidity, it is important to look back through time to the professional life as such, and to begin by saying that the ancestral jokes that ridicule the stereotypes of each calling can be disregarded: the clergyman is holier-than-thou and hypocritical; the professor is dry-as-dust and lives in an ivory tower—lucky man!; the lawyer thrives on complicating simple things and drags out the case to swallow up the estate in costs. These clichés go back thousands of years. Plato says: "The surest sign of bad government and social anarchy is to find many judges and many physicians." Four hundred years ago, Montaigne wrote that "lawyers and physicians are a bad provision for a country." The professionals themselves have occasionally turned on their own kind. Dr. Benjamin Rush, the famous physician of the American Revolution, wrote to a doctor friend: "Oh, that I were a member of any other profession, that I might not call physicians my brethren!" And Chekhov, who was a doctor as well as a playwright, makes one of his intelligent characters say, "Lawyers merely rob you; doctors rob you and kill you too."

These grumbles have counterparts outside the professions: the plumber always forgets his tools so as to increase his bill, and—in the Middle Ages—the miller adulterates the flour after taking more than his share for grinding it. The complaints only express the public's desire for workmanship and accountability; and obviously, the more advanced the profession, the less the public can judge of either.

But, however stupidly put, these demands by the laity should remind all professionals of the permanent clash of cross-purposes between even the best practitioner and his customer. In teaching, the professor's tendency is to expound the truth in all the particulars of interest to him. The student wants to be interested in another sense, besides learning what will get him past the examination and certified for his line of work. In the law, it is true, both client and advocate want to win, but the lawyer is trained to exercise caution amid confusion and often takes pleasure in a nice point of law. Both attitudes then seem to the client a pretext for additional expense of time and money. In medicine also, the case may claim the physician's absorption in the disease rather than in the patient, who will judge his comfort neglected or even his life endangered for the sake, presumably, of medicine in the abstract.

THE RULE OF DIVERSITY

This tension of opposites is inevitable; its ever-present reality should be a constant warning to the professional. It is his duty to reduce the difference as far as possible, and in any case not imagine that the human being he is serving ought to be grateful for whatever he gets. On the contrary, as Nicholas Pevsner, the great authority on architecture, has pointed out, many modern buildings are bad because the customer did not make enough demands, letting the architect follow his own professional bent. This is but another way of saying, with Abraham Flexner, that a profession has a practical purpose in view, which is to satisfy a precise human need. The profession does not exist for itself, as a game, or as a field of free, uncommitted activity, such as pure science, philosophy, art, or mathematics.

In this regard, a rereading of Molière's comedies about doctors and other professionals yields a healthy lesson. The plays show how, in order to daunt the patient, the doctors of the time dressed in black robes and pointed hats and talked a barbarous Latin, in which they argued whether the patient was suffering or dying according to the rules of the faculty. The treatment that followed was an automatic routine. From other sources we learn that in Molière's day there were about 100 doctors in Paris and only four new ones admitted each year. The cost of training was high, and by habitual nepotism the profession was practically hereditary. Surgery was of course rigidly prohibited as being a low manual trade. It was carried on, as we can see in the *Marriage of Figaro* a century later, by the barber, often an itinerant, who was expert in sharp instruments if in nothing else.

All this, no doubt, sounds remote from present-day practice, but in times of troubles one can profit from parallels. Traits connected with monopoly practice recur in altered guise. More than one modern specialist has in effect asked a patient: "Have you got *my* disease?"—which is not far from the Molière model of making the customer fit the rules. As for automatic treatment, what of the temptation to prescribe the current antibiotic, formerly the sulfa drugs? The present complaint that surgery is resorted to too readily and too often needlessly again suggests treatment without forethought. Modern practice also shows that since Molière's day the surgeon has been admitted to the profession, but the prejudice against manual work subsists in the still invidious position of the dentist.

In these and other ways, the professions exhibit their fatal tendency toward routine. Routine relieves the mind of the effort of thought, and it is protected by the secret and the monopoly of the art. Among public-school teachers, "methods" play the same role and, paradoxically, it is no less routine when the fad changes and new gimmicks replace the old. As in Molière, too, "the faculty" can turn in on itself and its back on the job. During the great demand for academic talent twenty years ago, its central duty of teaching was openly neglected, explicitly downgraded; it was left to those who could not devise a project or obtain a nonteaching university post. The student troubles of the next decade no doubt had additional causes, but not a better excuse. Now there are stirrings to put an end to academic tenure and to limit academic freedom so that it does not permit the use of one's specialty to lighten one's student load. Accountability has become a watchword throughout the world of education, and all this has come about without the aid of a great satirist like Molière.

If one existed and our subject engaged his mind, he would have a harder time making his point. In the seventeenth century, uniformity and small numbers made the target unmistakable. Today, diversity is the rule among physicians and other professionals. Their training, work arrangements, and reserves of mental and ethical strength vary widely. The tendency of an egalitarian age to turn every occupation into a profession has complicated the substance of ethics. What is right or wrong for a journalist, a marriage counselor, an optometrist—all now under distrustful scrutiny? Like their elder guilds, these new mysteries get little support from outside their numbers. Especially in cities, the public cannot readily tell the competent from the incompetent; there is no local opinion, and the mutual judgment of laity and professionals that used to be exercised easily within a small community is a thing of the past. The professional does not know how the client lives, whether he is honest and conscientious, and what he can understand. This information has relevance beyond the single case, for in the present outcry against the professions one must not forget the repeated proofs that the public, too, can be charged with corruption: shoplifting, bilking, padding accounts, cheating employers or examiners or the government are no longer specialties in the hands of dedicated crooks; they have become ecumenical avocations. And equally important—as is true of the high-school graduate claiming entrance to college—words, titles, grades, symbols, and diplomas no longer mean what they say.

This confusion, moreover, is occurring at a moment when the nation wants a huge supply of first-class services evenly distributed. Leaving aside the question of costs, there is little chance that the requisite number of able and devoted professionals can be found. Here again we must stop and think what a profession *as institution* is for. It is to turn people who are *not* born teachers, born builders, born advocates, or born healers into a good imitation of the real thing. Easier said than done. The decline of the public schools since the 1920s is patently due to the enormously enlarged demand for teachers, in keeping with the increase in population. There are not enough true teachers to go around, and the imitation ones, owing to the normally wretched training, are largely useless. In other words, it is not true that with increasing numbers of people you get a proportional number of every sort of talent. Many other opportunities disperse and absorb that talent; and even apart from this fact, the supply is unpredictable. The country cannot say that now it has four times as many writers of the first rank as it had 100 years ago.

CRITICISM FROM WITHOUT AND WITHIN

As a whole, every profession is always horribly average, mediocre. By definition it cannot be anything else. But the public expectation aims much higher than mediocrity, so that in a time of reckoning, when the laity is hot about its rights, general dismay and recrimination are inevitable. What is more, although any art should be judged by its best results, a democratic nation, bent on equality in all things, is sure to judge a profession by its worst exemplars. That is the condition we are in now.

Consider the malpractice suits. From one point of view it is just that a patient—or his heirs—should recover damages for careless or ignorant treatment. From another point of view it is absurd that after the best professional efforts failure should be a cause of complaint. Yet a customer cannot tell whether he has had the best. He always judges by gross results—kill or cure—and wants the reasons plain. On such points the public intelligence is weak. Certain groups have voiced demands that show they expect from the professions nothing less than divination and infallibility. Again the requirement, now enacted here and there, that the fairly exact language of the law be reduced to the common tongue foolishly ignores both the potential litigation over loose terms and the appalling prose now being written by laymen and professionals. Thus do carelessness and indifference on one side beget angry utopian claims on the other.

The subtleties of the predicament are even clearer in education, where the failure to "educate" a particular student is evident in the student, yet assigning blame is beyond human wit. Nor can our modern system follow the example of the old-time college president who said to the indignant parent: "Madam, we guarantee results—or we return the boy."

It is because of these intricacies behind the gross results—a cure, a good education, winning the lawsuit—that for centuries it has seemed best to let the professions police themselves. The assumption is that inside the shop merit and demerit are correctly judged. That is by and large true, and lately in medicine Professional Standards Review Organizations, which are local committees of physicians, have been set up to pass upon the performance of fellow practitioners. It is too soon to say whether this is doing the profession any good, but on general principles one may doubt it. Anyone who has seen at first hand the evaluation of educational institutions by committees of colleagues knows that except for cases that could have been decided at sight by a sensible bricklayer, the reports have been pointless and false. The criteria are mechanical (e.g., equipment, number of books) or else abstract, hence remote from what goes on day by day. How can practice be judged otherwise than *in* practice?

If policing passes from a local committee of peers to a government agency producing the usual sort of questionnaire, the gap between printed paper and real life is even wider. Through their own routines, regulatory agencies lag behind the facts. They also protect featherbedding, prevent innovation, and give inspectors a power that may invite and spread corruption.

But this generally true verdict is incomplete. The regulation of business came about because business did not regulate itself. It exploited labor and the buyer, under the motto "The public be damned." An alert professional today has the uneasy feeling that the professions are at the juncture where the same motto, unspoken on their part, is being imputed to them by a public ready to clamor for regulation.

There are other signs of a gradual demoting of the professions to the level of ordinary trades and businesses. The right of lawyers and physicians to advertise, so as to reintroduce money competition and break down the "standard practices," is being granted. Architects are being allowed to act as contractors. Teachers have been unionized. Laymen demand the right to sit on various professional bodies, on boards of trustees, and wherever "community representation" may be argued for, on the ground that internal management is unable to serve the public fairly without supervision. The great force of government money works to the same end, for bureaucracy follows the funds and while directing their use is bound to control the user. And where government does not intrude, unionism will.

Such moves, whether viewed as threats or as reforms, signify one thing: the modern professions have enjoyed their monopoly for so long that they have forgotten that it is a privilege given in exchange for a public benefit. The myth of the profession as an immortal being, forever young and strong, fills their minds. The example of the scientist, till lately unchallenged as oracle and servant of mankind, has persuaded the professional man that expertise in itself is a sufficient raison d'être. Occasional complaints are interpreted as envy or misunderstanding, instead of what they have turned out to be   suspicion, resentment at breach of faith, contempt of complacency.

Forgetting the great principle of reciprocity will ruin any profession. The scientists felt a touch of the menace when they too loudly claimed complete autonomy from social judgment. Nor can the workaday professions be saved when only a faithful remnant performs well and behaves ethically. An institution exists for use, not to be an object of wonder.

It may be, of course, that we are witnessing the evolution some have predicted— the drive toward a society collectivized through and through, that is, in which groups interlock in mutual control; the theory being that no individual *or* group can be trusted. That would mean the death of the very idea of a profession, which so far has been synonymous with a blend of individual and group self-governance.

The message for the professions today is that their one hope of survival with anything like their present freedoms is the recovery of mental and moral force. No profession can live and flourish on just one of the two. For its "practical purpose" it requires the best knowledge and its effective use. But since that purpose is to transfer the good of that knowledge from the possessor to another person, the moral element necessarily comes into play. *Moral* here does not mean merely honest; it refers to the nature of any encounter between two human beings. Nobody sues a vending machine for malpractice; if it fails to work, no blame attaches to it, even though one wants to get one's coin back. But as soon as a person serves another, ethical issues spring to life and get settled well or badly. Perhaps they are clearly felt by only one of the parties; they are in any case never easy to reason out. Such practices as experiments on poor patients, or operations by young residents while the patient thinks he is in the hands of the great surgeon, seem clear-cut matters that find their parallels in teaching and in the law. But more subtle situations arise from group practice, in any profession, where the client may be tossed about among several hands, having to reestablish his identity and need, losing confidence all the while, and in the end knowing that responsible attention has been denied him.

None of this means that there are not arguments for the systems complained of, advantages that would be lost with their abandonment—for example, the training of beginners directly on the body, or the mind, or the hearth and home of the customer who has paid high fees. But that conflict of good reasons will not resolve itself by simply waiting it out, any more than moral sensitivity will return to the individual practitioner by his guild's writing up a fresh code. A code only sets the limits beyond which behavior will be condemned, and the moral level is not high when all or most of those who live under it always act within a hairline of those limits. Codes, in fact, are for criminals and competitors, not for professions that want to be known as dedicated.

No doubt the codes now in force (though hardly the Hippocratic oath, so moving and yet so ineffectual) can benefit from thorough revision. But what the professions need in their present predicament is, first, the will to police themselves with no fraternal hand, with no thought of public relations. Any few scandals giving the group a bad name will soon convince the public that self-policing means what it says and confidence will return. The examples will also inspire—shall we say?— others who might otherwise be weak. Screening and disciplining from within must always continue, steadily and firm, or it will be taken over by public bodies and officialdom. A simple fact of language may help drive in this simple point. English has borrowed from French the phrase *esprit de corps* and uses it to mean something good—team spirit, loyalty. But in French, to this day, it means something bad: the huddling together of members of a guild to hush up their mistakes; it means, in short, Shaw's conspiracy against the laity.

Policing, being negative, is not enough. It will not effect moral regeneration, which can come about only when the members of a group feel once more confident that ethical behavior *is* desirable, widely practiced, approved, and admired. After a marked decline, it can only be a slow growth and only one force can start it on its way, the force of moral and intellectual leadership. What all the professions need today is critics from inside, men who know what the conditions are, and also the arguments and excuses, and in a full sweep over the field can offer their fellow practitioners a new vision of the profession as an institution.

For each profession, details such as these will have to be spelled out, for in human affairs the great and perpetual shortage is that of imagination. Nothing is so obvious that somebody won't manage to overlook it. Details must of course be embedded in general principles, and these summed up in maxims to strike the mind. The whole aim is to lift the critique from a set of complaints to a set of purposes, purposes no professional can gainsay because they are stated or restated by one who knows. That is what Flexner did when he riddled medical education in 1910. He changed American medicine, having made it impossible for do-nothing schools and filthy hospitals to continue in being. In an earlier day Bentham did likewise for English law and John W. Burgess for American higher education. When the problem is a failure of competence and morality, nothing will solve it but the work of an individual mind and conscience, aided of course by the many scattered men of talent and good will who are only waiting for a lead. Without some such heroic effort, we professionals shall all go down—appropriately—as non-heroes together.

2.2

Moral Blueprints

SAMUEL C. FLORMAN

Technethics: *The responsible use of science, technology, and ethics in a society shaped by technology.*
Britannica Book of the Year, 1973
"New Words and Meanings"

The joining together of *technology* and *ethics*, two of the most supercharged words of this decade, was not an etymological success (as was *bioethics*, for instance), but in all other respects the union appears to thrive. Most people agree that this is all to the good: ethics rejuvenated and applied to technology; technology tempered by ethical considerations. Technethics seems to be just what is needed, as the proliferation of sermons, seminars, grants, articles, newsletters, conferences, professional codes, and academic courses devoted to this theme solemnly testifies.

Of course, worries about the ethics of technology did not begin in 1973. Atomic weapons have troubled the public conscience since 1945. By 1960, when Vance Packard called us "wastemakers," we were already concerned about the world's diminishing natural resources. Soon Rachel Carson and Ralph Nader were expressing their anxieties, and the Club of Rome announced that environmental catastrophe was imminent. But it was not until the oil embargo and energy panic of 1973 that concern about the effects of technology reached pathological intensity.

That same year, the Agnew scandal embarrassed the embassies of technology with its revelation that several prominent engineers had been involved in illegal payoffs. Previous incidents had raised questions about the integrity of technological enterprise and were recalled with much wringing of hands. In 1968, at the B. F. Goodrich plant in Troy, Ohio, test results were falsified so that faulty aircraft brakes might be accepted for use on the U.S. Air Force A-7D. In March, 1972, three engineers were peremptorily dismissed by the management of BART (the San Francisco Bay Area Rapid Transit) for speaking to the board of directors about inadequacies in the train control system being furnished by the Westinghouse Electric Corporation.

In a different era these episodes might have been forgotten, but in the revisionist political climate of the 1970s they were resurrected to become the canon of a new folklore. Concerned engineers and observers of the engineering scene would point to these incidents repeatedly as evidence of the need for new ethical standards.

In October, 1971, an international symposium was held at the newly established Joseph and Rose Kennedy Institute for the Study of Reproduction and Bioethics at Georgetown University. The meeting concluded with a call for federal funds to support research and teaching in the area of science and values. Usually such statements disappear into dignified obscurity, but this one was reported prominently by the press and was cited at Congressional hearings. The message was heard at the National Science Foundation (NSF), where within a month a task force was at work studying the contributions the NSF might make. Similar concerns were voiced at the National Endowment for the Humanities (NEH), which joined with the NSF in 1973 to announce its eagerness to fund programs.

Interest and anxiety had been growing. Now the all-important final ingredient—government money—was at hand. At present the NSF's program of Ethics and Values in Science and Technology, funded at just under $1.5 million annually, awards about twenty grants a year, mostly to scientists and engineers. The NEH's Program of Science, Technology, and Human Values—which supports the work of philosophers, historians, and other humanists—has an annual budget of $3.5 million, and awards about fifty grants. Interdisciplinary activity is encouraged and funded by a number of joint NSF–NEH grants. Although the majority of recipients are colleges and universities, substantial grants are also made to museums, scientific and engineering societies, and other nonprofit institutions.

During the five years since the NSF–NEH program was announced, events have lent new urgency to the technethics crusade. The DC-10 catastrophe over Paris in March, 1974, was caused by a faulty rear cargo door that, according to falsely certified records, was said to have been repaired. During a fire at the Tennessee Valley Authority's Brown Ferry nuclear power plant in March, 1975, a reactor ran out of control for almost seven hours; safety improvements recommended by government inspectors had been deferred because of cost. In February, 1976, three General Electric engineers resigned in protest over what they considered to be inadequate safety provisions in the design of nuclear power plants.

Dismayed by such scandals and uneasy about the initiatives taken by academe and the government, practicing engineers hurried to join the crusade. In late 1974, the Engineers' Council for Professional Development adopted a new Code of Ethics of Engineers, which was quickly endorsed by most of the major engineering societies. The societies have long observed codes of ethics, but these have traditionally stressed gentlemanly conduct rather than concern for the public welfare. An engineer was to be honest and impartial; he was to avoid conflicts of interest; he was not to criticize a fellow professional; and mainly, he was not to compete for commissions on the basis of price (an injunction that the Supreme Court disallowed in April as a violation of antitrust laws). The first canon of the revised code now reads:

> Engineers shall hold paramount the safety, health, and welfare of the public in the performance of their professional duties.

The original code had enjoined the engineer to show "due regard" for the public. In revision, "proper regard" was proposed and rejected as a weak compromise; only "paramount" would do. Most engineers seem to agree that this revised statement is worthy of approval. Nobody, however, seems to be sure of its precise meaning, whether or not it can be taught, and if it can, how, and by whom.

Skeptics—both within academe and without—argue that moral character is formed in the home, the church, and the community, and cannot be modified in a college classroom or professional symposium. I cannot agree with the skeptics on this count. Most evil acts are committed not by villains but rather by decent human beings—in desperation, momentary weakness, or an inability to discern what is morally right amid the discordant claims of circumstances. The determination to *be* good may be molded at an early age, but we grapple all our lives with the definition of what *is* good, or at least acceptable. I see nothing inherently wrong with doing some of that grappling in the classroom.

An ethics program for engineers is no doubt a prudent step for society, akin to an inoculation of gamma globulin. Even so, I am apprehensive about the possible consequences of an uncritical acceptance of the new engineering ethics in articles and speeches, and worst of all in classroom lectures. The revised code, which accurately reflects the current mood of guilt and introspection, is based upon the deceptive platitude that the professional's primary obligation is to the public—"the ultimate client," according to one teacher of engineering ethics. The engineer is no longer to be guided by his employer's wishes or instructions, or by his own creative imagination, as constrained by laws, regulations, and technical parameters. He must answer first to what his conscience tells him is best for the common good. Technethics, born of public outrage, ends by seeking solutions in private virtue.

If this appeal to conscience were to be followed literally, chaos would ensue. Ties of loyalty and discipline would dissolve, and organizations would shatter. Blowing the whistle on one's superiors would become the norm, instead of a last and desperate resort. It is unthinkable that each engineer determine to his own satisfaction what criteria of safety, for example, should be observed in each problem he encounters. Any product can be made safer at greater cost, but absolute freedom from risk is an illusion. Thus, acceptable standards must be specifically established by code, by regulation, or by law, or where these do not exist, by management decision based upon standards of legal liability. Public-safety policies are determined by legislators, bureaucrats, judges, and juries, in response to facts presented by expert advisers. Many of our legal procedures seem disagreeable, particularly when lives are valued in dollars, but since an approximation of the public will does appear to prevail, I cannot think of a better way to proceed. It would be a poor policy indeed that relied upon the wisdom of individual engineers.

There have been a few cases in which scientists far in the forefront of some exotic field—splitting the atom or research in recombinant DNA—have had special problems of conscience. But such rare instances, interesting as they may be, have little pertinence to the responsibilities of the average technologist.

It will be argued that the pledge to serve the public is not intended to transform each engineer into an independent review authority, but simply to serve as an ideal. Yet even as an ideal the precept is insidious. Just as some lawyers dedicate themselves to protecting consumers or prosecuting criminals, so there are engineers in public service who do the research, write the codes, and make the inspections that keep technology in check. Engineering has a place for both the *creator* and the *guardian*; the dynamic tension between the two is crucial to social vitality and has been obscured by the shapers of the new engineering ethics. The existence of an adequate cadre of public-service engineers depends mainly upon the determination of the public, who must support it. This determination can become weak if there is too much reliance upon morality.

Engineers are obliged to bring integrity and competence to whatever work they undertake. But they should not be counted upon to consider paramount the welfare of the human race. An engineer who declines a commission in deference to his scruples will only pass it along to a colleague of less-refined sensibility. Vows of rectitude will not reform reality. The ethically sensitive engineer should welcome effective regulations, laws, and definitions of legal liability within which he can energetically pursue his calling.

The new ethics does not stop with considerations of public safety, but goes on to hold the engineer accountable for the quality of life in this technological age. Not wanting to be taunted as mere cogs in the social machine, engineers enjoy the messianic importance that comes with the title "shapers of culture." The new interdisciplinary courses they are taking, to quote an NSF-funded report, "attempt to develop skills to bring about value-sensitive social change." The president of the Polytechnic Institute of New York has proposed a new precept for the profession: "We shall in general design our systems so as to enhance and glorify man, not dehumanize him."

It is difficult to quarrel with this objective. But just as the ethicists have failed to distinguish between creators and guardians, so have they confused the functions of solving problems and establishing goals. The problem-solver cannot factor his personal fancy into each equation. He must operate within constraints and expectations set by those who commission his work.

An engineer designing a monorail car for a rapid transit system cannot become an expert in acoustics, urban planning, and the habits of woodland birds, and at the same time be an expert designer of monorails. Nor can he do his best work if he is excessively apprehensive about the consequences of his every move. He must have confidence that other members of society are doing their jobs—planning cities and protecting birds.

Engineers can (and should) contribute to public policy as citizens, but this is very different from filtering their everyday work through a sieve of ethical sensitivity.

As a class, engineers have neither the power nor the right to plan social change. If they did, we might be well on our way to George Orwell's *1984*. Fortunately, engineers are no more agreed upon how to organize the world than are politicians, novelists, dentists, or philosophers. Should we risk oil spills and increase our reserves by offshore drilling? Accept the hazards of pesticides in order to feed hungry people? Stop building a dam and thus protect an endangered fish? These are political questions; it is pathetic and a little frightening to see citizens abdicate their responsibilities by assigning them to the realm of engineering ethics.

Should professionals work only on projects that they as citizens approve? The new ethics implies as much. Conventional wisdom suggests that it is the duty of enlightened professionals to lead. But, paradoxically, it is essential that professionals should serve. I once heard Isaac Stern and Eugene Istomin discuss this paradox as it bedevils performing artists. Stern believed that he was obliged to use his musical art to further his political beliefs, promoting "good" causes and boycotting "bad" ones. Istomin argued that a musician has a responsibility to perform wherever people want to hear music. Relating this debate to engineering, I tend to side with Istomin. If each person is entitled to medical care and legal representation, is it not equally important that each legitimate business entity, government agency, and citizens' group should have access to expert engineering advice? If so, then it follows that engineers (within the limits of conscience) will sometimes labor on behalf of causes in which they do not believe. Such a tolerant view also makes it easier for engineers to make a living.

Liberals are not alone in their advocacy of technethics. People in the business and professional communities see in technethics a welcome opportunity to repudiate the government controls they detest and fear. Witness, for example, American Viewpoint, Inc., an industry-supported, nonprofit "educational corporation" located in Chapel Hill, North Carolina, whose directors include the president of the National Association of Manufacturers, the executive vice-president of the American Medical Association, the assistant to the president of the American Mining Congress, and other representatives of Establishment America. In 1976 this organization published a volume of esssays entitled *The Ethical Basis of Economic Freedom* and sent copies to all U.S. Senators, 1,000 banks, all companies on the "Fortune 500" list, many college presidents, and the deans of most graduate schools. The funding for this distribution came, in part, from William E. Simon, then Secretary of the Treasury, now chairman of American Viewpoint's board of directors. Mr. Simon wrote the book's concluding essay, "A Challenge to Free Enterprise."

In November of last year [1977] this organization took full-page advertisements in the nation's leading newspapers to announce the founding of an Ethics Resource Center. The essence of the American Viewpoint message is expressed in one sentence from that announcement: "When honesty and ethics sink down, centralized authority and coercive regulations rise up." A special message is addressed to the nation's professional organizations, proclaimed, one can almost imagine, with a wink: "The more trust earned, the fewer restrictions needed."

Thus, from both the Left and the Right we find zealous armies marching under the banner of technethics. (Every army marches under a holy banner, which is why I feel edgy when I see such banners on the horizon.) These forces are dangerous because they attack from different directions the policy that is our best hope—the formulation of specific rules and regulations equal to the complexities of our technology.

The regulations need not all be legislated, but they must be formally codified. If we are now discovering that there are tens of thousands of potentially dangerous substances in our midst, then they simply must be tested, the often-confusing results debated, and decisions made by democratically designated authorities—decisions that will be challenged and revised again and again. The Clean Water Act, passed by Congress in 1972, was modified last year [1977] by more than 100 amendments. Now the Environmental Protection Agency—under the watchful eye of industry, environmental groups, and local governments—is drawing up detailed regulations, scheduled for issue in late 1979. The announced EPA policy is to encourage local planning and enforcement of waste-treatment programs within federal standards that will eventually cover twenty-one categories of industries, and perhaps 400 subcategories. This is an excruciatingly laborious business, but it cannot be avoided by appealing to the good instincts of engineers.

If the multitude of new regulations and clumsy bureaucracies has made life difficult for corporate executives, the solution is not in promising to be good and eliminating the controls, but rather in consolidating the controls themselves and making them rational. The world's technological problems cannot even be formulated, much less solved, in terms of ethical rhetoric: especially in engineering, good intentions are a poor substitute for good sense, talent, and hard work.

No wonder that so many people—intimidated by complexity, impatient with detail, and jealous of privilege—have seized upon the false belief that technological utopia depends mainly upon moral reformation.

2.3

The Special Responsibility
of Engineers

FRANK COLLINS

The question that this conference is exploring is whether engineers have a special responsibility for the end results of technology, a responsibility that transcends that of ordinary citizens.

The general public is beginning to believe that engineers do indeed have such a special responsibility. Engineers design and build the industrial plants that pollute air, water, and the land, and they design the "unsafe-at-any-speed" automobiles that have become symptomatic of the growing environmental crisis. It is also the engineers who design and build the weapons for mass slaughter now being automated for the "electronic battlefield."

Unlike scientists, who can claim to escape responsibility because the end results of their basic research can not be easily predicted, the purposes of engineering are usually highly visible. Because engineers have been claiming full credit for the achievements of technology for many years, it is natural that the public should now blame engineers for the newly perceived aberrations of technology.

In considering the question of the general responsibility for the uses and effects of technology, it is clear that the responsibility exists on two levels, the first on the level of policy making and the second, the implementation of policy decisions.

Most engineers play little part in primary decision making. Decision making of this kind is the domain of high government officials and corporation executives. Here, with some notable exceptions, engineers generally participate only to the extent of providing technical background and data for the information of the actual decision makers. Later on, engineers may modify the technical execution of the decisions or implement them in ways unforeseen by the original decision makers but nevertheless the fact remains that engineers in their normal capacity can not change general policies nor reverse basic decisions.

The lack of independent decision-making power by most engineers is reflected in their status as employees, a position quite different from that of the practitioners of medicine and the law who are generally independent entrepreneurs. Typically, engineers are "management" only in the restricted sense of foremen and other supervisory employees. From their position as "professional" employees, it would seem to be difficult to ascribe to engineers any special responsibility beyond that assumed by the other employees of their companies.

What then is the meaning of the "professionalism" that is claimed by engineers? The dictionary defines *profession* as "a vocation or occupation requiring advanced training in some liberal art, or science, and usually requiring mental rather than manual work, as teaching, engineering, etc."

This formal definition says nothing about the socioeconomic status of professionals nor about any special responsibility. Nevertheless, there is a deeper meaning to professionalism that is worth exploring here. The definition implies that professionals, apart from any economic questions, have a definite responsibility for the quality and, more importantly, for the character of their work. The latter takes us beyond the question of workmanlike competence, which surely is basic to the practice of engineering, to the question of the general objectives of the work.

Because of their advanced training, engineers should be in a unique position to see the broad consequences of their technology. It must be conceded, however, that narrowness in technical education and a high degree of specialization handicaps many engineers in evaluating the relationship of their work to societal questions. Even so, by virtue of their training and their access to detailed technical information, engineers have a capacity for the evaluation of the uses and effects of technology that is far beyond that of most nontechnically trained citizens. From this ability arises the special responsibility of engineers we are considering here.

In rebuttal to this point of view, it is frequently urged that engineers possess some special quality of objectivity that would be sacrificed if they were to concern themselves actively with the social consequences of their technology. This argument is a subterfuge for inaction.

Although it is mandatory that engineers deal only in independently verifiable facts in the objective aspects of their work, this kind of objectivity carried with it no a priori necessity for moral neutrality with respect to the uses of engineering. There is no contradiction here. Facts are neutral; morality and social responsibility relate to the choice of alternative courses of action. Factual knowledge makes the choice more intelligent by indicating the probable consequences of the alternatives, but it does not dictate the choice. This choice can be made only on the basis of systems of human values incapable of validation by the scientific method. In this, engineers are no worse off than anyone else, and we can not escape responsibility for choice of action by alleging some kind of objectivity, not possessed by the ordinary citizen.

There are three ways in which the special responsibility of engineers for the uses and effects of technology may be exercised. The first is as individuals in the daily practice of their work. The second is as a group through the technical societies. The third is to bring special competence to the public debate on the threatening problems arising from destructive uses of technology.

The first responsibility of any engineer is the individual one. It is for the social consequences of his own day-to-day work. The rising storm of public controversy regarding the uses and effects of technology is creating doubts in the minds of some engineers as to whether their particular employment is in the public interest. But an engineer is fairly well trapped in his particular specialty and his job. He has few alternatives but to continue, particularly if he has family responsibilities. This is especially true in a time of engineering layoffs and difficulties in obtaining new jobs.

In this context, it is natural that some engineers develop rationalizations justifying the end results of their jobs. For example, it is likely that many automotive engineers pooh-pooh the environmental pollution of the automobile simply because it is more comfortable for them to do so. It is almost certain that any survey of aerospace engineers taken last year [1972] would have indicated that the preponderant majority favored going ahead with the SST. Likewise, it is probable that the majority of the engineers who are developing the "electronic battlefield" are in favor of continuing the Vietnam war.

In spite of all this, it is probably true that an increasing number of engineers are seeking jobs they regard as socially desirable. Further, a few engineer employees are beginning to speak out on their jobs and in their technical societies on the question of the end uses of technology.

Until quite recently, there was little evidence that technical societies took seriously any special responsibility for the uses and effects of technology. In fact, the general policy has been to avoid studiously open discussions of the social and political issues involved in the practice of engineering. This is not to say that the actual policy of most technical societies was scrupulously neutral on such questions. On the contrary, the prevailing policy of the technical societies has been to adopt established industry and government views as their own without membership debate.

There are many examples of the lack of neutrality of the technical societies where their disciplines have touched on social and political questions affecting the interests of the corporations connected with the given discipline. Only one of these need be cited here. The case of *Chemical and Engineering News*, published by the American Chemical Society, in the matter of the environmental effects of chemical pesticides is particularly interesting because it long preceded general public concern. When Rachel Carson's *The Silent Spring* appeared in 1962, the C. & E. N. review of the book was entitled, "Silence, Miss Carson." Its editorials,[1-4] such as the one entitled "Pesticides on trial: shortages of fact and excesses of emotion could deprive us of major benefits," represented more the concern of the pesticides industry than an objective examination of Miss Carson's thesis that DDT was harmful to wildlife.

This and other similar examples show clearly that lack of member debate within technical societies by no means assures neutral objectivity. More seriously, lack of debate constitutes an abdication of responsibility on the part of the engineering societies for the uses and effects of technology, even in areas in which some of their membership is particularly competent. For example, it was not one of the relevant engineering societies but Ralph Nader, a lawyer, who first raised the question of automobile design safety. Likewise the problem of environmental pollution by the automobile was first pursued outside the professional societies concerned with the design and manufacture of automobiles. Up to right now, there does not appear to be a single outstanding example of an engineering society being the first to raise a question about a major misuse of technology within its special area of competence.

That this situation has begun to change in the past several years is shown by the extended discussions of the uses and effects of technology and their relation to human values which are now appearing in the technical news press. Some of the technical societies are currently founding sections on the social implications of technology, and such symposia as the present one are no longer uncommon. The former view that engineers as a group should blindly practice engineering without professional concern over the critical social issues involved in their work is, hopefully, in the process of being discarded.

References

1. ________. 1961. Chem. Eng. News Oct. 1:6
2. ________. 1961. Chem. Eng. News July 23:5
3. ________. 1961. Chem. Eng. News July 30:5
4. ________. 1961. Chem. Eng. News Oct. 15:5

Public versus Client Interests— An Ethical Dilemma for Engineers

TERRY L. TURNICK[1]

The most obvious and inescapable effect of scientific technique is that it makes society more organic in the sense of increasing the interdependence of its various parts. (5, p. 11)

This observation by Bertrand Russell over 20 years ago well describes our present industrial-technological society. Our general knowledge has ascended to such vast proportions in so many areas that an individual is hard pressed to maintain a mastery of any specific field of endeavor. Society is therefore faced with groups of specialized people in a dynamic system. Each exerts an influence on society and likewise each is influenced by the remainder known as the public.

With the understanding gained through specialization usually comes a type of "tunnel vision" which somewhat limits perception of other areas. This breeds an acceptance of the remainder of society's functions without understanding the how and the why that produced them. Thus, the interdependence outlined by Bertrand Russell.

If society is to remain cohesive and to move forward by aim and not by accident, then each specialized group must apply its skills and knowledge to the benefit of the public. Today, it is generally agreed that greater knowledge brings greater responsibility (2). The educated professionals, in light of their specialized skills, carry a large burden of responsibility. Yet, in their specialization they also carry the consequences of limited comprehension of the other facets of society and other fields of knowledge. Perhaps in the forefront of these nearsighted professionals are the people greatly responsible for today's technical society, the engineers.

[1]Struct. Engr., Fossil Power Generation Group, Babcock and Wilcox, Barberton, Ohio.

With his technical expertise, the engineer has used the forces of nature and the properties of materials to produce some remarkable accomplishments. A list of these would encompass much of what is accepted as modern life. From the largest civil structures to the minute electronic circuitry of pocket calculators, almost all segments of society are touched in some way by the work of the engineer. Yet, for all his achievements, a change in public attitude toward the engineer has taken place.

As recently as the late 1950's the United States considered its engineering skills among the foremost reasons for national pride—as witness the great public disappointment that U.S. engineers were not allowed to build the Aswan Dam. Today, engineering faces increasing disdain, mistrust, and fear (6, p. 19). Ill conceived projects that damage the environment and far outweigh the benefits to society are darkening the engineer's reputation. Products which endanger life and health have been discovered long after their exposure to the public. But is the engineering profession to blame for its transgressions against society? The answer to this question requires an examination of the procedures of conception and design.

THE ENGINEER–CLIENT RELATIONSHIP

As previously stated, the engineer is a specialist and in specialization has somewhat limited his view of society and its needs. He therefore has come to rely on the dictates of others who, by virtue of their broader spectrum of knowledge, supposedly know the desires and needs of the public whom the engineer is ultimately to serve. These would include private entrepreneurs, corporations, governmental bodies and agencies, and anyone who would engage the engineer and his services. The client and engineer form a symbiotic relationship with the client providing direction and the engineer the means.

The strength of this relationship and the responsibilities that accompany it are not taken lightly by the engineering profession. An examination of the American Society of Civil Engineers Code of Ethics reveals that two-thirds of the articles guiding professional conduct outline the engineer–client relationship, principally the engineer's duty to his client. This is understandable, for, along with providing direction, the client assumes many risks and must be insured of the engineer's faithful service.

In the past, the engineer was able to concern himself narrowly with the technical aspects of meeting his clients' stated needs. His criteria included economy, functional efficiency, quite often style, if not esthetics. The engineer defined the limits of technical feasibility and strove to find the best solutions to problems within these bounds. Unless no solution presented itself, despite his most exhaustive and imaginative efforts, he seldom found a reason for questioning the wisdom of what the client wanted (1, p. 164).

Then the criticism began. Ralph Nader focused attention on every man's engineering marvel, the automobile, and made the public aware that the manufacturers were not giving all the safety that technology could provide and that common sense demanded.

The Sierra Club and other environmental protection groups began questioning the real benefits of the massive civil projects which had previously been such sources of pride. Ecology became a common word in the vocabulary as an anti-technology backlash swept the country. Advanced technology was no longer synonymous with progress and the engineer no longer a hero, for, in his devotion to serve the client, the engineer had failed the public who placed in his trust the power of technology. The engineer knows the full requirements and results of his technology better than any other segment of society. And yet, the profession is in no way fully competent to predict the full effect of its work on the environment and the remainder of society. Because of technology's ponderous effects, the engineer must seek the guidance of others in its administration.

Relying solely on his client for this responsibility must become a thing of the past. Lacking any system of public scrutiny, the client has a record of disregard or misunderstanding of public welfare. Where profit and economy are chief measures of success, public well-being is not always served.

The client–engineer bond also tends to promote economy over conscience. As "partners" in a venture which realistically should prove profitable to both, financial and personal pressures can be created which cloud issues of public welfare. By limiting discussion of these issues to the engineer and his client, society on the whole can suffer. However, too much public scrutiny can cause equal harm.

An undertaking of any magnitude will be met by some dissent. With public opinion being so diverse and public processes being so slow, a total referendum on any issue could cause unaffordable delays and cancellation of a beneficial proposition. Thus, the general public is not necessarily a good referral system for the engineer's work. The seeming contradiction of allegiance without respect is one of practicality and only adds to the engineer's dilemma.

PROBLEMS AND ALTERNATIVES

The dilemma for the engineer is principally this—he must serve two masters whose wishes do not always coincide and whose opinions are not always in the best interest of all. Like most dilemmas, this has no suitable single solution. However, steps can be taken to help with resolution of the problem and to avoid future conflicts of this nature.

For the benefit of all, the engineer must gain a greater social consciousness. As man begins to develop a more enlightened view of his world as a unified system, the engineer's preview also seeks broader dimensions. His works are far too numerous, too massive, too powerful, and too pervading to continue to be viewed individually as mere technical services. Collectively, they form a major part of the framework of our environment (1, p. 165).

By establishing a more meaningful dialogue with other groups such as professional, private, and governmental agencies who speak authoritatively in representing the social and environmental interests that now interface so closely with the work of the engineer (1, p. 166), he can choose with greater decision the solution to the conflict.

However, this does not resolve or redefine the conflict between client and public interest. Education only allows the engineer to see the questions more clearly and hopefully to select a proper course of action. By taking steps to aline public and client interests, future conflict could be avoided.

One means of doing this would be more engineering counseling at the upper management levels where the initial directions of a project are determined. To provide such advice would take the concerted efforts of many groups, among which the engineering profession must be prominently represented (1, p. 165). This team of professionals could be tailored to the specific problems of each project, and led by the discipline of who understands the total problem most completely and who can communicate most clearly the many aspects of the problems to technologists and laymen (4, p. 222).

Another direction would be greater public review of an undertaking. During conception, public sentiments could be expressed in the form of codes and laws that guide the client and engineer's design. Zoning laws, safety codes, and environmental impact studies are existing samples of this type of legislation. As the effects of technology become increasingly evident, it is conceivable that the future may require social impact studies covering a broad range of doctrines affected by a project. Boards of review could follow a project through execution and completion, settling any disputes that arise and insuring that all commitments be met. If these processes are to be workable programs that are truly beneficial and equitable, the engineer must play a major role in their conception and administration.

Finally a procedure of due process should be established so that engineers can voice dissent without undue consequence. Faced with possible reprisal from employers, the engineer is pressured into silence. At the other extreme, the engineer faces unnecessary project delays if his questions, made public, prove unfounded. A "disinterested party" could hear the grievances, review them, and make recommendations, without causing project delays, and still be responsible to the publc interest. Whatever action was recommended, the engineer could be protected against unwarranted actions from his superiors by legislation and his colleagues in the professional societies. To require an act of courage for stating perceived truth is to foster a system of self-censorship and the demise of individual conscience against the organized (3).

The Periclean dictum, "In Athens, we think that silent men are useless," still applies in a twentieth century democracy based on science and technology (2). The engineer has been silent for too long. It is time he realized the magnitude of his work while remembering its place in the framework of society.

The strength of the client–engineer relationship must remain, but also must be modified if it is truly to serve society's best interest. It must become more flexible and sensitive to issues other than dollar value and present technology.

For, while there are many guardians of "what might have been" or "what should be done," the engineer must bear a unique responsibility for "what was done" and "what can be done" (1, p. 167).

References

1.   Clarke, F. J., "The Engineer's Role," *Engineering Issues—Journal of Professional Activities*, ASCE, Vol. 98. No. PP2, Proc. Paper 8825, Apr., 1972, pp. 163–167.

2.   Mayer, J., "Science Without Conscience," *American Scholar*, Vol. XXXXI, Spring, 1972, p. 266.

3. Nader, R., "A Code for Professional Integrity," *The New York Times*, January 15, 1971, p. 43.

4. Novick, D., "The Civil Engineer in the Multidisciplinary Design Team," *Engineering Issues—Journal of Professional Activities*, ASCE, Vol. 98, No. PP2, Proc. Paper 8826, Apr., 1972, pp. 221–226.

5. Novick, D., and Schumaker, S., "The Civil Engineer of the Future—A Renaissance Man?," *Engineering Issues—Journal of Professional Activities*, ASCE, Vol. 99, No. PP1, Proc. Paper 9489, Jan. 1973, pp. 8–18.

6. Roberts, K., "The Public Role of Engineering: The California State Water Project," *Engineering Issues—Journal of Professional Activities*, ASCE, Vol. 99, No. PP1, Proc. Paper 9492, Jan., 1973, pp. 19–38.

2.5

The Professions and Government: Engineering as a Case in Point

RICHARD L. SCHOTT

Engineering, among the largest of the professions, has had a substantial impact on American government at nearly all levels. As the domain of government activity has increased and its programs become more complex, engineers have provided an important source of talent for the implementation and the management of wide-scale technical ventures. Engineering's oldest branch, civil engineering, has, since its beginnings, been closely associated with public technology; some of its newer branches, such as aeronautical engineering, depend heavily on governmental support. In this essay, I hope to provide a brief introduction to the engineering profession, examine some characteristics of engineering which influence its impact on government, and raise several issues which surround the role of engineers in the public service.

ENGINEERING AND ENGINEERS

Engineering, with roughly one million practitioners, is among the largest of American professions. Second in size only to primary and secondary school-teaching (which is largely a public function), it is the largest "general" profession. Since there is no commonly accepted definition of "engineer," their exact number in the American workforce is a matter of contention. Numerous individuals who identify themselves as engineers in census data, for example, do not hold even bachelor-level degrees in engineering. Although the 1970 census lists some 1,200,000 engineers, an analysis by the National Science Foundation suggests that the number is closer to 986,000.[1]

Engineering is predominantly a profession of white males from lower-middle and lower class backgrounds. Indeed, it is the traditional under-graduate profession for upwardly-mobile males, as nursing and school teaching have been for females. Women engineers make up less than two per cent of the engineering population, minority groups about one per cent.[2] Engineering offers a professional curriculum of four years' duration which has tended to attract students who cannot afford the extended preparation required by the more prestigious graduate professional fields such as law and medicine. During this century, there has been a tendency for engineering to recruit from increasingly lower socio-economic strata; and the percentage of students coming from blue-collar families has steadily increased.[3] "Clearly," one observer concludes, "engineering has become an avenue of upward mobility for the intelligent sons of working class families."[4]

Engineering is open, diverse, and pluralistic. Entrance to its ranks is subject to firm control neither by its practitioners nor by its institutions of education. Engineering is not a terminal profession: large numbers of its practitioners eventually leave technical engineering work for managerial and executive positions. Its knowledge base is fragmented into a variety of disciplines and sub-disciplines, and it lacks legal recognition in the sense that registration or licensing are not a requirement for most engineering practice. It lacks a generally-accepted code of ethics as well as effective peer control over the activities of its members. Its commitment to society and its concern for social responsibility traditionally have been weak. Engineering has yet to develop a guiding rationale for ideology; and its organizations, as we shall see, are numerous and diverse. And due partly to its large size, it lacks an essential characteristic of the older general professions—a sense of community.[5]

Sociological studies of the engineer depict him as an intelligent, stable, self-sufficient, conforming individual, rather introverted but not personally introspective. Engineers appear to be energetic and goal-oriented and to place a high value on advancement and on tangible (especially financial) rewards. Two scholars who have surveyed the literature concerning the characteristics of the typical engineer suggest he is "an authoritative person with few intimate relationships, strongly work-oriented, competent and energetic, and preferring to deal with objects and things rather than with people."[6] Studies of the occupational values held by engineers indicate that they place a high emphasis on jobs that provide a comfortable income and in which work is judged on its merits, and a low emphasis on jobs which involve working with people rather than things and which make a contribution to society.[7]

The professional model of the independent practitioner does not apply to engineers. The vast majority of them, over 95 per cent, work in and for organizations. Their dependence on large organizations has proved an historical impediment to full professional status, for it means that the organization rather than a client is the consumer of the engineer's services. And it also means that for the most part, the engineer's performance is judged by superiors in his organization rather than his colleagues. This being the case, one might assume that the engineer is an excellent example of the classic problem of the professional-in-the-organization—an individual frustrated by the restraints on professional activity which the organizational environment places on him. Empirical studies of engineers in a variety of settings, however, suggest that this is not the case. These studies have found that engineers place little importance on such traditional professional values as a collegial orientation, a sense of professional community, and the acquisition of knowledge. Such findings have led two scholars to suggest that the engineer is rarely in conflict with the demands of large organizations, but rather adopts them to his own advantage. "The B.S. degree engineer", they conclude, "is in many ways the essence of the organization man. For, although he is a professional, this is largely due to his occupation having status in the larger society."[8]

This seemingly symbiotic relationship between the engineering profession and the organizational environment in which it moves may help explain a phenomenon of particular interest to us—the tendency of engineers to move into management and administration. Those from engineering backgrounds figure prominently among the executive cadres of private industry and, as we shall see, of government. Surveys of engineers' career patterns suggest a pronounced tendency over time for them to move into supervisory and administrative work.[9] One study by the Engineers' Joint Council found that of the sample they surveyed, over a third had become managers of programs, departments, or whole organizations.[10] The seeming appeal of administration may reflect a certain congruence between the engineer's occupational values and the norms of large organizations, as well as the scope within such organizations for the application of "rational" management. But it may also reflect the potential of executive positions to help engineers realize such career objectives as advancement, a sense of power and control, and financial reward which are important to many of them. "For many career-minded engineers," one observer suggests, "the executive office rather than the senior engineer's office is the ultimate goal."[11] For such individuals, professional "success" means moving out of technical activities into administrative ones; the practice for which they were trained becomes not an end in itself but a stepping stone to executive careers.

ENGINEERING EDUCATION

One can gain a good deal of insight into the engineering profession and its practitioners by examining its professional schools, for it is here that attitudes, values, and the context of professional activity are largely determined. Engineering, like a number of the newer or emerging professions, differs from such well-established professions as law and medicine. Whereas the established professions developed a graduate curriculum on a base of undergraduate liberal or pre-professional courses, engineering education has, since its beginnings in the nineteenth century, offered professional training at the undergraduate level in a period of four years. This has placed considerable strain on the content of engineering curricula, which attempt to cover basic and applied science, mathematics, engineering design and tool courses, and liberal studies in a relatively short period of time. The result is a curriculum of great density, few electives, and only scant attention (less than a year) to courses in the humanities and social sciences.

Much of the debate over curricular matters in engineering education has turned on whether to give greater emphasis to the science underlying engineering practice (the approach most often favored by Ph.D.-trained engineering faculty) or to applied and technical courses which prepare the student for immediate employment (the approach favored by most industrial employers of graduating engineers). As a result, the social science and humanistic components of the curriculum ("socio-humanistic studies" in the argot of engineering education) are lost in the shuffle. Indeed, since the introduction of engineering education on a a large scale, stemming from the impetus of the Morrill Act in the 1860s, the average proportion of the curriculum devoted to liberal studies has fallen from a little over a year to one semester.[12]

From our perspective, the problem of socio-humanistic studies is pressing. They are often the only "general" education many engineers receive, offering the potential of expanding his horizons beyond the narrow confines of technical study to the social and political environment and to the ecology of the profession. In fairness, it should be noted that a number of engineering educators have been sensitive to this problem, and a number of self-studies have argued for greater attention to the socio-humanistic area. Recently, there has been a growing awareness of the role of such studies in giving the engineering student an appreciation of the social impact of his work. While traditional views of the engineer's societal role held that rational design and the "impersonal service" rendered by the engineer was sufficient, a recent self-study countered that "It is becoming increasingly clear... that this genial optimism is not well-founded. Solutions to problems often merely generate more problems, with no guarantee that the end result will be in the interests of mankind. . . . It is the background of urgency which constitutes the new challenge to engineering education and to the humanistic and social sciences as part of engineering education."[13]

Despite such affirmations by engineering educators of the importance of liberal studies, they have simply not grown substantially, if at all, within the engineering curriculum. One limiting factor, of course, is that the press of technical subjects forces them to take a back seat. A contributing reason may be the fact that only since the 1950s has the engineering accreditation process paid formal attention to the place of socio-humanistic studies. Though accreditation guidelines require that one-half to one year be devoted to such studies, there is evidence that a number of schools have not significantly increased the amount of liberal arts courses in the curriculum.[14]

A key factor, however, may be the attitudes of many engineering faculty toward such studies. Analyses suggest that engineering faculty are often hostile or at best indifferent to the liberal arts, and reluctant to see them expand as a proportion of the curriculum. One study of the views of the liberal arts held by faculty in various professional schools found that of nine undergraduate-level professions, engineering faculty ranked lowest in favorable attitudes toward liberal studies.[15] (There appears to be a tendency on the part of many faculty, I have been told on several occasions, to dismiss liberal arts courses as "cultural b.s.") Moreover, a survey of engineering schools found that majority of them were satisfied with the amount of socio-humanistic studies presently offered in the curriculum; only a minority felt more were needed.[16]

Of those liberal studies which *are* offered in engineering curricula, courses which would give the student an appreciation of the ecology of the profession often are neglected. Engineers tend to place greater emphasis on subjects such as English composition and speech which contribute directly to professional performance. The social sciences, including government and political science, appear to be at a particular disadvantage: those engineering educators who accept the role of liberal studies as a contribution to their students' intellectual horizons often stress humanities courses such as literature, philosophy, and history, rather than social science offerings.[17]

THE ENGINEERING STUDENT

What of the students who pursue and eventually complete the demanding and highly-technical engineering curriculum? Most research tends to indicate that engineering schools attract more than their share of bright students who rank high on aptitude tests, especially those which measure quantitative ability. High school male graduates near the top of their class often choose engineering as their initial major. As one scholar has observed, engineering "has not lacked for capable students, at any level, especially when compared to other areas."[18]

Recruits to engineering schools share a cluster of values which distinguish them from those entering other fields. Freshmen engineers studied by the National Opinion Research Center ranked high on measures of desire for financial reward and on measures of originality and creativity, but quite low on measures of "people orientation" and service to society. Student engineers appear to enjoy ordering and systematizing knowledge; to be pragmatic, practical, and "thing-oriented;" and to have a rather low tolerance for ambiguity.[19] A number of these characteristics are especially typical of students from lower socio-economic backgrounds. One observer suggests that their social origins influence strongly the values held by many engineering students, whom he found basically anti-intellectual, contemptuous of middleclass values, and suspicious of liberal education. "These students believe in efficiency and hard work. The narrow, workmanlike and anti-intellectual attitudes held by so many engineers are in part a function of engineering education [and] in part a function of social recruitment to the profession...."[20] As I will note later, these characteristics of engineering students and the curriculum they pursue may have serious implications for the ability of engineers to grapple with the policy and administrative roles many assume at the upper levels of the public service.

ENGINEERING ORGANIZATIONS

Engineering organizations, like the engineering profession itself, are diverse; the various national groups number well over one hundred. There is within engineering no one organization or society that can claim to speak for, or to represent, the engineering profession as a whole—certainly not in the sense that the American Bar Association, for example, claims to speak for the bulk of the legal community. Some engineering organizations serve engineering disciplines and technical specialties; others serve its institutions of learning; still others like the National Society of Professional Engineers (NSPE) attempt to increase the professional status of its members. In addition to these organizations there have appeared groups which have tried to act as "umbrella" organizations for the engineering profession as a whole.

The earliest, and still among the most influential discipline-oriented engineering organizations are the four "Founder" societies, formed around the dominant disciplines of the nineteenth century—the American Society of Civil Engineers (ASCE), the American Society of Mechanical Engineers (ASME), the American Institute of Mining, Metallurigical and Petroleum Engineers (AIME), and the Institute of Electrical and Electronics Engineers (IEEE). The numerous disciplinary and technical societies which exist today are spread over a spectrum which ranges from the highly-professional (in the sense of an orientation toward the individual, practicing engineer) at the one end, to those that are basically business or industry-oriented at the other. One clue to a particular society's place on this spectrum is the kind of membership which it allows. Those which admit only individual members fall toward the professional end; those which admit individual *and* institutional members somewhere in the middle; and those whose members are solely institutional toward the business or trade-association end. Another clue is the qualifications necessary for full membership. The more "professional" organizations stipulate that full members be qualified in engineering *design*, a requirement which limits full membership largely to practicing engineers. Other societies require that a full member be only in "responsible charge" of engineering activities—a term broad enough to include most managers and executives. A last category of membership requires only that members be engaged or interested in a particular field of engineering.

Two engineering organizations—the American Society for Engineering Education (ASEE), and the Engineers' Council for Professional Development (ECPD)—serve the interests of engineering education and its educators. The ASEE, formed in 1893 as the Society for the Promotion of Engineering Education (SPEE), was the creation of engineering educators. Its founding gave expression to a need for greater cohesiveness and planning among engineering schools and curricula and as well to a feeling that the existing technical societies seemed disinclined to give the educational process sufficient attention. The early members of the ASEE were educators, not practitioners, representatives of a newer "school culture" which gradually gained control over curriculum content and admissions standards for a majority of engineering schools.[21]

The Engineers' Council for Professional Development, formed in 1932, was a response to pressures for the formal accreditation of engineering curricula, generated largely by members of the "school culture." The ECPD was formed originally from members of the four Founder societies plus the American Institute of Chemical Engineers; the (then) SPEE, and the National Council of State Boards of Engineering Examiners—organizations representing the professional, educational, and legal aspects of engineering. Today, the ECPD accredits curricula leading to the first degree in engineering, usually the B.S., as well as programs in engineering technology, considered sub-professional.

A special role in the professionalization of engineering has been played by the National Society of Professional Engineers (NSPE). With membership open only to individual engineers, NSPE was formed in the early 1930s in support of then-current attempts to enact strict provisions for the registration and licensing of engineers. NSPE has since attempted to promote greater professional competence among engineers by encouraging them to take and pass examinations for registration. And as it has promoted registration at the upper levels of engineering practice, it has also attempted to combat unionism at its lower levels and has campaigned actively to defeat engineering unions.[22]

Though the engineering profession has proved too diverse and perhaps simply too large to enable the growth of a single, unified professional group representing all engineers, it has been successful in creating a succession of umbrella organizations made up of technical society, institutional, and trade group members. The current [1978] engineer umbrella organization, the Engineers' Joint council (EJC), was formed in the 1940s and today includes some two dozen member societies and associates. Since 1967, it has allowed the affiliation of industrial corporations as well (a move which led one of the Founder societies, the IEEE, to withdraw). The EJC, in the eyes of one of its critics, has demonstrated a special concern for the "interests of employers of engineers." It helped to fight the growth of engineering unionism in the 1950s, has generally supported increasing the number of engineers to ensure an adequate manpower pool for employers, and has worked to keep engineering "open" by supporting less restrictive registration and licensing laws.[23]

The newest arrival on the scene is the National Academy of Engineering (NAE), which received a charter from the National Academy of Sciences in 1964. Its establishment represented a counterweight to what many engineers considered an over-emphasis, institutionally, on pure science at the national level. Among the Academy's activities is the provision of advice to the federal government on the application of engineering techniques to problems encountered by government agencies. And its Commission on Education has paid a good deal of attention to new developments in engineering education, especially those related to an enhanced appreciation of the social role of engineering.

The organizational picture of the engineering profession is thus one of great diversity—a melange of technical and special interest groups, some looking toward the individual practicing engineer, some toward engineering education, others toward engineering as a profession, and still others to business and industrial interests. No single organization speaks for all of engineering, and few even for the majority of those they claim to represent. Historically and functionally, engineering organizations are as diverse as the engineering profession itself.

THE ENGINEERING PROFESSION AND GOVERNMENT

Engineering has a substantial influence on governments at many levels. Its practitioners perform not only a variety of technical tasks but also form an important managerial cadre at the upper echelons of the public service. Recent data indicate that roughly 16 per cent of employed engineers work for federal, state, or local government with the vast majority, some 73 per cent, employed by private business and industry. The balance are in college teaching or non-profit organizations. There is considerable variation, however, among the various branches of engineering. Government employs about 43 per cent of all civil engineers, for example, but only six per cent of chemical engineers.[24] Though the percentage of engineers working in government is small compared to their total population, the fact that engineering is such a large profession helps explain why its members constitute such a large segment of all professional employees of government. In the federal government, engineering—with some 84,000 of its practitioners employed in professional positions—is far and away the largest single professional group. Indeed, engineers account for nearly one-third of all professionals in federal employment.[25]

There is evidence that the engineering profession, as a whole, denigrates government and government employment, and that its sympathies lie with the private sector. Support for this assumption comes in part from the Kilpatrick studies of the image of the federal service, which included sample groups of engineers in industry and business, in college teaching, and in public service. Attitudes toward government and public service among the sample from business and industry (which employ the vast majority of engineers) were found to be quite negative. Business engineers, for example, rated their present job much higher than the same job in the federal government. Asked how their families would feel should they go to work for government, the majority of engineers who said it would make a difference indicated that their families would see such a move as "down the ladder." Engineers in business saw government employment as restrictive, involving red tape, wasted motion, lack of self-determination, and rigid supervision. And they tended to have unfavorable views of federal employees in general, seeing them as security-conscious and lacking ambition.[26] (Such feelings were also reflected in the observation made to me by the head of an engineering professional organization that only the bottom members of his graduating class went to work for government.) Engineers in federal employment, however, do not share these negative views of themselves and their work. They prefer their present job in the federal service to the same job in the private sector, and view government work overall as more appealing. They are attracted to the fact that government employment offers greater security, but agree with their colleagues in the private sector that business organizations provide a better oppportunity to become "really successful."[27]

In the federal government, engineers are employed in large numbers by those agencies whose missions are closely associated with applied science and technology. The Department of Defense is the single largest employer of engineers, followed by the National Aeronautic & Space Administration (NASA), the Department of Transportation, the Department of Interior, the Environmental Protection Agency, and the Department of Agriculture. Engineers are found in substantial numbers at the middle and upper levels of the federal service, their *median* grade being GS-13.[28]

Of greatest interest to us are not primarily those engineers (the bulk of the federal engineering population) who are journeymen pursuing technical work, but rather those—like many of their colleagues in the private sector—who have moved upwards into administrative and executive positions. It is these individuals through whom the influence of the engineering profession on government is most likely to be translated. Exact estimates of the number of engineers in the federal executive population (defined as GS-15 and above by the U.S. Civil Service Commission) are difficult to make, since those from engineering backgrounds hold a proportion of the jobs in the category "administration," and as well in the "public law" group. But sources suggest that the number of engineers and engineers-turned-administrators in the federal executive population is probably as high as 25 percent.[29] It is, however, interesting that during the past few years, the number of engineers in the engineering occupational category has been steadily decreasing, due at least in part to retrenchment in NASA, traditionally a large employer of engineer executives, and in the Navy. The legal profession, by contrast, has shown a corresponding growth over this same period, reflecting perhaps the growing influence of the legal profession at the highest levels of the federal service as the national government has placed more emphasis on regulation and less on the development of technology.[30]

A study of federal engineer executives conducted several years ago addresses a number of the concerns raised earlier in our discussion of the engineering profession.[31] Based on a sample of engineers in NASA, the Corps of Engineers, and the Federal Highway Administration, the study found that by the time engineers in these agencies reached grade GS-14 and higher (including thus a grade lower than the Civil Service definition of the federal executive population), their work had already shifted substantially from technical activities into administration and management. More than 80 per cent of the sample reported that their work was at least half administrative in nature, with over a quarter reporting that it was primarily or exclusively so. And as he advanced in rank, the engineer's work focused increasingly on administration.[32]

These federal engineers, as administrators, are working in areas far removed from their formal education, which, as we have seen, has little relevance to administration and broad policy concerns. This group of engineers were well aware of this situation, and agreed almost unanimously that their engineering curriculum had done little to prepare them for the broader executive roles they had assumed. In looking back on their education, they favored increasing substantially the portion of their curricula devoted to socio-humanistic studies, especially the social sciences. Had these engineers known they would have ended up in their present positions, they would have (as before) done undergraduate work in engineering; but of those who went on to graduate work, nearly half would not have continued in engineering, but would have taken degrees in public or business administration. Though they felt their continuing, in-service education had contributed substantially to their ability to cope with their new roles, more than half expressed a moderate-to-strong need for additional training in such areas as human relations, public administration, and personnel management.[33]

The sensitivity which these federal engineer executives display toward their educational needs is triggered, at least partly, by the demands of jobs for which they were not well-prepared. But why did they opt for administrative and executive positions? One answer may be found in the occupational values—financial reward, increased responsibility—discussed earlier in connection with the proclivity of engineers to move into management. However, a seminal study of the transition from professional to administrative positions made by engineers (and scientists) in several federal agencies suggests other factors also may be important. This study found that engineers fall into several different groups, depending on how they view the prospect of moving from technical to administrative positions. One group included those whose motivations and values were consistent with moving into management. A second group was comprised of those who were reluctant to do so, but who grew to enjoy administrative responsibilities. A third were inclined against such a move, and remained strongly opposed. But over two-thirds of those studied fell into the first group,[34] thus suggesting that the federal executive engineer cadre is drawn largely from those with an inclination to assume executive roles—an inclination no doubt buttressed by the values which appear to permeate the engineering profession.

CONCLUDING OBSERVATIONS

It is impossible in an essay of such brief compass to assess in any comprehensive way the impact of the engineering profession on government. There are, however, a few central themes or questions which can be raised. As we have seen, the engineering profession is oriented in the main to the private sector, its major employer. The need to produce engineering graduates for immediate employment in industry and business is reflected in the engineering schools, where the humanities and social sciences, not to mention the study of government and the "political" are given short shrift. In spite of some pressure for reform from within, engineering education remains a narrow, technical curriculum which allows the student little time for reflection or for exposure to vast domains of human knowledge. It is doubtful that this narrow training does much to broaden or inform the public service by those engineers—often second-class citizens in the eyes of their colleagues—who end up working for government. Granted, however, that the relationship of engineering to government varies with the field concerned. Civil and environmental engineering, for example, have generally demonstrated a deeper appreciation of public sector problems through their close association with government than have such branches as industrial and mining engineering, which traditionally have been closer to the private sector.

A second question concerns the *kinds* of individuals who traditionally have been recruited to engineering. As we have seen, they have come predominantly from lower socio-economic backgrounds and tend to share the provincialism and authoritarianism often typical of upwardly-mobile individuals. One long-time student of the engineering profession suggests that the social origin of engineers has foreclosed the development of a service ethic in the profession and has kept it from becoming concerned with human welfare and broad social goals.

> As the importance of engineers has increased to the point where major segments of the society are dependent upon their expertise, we have also observed a corresponding decline in the possibilities for the development of engineering as a service profession committed to the service of man.... The social origins and consequent mobility experiences of persons entering engineering lead to the reinforcement of business rather than professional values, thereby inhibiting the emergence of a service ethic which focuses upon human welfare.... [35]

Given what appear to be rather clear sets of characteristics of engineers and the lack of a social ethic among them, their impact on the social and human dimensions of government activity is quite possibly a negative one.

A special problem is created by the proclivity of engineers to move into management and executive positions. Such positions generally require a broad approach to analysis and decision-making and substantial interaction with the policy and political environment of the program or agency concerned. One must question whether government is necessarily well-served by placing in its highest career echelons individuals of rather provincial and technical backgrounds. It is likely that continued reliance on engineers for the highest administrative cadres will contribute to the narrowing tendencies which professions often have on public service—in the words of Frederick Mosher, "gradually but profoundly moving the weight toward the partial, the corporate perspective and away from that of the general interest."[36] Though one of the antidotes to such influence offered by Mosher is a reform of professional education, one cannot be sanguine about the possibility of such developments in engineering.

One should bear in mind, however, that the technical expertise of engineering has been invaluable to the number of complex tasks that governments have undertaken over the years: From water and roadways in the early nineteenth century, through sanitary engineering accomplishments in the rise of the cities later in that century, through the successes of the Manhattan Project in the 1940s and the Apollo space program in the 1960s, to the most recent developments in protecting the environment and developing new sources of energy. Engineering, moreover, had a certain rationalizing effect on the early development of a public service which was shaking off the vestiges of the spoils system, and has played a major role in designing information and other control systems which have contributed greatly to the administration of vast and complex public organizations. But the injection of engineers into the commanding heights of these organizations continues, justifiably, to give pause.

Notes

1. U.S. Bureau of the Census, *1970 Census of the Population. Vol 1, Characteristics of the Population: Summary Data* (Washington: USGPO, 1973), p. 1–718; U.S. National Science Foundation, *The 1972 Scientist and Engineer Population Redefined* (Washington: NSF, 1975), p. 5.

2. U.S. Bureau of the Census, *op. cit.*, pp. 1–718, 1–739.

3. Carolyn C. Perrucci, "Engineering and the Class Structure," in Robert Perrucci and Joel E. Gerstl, eds., *The Engineer and the Social System* (New York: John Wiley and Sons, 1969), pp. 282–284.

4. Robert L. Eichhorn, "The Student Engineer," in *Ibid.*, p. 144.

5. For an elaboration of these themes, see Robert Perrucci and Joel E. Gerstl, *Profession Without Community: Engineers in American Society* (New York: Random House, 1969).

6. Donald E. Super and Paul B. Bachrach, *Scientific Careers and Vocational Development Theory* (New York: Teachers' College of Columbia University, 1957), pp. 66–67.

7. See, for example, William K. LeBold *et al.*, "The Engineer in Industry and Government," *Journal of Engineering Education*, Vol. 56, No. 7 (March, 1966), p. 727.

8. Perrucci and Gerstl, *op. cit.*, pp. 118–119; 137; 161. It should be noted that there are variations in the degree of professional orientation among engineers, and that the level of professional activity as well as the importance of professional values appear to increase with a higher level of education. This relationship may show either that advanced education leads to more "professional" jobs or that professional values are imparted during the longer professional socialization which advanced study entails. *Ibid.*, p. 123.

9. See LeBold, *op. cit.*, p. 245.

10. Engineers' Joint Council, *A Profile of the Engineering Profession* (New York: EJC, 1971), p. 11.

11. Laure M. Sharp, *Education and Employment: The Early Careers of College Graduates* (Baltimore: The Johns Hopkins Press, 1970), pp. 54, 79.

12. Society for the Promotion of Engineering Education, *Report of the Investigation of Engineering*

Education, 1932–1929 (Pittsburgh: Lancaster Press, 1930–34), Vol. I, p. 552; American Society for Engineering Education, "Liberal Learning for the Engineer: Report," *Journal of Engineering Education*, Vol. 59, No. 4 (December, 1968), pp. 303–342.

13. American Society for Engineering Education, *op. cit.*, p. 310.

14. See *Ibid.*, pp. 332–333.

15. Edwin J. Holstein and Earl J. McGrath, *Liberal Education and Engineering* (New York: Teachers' College of Columbia University, 1960), pp. 97–103.

16. William K. LeBold, Robert Perrucci, *et al.*, "Educational Institutional Views of Undergraduate Goals of Engineering Education," *Journal of Engineering Education*, Vol. 56, No. 6 (February, 1966), p. 218.

17. Holstein and McGrath, *op. çit.*, p. 44.

18. Paul Heist, "The Student," in National Society for the Study of Education, *Education for the Professions* (Chicago: NSEE, 1962), p. 222.

19. James A. Davis, *Undergraduate Career Decisions* (Chicago: NORC, 1965), p. 180; Perrucci and Gerstl, *Profession Without Community*, p. 51.

20. Martin Trow, "Some Implications of the Social Origins of Engineers," in U.S. National Science Foundation, *Scientific Manpower: 1958* (Washington: USGPO, 1959), p. 72.

21. See the discussion in Monte A. Calvert, *The Mechanical Engineer in America, 1830–1910: Professional Cultures in Conflict* (Baltimore: Johns Hopkins Press, 1967), p. 48.

22. Joel Seidman, "Engineering Unionism," in Perrucci and Gerstl, *op. cit.*, pp. 225–226.

23. Edwin C. Layton, Jr., *The Revolt of the Engineers: Social Responsibility in the American Engineering Profession* (Cleveland: Case Western Reserve Press, 1971), pp. 249–252.

24. U.S. National Science Foundation, *1972 Scientist and Engineer Population Redefined*, Vol. 2, *Detailed Statistical Tables* (NSF76–306) (Washington: NSF, 1976), p. 2.

25. U.S. Civil Service Commission, *Occupations of Federal White-Collar Workers, October 31, 1974 and 1975* (Washington: USGPO, 1976), pp. xiv, 5–6.

26. Franklin Kilpatrick, *et al.*, *The Image of the Federal Service* (Washington: Brookings Institution, 1964), pp. 71–75, 221–237, 321–337. One should temper these findings by noting that the data were collected in the early 1960s before such projects as Apollo brought additional glamour to government engineering and also before the substantial salary increases of the middle and late 1960s.

27. Franklin Kilpatrick *et al.*, *Source Book of a Study of Occupational Values and the Image of the Federal Service* (Washington: Brookings Institution, 1964), pp. 71, 75.

28. U.S. Civil Service Commission, *Occupations of Federal White Collar Workers*, pp. 38–41, 86–89.

29. U.S. Civil Service Commission, *Executive Personnel in the Federal Service: October, 1976* (Washington: USGPO, 1976), p. 6; sources in the Bureau of Executive Manpower, U.S. Civil Service Commission.

30. U.S. Civil Service Commission, *Characteristics of the Federal Executive* (Washington: USCSC, 1969), p. 5; U.S. Civil Service Commission, *Executive Manpower in the Federal Service* (Washington: USGPO, 1971–1975); U.S. Civil Service Commission, *Executive Personnel*, p. 6.

31. Richard L. Schott, *Professionals in Public Service: The Characteristics and Education of Engineer Federal Executives* (Beverly Hills: Sage Publications, 1973).

32. *Ibid.*, pp. 16–24.

33. *Ibid.*, pp. 31–40.

34. James Bayton and Richard Chapman, *The Transformation of Scientists and Engineers into Managers* (Washington: National Academy of Public Administration, 1971), pp. vi., 103.

35. Robert Perrucci, "Engineering: Professional Servant of Power," in Eliot Friedson, ed., *The Professions and Their Prospects* (Beverly Hills: Sage Publications, 1973), p. 121.

36. Frederick C. Mosher, *Democracy and the Public Service* (New York: Oxford University Press, 1968), p. 210.

2.6

The Engineer and Public Policy-Making

MYRON TRIBUS

During the French Revolution, a priest, a lawyer, and an engineer were put on trial as enemies of the Revolution. The priest tried to defend himself by explaining how he had ministered to the poor, brought food to the hungry, and tended the sick. To no avail. He was sentenced to the guillotine. In those days, when a person felt he had been sentenced unjustly, he would demonstrate his feelings by making the supreme gesture—he would lie face up rather than face down in the guillotine. The priest did so. When the rope was tripped, it happened that the blade stuck in the channel about six inches above his neck. It was the rule that if ever the guillotine stuck, the victim was let free. So the priest was spared.

Then came the lawyer. He too argued for his life. He reminded them of the many criminals he had defended, of how he had served the poor without fee and had argued against unjust laws. But he too failed and was sentenced to die. And, like the priest, he insisted on looking up at the blade as it descended. And in his case, too, the blade stuck in mid-air, and he was let free.

Then came the turn of the engineer. He pointed out that he had built the water works. There were public buildings and roads he had made for the people. He had never been engaged in politics, nor had he become rich at the people's expense. But these arguments fell on deaf ears. And he too expressed his contempt by lying on his back, looking up at the blade. Then, just as the executioner was about to pull the rope, he let out a cry: "Look—I see your trouble. The rope has slipped off the pulley up there. No wonder the thing doesn't work."

The last few years in the United States have seen an increase in the political activity both of engineers as individuals and of the professional societies to which they belong. IEEE is a case in point. At this very moment, the Institute is engaged in considerable soul-searching to determine what its role should be in the public arena. The debate is raging not merely at the level of the Board of Directors or of the volunteer officers, but among the membership at large.

There are members who say IEEE should stay out of the public arena altogether; there are others who say IEEE should limit its public involvement to purely technical issues; and there are those who say IEEE should jump in with both feet. This variance in opinion says a great deal about engineers. At one and the same time, engineers may see themselves as society's saviors and as society's servants. And engineers may be both responsible citizens and inept in the exercise of that responsibility.

That all of this is true is aptly illustrated by the story that begins this article. That story points to the same rationale for IEEE and its members to try to understand and become involved in the public arena, and also to the pitfalls that await the Institute and engineers in general as they try to operate there.

In this article, I do not intend either to advocate IEEE or engineer support of public issues or to suggest that they should stick only to engineering. Rather, at what may be a crucial time in the escalating debate that is going on in the engineering community, I wish to offer some thoughts on the strengths and weaknesses of engineers faced with the public-policy-making mechanism.

LEARNING FROM HISTORY

Just after World War I, there was a brief surge of popular opinion in the U.S. that the engineers, who had been so instrumental in winning the war, could lead the nation to even greater peacetime victories. Had not Herbert Hoover, the engineer's engineer, fed the starving Europeans?

This period is well described in a book by Edwin T. Layton, Jr., called *The Revolt of the Engineer—a Revolt that Failed* (Cleveland, Ohio; Case Western Reserve University Press, 1971). Layton tells us how the engineers rallied to the cause of public service and turned their attention to energy conservation, materials conservation, material substitution, pollution, and the general improvement of the environment, including the working environment. For example, they prepared analyses to show the benefits to be expected if industry would adopt the eight-hour working day. Now, of course, we talk about even shorter working hours, but then the idea was quite radical. The engineers demonstrated conclusively that an eight-hour day would improve productivity. However, the owners of the factories considered this an intrusion, and they fought back. Among other things, they withdrew their support of the engineering societies and of their engineers' attendance at meetings. In the end, the engineers became divided and gave up. (More details on this can be found in *Spectrum*, May and June 1976.)

Today, things have come full circle, and we seem to be in the midst of a period of increased engineer activism in the public arena. To mention only one example, IEEE's Board of Directors recently took a stand calling for the registration of all U.S. engineers "in responsible charge"—even those in industry who have traditionally been exempt from mandatory registration. A storm of reaction has ensued that has induced IEEE's Board to suspend any implementation of this policy pending a report by a special *ad hoc* committee.

On a grander scale, U.S. engineers now flock to Washington to become involved in giving advice to legislators. The Government regulates more and more technical activities. There are laws on the books that specify technical details in consumer product safety, occupational safety, environmental protection, and health protection. The legislators often fancy themselves capable of doing engineering by law and their well-intentioned miscalculations have been evident to the engineers. So the engineers try to help. This may be a healthy attribute but, based on past experience, it is fraught with danger.

WHY ENGINEERS CAN FAIL

Up until Herbert Hoover was swept under by the tide of public reaction to the Great Depression of 1929, he was a much respected and admired public figure. In his book *The Shattered Dream* (New York: William Morrow, 1970), Eugene Smith analyzes how and why this great engineer was destroyed. Franklin Delano Roosevelt campaigned successfully against Hoover and, in the process, firmly fixed him, in the public mind, with the image of a helpless, hopeless president who simply didn't know what to do. Actually, many of the programs Roosevelt started were, in fact, designed by his predecessor. But my interest here is not to deal with Hoover's reputation and public image. Rather, it is to understand what went wrong. What shattered Hoover's dream? Why did the revolt fail?

My thesis is that Hoover failed partly because of the things that made him a great engineer. And the problem isn't restricted to Hoover. It starts with the prototypical characteristics of the engineering mind and it is exacerbated by the engineering education process.

The products of our schools of science and engineering have traits that reflect several factors.

1. *Self-selection.* The engineering student body is a self-selected group of young men and women. They see in science and engineering an outlet for their interests. These interests tend to be centered on things rather than people, facts rather than feelings, logic rather than emotion.

2. *Demands of the curriculum.* Emphasis on logical analysis, impatience with ambiguity, pride in the elegant solution, the delight in knowing and understanding, suppression of emotion, and the value of integrity—all are traits that may be considered typical of engineers.

3. *The image of the profession.* The role model presented by the professors is that of the disinterested professional—the "savior/servant" of society.

4. *Emphasis on solving the given problem.* Engineering students are usually given problems. They have very little experience in finding their own problems. And as they enter the world of work, they learn to solve more and more complex problems. However, they have little or no expertise in deciding what *ought* to be solved.

There are other traits, of course, that are reinforced by the schools and the canons of ethics of the engineers. And, in sum, these characteristics combine to produce a certain approach to solving life's problems—an approach called the "engineer's approach."

ENGINEERS AS PROBLEM-SOLVERS

Anyone with any experience in dealing with lawyers—particularly those in Washington, D.C.—will appreciate the vast differences between the lawyer's outlook and that of the engineer and scientist:

- Engineers and scientists are programmed to tell everything they know; lawyers tell only what they must.
- Engineers and scientists want to be sure the person they are informing really understands the situation, whether friend or foe; lawyers want to be sure the person is a friend.
- Engineers believe there is a "payoff function" that should be optimized for the system as a whole; lawyers don't bother looking for it.
- Engineers are team players—they are willing to subordinate themselves to the team, believing that if they do well, they will be rewarded; attorneys prepare for power struggles.
- Engineers and scientists believe the right solutions will inevitably win out, that time is on the side of truth; attorneys harbor no such Boy Scout illusions—they say they compete in a court of law, not a court of justice; they know that the guilty often go free and the innocent are often punished; and they know that the outcome is as sensitive to style and tactics as it is to fact.
- Engineers are taught to serve—to be on tap, not on top. Read again the "IEEE Code of Ethics for Engineers." The engineer's image is that of society's faithful servant. Attorneys believe they belong on top; power and decision-making are rightfully theirs.

These qualities often put the engineer at a disadvantage, especially when negotiating with an attorney. The principal business of Government is negotiation. People, in fact, do have real differences of opinion and real differences in their values. Engineers strive to find the happy solution that will satisfy all parties. But many social problems have no solutions, in the engineer's sense of the word—only adjustments achieved through negotiation. Since technical people seek "solutions," not adjustments, negotiation comes hard.

This obsession with "problems" and happy solutions comes from the mode of instruction used in science and engineering. In the beginning, students are assigned definite problems, and they are graded upon their ability to come up with the same answer as the instructor. Imagine a freshman saying to the professor of physics, "But sir, that's not the right problem."

As the students advance, they solve harder and more complex problems. After awhile, they learn to extract problems from their environment. Scientists adore unsolvable problems. These can provide them with a lifetime of support, even into the retirement years. The road to fame and fortune for the scientist is paved with problems that have definite solutions discernable only to the best and brightest. But the issues that concern the Government are often issues of choice—not answers, but choices. For the making of these choices the engineers look in vain to their training. Their self-image, developed over the years, tells them they are the advisors to those who make the choices—on tap, not on top.

This image of the faithful advisor impels engineers to come running to Washington to advise their Senators and Congressmen of, say, a threat to the environment. They have trouble understanding that their Senators may not want to hear about it. They see themselves as generous, selfless sources of knowledge, ready to be used in a battle against evil. The Senators would probably rather *not* hear about the evil until it has been established that the public cares, and that curing the evil won't hurt the Senator's constituents.

There is much that can be learned about the consequences of our decision-making by appealing to elaborate computer models of the world's energy needs, food needs, population growth, or even the cyclic nature of the paper industry. But the very things that usually are suppressed in the computer modeling may turn out to be the concerns of our fellow citizens—such as issues of equity, changes in the balance of power, and who gains and who loses prestige and influence and the impact of the process itself on family life, our children, and ourselves. These will have as much to do with what society decides to do (if "society" decides anything) as the "facts."

ENGINEERS AS SAVIORS

Engineers think of themselves as the saviors of civilization. They point as evidence to their handiwork, which is all about us. But in human affairs, what people *think* is true is as important a determinant of action as what really is true. So it matters not in public decision-making whether the engineers are or are not saviors of mankind. What matters is what people think. Engineers are outnumbered by about 200 to one in our society—and the other 199 emphatically do not think of the engineers as "saviors." Ipso facto, they are not.

To my way of thinking, the nontechnical people have completely erroneous views of the relationships of engineering to society. They endorse the "faithful servant" image and consider that they are the masters who make their wishes known and the engineers oblige. History seems to say otherwise:

- No one asked Rudolf Diesel to develop his machine.
- No one asked Henry Ford to develop mass production.
- No one asked Edison for light bulbs.
- No one asked Firestone for rubber.
- No one asked Carlson for xerography.
- No one asked Land for instant photography.
- No one asked for plastics, frozen orange juice, or even the telephone. Inventors, scientists, engineers brought them forth, unbidden—usually against the wishes of many.

But the myths remain. They please the egos on both sides—those who think they are the masters and those who think they are the saviors disguised as faithful slaves.

This self-image as faithful advisor to the decision-makers leads engineers to try to improve things by designing better decision processes. Reading the literature of the Decision Analysis Group in IEEE, for example, reveals this image with great clarity. Members of that Group consider that there exists a Decision Maker (always written with capital letters, too) whom they are to advise by "structuring" problems. In the structuring operation, they ask the Decision Maker for his or her preferences.

Imagine the scene: The advisors shout up to the top echelons where Decision Makers reside: "Tell us your risk aversions. Tell us your measure of impatience; your willingness to trade today's profits for tomorrow's security."

I see the Decision Maker shouting back: "Don't be crazy. Why should I tell *you*? Why would anyone need *me* if I told you? And besides, who says I'm supposed to know these things? If I told you what I really think, I'd be laughed out of the office. I didn't get here by thinking of those things and I know that I probably won't be able to stay here if I play the game your way."

The engineers persist in thinking that the task is to build mountains; the rest of society looks at it as a game called "King of the Mountain." No wonder the engineers tumble to the bottom so easily.

I have spoken of the disadvantages of a technical education. They are real. But then so are some of the advantages. One overwhelming advantage is the technically trained person's ability to see much further into the future, insofar as the physical world is concerned, than nontechnical people. Armed with the laws of conservation of energy, matter, momentum, and charge, the nonconservation law of entropy, the understanding of feedback, system dynamics, control, and information fluxes, an ability to do mathematics, to program computers, to codify natural phenomena, and to measure almost everything in the physical universe, the engineers can construct mental or computer models of the world. They can even make crude models of social behavior. By these means, they see just a little more into the future than their 199 neighbors. And within their own circles, shielded from the political world, they have often used this special ability to enlighten one another. In his book *The Existential Pleasures of Engineering* (New York: St. Martin's Press, 1976), Sam Florman has recounted instances.

For example, 25 years ago the American Society of Mechanical Engineers (ASME) appointed a full-time executive secretary to administer the activities of its committee on air pollution controls. This was 18 years before the creation of the Environmental Protection Administation. Twenty years ago, ASME sponsored the first International Conference on Air Pollution.

A full decade before Rachel Carson and Barry Commoner, Mark D. Hollis, Chief Engineer of the U.S. Public Health Service, addressed a convention of the American Society of Civil Engineers about the possible effects of new synthetic chemicals being spilled into our rivers and lakes, about toxic substances being introduced into the air, radioactive pollution, and new problems in food sanitation. As author Florman observes, the public was not interested. Instead, it spent billions for engineers to put a man on the moon—which they did.

WAITING TO BE CALLED

One consequence of the "faithful servant" image is that engineers stand around waiting to be invited into the decision-making process. Unfortuanately, the only time they are apt to be invited in is when the situation has deteriorated so badly that it has become desperate. In case of war, or of a natural or a man-made disaster, the nontechnical people will allow technical people to disrupt their lives. But in the absence of serious, immediately perceived threats, the average citizen is apt to consider the technical person as an unwanted busybody.

If you are constructed as I am, you will not be satisfied merely to write your reports, play with your computer models, and talk to other engineers about the coming catastrophes. You will want to do things that affect the outcomes. You will not want to wait until things become so desperate that people come to you; by then it may be too late. You want to see some action now, while there is yet time. I'm glad you feel that way. It is a healthy sign; it represents human qualities of which you can be proud.

But what can *you* do about it? In the public arena, the name of the game is always power. People will deny this, of course, but then when two sailboats are going in the same direction at the same time, they will also deny that they are racing!

People want power for lots of different reasons, not all of them bad. They want to be come effective and influence public events. Their reasons may be noble and selfless or base and selfish; but whatever their reasons, they cannot exert influence without affecting the power structure. And, of course, with power goes responsibility. And in the struggle for power, someone will be displaced. This means conflict—a central characteristic of events in the public arena. To contribute in the midst of conflict, the name of the game is winning.

When you are able to exert power, it is only because you have established a basis for it. As far as I can see, there are only five or so bases. For example:

1. You can be elected by the people.
2. You can be appointed by someone else who has been elected.
3. You can develop a constituency that will mobilize when you tell them to—as, for example, Mr. Nader has done.
4. You can head an organization that has a large membership—i.e., a big union or the American Legion.
5. You can have the facts and be a recognized authority, but this is the weakest position of all.

Nevertheless, most engineers opt for the last. And unless they are connected to a strong organization (or at least one with a good public relations officer), they are unable to get the public's attention or the ears of the officials of Government.

Yet it is surprising what engineers can do if they put their minds to it. The Rochester Engineering Society decided that as engineers they ought to play a larger role in the affairs of their city. Today, they operate a Technology Transfer Program for the City of Rochester and provide technical assistance to the local governments on a volunteer basis. One of their members became the chairman of the Pure Waters Rate Review Task Force, which proposed the presently used rate structure to the legislature, as well as organizational changes in the water systems of surrounding areas. They have become involved in the study of the transportation system and the approach the city uses to obtain consultation. They have helped the community establish policies regarding resource recovery. And it all started with a few engineers sitting around a table.

If you are willing to pick out something in your community that interests you and to which you can make a contribution, you can probably work your way into the system in such a way as to influence outcomes. You will do well to study the power structure of your community—and decide with whom you wish to join and with whom you wish to fight. If you go to work on something technical, you will have a great advantage—for you will not only know what the issues are, but you will be able to spot how little your opponents know about them. Don't hesitate to exploit that advantage. Your weakness will be in the list of things I outlined earlier. But you can learn to overcome these weaknesses a lot better than nontechnical people can learn to understand what you know.

Values and Professional Practice

MARVIN L. MANHEIM

INTRODUCTION

As public works professionals—engineers, planners, administrators—we do worry about "value" issues. Unfortunately, in the past, I think we have given "value issues" only a relatively narrow definition.

I find it important to distinguish two major types of ethical, or "value," issues.

The first type of issues is one that we do find addressed frequently in discussions of ethics in engineering and other professions. These are issues such as the giving and taking of bribes to obtain contracts, or the falsification or holding back of data, or allowing one's opinion to be sold to the highest bidder, or one's responsibility to speak out against unethical conduct of this type on the part of fellow professionals. These issues are widely recognized as important issues, and institutional mechanisms for reinforcements of appropriate behavior on these issues are continually debated (see for example Anderson (1) and references (3) and (5)). In my view, these are obvious enough issues and receive enough attention that each of us should know where he or she stands on these issues. Therefore, I will not address this type of ethical issue. Perhaps others at this workshop will wish to speak to some of these issues.

The second type of issues is one that is not addressed often. This set of issues concern the way we do our professional work: its substantive content, and our personal professional styles. Value issues in this broader sense are pervasive in public works; in our view, there are very important ethical issues in the content and style of our professional work. It is these issues which we will address in this paper.

Our discussion will be most relevant to those aspects of public works concerned with planning, design or analysis: project planning and evaluation, environmental impact assessment, etc. We will draw on transportation for examples, and will use the term "transportation analysis" as a shorthand label for all of the diverse activities that planners or engineers or administrators or operators undertake in working on transportation problems or in transportation organizations. We believe our comments will have applicability to other aspects of public works as well.

SOME ETHICAL ISSUES

We will examine a few selected issues: problem definition, technical methods, work program definition, policy issues, and the legitimacy of debate.

Problem Definition

This set of issues concerns what problems we work on and how we define them. Here, we have made tremendous strides in recent years; especially in transportation. Nevertheless, it is important for us to distinguish rhetoric from practice.

Options.   It is now widely accepted in most areas of transportation that a wide range of transportation options need to be considered: operating, maintenance, pricing and institutional strategies, as well as infrastructure construction. This acceptance by the profession was, of course, aided significantly by the pressures of environmental issues and fiscal restraints. In urban transportation, federal policies such as Transportation System Management and Alternatives Analysis appear to have institutionalized these concerns. Behind the rhetoric, however, can we be confident that, in a specific transportation analysis, a range of real options is being considered in a balanced, objective way?

Impacts.   Similarly, in terms of impacts to be considered we now attempt to deal with all significant social, economic and environmental impacts, as well as impacts on various user groups, and others. Of course, this is required of us anyway, by the National Environmental Policy Act, FHWA's Process Guidelines, and numerous other laws and regulations, and by political pressures in many situations. Again, how often do we pay only verbal attention to these issues—how often are the preparations of environmental impact assessments separate from, and after, the bulk of technical effort devoted to development of alternatives? How often do we allow the basic conception of a project to be changed significantly by what we learn in our environmental analyses? Do considerations of these issues really change what we do?

Objectives to be achieved.   It is an essential element of our professional rhetoric that transportation should serve other needs of society: urban development, regional and statewide development, etc.; and that transportation planning and decision-making should be integrated with "comprehensive" planning and decision-making. Here again, is there any substance behind our rhetoric? Transportation is a strong establishment, technically, professionally, and most important, politically (even with the fiscal crunch of recent years). Are we putting that political strength behind our rhetoric so that stronger and more comprehensive planning and decision-making processes can be established? Or do we only use the "comprehensive" argument when it suits us: urban land use objectives are suddenly fashionable again, perhaps because some advocates of more fixed-guideway transit may see this as an important justification for transit, and some advocates for highway construction may see land use as the primary determinant of air quality impacts and energy consumption so that the burden of meeting such objectives falls elsewhere than in transportation.

Client groups. Citizen participation and the expressions of concerns for transportation needs of the elderly, handicapped, poor, and minority group members are all accepted parts of our rhetoric. How deeply are they entrenched in our professional practice? Is citizen participation allowed to change the nature of what options we are considering, or the final decisions? Do we consider seriously the needs of special mobility groups only when we have funding for a study or service specifically tailored to these groups, or do we give serious and detailed attention to their needs even when planning a major new system or service? Most importantly, when it comes time to reach conclusions and recommendations on a major project, an areawide plan, or the Annual Element of the Transportation Improvement Program (i.e., next year's distribution of funds), whose values do we speak for?

Comment. I honestly don't know what are the answers to these questions about problem definition, viewed across cities and professions. I do have a fear that some of us, while faithfully responding to federal, state, and local policy pressures as they shift from year to year, do not really have a strong sense of conviction about these elements of problem definition: what *do* we believe—as individuals? as a profession? Do we always practice according to our beliefs? Do we have views about problem definitions which are strongly-enough held that we consider "unprofessional" a study or analysis which fails to define the problem consistent with these views? Are we willing to speak out when we see other professionals operate with problem definitions which we consider inappropriate?

Let me give an example. Let's assume that current policy lays out a potentially broad "problem definition" for a study on a particular type of transportation issue (e.g., UMTA's Analysis of Alternatives policy, FHWA's Process Guidelines, EIS requirements). Yet we observe a particular transportation analysis underway which does not appear to reflect this broad problem definition—perhaps in practice the options being considered are few, especially if a mayor or a governor is pushing for jobs and other benefits associated with a major construction project.

Some would argue that if there is a problem with a study that is in response to a federal requirement, then it is an administrative problem to be dealt with by the appropriate federal officials. Do any of us feel that it is our professional responsibility to speak out on the problem definition and scope of work of a planning study that others are conducting? Is there any mechanism through which we can do so, in a way that is professional, and hopefully constructive? Do we in fact believe that it is legitimate to criticize the work of others?

Technical Method

We like to believe that the methodologies we use are objective and value-free. This is especially true of those of us with engineering or systems analysis backgrounds. Yet this is clearly not true: choice of method has significant value biases. We will consider two specific aspects here.

Prediction methods.   The central concern of tranasportation professionals for many years, particularly in urban transportation, was to predict travel. This was done by exercising the standard battery of travel forecasting models consisting of trip generation—distribution—mode split—traffic assignment. Today, we have a broader view of prediction, in that we predict many other impacts in addition to travel flows, such as air quality, energy consumption, and so forth. Further, the state of the art in travel prediction methods has advanced several generations, in two key dimensions. The first dimension is that of the estimation methodology used—now we expect the highest standards of statistical estimation. The second dimension is in the theoretical clarity and practical power of the approaches used—for example, disaggregate models with explicit behavioral structure.

Are there ethical issues in the use of prediction methods? Most certainly so. A model is a construct, a representation of reality which can never be that reality. Even with the highest standards of behavioral reasoning and of statistical method, we never know for certain that we have captured in our model the ability to produce predictions which are 100% accurate. In fact, we know for certain that we have failed to capture reality perfectly; we know that we always must make judgements and compromises in developing practical models, and that our resultant models are imperfect.

Once we recognize this, then we realize the highly subjective nature of model development. Fancy mathematics, powerful statistical methods, complex computer models may help us to develop better models, but they never remove the art and judgement from the process. This means, of course, that models must be subject to the same scrutiny for hidden value assumptions as anything else we do: what have we included in the model? what is left out? what is approximated? what is modelled in greater detail? If I look at the model and what it predicts from the perspective of interest group A, would I find it acceptable, or biased, or irrelevant from the perspective of interest group B or of other interest groups?

Because of the high degree of judgement that goes into model development, predictive models are not objective but are themselves value-laden and need to be scrutinized carefully. This is true for all types of impacts, and especially in transportation for prediction of system performance, system costs and demand. For example, it is easy to shift levels of forecasted mode splits by varying assumptions about average access times and costs, or other seemingly objective technical elements. Changes in external assumptions can also affect results: e.g., central city employment assumptions can shift corridor volumes; inflation and interest rate assumptions can shift costs, etc.

Can we accept the predictions that come from models? Some would throw out the use of models altogether; do we believe that models can be useful for prediction in spite of these shortcomings? How do we overcome these problems?

Clearly, the first step is to recognize that it is unacceptable and unethical in transportation to use any model which is not transparent—i.e., understandable to others. Certainly, no proprietary "black boxes" should be used in any public-sector analyses. How transparent? To whom—with what degree of expertise? with what degree of effort? The second step is to recognize that fairly drastic actions may be needed. Greenberger, Crenson, and Crissey (8) suggest a term, "counter modelling," for professionals who take responsibility to critically appraise a model from different value perspectives. This may include building and exercising alternative models, as well as such modes of examination as sensitivity experiments to model structure and parameter values. (A second step which can be taken is to make models more transparent, simpler to use, and widely available. For example, Manheim, Furth and Salomon are working to develop pocket-calculator procedures for transportation analyses.)

Evaluation method.   We have gone through several intellectual evolutions on the subject of evaluation methods. Economic analysis, such as benefit-cost analysis, began replacing "engineering judgement" in the late 1950's. As social and environmental issues began to be perceived as important, cost-benefit was replaced by cost-effectiveness and various kinds of rating and scoring schemes. In parallel, various operations research methods such as multiattribute utility and mathematical programming were proposed. Some would argue that none of these quantitative methodologies is appropriate as the sole or dominant evaluation procedure in an era when social and environmental issues are important, and equity is as important, or more important, than efficiency as an objective. (We are among those who have argued strongly for this view (11, 13).)

It seems clear to us that the choice of an evaluation method is a value judgement in itself: which issues will enter into the decision will be influenced by the evaluation method chosen. Alternatively, the evaluation method may serve as a "smokescreen," making the decision appear to be rational and free of value biases and helping to obfuscate the real bases of the decision.

What do we believe should be an "appropriate" evaluation approach—as individuals; or as a profession? Should we criticize evaluations which do not meet our ideals?

Work Program Definition

Many of the ethical issues we are concerned with are manifested in the "work program" for a planning analysis or design activity. Rather, many of the ethical issues are manifested in what is *left out* of a work program. It is a fundamental law that there is never enough time or resources to do the analysis that really needs doing. It seems as if serious questions always emerge in a crisis atmosphere, and an analysis must always be done under very tight time constraints and budgets that are just short of unreality. Very rarely is there sufficient prior planning so that an analysis is planned to proceed at a judicious and balanced pace, to address the issues in a measured and deliberate way.

Is this an ethical question? We think so—the decision as to what is or is not included in an analysis is a value question.

Elements of problem definition are involved. Elements of "willingness to learn" are also involved. If a study is tight on budget or time, we cannot afford to let unanticipated issues emerge during the process, because then we risk running behind time or over budget if we shift gears to deal with these new issues. As a result, we strive to stick to our computer-generated critical path diagrams and work programs, and resist learning about new issues or ideas, because they may be too disruptive.

Whose responsibility is this? What roles should we play? As government officials trying to get the most value out of their budgets? or as consultants trying to get the opportunity to do at least some of the work "right"? Should we refuse to bid on an agency's jobs because it consistently underfunds projects? How can local agencies or consultants raise issues about state or federally-defined work programs when their receipts of funds depend on the continued goodwill of, and acceptance by, officials controlling the budget? Do we need some mechanism through which some disinterested professionals can raise questions about the work program of a study and probe the underlying value assumptions?

Policy Issues

So far we have discussed primarily "method". What about our readiness to address "policy" issues: substance? When do we actually express views on federal, state, or local policies, plans or projects? Public agency personnel often feel constrained from speaking out except in well-defined, well-muted internal channels. Local agency personnel seem to fear speaking out publicly on federal policies; local fears of alienating federal bureaucrats, who will review funding applications, tend to inhibit discussion. As for us academics, we may yell and shout at one another, but as for carefully-structured debates and the taking of strong policy positions—do enough of us do this? And consultants, well, the life of the consultant is hard enough—how many really feel they can afford to take positions on issues for which they have not been paid to do a study?

One way in which the public works arena is complex is because of the heavy role of government and the resulting complex interrelationships among agencies at different levels of government, private firms, and individuals. There are tremendous built-in inhibitions against speaking one's mind on the substantive issues. There is no encouragement of debate; there is no positive mechanism which stimulates us to feel it is our *responsibility* to raise and debate issues; there is no institutional mechanism to nurture such debates and protect our rights to speak out. Should there be?

Legitimacy of Debate

Many of the preceding issues come down to one central question: Do we believe in the legitimacy of debate?

There seems to be a prevalent attitude that it is inappropriate to criticize the work of other professionals; professional work must be so grossly incompetent as to require major disciplinary action before we can speak out. Even then, we should, it appears, speak out only through professional society channels—e.g. through the engineering societies.

This raises two problems. First of all, in my own profession of transportation, there are no channels for disciplinary proceedings because there is no institutionalized profession. While organizations such as the Transportation Research Board and the Transportation Research Forum operate as arenas for presentation of research and technical information, they do not operate to reinforce standards of professional practice (probably by design). There is a void in transportation. Should that void be filled?

The second problem is even more important. Should we criticize fellow professionals only when they do something which clearly violates standards of human decency or honesty, such as the first type of issues identified in the introduction (e.g., "kickbacks")? Or should we be free to criticize and debate the second type of issues just discussed? Are these important ethical issues which should be debated?

For example, consider these circumstances:

a. We become aware of a planning, analysis or design activity being undertaken which is operating with what we believe to be an inappropriate problem definition;

b. A report is published recommending particular actions based upon technical analyses utilizing models or other analysis methods which we believe to be inappropriate or invalid;

c. A policy is issued by an agency or official which we believe inappropriate;

d. As a consultant working on a study, we are faced with agency officials who do not want us to analyze certain issues which we feel are important.

Is it professional to express our views on these issues? through what mechanisms? publicly, through the press and in public meetings? Is whether we speak out up to us as individuals or is it our professional responsibility to do so? Regardless of the consequences?

ACTION PROPOSALS

We have touched on just a few issues, of many possible, in the preceding discussion. The implication is clear:

- We view our professional work in public works as rational, objective, "technical," value-free; it is not. What we do in our substantive professional, seemingly "technical" work is an ethical issue.
- We do not engage in enough self-examination nor enough open debate about substantive issues in our own work or in the work of others. We appear as a profession to feel uncomfortable with the notion that criticism of one's professional work is legitimate.

To achieve a high standard of professional practice, we must change the norms of behavior which guide our profession:

> Scientific objectivity is not a product of the individual scientist's impartiality but rather an outcome of participation in a particular scholarly community based on the inculcation of standards of discourse and investigation, as well as the public disclosure of methods and results.[°]

To achieve this, I would like to propose for consideration specific actions of two kinds: first, some general actions to increase the level of awareness of value issues in our profession; and second, a specific model for our roles as professionals.

[°]Karl Popper, *The Open Society and Its Enemies*, pp. 405–406, as cited in Hoos (9, p. 246).

Actions for Professional Awareness

Some of the actions we might consider taking are these:

- *Policy debates.* Let's have more substantive discussions—at panel sessions or in writing; about federal policies, state policies, local plans and projects. Let's argue, using formats which encourage professionals to express divergent views, and which allow several rounds of questioning and clarification so that the observer or reader can decide for him/herself where he/she stands on the issues.

- *Clarify professional responsibility.* We have proposed here that we have a professional responsibility to examine one another's work critically and to express our views. Do we agree on this view? Is this desirable? How far should we go in carrying this principle into practice? What institutional mechanisms, if any, do we need to encourage or to protect our responsibilities to criticize?

- *Counter-modelling.* Because modelling plays such a significant role in many analyses today, mechanisms for constructive critiques of technical methods used in a particular planning study are needed. What mechanisms might be developed? Is internal technical review sufficient? Under what circumstances should formal counter-modelling efforts be initiated? Under whose auspices? How public should debates about methods be? (Some would undoubtedly argue that extensive debates about methods would lose the confidence of decision-makers and the public.)

- *Code of ethics.* Do we need to formulate, in writing, a "code of ethics" or a statement of professional responsibilities, dealing not only with the traditional definition of ethical issues, but with the broader, more substantive view espoused here?

- *Scopes of work.* Do we need to address the complex problem of how the scope of work for a professional analysis is developed? Should we examine the issue of professional responsibility to critically appraise inappropriate scopes of work (inappropriate because of problem definition, resources or time limitations, institutional arrangements, etc.) or the issue of whether consultants should refuse to bid, or agency staff refuse to work on, an inappropriately-defined study? Is there a need for institutional mechanisms which would encourage such responsibility, and protect those who speak out?

- *Education.* Do we know what we want to do better, to increase our standards of professional practice? Are we ready to begin exchanging views, and educating ourselves and others as to our views on these and related issues? How should our university-level and mid-career educational programs deal with these issues?

We raise these questions because in our view ethical issues are not abstract but are here-and-now, practical, bread-and-butter issues. They concern the content and processes of planning or design activities, of Environmental Impact Statements, of Analyses of Alternatives, of developing Transportation Improvement Programs at the local, county, and area levels, county or local road programs, and almost everything else we are dealing with today. Shouldn't we confront these issues and attempt to deal with them?

A Normative Model for Our Roles as Professionals

We would also like to propose a specific model for how we should operate as professionals. The reasons for this model and its many practical ramifications have been described in detail elsewhere (11, 13).

We propose that public works professionals adopt the following principles to guide their professional work in planning, design, operations or analysis activities:

Principle I. Range of Alternatives. There must be a range of alternatives available. The range of alternatives must be sufficiently broad so as to represent real choices, and must include the option of not doing anything.

Principle II. Identification of Effects. There must be adequate information on the effects of alternatives. This information must include not only the beneficial effects of the alternatives, but also the adverse effects; must include the incidence of those effects; and must include particularly any effects that any particular interest thinks are important, whether or not those effects are readily quantified.

Principle III. Public Involvement. There must be full opportunity for timely and constructive involvement of affected interests in the process, such that every interest—individual or group—which may potentially be affected by the changes being considered has full and timely access to all relevant information and has full opportunity to influence the process constructively.

Principle IV. Equity. Where significant adverse impacts might result for some interests in order that benefits can be provided for others, those adversely impacted should be adequately compensated. Further, the benefits should be distributed broadly.

Principle V. Uncertainty. Uncertainties which may exist should be explicitly recognized. Uncertainties in predicted effects should be explicitly identified; and some alternatives should be designed for staged implementation so as to retain flexibility in the face of uncertainty. Careful consideration should be given to identifying future options foreclosed and future options left open by proposed alternatives; and the process should be flexible, allowing periodic reviews of prior decisions and possible revisions of prior choices.

Principle VI. The Decision Process. The process of reaching a decision should be open, participatory and decisive, resulting in decisions that are implementable and implemented.

If we do adopt such principles as these, then we find that our professional work must be organized and conducted in ways that are significantly different from past practices (see further the references cited). Such principles *can* be implemented, and do have significant impact on professional practices.

In the field of highway planning, the Federal Highway Administration of the U.S. Department of Transportation has issued regulations to the state highway departments which explicitly incorporate most of these principles. Numerous changes and innovations have taken place in response to this process (2, 14).

Under the sponsorship of the American Association of State Highway and Transportation Officials, a manual reflecting these principles has been developed. Entitled "Transportation Decision Making: A Guide to Social and Environmental Considerations" (13), this manual describes a number of specific techniques which transportation agencies can use in implementing these principles.

In the field of urban transportation, several recent federal regulations also reflect these principles. The Urban Mass Transportation Administration (UMTA) of the U.S. Department of Transportation has promulgated a policy for "Major Urban Mass Transportation Investments" (16), which explicitly requires that transit system planners plan their systems (1) for implementation in stages with periodic revision of long range plans to respond to changing conditions (Principle V), (2) with full opportunity for timely public involvement in the planning process (Principle III), (3) with consideration of a wide range of alternatives, including improved management of existing transportation systems and measures to reduce the use of automobiles in congested areas (Principle I), (4) with equitable distribution of transit service improvements throughout the metropolitan area (Principle IV) and (5) with identification and evaluation of all significant impacts (Principle II). In addition, UMTA and the Federal Highway Administration have issued jointly regulations on the urban transportation planning process which explicitly reinforce the emphasis on programming (Principle V) (7, 10, 15).

In the field of airport planning, the philosophy reflected here has been incorporated in an airport planning manual issued by the U.S. Department of Transportation (4).

CONCLUSIONS

We believe there are significant value issues in what public works professionals do. In this paper we have examined a few selected aspects: problem definition, technical methods, work program definition, policy issues, and the legitimacy of debate. As a result, we conclude that professional work is *not* value-free, and that we do *not* engage in enough self-examination. To change the norms of behavior, we have proposed a number of specific actions to promote professional self-awareness, and some principles to serve as a normative model for our roles as professionals.

References

1. Anderson, Roy W., "Professionalism, Ethics and the Public Welfare," in *Conference on Engineering Ethics* (3).
2. California Department of Transportation, *Action Plan for Transportation Planning and Development*, Sacramento, Calif.: Department of Transportation, June 1973.
3. *Conference on Engineering Ethics*, New York: American Society of Civil Engineers, 1975.

4. CLM Systems, Inc., *Airports and Their Environments: A Guide to Environmental Planning*. Report prepared for the U.S. Department of Transportation (DOT P 5600.1), 1972.

5. *Ethics, Professionalism and Maintaining Competence*, Preprints for the ASCE Professional Activities Committee Specialty Conference, New York: American Society of Civil Engineers, 1977.

6. Federal Highway Administration (FHWA), *Process Guidelines (Social, Economic, and Environmental Effects of Highway Projects)*, Policy and Procedure Memorandum 90–4, Washington, D.C.: FHWA U.S. DOT, 1973. Codified in Federal-Aid Highway Program Manual as Vol. 7, Chapter 7, Section 1.

7. FHWA and Urban Mass Transportation Administration (UMTA), U.S. DOT, Transportation improvement program 23 CFR Chapter I, Part 450, Subpart A; *in Federal Register* 40:181 (Wed., Sept. 17, 1975, Part II).

8. Greenberger, Martin, Matthew A. Crenson, and Brian L. Crissey, *Models in the Policy Process*, New York: Russell Sage Foundation/Basic Books, 1976.

9. Hoos, Ida R., *Systems Analysis in Public Policy: A Critique*, Berkeley, California: University of California Press, 1972.

10. Manheim, Marvin L., "The Emerging Planning Process: Neither Long-Range Nor Short-Range, But Adaptive and (Hopefully) Decisive," paper presented to Transportation Research Board (TRB), 1977.

11. Manheim, Marvin L., *Fundamentals of Transportation Systems Analysis*, Vol. I, Cambridge, MA: MIT Press, in press (1978).

12. Manheim, Marvin L., John H. Suhrbier, and Elizabeth D. Bennett, *Process Guidelines For Consideration of Environmental Effects*, Final report to FHWA, USL Report 72–11, Cambridge, MA: U.S.L., MIT, June 1972.

13. Manheim, Marvin L., et al., *Transportation Decision-Making: A Guide to Social and Environmental Consideration*, National Cooperative Highway Research Program, Report 156, Washington, D.C.: TRB, 1975.

14. Pennsylvania Department of Transportation, *Action Plan* September 1973.

15. Reno, Arlee T., Jr., Ben Schneideman, and Marvin L. Manheim, *Opportunities to Improve the Interrelationship of Urban System and Project Planning*, Report for FHWA, Report 73–1, Cambridge, MA: U.S.L., MIT, 1973.

16. Urban Mass Transportation Administration, U.S. DOT, Major Urban Mass Transportation Investments, *Federal Register*, 41:185, Part II, September 22, 1976.

2.8

The Consultants' Competitive Negotiation Act

STATE OF FLORIDA

The following is a complete text of the Consultants' Competitive Negotiation Act, as signed by the Governor on May 2, 1973 and revised 1975, and 1977.

287.055 Acquisition of professional architectural, engineering, landscape architecture or land surveying services; definitions; procedures, contingent fees prohibited; penalties.—

[1]   **Short Title.** This section shall be known as the "Consultants' Competitive Negotiation Act."

[2]   **Definitions.** For purposes of this section:

[a]   "Professional services" shall mean those services within the scope of the practice of architecture, professional engineering, landscape architecture, or registered land surveying as defined by the laws of the state, or those performed by any architect, professional engineer, landscape architect or registered land surveyor in connection with his professional employment or practice.

[b]   "Agency" means the state or a state agency, municipality or political subdivision, a school district or a school board.

[c]   "Firm" means any individual, firm, partnership, corporation, association, or other legal entity permitted by law to practice architecture, engineering, landscape architecture or land surveying in the state.

[d]   "Compensation" means the total amount paid by the agency for professional services.

[e]   "Agency official" means any elected or appointed officeholder, employee, consultant, person in the category of other personal

service or any other person receiving compensation from the state, a state agency, municipality, or political subdivision, a school district or school board.

[f] "Project" means that fixed capital outlay study or planning activity described in the public notice of the state or a state agency pursuant to paragraph (a) of Subsection (3). An agency shall prescribe by administrative rule procedures for the determination of a project under its jurisdiction. Such procedures may include:

- Determination of a project which constitutes a grouping of minor construction, rehabilitation or renovation activities;

- Determination of a project which constitutes a grouping of substantially similar construction, rehabilitation or renovation activities.

[g] "Continuing Contract" is a contract for professional services entered into in accordance with all the procedures of this act between an agency and a firm whereby the firm provides professional services to the agency for work of a specified nature as outlined in the contract required by the agency with no time limitation except that the contract shall provide a termination clause.

[3] Public Announcement and Qualification Procedures.

[a] Each agency shall publicly announce, in a uniform and consistent manner, each occasion when professional services are required to be purchased for a project whose basic construction cost is estimated by the agency to be more than $100,000 or for a planning or study activity when the fee for professional services exceeds $5,000 except in cases of valid public emergencies so certified by the agency head. Public notice shall include a general description of the project and shall indicate how interested consultants can apply for consideration.

[b] Each agency shall encourage firms engaged in the lawful practice of their profession who desire to provide professional services to the agency to submit annually a statement of qualifications and performance data.

[c] Any firm or individual desiring to provide professional services to the agency must first be certified by the agency as qualified pursuant to law and the regulations of the agency. The agency shall make a finding that the firm or individual to be employed is fully qualified to render the required service. Among the factors to be considered in making this finding are the capabilities, adequacy of personnel, past record, and experience of the firm or individual.

[d] Each agency shall adopt administrative procedures for the evaluation of professional services, including, but not limited to, capabilities, adequacy of personnel, past record and experience and such other factors as may be determined by the agency to be applicable to its particular requirements.

[e] The public shall not be excluded from the proceedings under this section.

[4] Competitive Selection.

[a] For each proposed project, the agency shall evaluate current statements of qualifications and performance data on file with the agency, together with those that may be submitted by other firms regarding the proposed project, and shall conduct discussions with, and may require public presentations by, no less than three firms, regarding their qualifications, approach to the project, and ability to furnish the required service.

[b] The agency shall select, in order of preference, no less than three firms deemed to be most highly qualified to perform the required services after considering such factors as the ability of professional personnel, past performance, willingness to meet time and budget requirements, location, recent, current and projected work loads of the firms, and the volume of work previously awarded to the firm by the agency, with the object of effecting an equitable distribution of contracts among qualified firms, provided such distribution does not violate the principle of selection of the most highly qualified firms.

[c] This subsection shall not apply to professional service contracts for a project whose basic construction cost is estimated by the agency to be $100,000 or less or for a planning or study activity when the fee for professional services is $5,000 or less.

[d] Nothing in this act shall be construed to prohibit continuing contracts between firm and agency.

[5] Competitive Negotiation.

[a] The agency shall negotiate a contract with the most qualified firm for professional services at compensation which the agency determines is fair, competitive and reasonable. In making such determination the agency shall conduct a detailed analysis of the cost of the professional services required in addition to considering their scope and complexity. For all lump-sum or cost-plus-a-fixed-fee professional service contracts over $50,000, the agency shall require the firm receiving the award to execute a truth-in-negotiation certificate stating that wage rates and other factual unit costs supporting the compensation are accurate, complete, and current at the time of contracting. Any professional service contract under which such a certificate is required shall contain a provision that the original contract price and any additions thereto shall be adjusted to exclude any significant sums by which the agency determines the contract price was increased due to inaccurate, incomplete, or noncurrent wage rates and other factual unit costs. All such contract adjustments shall be made within one year following the end of the contract.

[b] Should the agency be unable to negotiate a satisfactory contract with the firm considered to be the most qualified at a price the agency determines to be fair, competitive and reasonable, negotiations with that firm shall be formally terminated. The agency shall then undertake negotiations with the second most qualified firm. Failing accord with the second most qualified firm, the agency shall terminate negotiations. The agency shall then undertake negotiations with the third most qualified firm.

[c]   Should the agency be unable to negotiate a satisfactory contract with any of the selected firms, the agency shall select additional firms in order of their competence and qualification and continue negotiations in accordance with this subsection until an agreement is reached.

[6]   Prohibition Against Contingent Fees.

[a]   Each contract entered by the agency for professional services shall contain a prohibition against contingent fees as follows: "The architect (or registered land surveyor, professional engineer or landscape architect, as applicable) warrants that he has not employed or retained any company or person, other than a bona fide employee working solely for the architect (or registered land surveyor, professional engineer or landscape architect, as applicable) to solicit or secure this agreement and that he has not paid or agreed to pay any person, company, corporation, individual or firm, other than a bona fide employee working solely for the architect (or registered land surveyor, professional engineer or landscape architect, as applicable) any fee, commission, percentage, gift or any other consideration contingent upon or resulting from the award or making of this agreement." For the breach or violation of this provision, the agency shall have the right to terminate the agreement without liability and, at its discretion, to deduct from the contract price, or otherwise recover, the full amount of such fee, commission, percentage, gift or consideration.

[b]   Any individual, corporation, partnership, firm or company, other than a bona fide employee working solely for an architect, professional engineer, registered land surveyor or landscape architect, who offers, agrees or contracts to solicit or secure agency contracts for professional services for any other individual, company, corporation, partnership or firm and to be paid, or is paid any fee, commission, percentage, gift or any other consideration contingent upon, or resulting from, the award or the making of a contract for professional services shall, upon conviction in a competent court of this state, be found guilty of a first degree misdemeanor punishable as provided in 775.082 or 775.083.

[c]   Any architect, professional engineer, registered land surveyor or landscape architect, or any group, association, company, corporation, firm or partnership thereof, who shall offer to pay, or pay, any fee, commission, percentage, gift or other consideration contingent upon, or resulting from, the award or making of any agency contract for professional services, shall, upon conviction in a state court of competent authority, be found guilty of a first degree misdemeanor punishable as provided in 775.082 or 775.083.

[d]   Any agency official who offers to solicit or secure, or solicits or

secures, a contract for professional services and to be paid, or is paid, any fee, commission, percentage, gift, or any other consideration contingent upon the award or making of such a contract for professional services between the agency and any individual person, company, firm, partnership, or corporation shall, upon conviction by a court of competent authority, be found guilty of a first degree misdemeanor punishable as provided in 775.082 or 775.083.

[7] Authority of Department of General Services.

Notwithstanding any other provision of this Section, the Department of General Services, Division of Building Construction and Property Management, shall be the Agency of State government which is solely and exclusively authorized and empowered to administer and perform the functions described in subsections (3), (4) and (5) of this Section respecting all projects for which the funds necessary to complete same are appropriated to the Department of General Services irrespective of whether such projects are intended for the use and benefit of the Department of General Services or any other Agency of government; provided, however, that nothing herein shall be constructed to be in derogation of any authority conferred on the Department of General Services by other express provisions of law. Additionally, any Agency of government may, with the approval of the Department of General Services, delegate to the Department of General Services authority to administer and perform the functions described in subsections (3), (4) and (5) of this Section. Under the terms of the delegation, the Agency may reserve its right to accept or reject a proposed contract.

[8] State Assistance to Local Agencies.

On professional service contracts for which the fee is over $25,000, the Department of Transportation or the Department of General Services shall provide, upon request by a municipality, political subdivision, school board, or school district, and upon reimbursement of the costs involved, assistance in selecting consultants and in negotiating consultant contracts.

[9] Applicability to Existing Contracts.

Nothing in this section shall affect the validity or effect of any contracts in existence on July 1, 1973.

[10] Notwithstanding any other provision of this section, there shall be no public notice requirement or utilization of the selection process as provided in this section for projects in which the agency is able to reuse existing plans from a prior project; provided, however, subsequent to July 1, 1975, public notice for any plans which are intended to be reused at some future time shall contain a statement which provides that the plans are subject to reuse in accordance with the provisions of this subsection.

[11] Nothing in the amendment of this section by ch. 75-281, Laws of Florida, is intended to supersede the provisions of ss. 235.211 and 235.31.

History.—ss. 1-8, ch. 73-19; ss. 1-3, ch. 75-281; s. 1, ch. 77-174; s. 1, ch. 77-199.

2.9

Suggested Key Steps in Administering The Consultants' Competitive Negotiation Act

FLORIDA ENGINEERING SOCIETY

The 1973 session of the Florida Legislature passed a Consultants' Competitive Negotiation Act (House Bill 309). The ACT sets forth the procedures for contracting professional services to be used by the state, its agencies, municipalities or political subdivisions, school boards and school districts. The ACT took effect July 1, 1973 and was revised in 1975 and 1977. The ACT covers the selection of Professional Engineers, Architects, Registered Land Surveyors and Landscape Architects, and the negotiation of fees for their professional services in contracts where the compensation for these professional services exceeds $5,000.00 in cases of studies and where the cost of construction exceeds $100,000 in the case of construction.

It is important to note that these procedures do not vary significantly from those now used by many agencies of state and local governments.

In an effort to be of public service, the Florida Engineering Society has prepared the following explanation of the procedures required by the Consultants' Competitive Negotiation Act. The outline and explanation offers comments and suggested steps that might be a helfpul guide to be followed in contracting for professional services.

STEP 1   PUBLIC ANNOUNCEMENT OF THE PROJECT

Announcement. Section 3 of the Act states that the agency shall publicly announce in a uniform and consistent manner on each occasion when professional services are required. Public notice shall include a description of the project and shall indicate how interested consultants can apply for consideration.

127

Advertisement. It is suggested that your attorney advise you as to the proper method of publicly announcing your agency's requirement for professional services. A good method of reaching many registered professional engineers, architects, landscape architects and registered land surveyors in the State is through the publications of the Florida Engineering Society, the Florida Association of the American Institute of Architects and the Florida Society of Professional Land Surveyors.

The Florida Institute of Consulting Engineers, the private practice section of the Florida Engineering Society, publishes a weekly report, "The Florida Register," which is circulated to consulting engineers, architects and planners throughout the state. Agencies may use this publication, free of charge, by mailing public announcements for professional services to "Florida Register," Suite 106, 1311 Executive Center Drive, Tallahassee, Florida 32301.

In addition, the Florida Engineering Society, with headquarters at 1311 Executive Center Drive, Tallahassee 32301, publishes a monthly magazine, the *Journal,* which has a circulation of about 5,000 and is distributed to nearly all the registered professional engineers in the State of Florida. Agencies interested in publicly announcing their requirement for professional engineering services to the widest possible audience of registered professional engineers may wish to consider placing an advertisement in the Florida Engineering Society *Journal.*

Letters. Some firms may have submitted their qualifications to the agency at an earlier date prior to the advertising with the request that they be considered for any work in their specialization being considered by the agency. The agency using its own criteria may screen these qualifications at the time they are submitted. These firms may then be listed under the categories of engineering work which the government agency determines the firms are qualified to perform (Qualifications are more fully discussed under Step 2).

The law provides that those firms which have requested and which are qualified, shall be notified. Those firms could be notified by letter or other verifiable means.

Notice. The notice or letter should give a general description of the work to be performed, including the time element involved. Those firms interested should be invited to notify the local agency indicating their willingness and qualifications for providing the required professional services.

STEP 2   PRE-QUALIFICATION AND CERTIFICATION OF FIRMS

Qualifications. The firm must be certified as qualified to perform the work pursuant to law and to any regulations adopted by the agency. The government agency may wish to establish additional relevant criteria by which interested professionals would be certified. Some major qualification factors mentioned in Section 3, Paragraphs c and d are the ability of professional personnel, past performance, willingness to meet time and budget requirements, location and recent, current and projected work load of the firms. An example of another consideration would be the recommendations of prior clients.

It might prove useful for the agency to prepare and require all interested professionals to submit a standard qualification form which outlines the company's capabilities and experience. A qualification form to be completed and submitted by interested professionals may be devised by the agency. Most engineering firms are familiar with the U.S. Government Architect-Engineer Questionnaire, Standard Form 254. Submittal of this form is required by most state and federal agencies in evaluating firms. The form is a clear, concise, informational document that the agency may choose to adopt to assure uniformity for presentation and evaluation of qualifications. The need for the firms to submit qualification data should be clearly indicated in the advertisement or notification letter (Step 1 above).

Other means of obtaining qualification information may be deemed advisable by the agency, but uniformity for ease in evaluating the firms is considered highly desirable.

Certification.   If, after reviewing a firm's qualifications, the agency finds the firm qualified to render the required services, the agency must certify the firm as qualified. There is no limit on the number of firms which could be certified. A principle purpose of the ACT is to foster competition based on qualifications.

STEP 3   SELECTION OF CERTIFIED QUALIFIED FIRMS

Selection.   After public announcement and certification of qualified firms, the agency must then study and compare the qualifications of those firms who are certified as qualified by the agency (Step 2) and who indicate an interest in the project. The purpose of this process is to rate and select firms from the certified list and establish the order of preference by the agency based on the firm's qualifications.

The law states that the agency shall evaluate the qualifications, and shall conduct discussions relative to the qualifications, etc., with no less than three firms. Public presentations also may be required by the agency. The presentation may be before the full agency governing board or it may be before a selection committee. It is important to note that, in any case, the law stipulates that the public may not be excluded from the proceedings under this law.

Order of Preference of Firms Selected.   The ACT provides that the agency shall select no less than three firms in order of preference based on qualifications. The law provides that the firms be given a definite preference rating, first, second, third, etc., based on the firms' qualifications, prior to the commencement of contract negotiations.

STEP 4   NEGOTIATION OF PROFESSIONAL SERVICE CONTRACTS

Negotiation.   The law provides that the agency shall negotiate a contract with the most qualified firm, **i.e., the firm ranked first in the selection process set forth in Step 3 above,** at a compensation which the agency determines is fair, competitive and reasonable. Detailed discussions should be held between the firm and the agency to establish clearly the detailed scope of the project so that it is well understood by both parties. After the scope of services or the exact work to be done is fully established and understood, negotiations for compensation or fee can commence.

The Florida Department of Transportation and the Florida Department of General Services have developed successful procedures for use in arriving at reasonable compensation.

Many methods of arriving at compensation are used. In general, the most commonly used are:

1. Lump Sum.
2. The firm's payroll costs times a multiplier.
3. Total cost plus a professional fee.
4. Percentage of construction costs.
5. Per diem and expenses.
6. Retainer.
7. Combination of the methods.

Failure to Arrive at Agreement. Experience indicates that compensation negotiations with the top qualified firm are usually successful. Termination of negotiations with the top qualified firm is infrequent. However, should the local agency and firm fail to agree upon a compensation deemed by it to be fair, competitive, and reasonable, negotiations with the first firm shall, under the law, be terminated. The local agency shall then commence negotiations with the firm selected as being the second most qualified by the selection process outlined in Step 3 above. The full scope of the work to be done is again established and clearly understood by the agency and the firm designated number 2. Again efforts are made to negotiate a compensation that is fair, competitive and reasonable. If negotiations are successful, a contract can be executed. If not, negotiations are terminated with firm number 2 and the process is repeated similarly throughout the list.

ADDITIONAL CONSIDERATIONS

Truth in Negotiations.   For all lump-sum or cost-plus-a-professional-fee professional service contracts over $50,000, the agency shall require the firm receiving the award to execute a truth-in-negotiation certificate stating that wage rates and other factual unit costs supporting the compensation are accurate, complete and current at the time of contracting. The certificate can be included in the body of the contract.

Contracts for professional services for all lump sum or cost-plus-a-professional-fee type of compensation shall contain a provision that the original contract price and any additions thereto shall be adjusted to exclude any significant sums where the agency determines the contract price was increased due to inaccurate, incomplete or non-current wage rates and other factual unit costs. Such adjustments must be made within one year following the end of the contract.

Contingency fees.   Section 6 of the ACT speaks to contingency fees and is specific in the prohibition and penalties. Each contract for professional service is required to contain a statement by the professional warranting as follows:

> The architect, registered land surveyor, professional engineer or landscape architect (as applicable) warrants that he has not employed or retained any company or person, other than a bona fide employee working solely for the architect, registered land surveyor or professional engineer to solicit or secure this agreement, and that he has not paid or agreed to pay any person, company, corporation, individual or firm, other than a bona fide employee working solely for the architect, registered land surveyor, landscape architect or professional engineer any fee, commission, percentage, gift, or any other consideration contingent upon or resulting from the award or making of this agreement.

For the breach or violation of this provision, the agency shall have the right to terminate the agreement without liability and, at its discretion, to deduct from the contract price, or otherwise recover, the full amount of such fee, commission, percentage, gift or consideration.

Assistance to Local Agencies.   On professional contracts where compensation exceeds $25,000, the local agency may request aid from the Florida Department of Transportation or the Florida Department of General Services in the selection and negotiation procedure.

If such aid is used, the local agency must reimburse the state agency for the actual costs incurred in furnishing this assistance. Costs for this assistance may vary, but the Florida Department of Transportation and the Florida Department of General Services may be contacted for estimates of the cost of their assistance.

Existing Contracts.   Nothing in the ACT shall affect the validity of contracts in existence on July 1, 1973.

Continuing Contracts.   The ACT defines continuing contracts and does not prohibit same.

Selected Additional Readings

Ashkinazy, A., "We Didn't Know—," *Annals of the New York Academy of Sciences*, Vol. 196, Article 10, 1973, pp. 448–450.

Calkins, M. D., "Decision Making: Professional vs. Political Considerations," *APWA Reporter*, January 1974, pp. 25–26.

Carpenter, S., "Values and Technical Feasibility Studies," *Values and the Public Works Professional*, D. L. Babcock and C. A. Smith (eds.), University of Missouri-Rolla, Rolla, MO, 1980, pp. 37–41.

Florman, S. C., "Hired Scapegoats," *Harper's Magazine*, May 1977, pp. 26–29.

Lerner, M., "Watergating on Main Street," *Saturday Review*, November 1, 1975, pp. 10–25, 56–57.

Marine, G., *America the Raped*, Simon and Schuster, New York, 1969.

McCollough, T. J., "When is a Political Contribution Influence Buying," *Engineering Issues*, January 1976, pp. 1–6.

Morgan, A. E., *Dams and Other Disasters*, Porter Sargent Publishers, Boston, 1971.

Perucci, R., and Gerstl, J. E., *The Engineers and the Social System*, John Wiley & Sons, Inc., New York, 1969.

Price, W., "Values in Public Works Decision Making: the Distribution of Services," *Values and the Public Works Professional*, D. L. Babcock and C. A. Smith (eds.), University of Missouri-Rolla, Rolla, MO, 1980, pp. 23–35.

Ross, S. S., "The Data Dilemma: Using Figures in a Responsible Way," *Professional Engineer*, December 1930, pp. 32–33.

Zuesse, Eric, "The Love Canal: The Truth Seeps Out," *Reason*, February 1981, pp. 17–33.

Obligations to Employers and Peers

We cannot live for ourselves alone. Our lives are connected by a thousand invisible threads, and along these sympathetic fibres, our actions run as causes and return to us as results.

Herman Melville

If engineers have obligations to the public because of the effects of their activity upon the public, then engineers equally must have obligations to their employers and peers who are likewise affected.

The professionalization of engineering in this century and the role of employee that has evolved in the corporate setting complicate the situation by creating complex relations between engineers and other engineers and employers. This is the reverse side of the question discussed by Turnick in the preceding chapter. The nature of engineering obliges most engineers to work for and to pursue the interests of an employer, a client, or government while requiring the engineer to take into consideration the general interests of the public. In fact, we can go further and say that the essence of the engineering profession's dilemma lies in a triad of obligations. The individual engineer is caught between potentially conflicting sets of obligations and responsibilities: (1) to the public, (2) to the client or employer, and (3) to other engineers. Further, many times the solution to the dilemmas works counter to reasonable expectation of self-interest for the engineer.

John Ullmann in "The Responsibility of Engineers to Their Employers" tries to explore the full range of obligations and responsibilities to employers. If engineers have any responsibilities at all to their employers, he suggests these responsibilities go beyond slavish acceptance of the goals and means assessments of their employers and extend to a critical attitude toward both ends and means. Ullmann catalogs a dismal list of recent "engineering failures" that can be seen as just cause for complaint from the public. But, more importantly for Ullmann, they can also be viewed as failures of engineers to discharge their full responsibilities to employers. Thus, it is obvious that employer and public interests are not necessarily in opposition to each other.

Bruce Gordon and Ian Ross analyze the three roles that engineers fill in the corporate setting and try to determine which of them can be called "professional." Given the difficulties involved in defining "profession" and "professional," their definition of "professional" and their choice of the role that best fits that definition will probably not find overwhelming assent. What is of greater importance is the description and analysis of the three roles (artisan, protege, professional) and their differing relations to corporate management.

Conflict of interest is a large and ambiguous issue in engineering. It is usually understood to concern the employed engineer where the conflict is between the employer's interests and those of the individual engineer. The two articles "Conflict of Interest" by Donald Christiansen and "Conflicts of Interest Pose No Big Problem" attempt to determine more precisely the full extent and nature of the conflicts of interest that an engineer may encounter.

In "Can an Engineer Join a Union and Still Be Professional?" William Haga discusses the issue of unions with an argument that will be unsettling to many engineers. He analyzes both professions and unions and reaches the conclusion that neither form of organization represents a satisfactory way of formalizing the relations between engineers and their employers and clients.

The Responsibility of Engineers to Their Employers

JOHN E. ULLMANN

Indeed, the idols I have loved so long
Have done my credit in men's eyes much wrong:
Have drown'd my honor in a shallow cup
And sold my reputation for a song.

The Rubaiyat of Omar Khayyam

The engineering profession is currently in a state of ill repute more profound and widespread than at any time since it overcame the last-ditch opposition of the likes of the Duke of Wellington more than a century ago Engineers, it is widely accepted, spend most of their time producing weapons of ever-increasing fiendishness, befouling the air and waters, gradually poisoning our food, or designing and manufacturing a whole lot of dangerous and poorly functioning products—what has been aptly called "future schlock." They do all this, moreover, at exorbitant cost to their customers and to society as a whole. Such stereotyping is, of course, as unfair as most such exercises; there are many engineers whose lives, values, intellectual stature, and achievements belie this kind of dismal performance. There are many others, however, who are up to their necks in it and do their job in an environment of ethical myopia that leads them to carry on with their miserable work in unquestioning obedience to their employers or clients.

If doing one's job without question and without limits were all there is to anyone's responsibility to an employer, there would be no need for us to discuss this matter. But, of course, that is not the case. Not only does this simplistic dogma not work when there is the kind of opposition that engineering and its works have now generated, it doesn't seem to do engineers much good either and what doesn't work tends to become a sin in a country as pragmatic as the United States. Engineers are currently unemployed by the thousands and their training schools are perhaps the least fashionable sector of Academe. When they do work, they are given to a multitude of discontents, many of which raise the aggrieved question of why "professionals" should be treated no better than production workers.

Accordingly, engineers are confronted with three questions and it is to these that we must address ourselves. The first echoes the *cri de coeur* of the typical parent of a juvenile delinquent: Where have we gone wrong? This quickly leads to the second: What kind of improvements can be made? And to the third: What can be done to encourage these improvements and establish them firmly?

FAILURES, WASTE, DEATH

It is easy to cite chapter and verse in the matter of engineering failures and barbarities. In professional scope and financial extent alike they are unprecedented. Thus, the war in Indochina has been fought with weapons of mass destruction that break both spirit and letter of most of the international weapons conventions to which the United States is a party, including the limitations on the destruction of enemy territory that had been set at the war crimes trials of the 1940s. Weapons regularly used include the Daisy Cutter bomb which destroys vast areas of vegetation, pellets and flechettes able to kill every living thing over a wide radius, and specially built giant bulldozers and chemical defoliants that are busily wreaking virtually irreversible ecological damage. As we contemplate their effects we can but say: This is the way some engineers have discharged their duties to their employers.

But we must look beyond the Vietnam weaponry. Over the last years, equipment of doubtful military justification has been produced in programs so riddled with waste, overcharge, and functional deficiency as to have created a whole new dimension in blundering. The F-111, the C-5A, the F-14, and the Main Battle Tank are only four of the more prominent examples. The details are abundantly covered in the literature and what makes matters even worse is the fact that not just ordinary engineers flocked to these projects, but those who described themselves as "working at the frontier." And why not? For a long time there were copious job opportunities that offered much higher pay than comparable commercial work. There was a leisurely work pace for the most part and ample support services had been laid on; it all served to create a new generation of engineers more oblivious than ever before to costs and professional independence. Discharging one's "responsibility" to one's employer in such an environment can be quite personally pleasant and there is always technical hair-splitting and overspecialization to blunt any sensitivity to the exact purpose of the device, let alone the policies, which it is supposed to implement.

It has by now obtruded itself on the consciousness of even fervent patrons of that branch of engineering that there has to be a limit to all of this. In recent months the Senate Armed Services Committee particularly has taken a lead in questioning and criticizing the quantity and quality of the engineering work that has escalated the cost of weaponry as shown below:[1]

Relative Cost Index

Item	World War II	1958–1963	1971 or in Development
Strategic bombers	1	11	43
Air Force fighters	1	25	100
Aircraft carriers	1	96	191
Attack submarines	1	16°	37

°1968

The activities of engineers and their paraprofessional fellow-travelers are major contributing factors to these soaring costs. They too, no doubt, felt that they were discharging their responsibilities to their employers with no more than the average degree of cynicism that this process generally involves.

But there are other troubled areas, as well. Consider, for example, the state of communications in the United States, circa 1972! It appears to be in the throes of a true "technical collapse scenario." For practical purposes, the railroads are gone because one train a day, taking second place to freight movements, is not a service. Planes fly from one inadequate airport to another, charging high fares, yet unprofitable and with such obvious engineering problems as a warning system for clear air turbulance still unsolved. The postal service first introduced "business methods" and new technology, and nevertheless went into a functional tailspin; the ultimate judgment on it is that government departments with proposal deadlines no longer take mailing receipts as sufficient proof of compliance, giving instead elaborate instructions for hand delivery. The telephone system is an increasing shambles, with inadequate investment and reliability and again with lagging technology. The sending of individual telegrams may soon be a thing of the past, and if anybody wants to brave the congested roads, they are sprouting ever more potholes. Although it would be excessively harsh to blame engineers for all of this, there is no evidence of their sustained concern over the distorted priorities in research and investment that *are* responsible. But such concern should be very much a part of their responsibilities *toward their employers*: What is now happening bedevils and complicates their work and, with an eye to the massive layoffs and bankruptcies in the field, it is surely part of the engineers' job to help keep an employer solvent, in the absence of more crucial criteria.

Engineers also help their employers despoil the environment. As their faithful messengers, they then come to conferences where they promptly develop the fashionable hypocrisies of the conventional wisdom. Exhibit A is usually a curve which has percentage pollution abatement on the horizontal, and cost on the vertical axis and shows a gradual cost increase up to a certain percentage, after which the cost rises through the roof. Greater accuracy or better quality does often produce a sharp rise in costs, but by an odd coincidence we are always precisely at the point in abatement before the big rise. In short, it is an argument for doing nothing further, and what renders it suspect is that the industry concerned fought the antipolluters all the way up to the steep climb as well.

Another technical-economic angle is the one that unless we develop ever more intensively, without putting undue antipollution burdens on industry, there is no prospect of helping the poor. Compassion like that is always welcome, especially in a time when the poor have been made scapegoats, but it is perhaps not too ungracious to detect a touch of the crocodilian in it.

A third major point is that we must not encumber our industries with extra abatement costs because we would then be unable to compete in the international market. But the precipitiously unfavorable trade balance, which the United States acquired in 1970 for the first time since the 1890s, is largely due to the failure to produce the kind of sophisticated machinery that for a long time accounted for much of the favorable American balance. The export losses are in machine tools, instruments, electrical machinery, appliances, and transportation equipment and have not been balanced by gains in large computers and a few other still good products. These are not industries that are conspicuous polluters. Some of our chemical process plants do have major problems and they also still contribute a sizable share to our exports, but all their case really shows is the international character of pollution as well as of engineering responsibility. Developing countries may be expected to be more "tolerant," just as American local officials have often shown similar toleration, with and sometimes without tangible encouragement; that, in fact, is a major reason for Federal standards. Americans know they have environmental problems but who would exchange ours for those of Japan, Germany, France, Italy, and the rest? And whatever happened to the Russian caviar industry? It isn't only American engineers who find it compatible with their responsibilities to their employers and everyone else to pollute.

Clearly then, engineers have done things of which many others would disapprove and which, in simple business terms, have often left their employers ill-served as well. Obedience, as Shelley put it, "makes slaves of men, and of the human frame a mechanized automaton"—which, of course, is a fate that engineers should reserve for their output instead of themselves.

THE POWER OF ENGINEERS: EARLY CONCERNS

What, then, should be the nature of "responsibility?" The dictionary defines it in part as "accountability, including moral accountability;" it thus becomes a set of values or system of beliefs that delimits what engineers are willing to do for their employers or clients. Furthermore, the humanist critique of engineering presupposes a *particular* set of values, encompassing, for example, a refusal to do work for the destruction of man and his environment. It is not, of course, a universally accepted view and even engineers who took a broad view of their professional skills and sought to apply them to other purposes did not and do not accept them. Neither an acceptance nor a rejection, nor neutrality, assure security of employment, however. Thus, many engineers accept the view that "engineering has no goals of its own but is charged with the implementation of the goals of others by proven know-how and technical skills."[2] Engineers, accordingly, did not decide and did not choose to build the atomic bomb although but for them it would not exist. Leonardo da Vinci, of course, knew better than to take so disingenuous a stance:

> How by an appliance many are able to remain for some time underwater. How and why I do not describe my method of remaining underwater for as long a time as I can remain without food; and this I do not publish or divulge on account of the evil nature of men who would practice assassinations at the bottom of the seas, by breaking ships in their lowest parts and sinking them together with the crews who are in them.[3]

The decision not to participate further in weapons design was a very important one in Leonardo's life but his talents were so ubiquitous that his fame rests on them rather than his engineering achievements and the moral resolution of his conflicts.

By contrast, in Louis XIV's time, there was a French military engineer, Marshal Sebastien de Vauban, who tried to apply engineering principles of efficiency to wider areas and failed.[4] He had made his fame by writing an elaborate treatise on how to storm fortresses, including the necessary hardware, and, when Louis XIV wanted to preserve what he had conquered, de Vauban wrote a second (much less uesful) work on how to defend them. In retirement he developed a new system of taxation. Essentially, it was a value-added tax collected directly by the central government which, though considered regressive now, was then a vast improvement on the rakeoffs and arbitrary extortions of the indirect local collectors. His suggestions were later followed in the French Revolution. Louis XIV, however, listened to those who sought to keep the technocratic meddler in his place and their pockets lined as well, and de Vauban died shortly thereafter of what was called a broken heart.

De Vauban had his problems in part because, as Malcolm Muggeridge once observed, a ruling class in trouble cannot tell its friends from its enemies, and may actually prefer the latter. As an engineer, however, he encountered some of the very same problems facing the profession now. He thought efficiency in collection and equity in the levying of taxes were important and so fell afoul of the real masters of France who were busy consummating the Pyrrhic ruin of Louis XIV's wars. He was "right" in the long run but wrong in the short run. He applied what he know of military logistics and, as part of his tax plan, also proposed new kinds of expenditures and thereby also encountered the priorities problem. As he might have said, *plus ça change . . .*, but he did try to broaden his responsibility in the *direct* interest of his employer as he conceived it and not merely by doing the jobs he was given.

The long-run versus short-run problem is another area on which engineers have often disagreed. In some notable instances, the decision made was really quite wrong. For example, at 4 feet, 8½ inches the standard railroad gauge is too narrow for modern use; the 3 foot, 6 inch or metre lines used in much of Africa, Asia, and South America are even worse. I. K. Brunel's 7-foot gauge of the 1850s would have produced a much speedier system with better stability and capacity. Electric motors would be faster and smaller and fluorescent lights would work better with a higher AC periodicity, as B. G. Lamme had advocated, only to be overruled by George Westinghouse and the other pioneers, for mechanical reasons. Most recently, the United States standardized the lines on a TV picture at a relatively low 525, thus permanently impairing picture quality in comparison with European practice.

THE POWER OF ENGINEERS: PRESENT PIPE DREAMS

It is hard to ask for enough clairvoyance to have avoided these mistakes, but the assessment of current and future technology is not only very much the responsibility of the engineer, but the source of much of his conflict with the rest of society. Unfortunately, much technological forecasting is still being done, science fiction style, by simply extrapolating into the future certain current trends and neglecting "cross impact" of other factors that would, in fact, prevent them from being followed. The late American SST failed in large part because the analysis was faulty in this way and so Congress found the critics more persuasive.[5]

Thus, at a time when engineers should be sensitive to their credibility, one still encounters variants of the old Pollyanna, as for instance, in a recent book prepared under the auspices of the American Trucking Associations and the U.S. Steel Corporation.[6] Even discounting the fact that with such sponsorship, eternal life to the internal combustion engine must be expected to emerge as a central conclusion, it is worth a more detailed look because, unless this is merely a public relations exercise for laymen, this is the sort of advice that the sponsors are getting from their engineers; such advice, in turn, to be worth paying for, must be good enough to serve as a planning input.

The book first leans heavily on the myth of military-space spinoff. (Fortunately we are no longer regaled with tales of how ovenproof cookware has been improved by the technology of space-capsule heat shields; it was actually the other way around and besides not even the most benighted home economics dropout would feel a need to cook a casserole at 1,500 degrees.) In the brave new world we have, of course, a home computer terminal able to "print out" a complete list of party supplies and, when asked, order them from a supermarket and have them delivered to the house—by truck, of course; there is no comment on why truck deliveries of this kind are impossible altogether in the suburbs and going the way of the passenger pigeon in the city. A cynic would, moreover, point out that in the ordering of party supplies such logistics are of less importance than the more melancholy fact that an acceptable salami may well cost more than three dollars a pound.

The computer next prints out the best road to take to the beach, given the miserable traffic conditions of that particular day. One travels there, of course, by an "automated beach highway" where one can relax and let the car move forward on its own; it is a notion not without its bizarre humor to anyone living, as the writer does, in the vicinity of New York's Jones Beach.

But there is trouble ahead:

> You leave the automated highway, take over the controls and hit the gas a little too heavily. A state trooper stops you and asks you to accompany him to a patrol car. He has you place your right thumb on a little screen on the dashboard. Instantly the FBI computer in Washington puts a check through on that print and in a few minutes a voice tells the trooper who you are, that you were honorably discharged from the U.S. Marine Corps, and that you have no record. The trooper lets you off with a warning.[6]

We can let constitutional lawyers debate the legal niceties of this procedure but one could specify other simple systems well within the current state of the art, like having an optical scanner read political contribution checks and then program an electric typewriter in the Department of Justice to prepare antitrust consent decrees favorable to the contributor.

The brave new day concludes with the above-mentioned party, which culminates in a quiz game between men and women judged, of course, by the ubiquitous computer; and finally the host has pains in his stomach, which by attaching various electrodes to his person he can have telemetered to "his doctor" who tells him that he is only suffering from "too much party." We are not told what doctors might be persuaded to answer the telephone at that hour on a weekend but that is a not untypical weakness of all these projections of a happily telemetered future. We may know all about how to transmit things but the real problem is still who takes the remedial action at both ends. The shortage lies in medical and paramedical personnel, not in electronic hardware. Again, one would laugh, were it not for the fact that electronic engineers have often waxed enthusiastic over gadgets that their employers accepted, only to be consigned to the boneyards of bankruptcy statistics as a result. Thus, there were "point of sale" input devices that were to produce "real time, on line" inventory data. Instead, many department stores use a more time-honored method of collecting the prepunched inventory tickets, to wit a cardboard box with a slot in the cover. To expedite the work at Christmas, they take the cover off.

RESPONSIBILITY, ADVICE, AND THE LAW

It is the essence of responsibility or accountability that one is being held personally responsible or accountable for something. In general, business law makes an employed engineer an agent of the employer who is responsible for the result. It has followed from this that the sole judge of how well the employee's responsibility is carried out is the employer (who might be the state) and he also determines the nature of the work. This means that he is able, if he wishes, to predetermine the result and to keep trying until he finds underlings pliable enough to do his bidding. The individual conscience thus faces formidable adversaries. Certainly there does not exist now a mechanism of accountability of the kind, which in the immediate postwar atmosphere of the Nuremberg trials, resulted in the execution of some of the manufacturers of *Zyklon B*, the gas used in the death camps; Walter Dejaco and Fritz Erpl were much more fortunate when, in February 1972, they were acquitted by an Austrian jury of having been accessories to murder by designing the gas chambers and crematoria of Auschwitz-Birkenau. There is no comparable legal responsibility for the designers of the Daisy Cutter bombs.

Nevertheless, the legal responsibility of engineers seems certain to be increased beginning with some more mundane problems of quality and product safety. Recently, a meeting of the Society of Automotive Engineers[7] heard advice from a lawyer that since the files of engineers had recently been subpoenaed in cases of this sort, engineers should watch carefully in the future before they "put pen to paper." He was arguing in effect that the same sort of "chilling" effect on candor would exist as that cited by the government in the case of the Pentagon and Anderson papers. The speaker pointed to the special hazards of reports from field representatives that might contain "something absolutely horrendous about the product" which, he notes, often serves the purpose of attracting the speedy attention of the service department.

Even a general and quite unexceptionable formulation of responsibility such as the following[8] shows potential conflict between employee and employer:

1. Is the product safe for the user under all feasible conditions of usage?

2. Does it adequately perform its function within the framework of acceptable standards of performance, reliability, durability, and cost?

3. Does the product truly serve a consumer need or does it have reasonable probability of providing customer satisfaction?

4. Is the product designed for efficient serviceability?

Consider this last point, for instance: Servicing costs of cars and appliances have risen so much faster than their sales volume as to raise at least the suspicion that some of that servicing pays for sloppy design and manufacture, rather than honest wear and tear.

Can engineers do better? The evidence is contradictory, and engineers are especially reluctant to assert potentially useful views that may offend their employers. Thus, a poll conducted among chemical engineers on various pollution issues,[9] shows majority opinions, anonymously expressed, that might have emanated from the board of directors of the Sierra Club or Friends of the Earth. But one respondent was perceptive enough to point out that when it came down to practice, the result would be determined by such factors as whether the engineer was young, single, and mobile, married and in debt, or old and well provided for. Another aspect of the gap between advocacy and practice is probably to be found in the complaints voiced about "increasingly emotional meetings and slanderous accusations," and the labeling of expert opinions as having ulterior motives due to conflict of interest.

Unfortunately, it appears from other writings on the subject that employers expect engineers to render only some rather specific kinds of advice, much as Louis XIV expected from de Vauban. When, for example, an executive of General Motors suggests that engineers participate in local government and politics as "concerned citizens" who can "assimilate facts and make decisions objectively," the suspicion is strong among engineers that what is meant is an unhesitating defense of company interests.[10] It is a well-founded suspicion grounded in part in observations such as those of Packard[11] or Stryker,[12] that only some variant of extreme conservatism was a suitable political orientation for an executive.

It is part of a broader problem which, in turn, relates to the status of engineers. The comment of an executive of Western Electric Company is not untypical: "Fundamentally, engineers and managers share in the same responsibilities. Mutual confidence is endangered and effectiveness blunted when a union seeks to drive a wedge between them."[13] There is, however, another view that holds that the so-called professionalism of engineers is far too flimsy to deserve the name. As one engineer puts it: "How can I be a true professional when my paycheck comes from one source only and therefore I am not really free to speak my mind? What can I tell my boss when he gives me the go-ahead to pollute a river?"[14] Perhaps there is something of a semantic confusion here; to the employer, professionalism means a nonunion shop, but surely a profession means adherence to standards and *outside* standards at that. But then this would (and should) demolish the notion that *all* engineers are part of management in some way, even if they have neither the title nor the perquisites.

Although he winds up as a second-class citizen on the management team, he is supposed to accept that "engineers must be encouraged to understand that they have a stake in the company's profit or loss picture, its competitive position and its future. They must be *fully informed* (emphasis added) of management's goals, views, plans, and projects."[15] That's fine, but what does "informed" mean? Unless it means that engineers must closely work with management in developing all these hoped for characteristics of the future, the position as part of the management team is an illusion.

"HEAL THYSELF?"

The prospects of engineers doing better in the future necessarily depend on a combination of their inner resources and outside safeguards. The inner resources, in turn, consist of certain crucial elements in the education and background of engineers. It is significant that the government has been sufficiently concerned with the characteristics and attitudes of engineers to conduct a post-Censal sample survey of about 150,000 of them in cooperation with the Engineering Manpower Commission. Unfortunately, the results will not be available until 1973 and in the absence of comparably thorough studies the following observations must necessarily be somewhat subjective.

First, at the university level, engineering students generally tend to be the least critical, the least involved, and among the most conservative on their respective campuses; with rare exceptions their teachers conform to similar views. Of course, all such statements are necessarily relative; here the comparison is with liberal arts departments and such professional schools as law, medicine, social work, and the graduate faculties of the natural and social sciences.

In an article published shortly after Cambodia, Susan Lefcourt, an engineering student in New York, details some of the reasons "why engineering students don't riot."[16] She points to their silent neutrality, indifference, or hostility to such actions as the "open enrollment" strike at City College, and the Cambodian closings; in the former case the engineering students insisted on keeping their school open. Another factor is the rigid routine in the professional training of engineers, comprising multitudes of courses connected in series with much time-consuming laboratory work, and often more than a suggestion of make-work. In fact, the process has been called brutalizing by many of its critics.

Faced with that kind of heavy professional load, moreover, engineers have little taste for nonengineering or nonscientific work. Even industrial engineering or engineering economics face a rather frosty atmosphere generated by their more literal minded technocratic colleagues. These attitudes have, in fact, led to the eclipse of much of the training that American engineering schools once routinely offered in the processes of manufacturing and commercial product design; their replacement with more work at the "scientific" frontier bears a heavy responsibility for the competitive decline of American manufactures mentioned earlier. The result has been to relegate precisely that subject area that could generate informed social responsibility to a diversion reluctantly suffered. The humanities courses required are frequently still of the instant and concentrated variety which used to be called "Western Civ" or "source studies." These offerings have declined to "service courses" of liberal arts departments that are offered only to engineers who are thus segregated from students with different backgrounds, values, and ambitions.

Finally, Ms. Lefcourt speaks of "alienation," but insofar as this word denotes the decline or destruction of something that once was there, the term is probably misleading. There is evidence from an older study that engineers are notably apolitical and score much lower in their desire to participate in community, national, or international affairs than their fellow students in other fields.[17] The study cited goes back to the late 1950s but that still makes it relevant because it not only accords with empirical observations in the years since, but also reflects the attitudes of the then students who are now practitioners.

As to the social background of engineering students, the great majority came from working-class and lower-middle-class families. Less than half the parents studied in the sample had completed a high school education and a significant proportion had gone no further than the eighth grade. The median family income was substantially less than that of students in other major fields. In such circumstances, a desire not to rock the boat so as not to jeopardize upward social mobility is probably combined with a background not noted for its questioning of sociopolitical conditions; in fine, we are dealing here to a great extent with the children of hardhats. It has, incidentally, been the author's observation that the children of successful engineers in high corporate management almost never become engineers themselves, preferring other professions instead.

All this does not bode well for internal efforts at change and yet, as the survey[9] indicates, the conscience of engineers is there, but doth make a coward of him, as it did of Hamlet. Engineers are not likely to go on the barricades and at present, and for the next few years, their bargaining position is not likely to be good. In part, this is because so many engineers have allowed the technical content of their work to sink to the paraprofessional level, thus allowing them to be replaced, if need be, by people with far less training. They will, therefore, need outside help.

GUARDING THE GUARDIANS

The proposed reorganization of engineering is drawn, without apology, from a much maligned type of management, that of the university. It does not seem to have dawned on the many writers who bemoan the status of the engineer and the way an ungrateful world treats him that some of the things regularly and wistfully asked for already exist in full measure and tradition in universities. There is, for example, the matter of parallel path advancement, i.e., engineers can move upward on professional merit alone without having their position and earnings tied to an arduous climb up the pyramid of line advancement. In engineering firms this is enough of an exception to merit special articles[18] but in a university one can be successful and, as a matter of fact, usually happier by not bothering with administrative chores.

Academic freedom, as it is now practiced—may it long resist all attacks!—materially reduces such problems as recently stated in a proposed ASME position on public issues:

> The individual engineer may find his professional obligation to the public in conflict with the demand of his employer or his client. Without the support of established and currently applicable standards of ethical practice that have the endorsement and practical backing of his professional society the engineer may well jeopardize his position or even his career by standing for what his engineering conscience tells him is right.[19]

Again such professional independence is second nature to academicians in all but the more benighted institutions.

The key to the difference is, of course, tenure. It provides security of employment after a probationery period and can terminate only for cause or under extreme financial stress. Like a lot of worthy institutions, that of tenure is now under attack and, as befalls only the worthiest of all, the attack comes both from the extreme left and the right. We need not go into the details of this debate, which has recently been most ably summarized,[20] but a comment by Prof. Fritz Machlup is germane. "With respect to some occupations, it is eminently in the interests of society that men can speak their mind without fear of retribution"—and there surely is the nub of the issue. Machlup continues: "The occupational work of the vast majority of people is largely independent of their thought and speech. The professor's work *consists* (emphasis in original) of his thought and speech." But it is *precisely* the duty of engineers to feel free to give their own judgment without fear or concern; their work too consists of thought, plans, and writings.

Would such an arrangement protect mediocrity and time serving? Of course it would. But these qualities are hardly in short supply in the engineering industries as they exist today; they seem a small price to pay for getting some good judgment as well.

What are the prospects for such a reorganization of the profession? Of course, the impediments are many. The concept of tenure in commercial employment as distinct from civil service and academia is unknown in the United States where profligate hiring and firing still remain the rule. But the system is firmly entrenched in other countries. It exists in various forms in western Europe and most directly in Brazil and especially in Japan; to use a conspicuous understatement, it doesn't seem to have hindered that country's technical success. In the American context, however, there appear to be two ways in which professional independence of this sort is likely to be furthered. The first is the result of an increasing demand by industries that government sponsor and directly pay for such parts of their operation as product design and environmental protection. This is already leading to a perceptible reversal in what had been a previous trend toward the conversion of social costs into business costs.[21] It suggests that, in return, the government should demand that engineers giving their professional judgment on matters of this sort should consider themselves as acting in the interests of the public that paid them rather than the employer who takes care of them in line with what is sometimes called the "trickle down theory."

So far, the government itself has not taken kindly to such advice as, for instance, in the case of A. Ernest Fitzgerald, who blew the whistle on the C-5A mess and was promptly eased out. However, it is exactly such cases that argue in favor of a more receptive attitude in the future. Anyway, it is the height of illogic for management to pay for professional advice and then restrict what it wants to hear—like J. P. Morgan telling his lawyer: "I did not ask you whether to do it, but how to do it." It is also an oversimplification to paint engineers as suppressed dissenters who are just aching to make their true feelings known. There are many employment situations which are anything but despotic and in which the conditions more extensively advocated above already prevail to a significant extent. Even as to tenure, there are enterprises where few people, if any, are ever fired, but even so, there is extra security in a formal codification of these arrangements.

A second opportunity for making opposition known is in the growing Public Interest Research Groups (PIRGs), which increasingly deal with technical matters. Engineering managements certainly have recognized their potential. When *Chemical & Engineering News* reported on Ralph Nader's Clearing House for Professional Responsibility, in which he sought information on alleged unethical or wasteful practices of industrial firms (if necessary, submitted anonymously), "fink tank" was one of the more respectful epithets in the managerial reaction. It was conceded that there was a problem here, but that engineers should go through channels to get things changed, and if they did not like it they should quit.[22] Nader countered that this "love it or leave it" scheme only entrenched the status quo and asked why a man should quit or be fired merely because he is doing a competent job.

In matters affecting the public, there should indeed be possibilities for making objections widely known; once the damage is done, the subpoenas increasingly issued have the same effect but it would be better to know earlier. The sub-servience of engineers rests on the notion that even those who have devised something are no better qualified than the average "citizen" to judge its use. But obviously only the engineer or other professional who knows what weapon is being made or what river polluted knows enough to alert the public in a meaningful way. It is part of the great American battle between the document shredders and the Xerox machines.

If reporting to Mr. Nader or the press, or the police, for that matter, is somehow considered unseemly, then how about engineers being able to report the blun-dering of the management to the board of directors? But then, that would require a real board which would take its supervisory responsibilities seriously and be more than a rubber stamp for the existing management. In the case of large companies, it might even include a public director with an internal audit staff of the kind suggested by Robert Townsend.[23]

There is also a strong case for the more generous sponsorship by legislative and regulatory bodies of independent public-interest research which has as its objec-tive the prevention or early elimination of technical blunders. Drs. A. R. Tamplin and J. W. Gofman of Lawrence Radiation Laboratory have proposed the insti-tutionalization of an adversary system of scientific inquiry, as they call it, in which all public projects with technology content would have to face some technically sophisticated devil's advocates.[22]

It is thus clear that there is no shortage of ideas and indeed of precedent to expand the scope of professional responsibility of engineers in relation to their employers and to their society. The trouble is still, however, that they act like workers and very unmilitant ones at that; they are certainly not up to the stage of the well-known Lordstown workers at General Motors who, in early 1972, struck against the drudgery of the assembly line.[24] Engineers will, no doubt, have to take the lead in solving the technical problems that now beset us. However, until they doff their servility and learn to act like professionals they will neither merit nor receive the accolade set forth in the Gospel of St. Matthew: "Well done thou good and faithful servant: Thou hast been faithful over a few things, I will make thee ruler over many things."

References and Notes

1. ________. 1972. Stopping the incredible rise in weapons costs. Bus. Wk., Feb. 19: 60–1. See also U.S. SENATE, COMMITTEE ON ARMED SERVICES. 1971. Report on defense appropriations bill. Report 92–359 92d Congress, 1st Session, Sept. 1971 : 16–19 and U.S. SENATE, COMMITTEE ON ARMED SERVICES. 1971. Hearings on Weapons Acquisition Process, 92d Congress, 1st Session, December 3–9. In the latter, see especially the comments on this point by Sens. Symington and Goldwater.

2. Kolenda, K. 1971. The new consciousness and the engineer. Hydrocarb. Proc. 40(5): 165–73.

3. MacCurdy, E., Ed. 1954. The Notebooks of Leonardo da Vinci ; 850–851. Braziller. New York, N.Y.

4. Saint-Simon, L. De R., Duc De, 1913. Memoirs. Vol. 1: 353–355. Allen & Unwin. London.

5. U.S. Congress, Joint Economic Committee. 1970. Economic analysis and the efficiency of government; Part 4. Supersonic Transport Development. U.S. Govt. Printg. Off. Washington, D.C.

6. Fuchs, W. G. 1972. Marketing at a Crossroads. Ch. 3. American Trucking Associations. Washington, D.C.

7. ________. 1971. Engineers play bigger role as lawsuits go after poor design. Ind. Wk. 169: 15–16.

8. Staebler, L. A. 1970. Engineering ethics and consumerism. ASHRAE Jl. 12(9): 70.

9. Hughson, R. V. & H. Popper. 1971. Engineering ethics and the environment: The vote is in! Chem. Eng. 78 (4): 106–112.

10. ________. 1971. Engineering mentality is label nobody wants. Prod. Eng. 42: 22.

11. Packard, V. 1962. The Pyramid Climbers : 220–222. Crest. New York, N.Y.

12. Stryker, P. 1960. The Men from the Boys. Ch. 3. Harper. New York, N.Y.

13. Kleingartner, A. 1971. Professionalism and engineering unions. Ind. Rel. 8: 224–235.

14. Garcia-Borras, T. 1971. The estranged engineer. Hydrocarb Proc. 50 (6): 157.

15. Simrall, H. C. 1971. If you are a concerned.... IEEE Spectrum 8(2):69–71.

16. Lefcourt, S. 1969. Why engineers don't riot. Chem. Eng. 76(17): 108–110.

17. Krulee, G. K. & E. B. Nadler. 1960. Studies on education for science and engineering: Student values and curriculum choices. IRE Trans. Eng. Mgmt. EM–7(4): 146 ff.

18. Kahle, H. 1969. Parallel path advancement: Plan promotes productivity. Mach. Des. 41: 98–105.

19. American Society of Mechanical Engineers. 1970. Goals: A proposed statement. Mech. Eng. 92(4): 18–23.

20. Bierstedt, R. & W. P. Metzger. 1972. For tenure: Safeguarding academic freedom. Civil Liberties 285: 5–7.

21. Ullmann, J. E. 1969. Waste disposal and social costs. In Waste Disposal Problems in Selected Industries. J. E. Ullmann, Ed. : 271–284. Hofstra Yearbook of Business. Hempstead, N. Y.

22. ______. 1971. Groups urge advocacy roles for professionals. Chem. Eng. News 49: 15–16. For the earlier report on the Nader organization, see Chem. Eng. News 49: 3. For a later status report, see ______. 1972. Student interest in advocacy groups grows. Chem. Eng. News 50: 20–23.

23. Townsend, R. 1971. The wreck of the Penn Central. N.Y. Times Book Rev. Dec. 12.:3.

24. ______. 1972. The spreading Lordstown syndrome. Bus. Wk. Mar. 4: 69.

3.2

Professionals and the Corporation

BRUCE F. GORDON
IAN C. ROSS

One cannot escape exposure to frequent use of the word *professionalism* on the contemporary industrial scene. It is a word used widely by engineers, scientists, and the management responsible for them. It is a word that has a positive affective value on those that hear it. It is a vague word with meanings that range from workmanlike to nonamateur, from independent to belonging to professional societies, and from educated to certified. The vagueness of the word leads to the propagation of a great many statements about professionalism which may be true in some sense or in some situation but do little to communicate much about the central meaning of the concept of professionalism. The favorable image of the word, together with its vagueness, has made it a useful base for many statements that must in all honesty be viewed as propaganda. Professionalism has been advanced as a reason for engineers to join unions and as a reason to refrain from joining unions; it has been used to promote the solidarity of employees as well as to distinguish some employees from others; it has been used to promote the identification of some employees with management as well as to distinguish them from management. In short, confusion has been generated by the use of professionalism to justify positions which interest groups would espouse even if there were no concept of professionalism available.

Much of the discussion about professionalism has concerned scientists and engineers in corporate employment. The statements have often come from sources which are explicitly management or explicitly representative of groups of scientists or engineers. Statements from these sources generally do not do much more than exploit the favorable image of professionalism for the propaganda purposes of these groups. Indeed, as we shall later show, they cannot analyze the basic implications of professionalism without creating large problems for their own groups. However, it is our belief that such an analysis is necessary. The research and development effort of our nation is dependent upon corporately employed scientists and engineers. Understanding the position of these technical people may aid in maximizing their contribution to our effort and posture. Very little understanding can come from listening to the propaganda on the subject and the authors present the following analysis in the hope that a reasoned inspection of the roles of technical people will begin to receive the attention is merits.

ANALYSIS OF WORK RELATIONSHIPS

We shall now develop our analysis of the place of professional employees in a corporate organization. To do so we shall first distinguish among some classes or working relationsihps and then examine one of them, the professional, in detail.

The employment relationship is often thought of as one in which the employee does what the employer wants him to do. Theoretically and typically the relationship can be terminated when the employer is no longer satisfied with the performance. Exceptions, however, exist, particularly in the cases of employees protected in their jobs by features of collective bargaining agreements. Where the employee is required to do in detail what the employer prescribes, we shall refer to the relationship as that of *artisan–master.*

Less widely recognized is that the autonomy of the employer to judge the production and quality of his employees has been more limited in two classes of employment.

In some cases, it is the employee's duty to give his employer what the employer ought to have, regardless or whether or not the employer says he wants it, regardless of whether or not the employer likes it, and regardless of whether or not the employee likes to give it to him. To deliver service under this kind of obligation has been the essential meaning of *professional* status.

A preacher should not care whether his parishioners want to hear of the eternal damnation that awaits them in retribution for their sins. If the word of God so states, he is obliged to tell them even though he himself dislikes to utter the word "hell." A doctor may be obliged to prescribe bitter medicine. A teacher is obliged to try to teach a child no matter how much he lacks interest in arithmetic. A lawyer is obliged to make his employer's purchase of property effective no matter how much the employer hates to pay for registering the deed.

When an employee is obliged to consider the need of his employer or client, his own desires and the employer's desires notwithstanding, he is functioning in a professional capacity. The motto of a professional, therefore, might be, "Give him what he ought to have!"

The second class of employment in which, historically, the autonomy of the employer is limited, is that of employees whose mandate is to give the employer what he, the employee, would like his employer to have. Such employees are paid for doing what they would like to do anyway.

To be able to employ people in this relationship was, in the past, a prerogative of royalty or at least nobility. Only recently have corporations been aware of the advantage of employees with such commissions. This is the kind of employment which was enjoyed by Michaelangelo, Hadyn, and Bach. Each of them had patrons and may, therefore, be referred to as a "protege." This protege–patron relationship is a difficult one to describe. Perhaps we can get a better understanding of it by examining the relationship as it has occurred between the artist as a protege and his patron.* Clearly, the artist cannot elect to give his employer nothing. Indeed the patron must be pleased with the production after he gets it, though he may not have been able to ask for it. The patron paid the artist and provided him with opportunities in which to execute and display his creations. The patron was willing to support the artist because he obtained aesthetic pleasure and glory from the artist's products. In those days glory was a very practical outcome. In a sense the artist's production provided the means by which the patron could demonstrate for all the world his political power, wealth and status. There are analogous reasons for the corporate employment of protégés. The products of protégés are hedges against an unknown future; they increase the likelihood of corporate stability and may enhance the future prospects of the organization. Stability and prospects are essential to a corporation in the marketplace of investors.

We cannot be sure that those who usually were proteges always created and produced what they would like to give the patron. In some cases the protege may have performed skilled work as an employee in which the patron prescribed how the work should be performed in detail. (In such cases, the artist had neither responsibility for deciding whether the product was one which the patron needed nor responsibility for whether the product was what the artist wanted to give the patron.) However, patrons able to give this direction successfully were rare and had some claim to artistic competence themselves.

It is our contention that the modern industrial research organization may employ people in each of the three employment relationships which we have defined. One would never get this impression from the contemporary propaganda because employers and employees have reasons for suppressing the distinctions. There are different needs and aspirations associated with each of these categories and these may not be discussed because to do so would raise all sorts of invidious comparisons and would contradict the notion that all scientific and engineering employees are professionals. This notion is one of the functional fictions of industrial research. This fiction is used in many morale building speeches and even actions. It attracts scientists and engineers to the corporation and enlarges their place in the organization. Let us now look in more detail at this fiction and why it is used by spokesmen for both management and employees.

There are basically two reasons why spokesmen for technical employees should not raise the distinctions. First of all the distinctions imply that technical employees do not have a homogenous set of goals, objectives, and situations. To recognize the distinctions would reduce the solidarity of the workers and consequently the constituency which employee spokesmen may represent. The second reason for silence is that the classification implies a certain amount of power on the part of the management to make some technical employees professional and to make others something else. If management makes some employees artisans by virtue of their assignment, there may be no apparent reason why such people need special representation as technical employees but might be bargained for by regular employee organizations which also represent machinists, janitors, or clerks.

---

*The artist–patron relationship as an ideal role model for the basic research scientist employed in industry had previously been suggested by William M. Evan in a paper presented before the annual meeting of the American Association for the Advancement of Science, New York City, December 1960.

On the other hand the notion of protégé is hard to reconcile with the notion of being represented by an organization of technical employees. The protégé concept implies that protégés exist and do their work because they have been given gifts (called stipends, fellowships, grants, and other synonyms for gifts) which carry a limited commitment on the part of the patron for continuity. Protégés have been subject to variations in the taste of their patrons and have suffered from a limited market for the service which they render. There is also the implication that protégés do not pay their own way but produce decoration, memorials, and objects that, while admirable, are not functional, at least in any short term and accountable sense. The long term is different. We remember an age by the products of its artists and its discoverers. These people in their day generally were unappreciated, were accused of fooling around and of being unproductive of the stuff which society required then and there. The notion of protégé, regardless of what it is called, is not a particularly attractive one to a work and sales oriented culture. One of the troubles with the protégé concept is that it violates contemporary notions of work. Protégés are either doing something for which there is no recognized need or they are fooling around for their own enjoyment. However, if one calls them professionals, these nasty features are suppressed.

Neither side believes it may gain by recognizing that some scientists and possibly some engineers are protégés The representatives of technical employees would create invidious distinctions within their constituency were they to do so. Management would find itself in trouble with stockholders for wasting the corporate treasure on people of questionable productivity. These difficulties are avoided by asserting that people in the protégé role are professionals.

To recognize that some scientists and engineers are artisans would also get both sides into trouble. Here again one of the troubles for the employee side is the reduction of its constituency.

The problem with the artisan notion for the management side is somewhat different. Management needs to recruit engineers and many of them have to be recruited for the mountains of routine, skilled work which has to be conquered between the conception and fruition of the scientific breakthroughs and the pioneering engineering achievements. There has been some tendency to talk about the flag planting ceremonies at the top of the mountain and to ignore all the drilling that must precede the installation of one piton.

It is also clear that the notion of artisan is not completely compatible with the notion of a college education. There is still some carry-over from the days when higher education was generally a preparation for work in law, medicine, teaching, or preaching. If an engineer works as an artisan, it sounds as if a trade school education would have been sufficient. In order to give appropriate recognition to the arduous instructional process to which he has been subjected, it is desirable to accord him the dignity of being called a professional.

The suppression of the fact that some engineers work as artisans is believed to be useful in mobilizing talent for the hard, skilled, and often routine work that undergirds engineering progress. Both sides want this talent attracted into engineering and no one should be surprised that the possibilities of an artisan career for an educated engineer are seldom mentioned.

Once one has split off the concept of professional from those of artisan and protégé, it is possible to state analytically a set of criteria defining a professional and describing the manner in which he works.

The analysis consists of criterion characteristics for the professional–client relationship of three sorts: prerequisite, continuing, and evaluative.

All of the characteristics to be given are considered to be necessary to the essence of the professional–client relationship. The characteristics are neither mutually independent nor is any one of them sufficiently unique to differentiate the professional–client relationship from other possible relationships.

CHARACTERISTICS OF PROFESSIONAL STATUS

Prerequisite characteristics

1. The professional offers to provide a service for the client which is superior to attempts by the client, or by any other individual without special qualification, to perform such services.

2. The authority and responsibility for service which is to be performed passes from the client to the professional. Behind this transfer of authority and responsibility are the assumptions that (a) the client believes the professional has superior knowledge and judgment and (b) the client believes that the professional will not abuse the trust and confidence placed in him.

Continuing characteristics

3. The service performed for the client is what the professional, in his judgement, believes the client should have.

4. The service provided by a professional is intellectual, unique, and personal. All professionals in a given field have access to the same formal resources, techniques, and knowledge. The manner in which these are drawn upon and the judgment which is exercised in selecting the particular treatment for a client's problem, may be unique to each practitioner in the field.

5. The professional must continue to increase his knowledge to provide assurance that all available alternatives are considered in giving the client what he should have.

6. The professional has the obligation of (a) advising the client as to the nature of the service; (b) keeping the client informed of his progress and expectation of solutions to problems.

Evaluative Characteristics

7. The competence of a professional on a single performance may only be judged by professional colleagues.

8. The client can evaluate the professional's services over time by reference to (a) the extent to which such services have met his needs in the past; (b) the comparison of the services of several professionals all of whom he employs;

(c) the reputation of a particular professional among other clients and; (d) the reputation of the professional among other professionals.

ROLE OF THE PROFESSIONAL IN THE CORPORATION

Let us consider the total impact of all these characteristics of a professional. In sum, they say that the professional has great responsibilities. He is responsible for the quality of the services which he produces. He is responsible for giving his client an understanding of what the service is and what it can do for the client. He is responsible for making the service as good as possible and up to an ever-increasing standard of excellence, as the body of knowledge he is marketing is fortified by new discoveries of proven worth. To meet these responsibilities the professional is given considerable authority and protection. He has the authority that will lead him to expect a certain amount of obedience from the client who is employing him and he is protected from precipitous judgment.

The authority and protection accorded to the professional serves the professional in the discharge of his responsibilities. They are functionally justified by his duties towards his client. It would indeed be difficult for him to give the client what he should have without them and it might be even more difficult to maintain the professional–client relationship.

Clearly, failure to recognize the reality of the professional–client, or protégé–patron and artisan–master types of relationships precludes specifying the nature of such relationships. It is our contention that all three role relationships do exist in corporations. There is evidence that recruiting practices tend to promise the role of protégé to the scientist and the role of professional to the engineer. In practice a substantial portion of scientists work as professionals for the corporation and a substantial portion of engineers work as artisans.

The reality and necessity of all three role relationships within the corporation are implicitly recognized by some corporate managements. Recognition is not tantamount to evaluating the corporate requirements for artisans, professionals, and protégés. Evaluation of needs should proceed from a theoretical understanding of the role relationships and their attendant obligations and responsibilities. These have to be coordinated to the requirements of each research and development organization.

The proportion of protégés, artisans, and professionals that a corporation will need, of course, depends on the nature of the business and the size of the corporation. The small corporation may not be able to afford scientists in the protégé role. Such positions are a privilege of the large corporation. It may even have an obligation to the economy to employ such people. The reason for these assertions will become clearer when we examine the functions of the protégé, professional, and artisan in the development of new products.

In order to help the individual find the role in which he performs best, some flexibility should be provided so that there is no unnecessary rigidity about the role that a particular individual is expected to perform. Some shifting from one role to another will help to reveal the resources which might otherwise remain hidden in the staff of scientists and engineers. However, it probably should be made clear to the individual what role he is performing at any given time and the reciprocal behavior of his supervision should always be coordinated to the present role of the individual.

The balance of the paper is concerned with what corporate management can do toward maximizing the contribution of the protégé, professional, and artisan to the well-being of the corporation and society.

The function for the scientist as a protégé is to make inroads on the unknown according to his interest, knowledge, and skills. Utility of the knowledge derived is of secondary importance. The protégé, knowing the most about his specific area of competence, is best able to select the segment of the unknown that he wishes to bring into the known. The motivation of the protégé is in large part generated by a desire to unravel the portions of the particular mystery in which he is most interested.

The expectation of the corporation for the protégé is never on an individual basis in that the corporation has the right to demand utility from the knowledge that the individual protégé uncovers. The utility for the corporation is probabilistic over all protégés. Some of the knowledge gained by protégés will never have utilitarian value. Neither the protégés nor the corporate management can specify either the individuals or areas worked on that will have utility. However, selection of the protégé to work with complete autonomy in a given area should never be done capriciously. The corporate management has the obligation and duty to select a man for the protégé role whose interests, propensities, and previously demonstrated capacity provide evidence that important knowledge in areas of concern to the company (aside from the question of utility) will be forthcoming in the best tradition of science.

The function of the professional is utilitarian in that the professional is providing the client with what it ought to have. The professional is interested in converting known knowledge into a utilitarian form that will benefit his client, the corporation. Indeed much of the satisfaction of the professional is derived from observing the results of his recommendations. He is like an authorized distributor.

Unlike the expectations for protégés, the expectations for professionals are individual rather than group. As specified in the characteristics, the evaluation of the utility of an individual professional must be over a long enough time period to permit corporate management to strike a fair average for comparison with other professionals. Selection of the professional for a task is not difficult where the individual has demonstrated his professional capabilities previously either within or outside the corporation. Unfortunately, the vast majority of engineering school graduates hired upon completion of their bachelor degree have not yet demonstrated their professional capability. The result is assignment to artisan type work. This places a heavy responsibility upon corporate management to develop at least a portion of engineering artisans into professionals.

In order to carry out their responsibility, corporate management must create job assignments which provide a bridge whereby engineering artisans may become engineering professionals. The crossing of the bridge in many corporations traditionally has been haphazard. It has been crossed most often by the initiative and drive of individual engineers who aspire to become professionals rather than by any formalized help or procedures from corporate management.

Artisans are oriented toward performing a given specified task for value received. The value received (money) has a relatively more prominent value to the artisan than to the protégé or professional in that it is used to satisfy desires and needs that cannot be found in the nature of the task. The contract is between the individual artisan and the corporation. It is further identified by the fact that the task is specified and can be quantitatively measured over a short time period.

Selection for artisan positions is based on the individual possessing the necessary skill to carry out the specified task. In this sense, the artisan engineer is more like the blue or white collar worker than like the professional engineer.

SUPERVISION

The kind of supervision and amenities which should be given the protégé, professional, and artisan differ in some important respects.

It may well be that supervision of the protégé should give patient encouragement, physical facilities and personnel aides, statements about the importance of the phenomena, and protection from diversionary demands. It may also include statements that reduce any skepticism the protégé may have that the corporation wants him to do the kind of work that he has chosen to accept.

The supervisor of the professional, after transferring the responsibility for the problem and its outcome to the professional, has the obligation to make sure that no necessary relevant information is withheld. Assistance, whether personnel or material, that is functional and necessary should be supplied. He also is obligated as the client's representative to see that progress is being made toward a utilitarian solution. It is also important that the solutions of the professional be implemented. In instances where it is not possible to use the solutions of the professional, the supervisor has an obligation to demonstrate in a reasonable fashion why such solutions cannot be implemented. Failure to do this can undermine the very essence of professional motivation and performance.

The artisan, of course, requires the necessary tools and materials to exercise properly his skill. Clear instructions, fair measurement of output, fair compensation and fair human treatment in accordance with the dignity of man are all desirable and probably necessary.

SUMMARY

Appropriate supervision is only one way of recognizing the diversity of employment relationships which a research and development organization requires. It is also necessary to have firm notions of which developmental requirements call for which kind of employment relationship. Management should structure the whole role of the employees accordingly. Most important of all, it must not let itself be diverted from the requirements by the occasions for public discussion of professionalism which give opportunities to use the fictions of professionalism for propaganda purposes.

The authors have attempted to provide a more meaningful definition of the concept of professionalism than that in current use by the spokesmen for engineers, scientists, and management. Through the development of the role distinctions of protégé, professional, and artisan, the authors have attempted to clarify the kind of role relationships that exist for engineers and scientists within the corporation. In conclusion, we have made suggestions to corporate management in the belief that implementation of such suggestions cannot help but improve the contribution that engineers and scientists will make not only to the corporate well-being, but to the progress of man.

3.3

Conflict of Interest

DONALD CHRISTIANSEN

In today's business and professional world, mention of the term conflict of interest gains instant attention. It denotes something to be avoided at all costs, when, indeed, in its general sense, conflict of interest is unavoidable. Governments, corporations, and institutions are prey to it. Whole professions exist to adjudicate conflicts of interest; lawyers do it, judges do it, managers do it, and yes, even engineers do it.

In the case of Governments, it seems almost superfluous to note that they are rife with conflicts of interest (among constituents, to cite one example, as well as between one another). The corporation encounters conflicts galore in serving its many publics—its stockholders, managers, workers, customers, and the community. And even in the case of the individual, each of us faces his own internal conflicts of interest—some of them not as simple as whether to watch baseball on TV or mow the lawn. In the case of individual conflicts, sometimes we have only ourselves to answer to. If we find ourselves in conflict with others, it may be adjudicated on a custom basis, or, in cases where we conflict with an organization (perhaps the firm for which we work), the likelihood exists of some precedent to guide the resolution.

But the "rule of conflict" is that the more varied our interests, the more frequently will our conflicts occur. The same applies to the individual organization with which we are "conflicting." So we can readily postulate an exponentially escalating incidence of conflict as a function of the activity and interests of both parties.

A classic conflict of the engineer is between "things technical" and societal concerns. In this day of rapid change, in which societal or "professional" concerns are seen as valid by a greater number of colleagues, we even recognize a conflict between "inward" and "outward" directed professional concerns. (Inward concerns are related to such items as portable pensions, programs to limit the number

---

158

of available engineers, and the like, while outward concerns relate principally to interests of the public.)

If individual engineers are conflict-prone, it is not surprising that conflict of interest is no stranger to the IEEE itself. (The phenomenon of conflict escalation clearly applies to IEEE, as both its individual members and the Institute itself are broadening their interests and concerns.)

A perennial conflict of interest that IEEE has had involves the fact that some of its members, in their daily professional duties, are more accurately describable as employees, whereas others are better described as managers—or part of "management." (This is not to imply that even the company president himself is not usually an employee.) Thus, the Institute, in addressing matters of, for example, portable pensions, may encounter conflicting viewpoints among its own members. Individual members of the Institute's Board of Directors face dilemmas stemming from such conflict. Furthermore, when the Institute finds itself acting as spokesman for the profession, members of the Board ask themselves, "Should I support the consensus among my constituents, or should I vote my own conscience and belief, independent of such consensus?" The problem is not unlike that confronting any elected official in a representative democracy.

To close one's mind to conflict of interest or to "sweep it under the rug" is folly. There is little question that such an approach may work occasionally in matters of little or no consequence, but rarely does it aid in reaching decisions in complex situations. Furthermore, it seems that, in enlightened professions such as ours, progress involving resolution of conflict is more rapid if based on advocacy techniques as opposed to adversary techniques. The point may seem a fine one to those who view the two as opposite sides of the same coin. Yet adversary techniques often tend to polarize, while advocacy tends to resolve differences. The IEEE, for example, espouses the wide publication and dissemination of information concerning engineering employment practices, to the end that it will help the profession—both employees and employers. Sometimes, "advocates" for the employees' interests see defects in existing practices that endanger their careers or earning power, while, simultaneously, "advocates" for management see easily correctable shortcomings that put them in an unfavorable position in attracting and keeping high-calibre engineering talent. In any event, enlightened decisions cannot be made in a vacuum—and changes made unilaterally may be counterproductive.

In summary, conflict of interest is unavoidable. It is neither all good nor all bad. It is bad in certain business situations, as, for example, when an individual's interests are so evenly divided that he is not sure of his own loyalties. It is bad when he intentionally misrepresents his interests, or when he gains financially because of hidden influence. However, it can be positive when individuals or organizations can openly debate the interests they advocate. The result can be clearer delineation of the tradeoffs, solutions openly arrived at, and fair exposure of minority concerns in the process.

3.4

Conflicts of Interest
Pose No Big Problem

Conflicts of interest, by and large, pose no major problem now for the chemical industry. Not, at least, in the opinion of most companies with whom *C & EN* [*Chemical & Engineering News*] has talked during the past couple of weeks.

Nevertheless, most company managements recognize that problems could arise. And most leading companies have prepared formal policy statements dealing with conflicts of interest to guide their employees.

In many cases, these policies have been established, or at least reviewed and strongly reiterated, since the conflict of interest controversy at Chrysler Corp. made headlines last year and was followed by a drastic shake-up of Chrysler's top management.

C & EN was unable to find a single industry executive who felt conflicts of interest were a serious problem, either in his company or for the industry as a whole. A few cases may have cropped up, but only one company would admit to taking action against an employee because of a conflict within the past year or so— and that one case was "minor" and was handled simply by asking the employee to rid himself of the conflict.

But there is little doubt, judging from the number of formal policy statements or memoranda issued from executive suites in recent months, that companies have taken steps to guard against conflicts. Says one company: "It is an area where there should be sound policy and administrative procedure."

For many companies, too, it is a sensitive matter. Several seem to agree with the midwestern firm that feels "the entire subject is far too delicate and far too complex" to talk about with outsiders.

Most companies that have set forth a policy on conflicts have done so in a formal written statement worked up by top management, such as the president or the chairman of the board. Most often the policy applies to all employees, although for many companies it involves only top management or other key employees, such as technical and professional people, purchasing agents, or those involved with setting specifications. Although company policies often have been in force for many years, several companies, like Goodrich Chemical, have reviewed theirs and brought them up to date within the past year or so.

WHAT IS A CONFLICT?

Most chemical companies, in their policy statements, attempt to combat a broad area of business activity which would be unethical, illegal, or dishonest; where the employee's personal interest might injure the company's interest; or where the employee might use his position for personal gain. FMC Corp. in its statement, for example, says, "The fundamental principle to keep in mind is that there cannot be any compromise with high standards of business ethics. If there is room for suspicion that an employee's interest or connection with another venture might affect that employee's judgment in acting for FMC, he must dispose of that interest or sever that connection."

Frequently, a single policy is laid down to cover both conflicts of interest and fixing prices with competitors. More often, though, price fixing or other violations of antitrust laws are dealt with by separate policies.

In general, companies do not restrict their policies by attempting to spell out all potential conflicts. But typical examples of conflicts include:

- Owning a financial interest in or holding a position with a supplier, agent, customer, or competitor.
- Accepting fees, gifts, entertainment, loans, or other favors from suppliers, customers, or competitors.
- Acquiring an interest in a business, real estate, or other facilities in which the company may be interested.
- Having an outside interest or job which prevents the employee from performing his duties fully or which is detrimental to the company.
- Divulging confidential information.
- Speculating on the basis of inside information in stock of the company or of companies with which it deals.

These restrictions usually extend to all members of the employee's immediate family as well.

Not Clear-Cut

Conflicts, however, often are not clear-cut, and most company policies have a certain amount of flexibility. Companies usually do not object to employees' owning "nominal investments" in other firms, especially if the interest is in a publicly owned company.

Monsanto, for example, feels that "a conflict is unlikely if the financial interest consists of holdings of less than 5% of any type of a company's securities which are listed on a stock exchange or regularly traded on an over-the-counter market." Du

Pont requires its top management to report interests exceeding 1% in any company with which it does $10,000 worth of business or more a year. Goodrich Chemical questions its management on ownership of more than 1% of stock of any company not listed on a stock exchange and with which it does business. Other companies link investments to the employee's total assets.

Likewise, most companies do not object to employees' accepting "conventional business courtesies," especially if they can return the favor in kind. "Lunch is okay," says a West Coast company, "but not a yacht voyage or a hunting trip."

Digging Out Conflicts

Many companies rely on internal questionnaires to ferret out potential conflicts instead of turning to outside agencies, such as their law firms or auditors. Du Pont, for one, requires its officers, board members, department and assistant department heads, and employees involved with purchasing to answer a questionnaire each year regarding outside financial interests. Early this year [1961], Union Carbide sent a questionnaire to over 2000 key people, [and] had its auditors check replies for conflicts. Atlas Chemical similarly questioned 750 of its employees and had the responses reviewed by its auditors.

Other companies use personal interviews. Some require signed statements disavowing conflicts. But several merely circulate policy statements and rely on employees to report possible conflicts to their superiors. Division or department heads often are responsible for checking on employees under them.

Most companies say that the way in which they handle a conflict depends on how serious it is. In general, though, employees would be required either to end the conflicting activity or to leave the company. In addition, the company might take the offending employee to court to recover his gain.

Renewed alertness to conflicts in the chemical industry checks with what the National Industrial Conference Board has found for all U.S. manufacturers. An NICB survey this year shows that one company in three has a written policy on the subject now, compared with only one in 10 last year, and that more than a fourth of the companies use questionnaires to uncover conflicts. NICB found, however, that some executives question the value of questionnaires on the basis that a person would be no more ethical in filling out such a form than he is in his personal conduct.

CONFLICTS OF INTEREST

These are examples taken from chemical company policy statements:

Where an employee has an outside interest which materially encroaches on time or attention which should be devoted to the Company's affairs or so affects the employee's energies as to prevent his devoting his full abilities to the performance of his duties.

Where an employee, or close relative, has a direct or indirect financial interest in or is engaged in the management of an organization which does business with or is a competitor of this Company.

Speculation in materials, equipment, supplies, or property purchased by the Company.

Borrowing money from customers or from individuals or firms from which the Company buys services, materials, equipment, or supplies.

Purchase or lease of real estate or other facilities which the employee knows the Company may be interested in, or may need in the future.

Acceptance by an employee, or his immediate family, of money, loans, goods, services, or rebates in return for employment, leases, contracts, or purchase orders, or in return for the supplying of confidential Company information, or in return for otherwise conducting his prescribed Company responsibilities in a manner that is contrary to the Company's interests.

Acceptance by an employee, or his immediate family, of extravagant gifts, entertainment, or travel.

Where an employee uses or releases to a third party any data on decisions, plans, competitive bids, or any other information concerning the Company which might be prejudicial to the Company.

Speculative dealing in the Company's stock on the basis of information gained in the performance of the employee's duties and not available to the public.

The acquisition of an interest in a firm with which, to the employee's knowledge, the Company is carrying on or contemplating negotiations for merger or purchase.

If an employee learns that the Company is about to make or substantially increase a contract with a particular company, it is wrong for that employee to employ that inside information as a basis for making an investment, even if the employee has other reasons for so investing. It is equally wrong for the employee to disclose that information to anyone else for such person's personal gain.

3.5

Can an Engineer join a Union and Still Be Professional?

WILLIAM JAMES HAGA

Can an engineer embrace trade-union membership without compromising his status as a professional? This question has arisen anew as an issue in engineering circles. As in past epochs of engineers' interest in unions, many who debate the question assume that engineers are, indeed, members of a profession. I will accept that assumption for the time being, while I discuss engineering unionism; but I will come back to it in a negative vein.

A BRIEF HISTORY OF ENGINEERING UNIONISM

The first collective-bargaining effort by engineers in the United States came through the American Federation of Technical Engineers (AFTE), which affiliated with the AFL in 1918. AFTE originally was comprised of marine draftsmen, the vast majority, somewhere between 11,000 and 12,000, being what we would euphemistically classify as subprofessionals or preprofessionals, but it also included about 1000 graduate-engineer members. This original mixing of engineers and non-engineers in the same bargaining units had an impact on engineering unionization, for better or worse, from that time on.

Another early engineering-bargaining organization, the Federation of Architects, Engineers, Chemists, and Technicians (FAECT) came together during the Depression years of the 1930s. FAECT affiliated with the CIO in 1937 and later merged with the United Office and Professional Workers of America (UOPWA). AFTE and the FAECT were the only significant organizing efforts among engineers before the 1940s. Expelled by the CIO on charges of Communist domination in the 1950s, UOPWA eventually was disbanded.

The decade of the 1940s saw considerable organizing activity among engineers, most of it leading to the formation of small independent bargaining units. The impetus for unionization in that particular organizational form was largely institutional rather than economic. In the years before Taft-Hartley legislation, the large industrial unions attempted to include engineers within their jurisdiction in the course of organizing Blue Collar production workers. At that time, engineers confronted with these apparently unwelcome affiliations, regarded them as a threat to their occupational status. They responded by forming myriad collective bargaining units of their own to preserve their professional distinction. Engineers formed unions, ironically, as a means of avoiding the labor movement.[1]

Many of those independent units were organized with the help of the various engineering societies. Engineers at the Sunflower Ordnance Works, in Eudora, Kansas, protested to the War Labor Board in 1943 at being represented by a local of the International Federation of Technical Engineers, Architects and Draftsmen's Unions, AFL. The American Society of Civil Engineers (ASCE) assisted the Sunflower engineers in presenting their case to the Board. The protest was denied, partly on the grounds that the ASCE could not define "professional engineers" in a way that clearly set them apart from the nonengineers in the union.

As a result of losing the Sunflower Ordnance case, ASCE began to work with local sections to help convert them into collective-bargaining units. Because of possible legal problems, ASCE subsequently removed those units from its affiliation to let them operate as self-administered, autonomous shelters against the organizing drives of the industrial unions.

The response of engineers to industrial unionization in the 1940s continues to be felt. Engineers who are today represented by production-worker internationals, such as the IUE or UAW, are often organized into separate, autonomous bargaining units within the same firms. A disdain for sharing representation with draftsmen or technicians remains a mark of engineering unionism.

The economic demand for engineers in the late 1950s and early 1960s gave them a supply-and-demand advantage, raising their incomes in dollar terms, and, more importantly, in relative terms compared to other occupations. This economic reality mitigated organizing drives. The reversal of these same economic forces now favors the success of renewed unionization efforts among engineers.

It is a striking fact, in this regard, that unions have not been notably successful in organizing engineers, just as they have not been with White Collar people generally. A number of factors stand in their way.

Would an engineers' collective-bargaining unit find greater material success and a happier self-concept within an exclusively White Collar international such as AFTE or Office Employees International Union (OEIU) or as a special adjunct of large industrial unions? While recognizing the opportunity for organizing additional members among growing white collar staffs, industrial unions have been almost indifferent in their actions.[2] One reason is that a White Collar staff, and this would be especially true of an engineering function in many cases, may not number in the hundreds, let alone in the thousands. The dues income to an international may not justify an expensive organizing campaign needed to break through traditional White Collar resistance to the labor movement. Moreover, upward-striving White Collar workers usually want their identity separate from production workers. Virtually every union that has pursued White Collar organizing recognizes this. It is even truer for employees who see themselves in a professional role partaking of management, as engineers predominantly do.

Thus White Collar locals are exercises in euphemism where "district representatives" replace stewards, "guilds" replace unions, and "employees" replace workers. Although separation and special treatment has seemed a requisite to White Collar organization, it is also a hindrance, conveying to the potential new member that, indeed, an industrial union is not his true home precisely because it dares not treat him as its majority membership and still expect to gain his interest.

ECONOMIC CYCLES AND ENGINEERING

The growth of the engineering unions has been preceded historically by profound change in the nature of engineering employment. By the middle of the Nineteenth Century, there were only some 2000 engineers in the United States, and most of them worked as self-employed practitioners or in consulting partnerships.[3] At that time American engineers came closest to realizing the character of a genuine profession. Today our country has over a million engineers, virtually all of them working for large, formal organizations as salaried employees in jobs remote from the character of a profession (as it will be defined later in this article). Management sets their working hours, even to the point of some punching time-cards. Management closely controls their work.[1] Their employment security rides up and down on the same cycles that affect Blue Collar workers. And bureaucratic setting of engineers' employment, in which they have become narrowly specialized, renders them increasingly vulnerable to economic swings because of an inability to transfer their limited job lore to other opportunities.

Specialization has also fostered alienation in work settings where any one engineer contributes only a small part to a large, complex total project, much as production workers individually contribute but a small part to the manufacture of a total product. Bureaucratic employment has also sharply restricted the latitude an individual engineer can exercise in choosing problems. And it has constrained recognition of outstanding work because secrecy in its operations often prohibits publication of research findings or techniques and because employers hold patent rights on inventions by their engineers.

As the nature of engineering employment shifted from individual practitioner to bureaucratic employee, and as the number of graduate engineers expanded faster than the national population growth, engineers came to a position of economic disadvantage. They found more money in their paychecks after World War II, in absolute dollar terms, but their median salaries declined relative to other occupations.[4] Relative income slid downward until 1953 when sharply expanding demand ran ahead of the supply of engineers. The percentage changes in income were greatest at the junior end of the experience spectrum, resulting in an inverse relation between years of experience and percentage gains in income. In the economist's language this "compression in salary structure" means the dollar distance between the bottom and the top of the engineering salary scales became progressively smaller, leaving senior engineers at a further disadvantage in relative income.

These economic forces weigh heavily upon the success of an effort to organize engineers. After 1953, as engineers' incomes rose faster than those of other occupations and the extraordinary demand for engineers made them the momentary elite of the work world, unionization efforts fell on tough times. When engineering graduates, even at the baccalaureate level, were being figuratively, and often literally, wined and dined by prospective employers, trade-union organizers were left without their major appeals: financial gain and getting management attention for grievances.

From the late 1960s to the present engineers have experienced a turn for the worse in their economic and employment fortunes. Three major factors have been at work: big slump of the aerospace industry in the late 1960s, the acceleration of price inflation in the early 1970s, and the consequent severe recession now with us. The aerospace slump put thousands of engineers out of work while engineering schools still produced new graduates at a record pace. The result, an economic impact of textbook obviousness: supply exceeded demand on a grand scale and engineering salaries either dropped or were severely slowed in their annual gains. Because the salaries of other occupations generally were still rising in the late 1960s, engineers again suffered losses in relative incomes.

By the early 1970s, when the doctrines of altruism and careers for social service were peaking on U.S. campuses, while student interest in technology waned, the output of graduate engineers was slowed. Many undergraduates stayed away from engineering colleges because the specter of the aerospace collapse was still before them. Unemployment among engineers was still high, high enough to garner research grants for academic entrepreneurs to study alternative ways of employing the available engineering manpower. But even the slight moderation of the supply-demand imbalance arising from lessened graduate output was offset by the beginning of price inflation, which approached double digit levels in 1974. The fuel shortage and the parallel shortages of key petrochemicals, coupled with consumer resistance to higher prices, contributed to the economic downturn. This, in consequence, has further clouded employment opportunities for engineers.

Recent relative-income and salary-compression statistics describe a psychological trauma for occupations such as engineering. When other occupations, which involve much less in the way of education and skill, begin to close the income gap with a high-status occupation such as engineering, then engineers feel a generalized uneasiness. As the difference in pay between an engineer and, say, a welder narrows within a span of ten years, the engineer believes he deserves better.

This response to declining relative income is not limited, of course, to engineers. Consider the situation confronted by professors, another high-status occupation. A junior faculty member typically has spent up to six years in graduate school to obtain a Ph.D., sometimes longer. He leaves school with a large debt. If he can find a job at all, in the current depressed academic market, he celebrates by trying to buy a house for his growing family. Only he finds that he loses out in the price bidding to the proverbial truck driver. In the late 1950s, a tenured professor made about $10,000 in a year. But he also lived in a roomy old house in the nice part of town. His children went to private schools. Today, a beginning assistant professor may make as much as $16,000 in his first year of work. With that he can scarcely manage to finance a mobile home. He was attracted to the professorial life by the lingering image of a genteel life-style that provided a comfortable, secure, middle-class existence even if it was not a road to riches. In the 1950s, faculties were reputed to dwell near poverty but a nice sort of poverty. Today, the young academic finds that he receives more dollars than faculty people have ever been paid, but he is still left in a poverty that is closer to the real thing than his predecessors ever experienced. To keep the wolf from the door he has to hustle for supplemental income: consulting, quickie textbooks aimed at junior-college markets, extra teaching at nights and during the summers. These distractions keep him from the contemplative, scholarly life that was his original attraction to Academe. He finds himself caught up in a rat race for promotion, tenure, and security as hectic as any other occupation. Something is wrong. The world has not turned out at all as he expected. He is disappointed and alienated. That well over 300 American campuses have faculties recently organized into collective-bargaining units therefore comes as no surprise. How much different is the situation of the engineer?

COMPARATIVE VALUES OF PROFESSIONS AND TRADE UNIONS

White Collar workers, as a group, are known to be conservative in their outlook, embracing the proverbial Protestant Ethic work perspective. They believe in career advancement on the basis of individual achievement. They hold their employing management as their major occupational reference group. This is a sociologist's way of saying they know (a) what to do and (b) how well they are doing it according to what management wishes. Their work identity is based almost exclusively on their affiliation with management. They dislike collective action because it subordinates individual initiative and capacity to the will and mentality of a group. They dislike it because it threatens the thread of identity tied to management. Unionization, therefore, threatens the aspirations and images of a career that White Collar people expect to lead to a post in management.

Occupations that like to think of themselves as "professions" embrace some but not all of those White Collar values. As one difference, an occupation in pursuit of professionalism may identify with management while at the same time building separate occupational identity that cuts across employment ties. Nurses, for example, usually think of themselves primarily as members of the nursing profession and only secondarily consider themselves as part of an employing hospital organization. Engineers, however, for all their professionalism, seem to maintain a tie to their managements that supersedes even their ties to engineering.

Occupations that aspire to be known as professions, as Kleingartner at UCLA has noted,[5] embrace a work ethic stronger than that of other White Collar occupations. To the professional, his work manifests his identity. The work itself is his driving interest. He expects to be evaluated solely on the quality of his work performance. He places great store by autonomy in deciding what problems he will work on and in the choice of tools and means to do his work. He expects his skill and contributions to be judged by his final product, not to be directed and constrained by administrators. Moreover, he deems only his professional peers to be fit to judge the quality of what he has done. Kleingartner points out that a professional will tolerate many indignities and petty irritations at work so long as he maintains control of his core professional function. A professional cannot tolerate an infringement upon that core function. Many so-called professionals, however, do tolerate violations of their occupational autonomy, especially engineers.

Professionals think in terms of careers rather than simply as being employed at jobs in a purely economic *quid pro quo*. This leads to a corresponding value upon personal development throughout a working lifetime that has priority over simple efficiency in work output. Along the same line, professionals expect to make the fullest possible use of their skills and education in doing their work. They are acutely sensitive to being underemployed or misused by their employers. They expect their work to be personally satisfying and fulfilling of their selves. By and large it is. Occupations that aspire to professionalism usually are populated by people who are happier at work than the labor force as a whole.

Union organizers of professionally inclined work groups know these values and have been sensitive to them in their organizing approaches. The IUE, in fact, has turned the individualism value to advantage by arguing, in its organizing literature, that only through collective bargaining can individual ideals and identities be shielded from increasingly impersonal supervision and regulation by large organizations where the decision making about local matters is centered in remote layers of management.

The values of professional occupations are embodied in the conduct of professional associations, which, in Kleingartner's view, have "traditionally tended to accentuate aspects and values of professionalism over which they could exercise direct control and have rationalized away the importance of values which required direct confrontation with employers."[5] Professional associations embrace the view that their members and their managements share an appreciation and responsibility for doing work of a professional quality. Professionals prefer to believe that they will be rewarded by management in response to the quality of their work. They look upon any direct confrontation with their employers over matters of promotion and pay as beneath their professional dignity. Since it seldom takes the same exalted view of a profession's contribution as does a profession itself, management tends to take advantage of professional reluctance to quibble over economic concerns. The professionals respond by rationalizing away the importance of economic rewards while emphasizing a compensating devotion to service for a larger public.

Unions, on the other hand, begin with the assumption that matters of genuine difference separate employers and employees, even professional employees—differences deeper than a mere lack of communication. They contend that a truly professional group would be prepared to take any steps necessary to insure control over its work sphere and protect its professional identity from management infringements. Physicians, they point out, have usually taken whatever steps necessary to maintain control over their work, even to striking, without being regarded as any the less professional for having done so. Yet engineers are still inclined to regard such behavior among themselves as unprofessional.

Kleingartner found an inverse relationship, however, between measures of an engineer's professionalism and his rejection of engineering unions as unprofessional.[6] That is, the more professional an engineer, the more likely he was to see no conflict between unions and professionalism. This finding contradicts the folk belief among managers that only marginal or so-called "weak sister" engineers would be attracted to unions.

An explanation for Kleingartner's findings is that competent engineers are likely to be secure enough in their professional self-image to entertain the idea of union membership as a way to protect autonomy. Conversely, the least competent may be inclined to cling to fictions of professionalism and to identity with management while viewing union affiliation as a threat to fragile self-image. Walton supports this view.[7] He encountered a paradox in looking at which engineers actively participated in union activities. Inferior engineers, he found, often were insecure and therefore feared management disapproval of union affiliation. In reality, Walton found that the technically most competent engineers in a firm also held leadership positions in their union locals. Because engineers adore technical ability, they reward their competent colleagues with political support.

Walton distinguished four categories of engineers in their relationship to union activities:

1. *The middle-class-conscious engineer.* Extremely management-oriented, sees a union as an obstacle to this orientation and to reaching or keeping his class status.
2. *The traditional-professional engineer.* Views engineering work as a genuine profession and the union as a negation of individualism, which he regards as integral to professionalism.
3. *The socially aware engineer.* Sees the union as a means of expressing a general interest in social problems, a means of influencing his work environment.
4. *The leader aspirant engineer.* Enjoys positions of authority, sees a union as a way to exercise leadership skills, gain visibility for management positions.

THE HINDERED EFFECTIVENESS OF ORGANIZED ENGINEERS

For reasons at once intrinsic to the social nature of engineers as well as to the structure of trade unions, Walton finds that engineers usually make rather poor unionists. Even after engineers have joined a collective-bargaining unit, they seldom shed their identity with management. The early independent engineering bargaining units were careful to indicate to their managements that they meant no ill will. Clinging to the belief that the only thing separating them from management was a lack of communication, early unionized engineers saw their task as one of educating management about patent rights and salaries; and believing management to be comprised of reasonable men, much like themselves, they were certain their views would be understood.

This soft approach, however, led to frustration. Managements usually regarded engineering bargaining units casually. They did not approach them with the seriousness they accorded production-worker locals. In some engineering units there was an evolution from reasonable leaders to representatives willing to take a tougher stand. Still, management took its contract negotiations with organized engineers lightly, often sending junior labor-relations staffers, failing to prepare salary data for bargaining sessions, and firing engineering unit leaders at will.

Where militancy grew among engineering unions, the results were mixed. Some were able to get their managements to pay heed to their grievances and demands.

In other locals, a militant stand caused enough dissension among their professional-image-conscious members that they were eventually decertified as representatives.

In Walton's analysis, engineering unions lack a decisive element in any collective bargaining: *power.* A great deal of engineering work can be delayed without dire consequences to a firm. As a result, strikes by engineers alone have seldom been effective. Engineering bargaining teams, moreover, seldom maintain a bond of unity and the necessary fortitude for the frustrations and in-fighting of negotiation sessions. As Walton puts it, "... the typical engineer cannot resist the temptation to improve his standing with his supervisor even at the expense of his fellow engineers."[7]

A meticulousness about democratic procedure, to the point of tedium, also has hampered the effectiveness of engineers as collective bargainers. Seldom does an engineering negotiating team act on its own. Instead, its members prefer to return to their constituency for approval of each offer and consideration that arises at the bargaining table. Another hindering factor is that engineers are inordinately mobile in their employment. This turnover of members also weakens the leadership of many engineering locals.

One cure tried for these built-in problems has been to accustom organized engineers to taking strike action without seeing it as a threat to their self-image. Another attempted remedy has been to ally an engineering local with adjacent industrial-union locals that do have collective-bargaining power, in order to have engineering picket lines honored by production workers. Often the threat of affiliation or cooperation with a production local has strengthened an engineering local's power. However, the willingness of a production local to entangle its own bargaining position with that of an adjacent local representing a comparative handful of high-status, well-paid engineers cannot be automatically assumed.

Even when not hampered at the bargaining table, engineers create obstacles for themselves in their day-to-day relations with management by their reluctance to file grievances. This severely limits the effectiveness of their local's bargaining efforts. A union representative needs grievances to support charges of inadequacy of management procedures and policies. Union officials argue that engineers' immediate supervisors seldom possess the organizational power to modify existing practices and are simultaneously reluctant to approach their own supervisors. A formal grievance gives the immediate supervisors a compelling reason for bringing a shortcoming to the attention of upper managers. Walton notes that engineers usually want management to change facets of their work situation but are reluctant to be personally identified as malcontents. This timidity often drives an individual engineer to withdraw a grievance after it has been filed.

Finally, the engineers' insistence, already noted, upon professional purity in the membership of their union undermines their bargaining. One autonomous engineering union, the Engineers and Scientists of America, dissolved into small splinter locals in the late 1950s over the issue of mixing graduate engineers with subprofessional technicians in the same bargaining units. The extent of such mixing in a particular local, Walton found, had a clear impact on its willingness to stand up to management. Engineer bargaining units that included a high proportion of technicians were inclined to act very much like industrial-union locals; such heterogenous units were more likely to strike. In some cases, the so-called subprofessional people have been placed in a separate bargaining unit by the engineering leadership of a local. Union members who also think of themselves as professionals strive to keep technicians out because they are seen as a hindrance to making the union acceptable to other professionals. Employers seek a membership composition of a professional local that will either defeat itself in a certification election or dampen any militant tendencies in an existing local. For this reason, they generally favor a professional union devoid of subprofessionals.

DROPPING THE ASSUMPTION OF ENGINEERING PROFESSIONALISM

Up to this point I have gone along with the assumption that engineers are indeed members of a profession and therefore may feel a real conflict between professionalism and union membership. But by simply *dropping the assumption* of engineering professionalism I can dismiss the conflict between professionalism and unionism as false and distracting.

Certainly, employers of engineers have used the professionalism issue as a counter attraction to the economic and work-condition appeals of unionism. Most managements construe engineering professionalism as a synonym for loyalty to them. To engineers, of course, professionalism is a flattering label that their conceits can scarcely resist. It has been a potent force in the resistance of engineers to union organization.

Most students of the relationship of engineers to the labor movement, however, begin with the conventional assumption that engineering is a profession. The assumption itself has seldom been examined. This follows from the fact that sociology has not provided clear distinctions between occupations and professions. Discussants of engineering unionism versus engineering professionalism have been compelled to assume engineering is a profession for lack of a means by which to assert that it was not.

On Being a Profession

Most talk of professionalism bogs down in a welter of confusing terms: *profession, professional, professionalism,* and professionalization. I propose these working definitions for each of the terms:

- *Profession.* A field of work that enjoys a higher social status than occupations not so ordained. Why a profession is different from lesser occupations is a distinction noted for its elusiveness; yet it is a distinction pivotal to any understanding of professionalism. (A burden of this discussion is to offer a clear model of that distinction.)
- *Professional.* A person who sees himself as a member of a profession. He is a self-conscious about professionalism and occupational status. Many engineers are professionals, by this definition, even though engineering itself is not a profession. Related to this meaning of professional is the fact that any occupation contains some practitioners who stand out from the rest. In folk usage, they are the "real pros" in their work—e.g., the "engineers' engineer." This is not what is meant by "professional" in most analyses of professional-

ism. Such a usage would simply make professionalism synonymous with competence. Most self-conscious professionals, including many engineers, are not content to leave it at that. They mean something more by professionalism, although they are at a loss to specify just what it is.

- *Professionalism.*　This term refers to a behavior, acting like a member of a profession. It is an affectation of the outward symbols of profession.
- *Professionalization.*　The deliberate, programmed process that members of an occupation undertake in trying to convert to a profession. Not knowing just what makes professions different from other occupations, they mount predictably futile professionalization campaigns.

From its earliest treatments of the question,[8] sociology has been content to set genuine professions off from occupations of lesser status according to a "shopping list" of necessary conditions. Each observer of professions has developed his own version of the attribute list. Typically, a sociologist's list includes many of the following traits:

1. Altruistic service to the public.
2. Long, specialized training for entrants.
3. A code of ethics.
4. Associations and professional meetings.
5. Learned journals aimed at upgrading the practice.
6. Examinations as barriers to entry into practice.
7. Tacit prohibitions against advertising of services.
8. Practice limited to members licensed by the state or certified by association boards.
9. Symbolic costumes (e.g., black robes, white coats) and control of the access and behavior of nonmembers in work places (courts, operation rooms, sanctuaries).

Many occupations have been surprised by their success in fulfilling most of the prescribed attributes and simultaneously dismayed at *what little difference it made in genuinely enhancing their status.* This follows since such lists contain only traits that are the observable results of being a profession. These traits are not the *causal factors* that can convert any occupation into a genuine profession.

THE PROFESSIONALISM PLIGHT OF PUBLIC RELATIONS

Consider the travail of public relations, a White Collar field sufficiently remote from engineering to permit objective examination. It is also an occupation that few engineers would mistake for a genuine profession. A PR man typically feels that his occupation doesn't command sufficient respect from others with whom he interacts, especially those in his immediate working world, such as employers, suppliers, members of genuine professions (accountants and lawyers are the bane of PR men). He gets together with other PR men to do something about it. They collectively conclude that their lot would be greatly improved if only public relations were perceived by others to be a profession. Then they could command respect and deference on the basis of professionalism.

Thus, concerned PR men swear to work toward professional status for public relations. First they consult a sociologist, preferably one specializing in the study of occupations. He, of course, provides them with a typical list of professional attributes. List in hand, this group (now named "professional practices committee") sets out to achieve as many of those professional characteristics as they can. After only a few years, they prove quite successful, at least in terms of their *list*. To their pleasant surprise, public relations has come to possess most of the attributes of profession named by the conventional sociologists:

- It has a national public-relations association.
- It publishes a journal.
- It holds annual conventions and regional meetings.
- It bestows coveted awards for outstanding examples of "professional" public-relations practice.
- It writes a code of professional ethics and gives each member a handsome copy, suitable for framing.

That is only the beginning; still more rigor is injected into the professionalization campaign. A judicial board is established to "try" PR practitioners who violate the now-detailed rules of practice. Moreover, a new class of "certified" practitioners is developed. These superprofessionals have passed a day-long examination on the ethics and practice of public relations. They are entitled to place "Certified" after their names. Talk begins to circulate among the true believers of the need for state licensing of public-relations men.

In the terms of sociologists' traditional *lists* of professional attributes, public relations has moved well down the road to becoming a profession. But, has it? Has the daily working life of most public-relations men really changed, after years of professionalizing? *Not at all.* In the eyes of the media people, they are still "flacks." In the eyes of their employers or clients, they are still just "nice guys who are not profit-oriented. To CPAs and lawyers they are still "pretentious irritants." In achieving respect and deference, the whole professionalization effort was a flop. If anything, more PR men are now demoralized for having let themselves pretend to be a profession.

Failure of the attribute-list approach to professions can be read in the misfortunes of several occupations that have taken the lists seriously enough to use them as a foundation for energetic professionalization campaigns. Public relations' futile professionalization is not unique. Much the same frustrating story can be told by candid teachers, librarians, nurses, social workers, hospital adminstrators, insurance salesmen, urban planners, stockbrokers, and policemen. They have all tried to become professions and they have failed. Why, then, do so many occupations persist in chasing the Holy Grail of genuine professional status when so many before them have not succeeded? There are two reasons:

1. Professional status offers something they want very much.
2. They have no idea what a profession really is.

WHAT THE PROFESSIONALIZERS REALLY WANT

The chief advantages that occupational groups seek when they undertake professionalization are these[9]:

1. Above all, they want autonomy; they want freedom from supervision in their jobs.
2. Individually they want to be recognized for their particular field of work, not for the name of their employer.
3. They want to have the power of judging who is "in" their occupation and who is "out." They want monopoly over a certain line of work, freeing it from interference by outsiders (usually employers, but also clientele or the general public) who do not share their work ideology.
4. They want exclusive power to discipline wayward colleagues who deviate from their work ideology. They may not exercise this power but they do not want anyone else to have it.

The desire for *autonomy motivates a professionalizing occupation*: its members want to get others off their backs. They want to perform their jobs as they, and *they alone*, deem to be the right way.

INTIMIDATION: THE ESSENCE OF PROFESSION

What gets and keeps autonomy in the work world? To put it bluntly, *intimidation*. That is the essence of truly professional behavior—the ability to intimidate *whatever* audience threatens a profession's autonomy—employers, clients, other outsiders.

Authority conditions the essential relationship between a member of a profession and the others in his world. *A profession has authority over others*. It is, in turn, free from the authority of others.

When members of an occupation engage in argument or resort to persuasion in dealing with employers or clientele, it implies an equality of status between the parties involved. Argument or rational persuasion does not suggest a relationship characterized by authority. When a practitioner must rely upon argument or persuasion in dealing with his clientele or employer, this is *prima facie* evidence that he is not a member of a profession.[10]

Note carefully that a genuine profession need only fall back upon its power to intimidate when all other means have failed. In most daily interactions, members of a profession will likely employ both argument and persuasion in dealing with significant others at work. A good doctor will cultivate a "bedside manner" for cajoling patients; but cajolement is not his basic means of control. Persistently skeptical clients are labeled as "difficult" by professions. In the marketplace, these same clients would be admired as rational consumers. But the status of profession is only slightly founded on rationality. As a last resort, a member of a profession handles "difficult" others by an ultimatum threatening to withdraw services. If an occupation is truly a profession, the prospect of losing its services will be quite intimidating to its clients or employers.

An occupation has become a profession when it can rely upon intimidation as its ultimate, if unseen, means of controlling others at work. Members of a profession get their own sweet way precisely because they can go beyond cajolement, arguments, appeals to logic and persuasion. Intimidation works. Many occupations thus seek its power.

Many engineers who aspire to the status of a profession feel uncomfortable with the "unprofessional" connotations of intimidation. They prefer to picture the true professional as dealing with others on a more logical basis. To repeat, most members of genuine professions do not spend their days openly intimidating others. However, when a critical issue arises that threatens their work autonomy, they have the power to preserve that autonomy through intimidation, however latent it might be. Failure to grasp that distinction contributes to the spectacle of so many well-intentioned people wasting themselves in pursuit of a professional status they can never achieve.

Being a Profession—Two Conditions

The canons of logic dictate two elements to delineate an entity: its *genus* (which tells the category to which it belongs) and its *differentia specifica* (which tells how one entity differs from other entities of the same *genus*).[11]

Missing from sociology's conventional "shopping lists" of attributes are the *differentiae* needed to distinguish a genuine profession from other occupations. Medicine or law, for example, would still be treated with deference, if not awe, regardless of whether they possessed codes of ethics, held conventions, or published journals. Even such supposedly defining attributes as restricted entry or specialized training, often cited as bedrock "requirements" of professional status, are actually consequents. Medicine, law, and the clergy were historically accorded superior status as occupations long before they formalized their educational entry requirements to the standards familiar today. Entry barriers are a great help in maintaining a profession's status once it is achieved, but they are not its source.

The crucial distinction between professions and ordinary occupations lies at the intersection of two *differentiae*: cruciality and mystique. These are the joint *differentiae* needed to delimit profession from the *genus* occupation.

A combination of cruciality and mystique defines a true profession.

This definition can be stated as a hypothesis: the greater the degree of both an occupation's cruciality and its mystique, the greater its status as a profession.

Cruciality. An occupation possesses cruciality when some group of significant *others* define it as being crucially necessary to their prosperity or survival. Cruciality means an occupation has an almost life-or-death, fateful relationship to its clientele, publics, or employers. "Significant" others means it is not necessary that everybody consider it crucial, but that it *is* considered crucial by some people. Moreover, it must be a group of people that the occupation itself *needs* as an audience for its presentations of professionalism.

Before an individual will define an occupation as crucial, he must first perceive himself as being in need of a service or counsel that deals with a matter vital to his life, health, purse, or freedom. What is "crucial" to him arises out of the meanings he gives to his life situation and the judgments he makes about the demands, threats, or opportunities he sees in that situation.

Cruciality is not to be mistaken for indicating matters that are merely convenient or of marginal importance. The literal meaning of crucial is embraced here: Crucial means *deadly serious*. This meaning applies whether the situation is valued positively, such as the fateful opportunity for increased wealth, or valued negatively, such as overcoming a potentially fatal illness.

The threshold of perception of a matter as crucial is a function of individual interpretations of, and often of interaction with, members of the occupation involved. Some citizens will plead their own case in a traffic offense, others will hire a lawyer; but few would attempt to defend themselves against a capital indictment without legal counsel. The more crucial a matter is perceived to be, the greater the tendency to seek the aid of the profession that holds an occupational monopoly over it.

Mystique.　　The second *differentia* of profession arises out of the first. The person with a crucial problem seeks help from a particular practitioner he perceives as best able to solve it—a practitioner who seems to know a great deal more about the problem than the lay person can even begin to grasp. He sees this practitioner as in possession of a mystery. Mystique includes a great deal of what is meant by expertise, technical know-how, or having specialized, advanced training—but it is more than competence or skill alone. Numerous occupations require extensive training and technical skill for their performance, e.g., tool and die making, hand lettering of Biblical parchments, and steamfitting. But they are not professions. The vital ingredient in mystique is that the work is seen as esoteric, difficult, and consisting of doing things that ordinary mortals, in the layman's perception, cannot do.

The mystique of a profession creates a form of interaction that is fundamentally an authority relationship between the persons involved. Friedrich defines authority as the suspension of judgment in the presence of a significant other.[12] First, an individual perceives himself as incapable of understanding the solution to a crucial problem. He sees himself as confronting a body of knowledge largely impenetrable by his own mental capacity. Second, he defers to the judgment of a significant other whom he perceives as understanding the crucial situation at hand. The significant other then has effective authority over the layman in such a situation: he acts in the manner of a member of a profession. This particular sort of authority, arising out of the layman's suspension of judgment, is the difference mystique makes in creating profession.

In answering the basic question, "What is a profession?", the focus is shifted from the professionalizing occupations themselves to the relevant audiences before whom they work.

How well these two necessary *differentiae* distinguish occupations according to their professional status can be evaluated, at least impressionistically, by looking at their four possible combinations.

1. *Low Mystique and Low Cruciality.*　　This category of occupations faces the maximum constraint upon aspirations to the status of a profession. Nonetheless, it includes occupations that have been diligent in pursuing pro-

fessionalism, e.g., public relations, purchasing agents, personnel specialists, librarians, and association executives. None of them have made it. Based upon the model developed here, none of them will gain it.

2. *High Mystique and Low Cruciality.*   This category embraces such endeavors as computer operations, watchmaking, diamond cutting, astrology, the fine arts, so-called healers, and most occupations that deal with science at the technician level. Not all of them are among the irrevocably damned. They are in possession of a mystery—in the perception of a large public. That is no mean feat. Should any of them manage to develop an accompanying aura of cruciality they could move into the charmed circle. Computer operations is the best candidate from this group for becoming a profession. For many of us, it already is.

3. *Low Mystique but High Cruciality.*   This class includes still more occupational groups that have been immodest about their self-conscious claims to professional status, e.g., teaching, journalism, farming (yes, farming), military leadership, business management, and social work. Having one of the necessary aspects of profession offers some promise for these groups. However, their extravagant expenditures of effort, discussion, planning, money, time, and rhetoric have not moved them a bit closer to being professions. Of course, they have been unaware that they also had to possess mystique. Objectively, it is doubtful that the work settings of many of them provide an opportunity to present a mysterious aspect.

The fate of the professionalization campaign by school teachers presents a case in point.[13] By all measures, their effort has failed; and consequently most teacher groups have turned to unionization (so have nurses, and social workers). They failed because they followed the advice of sociologists who overlooked the essential element of mystique in creating a profession. But, even had they been aware of the need for mystique, they could not have developed it. Teachers had two major audiences to which they directed their professionalization efforts: school administrations and the community. Despite such classic professionalizing as requiring a special college curriculum for entry, powerful associations, state evaluation of credentials, and an amusing jargon, teachers have not become members of a profession. The point against them is that their relevant audiences have been too close to the teachers' everyday work roles to be mystified by them. We have all spent years in first-hand participation with teachers in their daily work. Where's the mystery? Such familiarity leaves little ground for building mystique. Much the same is true for managers, librarians, social workers, and military officers.

4. *High Mystique and High Cruciality.*   The occupations in this group possess both mystique and cruciality. They are the only true professions in the four categories. Law, medicine, and the clergy are obvious examples. So are dentistry, veterinary medicine, architecture, and accounting. Other occupations that possess both *differentiae*, but to a slightly lesser degree, might also qualify as professions, e.g., pharmacy, surgical nursing, and clinical psychology.

PERCEPTIONS AND INTERACTION—THE DEFINING PROCESS

If the cruciality and mystique *differentiae* separate the genuine professions from the pretenders, into which category does engineering fall? How does engineering measure up on the mystique and cruciality dimensions? These questions beg a further one: to which of an engineering field's audiences is the question directed? How a field scores on mystique and cruciality varies with the particular audiences at which it aims its work performances: employers, clientele, other occupations, the general public. There is an unavoidable relativity to the process of defining professions. Engineering may possess sufficient mystique and cruciality in combination to appear as a profession to some audience. However, it may not be a *significant* audience in the work sphere of engineers.

To the man in the street, many occupations appear to be professions—for example, pharmacy. The pharmacist works behind an elevated counter (above the customer's eye level, compelling him to look upward), surrounded by a bewildering array of bottles, and speaking an alien tongue. If anyone were ever in possession of a mystery (taking his cruciality for granted here), it is the pharmacist. Thus, most of us defer to him, indulge him as the superior person in our fleeting interactions, and find that he could be intimidating should we not accept his presentation of a superior role. Between the pharmacist and his public customers the relationship certainly describes a member of a profession and his clientele. At least one of his relevant audiences supports his claims to the status of profession.

But the pharmacist's customers do not make up his most important audience.[14] Who does? Physicians, of course. In the pharmacist's world, physicians control and direct and *have the last word*. The physician occupies a role superior to that of the pharmacist, and so doing severely constrains the pharmacist's autonomy to direct his own work. (This example is limited to the interactions of hospital pharmacists and staff physicians: the professional status of the retail druggist is a related but distinct matter.) Physicians constitute the most important audience for pharmacists. Physicians are also the audience least apt to buy the pharmacist's presentation of professionalism. Pharmacists have little hope of developing a mystique in the perceptions of physicians, nor much cruciality. Physicians simply know all about the pharmacist's work, what he does and how he does it. There is no room for mystery in the face of such familiarity ("pill rollers"). And so pharmacists, as a group, suffer frustrated aspirations to professional status.

Engineering suffers similarly and for much the same reasons. The proverbial man in the street may see a graduate engineer as a person clearly in possession of a mystery; but since the general public rarely has any direct need for the services of an engineer, the field lacks cruciality. In the perspective of the man in the street, it falls in the high-mystique-but-low-cruciality group along with astrologers and diamond cutters.

Such a characterization offers little solace to engineers, but it is also of little importance. Since engineers, indeed, seldom have a clientele from among the general public, the perceptions of the man in the street are of small consequence. Engineers overwhelmingly work in large, formal organizations as employees. It is their *employers* that are their vital reference group for the presentations of professionalism. Unfortunately, it is the employer audience that is least inclined to buy such presentations. The rub is that, predictably, the managers and supervisors of employed engineers are engineers themselves. So the work of other engineers holds no mystery for them. That is not to say they do not recognize nor admire the skill and competence of other engineers. Indeed they do. But that is a quite different thing from mystique. They are not awed or dumbfounded by the mysteries of engineering. That is a crucial difference.

Indeed, it is the very lack of mystique that forecloses any possibility of engineering becoming a genuine profession in the eyes of employers. Lacking mystique, engineers lack a basis for effectively intimidating organizational others and getting away with it. As a result, engineers work mightily at professionalization while agonizing greatly about their inadequate status, not fully grasping what it is they lack and cannot attain.

THE PROBLEM OF COMMITMENT

To repeat some of what has been said elsewhere,[15] engineers face a further difficulty in any professionalization efforts (as if a lack of perceived mystique were not enough): failure of commitment by engineers to *engineering.* An early study of Trow[16] found that practicing engineers and engineering students see their field as a means rather than end. The end they were usually pursuing was a management position. Trow theorized that since engineering students tended to come from farming or industrial Blue Collar families, they choose their field with an eye toward advancing their family's social status in their generation. Engineering therefore serves upward social mobility. In a recent study, I found a similar pattern of social backgrounds among engineering undergraduates.[17] To the extent that these studies are valid indicators, engineering fields will never be able to summon the commitment from their members needed to gain control over their work worlds, either through professionalization or through union action.

In Conclusion

In terms of mystique and cruciality and their latent weapon, intimidation, engineering fails the claims of a genuine profession. At most, it is a pretender to profession. *This renders the alleged conflict between professionalism and union membership a false issue.*

Without a basis for effective intimidation, engineers only entertain wishful delusions by asserting professionalism as a rationalization against unions. This does not make the delusions less strong. Thomas observed that "If men define situations as real, they are real in their consequences."[18] That engineering, in objective fact, is not a true profession, and never will be, does not diminish the attitudinal obstacles faced by union organizers among engineers. Nor does it prevent professionalism from being a hindrance to engineers in finding means to gain some measure of control over their work realms that accounts for the reality of their bureaucratic employment.

Whether trade-union membership represents the only such means or the best means is open to debate, even setting aside the distraction of professionalism. Unions are essentially political organizations. Union leadership at the international level, and at the local level, operates under the pressure of dealing with the question, "What have you done for us lately?" In response, at the next bargaining session they aim to "one up" the gains of the last round. The question also makes internationals competitive among themselves. What the UAW just won the Steelworkers must appear to exceed. The unions thus force up wages in absolute dollar terms; but this same process also progressively erodes relative income differences.

As discussed earlier, shrinking relative income differences undermine engineers' satisfaction with their employment situation (a force, ironically, leaving them increasingly amenable to the idea of union membership in the first place).

What organizational means will aid engineers best in gaining control over their work realms, curbing capricious behavior of management, and securing economic benefits that will widen, rather than compress, relative income differences? Professor Kleingartner, one of the closest students of the employment conditions of engineers, has concluded that engineers appear to require an organizational form unique to their situation.[5] In his view, neither professional associations nor industrial unions presently fits those needs.

References

1. Dvorak, Eldon J., "Will Engineers Unionize?," in *Professionalism*, edited by Howard Vollmer and Donald Mills, Prentice-Hall, Englewood Cliffs, 1966.

2. Blum, Albert A., *Management and the White Collar Union*, American Management Association, New York, 1964.

3. Chamot, Dennis, "Scientists and Unions: The New Reality," in the September 1974 *AFL-CIO American Federationist*.

4. Blank, David M., and Stigler, George J., *The Demand and Supply of Scientific Personnel*, National Bureau of Economic Research, New York, 1957.

5. Kleingartner, Archie, *Professionalism and Salaried Worker Organization*, Industrial Relations Research Inst., Univ. of Wisconsin, Madison, 1967.

6. Ibid.

7. Walton, Richard E., *The Impact of the Professional Engineering Union*, Harvard Univ. Graduate School of Business Administration, Boston, 1961.

8. Flexner, Abraham, "Is Social Work a Profession?," proceedings of the National Conference of Charities and Correction, Hildman Printing, Chicago, 1915; Carr-Saunders, A. M., and Wilson, P. A., "Professions," *Encyclopedia of the Social Sciences*, Edwin Seligman (ed.), Macmillan, New York, 1930; Greenwood, Ernest, "The Elements of Professionalism," *Professionalism* (see Ref. 1); Glaser, William, "Nursing Leadership and Policy," *The Nursing Profession*, edited by Fred Davis, John Wiley & Sons, New York, 1966; and Vollmer, Howard M., and Mills, Donald L., "Professionalization and Technological Change," (see Ref. 1).

9. Hughes, Everett C., "Professions," *The Professions in America*, edited by Kenneth Lynn, Beacon Press, Boston, 1963.

10. Friedson, Eliot, "The Impurity of Professional Authority," *Institutions and the Person*, edited by Howard S. Becker, Blanche Geer, David Riesman, and Robert S. Weiss, Aldine, Chicago, 1968.

11. Barber, Bernard, "Some Problems in the Sociology of Professions," (see Ref. 12).

12. Friedrich, Carl J., "Authority, Reason, and Discretion," *Authority*, edited by Carl J. Friedrich, Harvard Univ. Press, Cambridge, 1958.

13. Becker, Howard S., "The Teacher in the Authority System of the Public School," *Complex Organizations, A Sociological Reader*, edited by Amatai Etzioni, Holt, Rinehart, and Winston, New York, 1961.

14. Denzin, Norman K., and Mettlin, Curtis J., "Incomplete Professionalization: The Case of Pharmacy," *Social Forces 46*, March 1968, pp. 375–381.

15. Haga, William J., "Haga on the Secret Life of the Engineer," November 1974 *Astronautics & Aeronautics*, p. 79.

16. Trow, Martin, "Some Implications of the Social Origins of Engineers," *Science Manpower 58*, NSF–59–37, Washington, 1959.

17. Haga, William J., "Who Goes to Engineering School? Who Graduates?," *The Regional Role of Engineering Colleges*, edited by Bugliarello et al. Pergamon Press, New York, 1974.

18. Thomas, William I.: Quoted repeatedly without specification of the source in Robert Merton's *Social Theory and Social Structure, Revised Edition*, The Free Press, New York, 1957.

Selected Additional Readings

Carter, J. T., Letter to the Editor, *New Engineer*, September 1976, pp. 7–8.

Computer Practices Committee of the Technical Council on Computer Practices, "Ethical Considerations in Computer Use," *Issues in Engineering*, American Society of Civil Engineers, December 1979, pp. 415–427.

Goodkind, D. R., "The Impact of Bidding on Client Engineer Relationships," *Consulting Engineer*, April 1979, pp. 90–91.

Horlick, R. J., "Trade Unions and the Professional Engineer," *The Chemical Engineer*, April 1976, pp. 253–254.

McLaughlin, R., "The Oil Industry and the Engineers," *New Engineer*, June 1976, pp. 21–33.

McLaughlin, R., "Unprofessional or Unfairly Fired," *New Engineer*, September 1978, pp. 22–30.

Millman, P., and Ostrower, D. A., "The Engineer as Judge: Conflict of Interest?" *Consulting Engineer*, May 1979, pp. 50, 53–55.

Parker, D., *Ethical Conflicts in Computer Science and Technology*, AFIPS Press, Arlington, Va., 1977.

4

The Ethical Dilemma

*Never has humanity known so much power and so much con-
fusion, so much worry and so much play, so much knowledge and
so much uncertainty.*

Attributed to Paul Valery,
by Andre Cournand, *Science,*
Nov. 1977

In "The World of Epictetus" Vice Admiral Stockdale says, "the linkage of men's ethics, reputations and fates can be studied in . . . vivid detail in prison camp." His article is the report of a personal study of this linkage—a study made in a North Vietnamese prison camp. Stockdale did not write the article because he thought it likely that his readers would one day find themselves prisoners of war. Stockdale believes that equally real, but less clear-cut, ethical dilemmas face every individual in ordinary twentieth-century civil life. He analyzes the "extraordinary" ethical dilemmas that face a prisoner of war and reflects upon what a person needs in order to face this challenge and deal more effectively with the "ordinary" ethical dilemmas of life.

A dilemma is a choice between A and B when one would prefer to have both. The person who faces an ethical dilemma has a real problem requiring immediate action.

Descriptively the concept, "ethics," is very easily defined. Ethics deals with our relations with other people and the things and behavior we value and do not value, should and should not value, in our relations with others. The concepts of ethics are duties, rights, responsibilities, obligations, etc.—that is, standards of behavior. Most people agree with this definition in their actions, if not in their words. That is, most people, even those who say there is no such thing as ethics or that there is no point in discussing ethics, habitually make judgments about their own behavior and that of other people, judging that behavior by some set of standards. Moreover, the individual who simply judges, who simply says "that behavior is wrong," is very rare. More often people say, "that behavior is wrong, *because*—". What this means is that beyond agreeing on the definition of ethics, most people, no matter what they say, agree in their actions that ethics is an area that can be discussed rationally—an area where it is appropriate to give reasons.

What makes ethics difficult to discuss and makes many people despair of any fruitful discussion is that people apparently disagree on what exactly constitutes an acceptable reason and on what specific duties, rights, responsibilities, and obligations people have. Part of this seems to be that circumstances play a role here, and circumstances vary from individual to individual, from time to time, and from place to place. Hans Jonas in "Technology and Responsibility" discusses how changes in circumstances bring about changes in ethical concepts and, in particular, how man has gained through technology the ability to alter both his circumstances and himself. Jonas argues that these alterations create new ethical dilemmas and the need for alterations in our ethics.

Beyond simple disagreement there is another related reason why people find ethical discussions so apparently fruitless. These disagreements, in many cases, appear to come down to people's holding different values. The involvement of values in ethical discussions and judgments is often taken to mean that these disagreements are subjective and therefore not susceptible to objective, rational resolution. Science and technology for a time appeared to offer a way out. It was thought that science took us out of the realm of values and away from these disagreements and the need to discuss them and our behavior. Robert Baum in "A Philosophical/Historical Perspective" argues against this view. He demonstrates that science is not "value-neutral" and suggests that the principal lesson of the twentieth century may be that there is not much difference between scientific systems and value systems.

A third reason for the difficulties encountered in ethical discussions is that ethics deals with our lives in their interrelationships and in a scientific-technological society our lives have grown very complicated and interrelated. Not only do ethical problems become more complicated, they also become harder to see—a case of failing to see the tree for the forest—and easier to dismiss—"mistaking" a problem to which we do not have a ready answer for one to which we do. It is only when our lives suddenly become very simple, for example, in a prison camp, that we clearly see the importance of ethics in clear but hard choices. K. R. Pavlovic in "Autonomy and Obligation" describes how ethics permeates our every action and how we create obligations by pursuing our freedom. In this respect one should recall and consider Florman's argument, in "Moral Blueprints," that it is not so much a lack of knowledge of ethical principles that makes ethics difficult as it is a lack of sophistication in dealing with the complex interrelations of contemporary life. With these reflections we have returned to Stockdale's point. In fact, all these reflections suggest two things: (1) ethics is important because it is intimately bound up with our reputations and fates and (2) ethical discussion, in order to be meaningful and fruitful, must be based on the consideration of real individuals and real ethical dilemmas. What the precise formal definition of "ethics" should be, how the real or apparent differences between various systems of ethics can be resolved, and even whether they should be resolved are all questions that are secondary to the primary question, "What should I do in this situation?"

The three articles, "I Gave Up Ethics—To Eat," "The Story behind the Recent National Scandals Involving Engineers," and "Three Leave G.E." and the two articles dealing with ethics within a context of environmental concerns are focused on ethical dilemmas engineers are likely to encounter. The first two articles are primarily of interest for what the engineers involved did not consider and the ways they perceived their problem. Would they have chosen a different course of action had they considered their dilemmas in a wider context? It does not appear to have even occurred to them that they had the option of moving to a different area, consulting with their professional organizations, or contacting law enforcement agencies. Did they display a systematic avoidance of confrontation in their actions? The third article brings to light very clearly the potential for conflict between personal values, corporate values, and social values and illustrates well a variety of individual and institutional responses to such conflict. The two articles by Hughson and Popper raise some of the issues facing the engineer caught between the imperatives of business and the quality of life provided by the environment.

The truth that has become manifest in the twentieth century is that no one acts in isolation from others. What we should do in the face of this truth, however, will remain a mystery until thought is given to the myriad ways in which we are all interrelated and in which our actions affect others. This is particularly true of engineers and their work in a society that is so greatly dependent upon technology.

4.1

The World of Epictetus

VICE ADMIRAL JAMES BOND STOCKDALE, USN (ret.)

In 1965 I was a forty-one-year-old commander, the senior pilot of Air Wing 16, flying combat missions in the area just south of Hanoi from the aircraft carrier *Oriskany*. By September of that year I had grown quite accustomed to briefing dozens of pilots and leading them on daily air strikes; I had flown nearly 200 missions myself and knew the countryside of North Vietnam like the back of my hand. On the ninth of that month I led about thirty-five airplanes to the Thanh Hoa Bridge, just west of that city. That bridge was tough; we had been bouncing 500-pounders off it for weeks.

The September 9 raid held special meaning for *Oriskany* pilots because of a special bomb load we had improvised; we were going in with our biggest, the 2000-pounders, hung not only on our attack planes but on our F-8 fighter-bombers as well. This increase in bridge-busting capability came from the innovative brain of a major flying with my Marine fighter squadron. He had figured out how we could jury-rig some switches, hang the big bombs, pump out some of the fuel to stay within takeoff weight limits, and then top off our tanks from our airborne refuelers while en route to the target. Although the pilot had to throw several switches in sequence to get rid of his bombs, a procedure requiring above-average cockpit agility, we routinely operated on the premise that all pilots of Air Wing 16 were above average. I test-flew the new load on a mission, thought it over, and approved it; that's the way we did business.

Our spirit was up. That morning, the *Oriskany* Air Wing was finally going to drop the bridge that was becoming a North Vietnamese symbol of resistance. You can imagine our dismay when we crossed the coast and the weather scout I had sent on ahead radioed back that ceiling and visibility were zero-zero in the bridge area. In the tiny cockpit of my A-4 at the front of the pack, I pushed the button on the throttle, spoke into the radio mike in my oxygen mask, and told the formation to split up and proceed in pairs to the secondary targets I had specified in my contingency briefing. What a letdown.

The adrenaline stopped flowing as my wingman and I broke left and down and started sauntering along toward our "milk run" target: boxcars on a railroad siding

188

between Vinh and Thanh Hoa, where the flak was light. Descending through 10,000 feet, I unsnapped my oxygen mask and let it dangle, giving my pinched face a rest—no reason to stay uncomfortable on this run.

As I glided toward that easy target, I'm sure I felt totally self-satisfied. I had the top combat job that a Navy commander can hold and I was in tune with my environment. I was confident—I knew airplanes and flying inside out. I was comfortable with the people I worked with and I knew the trade so well that I often improvised variations in accepted procedures and encouraged others to do so under my watchful eye. I was on top. I thought I had found every key to success and had no doubt that my Academy and test-pilot schooling had provided me with everything I needed in life.

I passed down the middle of those boxcars and smiled as I saw the results of my instinctive timing. A neat pattern—perfection. I was just pulling out of my dive low to the ground when I heard a noise I hadn't expected—the *boom boom boom* of a 57-millimeter gun—and then I saw it just behind my wingtip. I was hit—all the red lights came on, my control system was going out—and I could barely keep that plane from flying into the ground while I got that damned oxygen mask up to my mouth so I could tell my wingman that I was about to eject. What rotten luck. And on a "milk run"!

The descent in the chute was quiet except for occasional rifle shots from the streets below. My mind was clear, and I said to myself, "five years." I knew we were making a mess of the war in Southeast Asia, but I didn't think it would last longer than that; I was also naive about the resources I would need in order to survive a lengthy period of captivity.

The Durants have said that culture is a thin and fragile veneer that superimposes itself on mankind. For the first time I was on my own, without the veneer. I was to spend years searching through and refining my bag of memories, looking for useful tools, things of value. The values were there, but they were all mixed up with technology, bureaucracy, and expediency, and had to be brought up into the open.

Education should take care to illuminate values, not bury them amongst the trivia. Are our students getting the message that without personal integrity intellectual skills are worthless?

Integrity is one of those words which many people keep in that desk drawer labeled "too hard." It's not a topic for the dinner table or the cocktail party. You can't buy it or sell it. When supported with education, a person's integrity can give him something to rely on when his perspective seems to blur, when rules and principles seem to waver, and when he's faced with hard choices of right or wrong. It's something to keep him on the right track, something to keep him afloat when he's drowning; if only for practical reasons, it is an attribute that should be kept at the very top of a young person's consciousness.

The importance of the latter point is highlighted in prison camps, where everyday human nature, stripped bare, can be studied under a magnifying glass in accelerated time. Lessons spotlighted and absorbed in that laboratory sharpen one's eye for their abstruse but highly relevant applications in the "real time" world of now.

In the five years since I've been out of prison, I've participated several times in the process of selecting senior naval officers for promotion or important command assignments. I doubt that the experience is significantly different from that of executives who sit on "selection boards" in any large hierarchy. The system must be formal, objective, and fair; if you've seen one, you've probably seen them all. Navy selection board proceedings go something like this.

The first time you know the identity of the other members of the board is when you walk into a boardroom at eight o'clock on an appointed morning. The first order of business is to stand, raise your right hand, put your left hand on the Bible, and swear to make the best judgment you can, on the basis of merit, without prejudice. You're sworn to confidentiality regarding all board members' remarks during the proceedings. Board members are chosen for their experience and understanding; they often have knowledge of the particular individuals under consideration. They must feel free to speak their minds. They read and grade dozens of dossiers, and each candidate is discussed extensively. At voting time, a member casts his vote by selecting and pushing a "percent confidence" button, visible only to himself, on a console attached to his chair. When the last member pushes his button, a totalizer displays the numerical advantage "confidence" of the board. No one knows who voted what.

I'm always impressed by the fact that every effort is made to be fair to the candidate. Some are clearly out, some are clearly in; the borderline cases are the tough ones. You go over and over those in the "middle pile" and usually you vote and revote until late at night. In all the boards I've sat on, no inference or statement in a "jacket" is as sure to portend a low confidence score on the vote as evidence of a lack of directness or rectitude of a candidate in his dealings with others. Any hint of moral turpitude really turns people off. When the crunch comes, they prefer to work with forthright plodders rather than with devious geniuses. I don't believe that this preference is unique to the military. In any hierarchy where people's fates are decided by committees or boards, those who lose credibility with their peers and who cause their superiors to doubt their directness, honesty, or integrity are dead. Recovery isn't possible.

The linkage of men's ethics, reputations, and fates can be studied in even more vivid detail in prison camp. In that brutally controlled environment a perceptive enemy can get his hooks into the slightest chink in a man's ethical armor and accelerate his downfall. Given the right opening, the right moral weakness, a certain susceptibility on the part of the prisoner, a clever extortionist can drive his victim into a downhill slide that will ruin his image, self-respect, and life in a very short time.

There are some uncharted aspects to this, some traits of susceptibility which I don't think psychologists yet have words for. I am thinking of the tragedy that can befall a person who has such a need for love or attention that he will sell his soul for it. I use tragedy with the rigorous definition Aristotle applied to it: the story of a good man with a flaw who comes to an unjustified bad end. This is a rather delicate point and one that I want to emphasize. We had very very few collaborators in prison, and comparatively few Aristotelian tragedies, but the story and fate of one of these good men with a flaw might be instructive.

He was handsome, smart, articulate, and smooth. He was almost sincere. He was obsessed with success. When the going got tough, he decided expediency was preferable to principle.

This man was a classical opportunist. He befriended and worked for the enemy to the detriment of his fellow Americans. He made a tacit deal; moreover, he accepted favors (a violation of the code of conduct). In time, out of fear and shame, he withdrew; we could not get him to communicate with the American prisoner organization.

I couldn't learn what made the man tick. One of my best friends in prison, one of the wisest persons I have ever known, had once been in a squadron with this fellow. In prisoners' code I tapped a question to my philosophical friend: "What in the world is going on with that fink?"

"You're going to be surprised at what I have to say," he meticulously tapped back. "In a squadron he pushes himself forward and dominates the scene. He's a continual fountain of information. He's the person everybody relies on for inside dope. He works like mad; often flies more hops than others. It drives him crazy if he's not liked. He tends to grovel and ingratiate himself before others. I didn't realize he was really pathetic until I was sitting around with him and his wife one night when he was spinning his yarns of delusions of grandeur, telling of his great successes and his pending ascension to the top. His wife knew him better than anybody else; she shook her head with genuine sympathy and said to him: 'Gee, you're just a phony.'"

In prison, this man had somehow reached the point where he was willing to sell his soul just to satisfy this need, this immaturity. The only way he could get the attention that he demanded from authority was to grovel and ingratiate himself before the enemy. As a soldier he was a miserable failure, but he had not crossed the boundary of willful treason; he was not written off as an irrevocable loss, as were the two patent collaborators with whom the Vietnamese soon arranged that he live.

As we American POWs built our civilization, and wrote our own laws (which we leaders obliged all to memorize), we also codified certain principles which formed the backbone of our policies and attitudes. I codified the principles of compassion, rehabilitation, and forgiveness with the slogan: "It is neither American nor Christian to nag a repentant sinner to his grave." (Some didn't like it, thought it seemed soft on finks.) And so, we really gave this man a chance. Over time, our efforts worked. After five years of self-indulgence he got himself together and started to communicate with the prisoner organization. I sent the message "Are you on the team or not?"; he replied, "Yes," and came back. He told the Vietnamese that he didn't want to play their dirty games any more. He wanted to get away from those willful collaborators and he came back and he was accepted, after a fashion.

I wish that were the end of the story. Although he came back, joined us, and even became a leader of sorts, he never totally won himself back. No matter how forgiving we were, he was conscious that many resented him—not so much because he was weak but because he had broken what we might call a gentleman's code. In all of those years when he, a senior officer, had willingly participated in making tape recordings of anti-American material, he had deeply offended the sensibilities of the American prisoners who were forced to listen to him. To most of us it wasn't the rhetoric of the war or the goodness or the badness of this or that issue that counted. The object of our highest value was the well-being of our fellow prisoners. He had broken that code and hurt some of those people. Some thought that as an informer he had indirectly hurt them physically. I don't believe that. What indisputably hurt them was his not having the sensitivity to realize the damage his opportunistic conduct would do to the morale of a bunch of Middle American guys with Middle American attitudes which they naturally cherished. He should have known that in those solitary cells where his tapes were piped were idealistic, direct, patriotic fellows who would be crushed and embarrassed to have him, a senior man in excellent physical shape, so obviously not under torture, telling the world that the war was wrong. Even if he believed what he said, which he did not, he should have had the common decency to keep his mouth shut. You can sit and think anything you want, but when you insensitively cut down those who want to love and help you, you cross a line. He seemed to sense that he could never truly be one of us.

And yet he was likable—particularly back in civilization after release—when tension was off, and making a deal did not seem so important. He exuded charm and "hail fellow" sophistication. He wanted so to be liked by all those men he had once discarded in his search for new friends, new deals, new fields to conquer in Hanoi. The tragedy of his life was obvious to us all. Tears were shed by some of his old prison mates when he was killed in an accident that strongly resembled suicide some months later. The Greek drama had run its course. He was right out of Aristotle's book, a good man with a flaw who had come to an unjustified bad end. The flaw was insecurity: the need to ingratiate himself, the need for love and adulation at any price.

He reminded me of Paul Newman in *The Hustler*. Newman couldn't stand success. He knew how to make a deal. He was handsome, he was smart, he was attractive to everybody; but he had to have adulation, and therein lay the seed of tragedy. Playing high-stakes pool against old Minnesota Fats (Jackie Gleason), Newman was well in the lead, and getting more full of himself by the hour. George C. Scott, the pool bettor, whispered to his partner: "I'm going to keep betting on Minnesota Fats; this other guy [Newman] is a born loser—he's all skill and no character." And he was right, a born loser—I think that's the message.

How can we educate to avoid these casualties? How can we by means of education prevent this kind of tragedy? What we prisoners were in was a one-way leverage game in which the other side had all the mechanical advantage. I suppose you could say that we all live in a leverage world to some degree; we all experience people trying to use us in one way or another. The difference in Hanoi was the degradation of the ends (to be used as propaganda agents of an enemy, or as informers on your fellow Americans), and the power of the means (total environmental control including solitary confinement, restraint by means of leg-irons and handcuffs, and torture). Extortionists always go down the same track: the imposition of guilt and fear for having disobeyed their rules, followed in turn by punishment, apology, confession, and atonement (their payoff). Our captors would go to great lengths to get a man to compromise his own code, even if only slightly, and then they would hold that in their bag, and the next time get him to go a little further.

Some people are psychologically, if not physically, at home in extortion environments. They are tough people who instinctively avoid getting sucked into the undertows. They never kid themselves or their friends; if they miss the mark they admit it. But there's another category of person who gets tripped up. He makes a small compromise, perhaps rationalizes it, and then makes another one; and then he gets depressed, full of shame, lonesome, loses his willpower and self-respect, and comes to a tragic end. Somewhere along the line he realizes that he has turned a corner that he didn't mean to turn. All too late he realizes that he has been worshiping the wrong gods and discovers the wisdom of the ages: life is not fair.

In sorting out the story after our release, we found that most of us had come to combat constant mental and physical pressure in much the same way. We discovered that when a person is alone in a cell and sees the door open only once or twice a day for a bowl of soup, he realizes after a period of weeks in isolation and darkness that he has to build some sort of ritual into his life if he wants to avoid becoming an animal. Ritual fills a need in a hard life and it's easy to see how formal church ritual grew. For almost all of us, this ritual was built around prayer, exercise, and clandestine communication. The prayers I said during those days were prayers of quality with ideas of substance. We found that over the course of time our minds had a tremendous capacity for invention and introspection, but had the weakness of being an integral part of our bodies. I remembered Descartes and how in his philosophy he separated mind and body. One time I cursed my body for the way it decayed my mind. I had decided that I would become a Gandhi. I would have to be carried around on a pallet and in that state I could not be used by my captors for propaganda purposes. After about ten days of fasting, I found that I had become so depressed that soon I would risk going into interrogation ready to spill my guts just looking for a friend. I tapped to the guy next door and I said, "Gosh, how I wish Descartes could have been right, but he's wrong." He was a little slow to reply; I reviewed Descartes's deduction with him and explained how I had discovered that body and mind are inseparable.

On the positive side, I discovered the tremendous file-cabinet volume of the human mind. You can memorize an incredible amount of material and you can draw the past out of your memory with remarkable recall by easing slowly toward the event you seek and not crowding the mind too closely. You'll try to remember who was at your birthday party when you were five years old, and you can get it, but only after months of effort. You can break the locks and find the answers, but you need time and solitude to learn how to use this marvelous device in your head which is the greatest computer on earth.

Of course many of the things we recalled from the past were utterly useless as sources of strength or practicality. For instance, events brought back from cocktail parties or insincere social contacts were almost repugnant because of their emptiness, their utter lack of value. More often than not, the locks worth picking had been on old schoolroom doors. School days can be thought of as a time when one is filling the important stacks of one's memory library. For me, the golden doors were labeled history and the classics. The historical perspective which enabled a man to take himself away from all the agitation, not necessarily to see a rosy lining, but to see the real nature of the situation he faced, was truly a thing of value.

Here's how this historical perspective helped me see the reality of my own situation and thus cope better with it. I learned from a Vietnamese prisoner that the same cells we occupied had in years before been lived in by many of the leaders of the Hanoi government. From my history lessons I recalled that when metropolitan France permitted communists in the government in 1936, the communists who occupied cells in Vietnam were set free. I marveled at the cycle of history, all within my memory, which prompted Hitler's rise in Germany, then led to the rise of the Popular Front in France, and finally vacated this cell of mine halfway around the world ("Perhaps Pham Van Dong lived here"). I came to understand what tough people these were. I was willing to fight them to the death, but I grew to realize that hatred was an indulgence, a very inefficent emotion. I remember thinking, "If you were committed to beating the dealer in a gambling casino, would *hating* him help your game?" In a pidgin English propaganda book the guard gave me, speeches by these old communists about their prison experiences stressed how they learned to beat down the enemy by being united. It seemed comforting to know that we were united against the communist administration of Hoa Lo prison just as the Vietnamese communists had united against the French administration of Hoa Lo in the thirties. Prisoners are prisoners, and there's only one way to beat administrations. We resolved to do it better in the sixties than they had in the thirties. You don't base system-beating on any thought of political idealism; you do it as a competitive thing, as an expression of self-respect.

Education in the classics teaches you that all organizations since the beginning of time have used the power of guilt; that cycles are repetitive; and that this is the way of the world. It's a naive person who comes in and says, "Let's see, what's good and what's bad?" That's a quagmire. You can get out of that quagmire only by recalling how wise men before you accommodated the same dilemmas. And I believe a good classical education and an understanding of history can best determine the rules you should live by. They also give you the power to analyze reasons for these rules and guide you as to how to apply them to your own situation. In a broader sense, all my education helped me. Naval Academy discipline and body contact sports helped me. But the education which I found myself using most was what I got in graduate school. The messages of history and philosophy I used were simple.

The first one is this business about life not being fair. That is a very important lesson and I learned it from a wonderful man named Philip Rhinelander. As a lieutenant commander in the Navy studying political science at Stanford University in 1961, I went over to philosophy corner one day and an older gentlemen said, "Can I help you?" I said, "Yes, I'd like to take some courses in philosophy." I told him I'd been in college for six years and had never had a course in philosophy. He couldn't believe it. I told him that I was a naval officer and he said, "Well, I used to be in the Navy. Sit down." Philip Rhinelander became a great influence in my life.

He had been a Harvard lawyer and had pleaded cases before the Supreme Court and then gone to war as a reserve officer. When he came back he took his doctorate at Harvard. He was also a music composer, had been director of general education at Harvard, dean of the School of Humanities and Sciences at Stanford, and by the time I met him had by choice returned to teaching in the classroom. He said, "The course I'm teaching is my personal two-term favorite—The Problems of Good and Evil—and we're starting our second term." He said the message of his course was from the Book of Job. The number one problem in this world is that people are not able to accommodate the lesson in the book.

He recounted the story of Job. It starts out by establishing that Job was the most honorable of men. Then he lost all his goods. He also lost his reputation, which is what really hurt. His wife was badgering him to admit his sins, but he knew he had made no errors. He was not a patient man and demanded to speak to the Lord. When the Lord appeared in the whirlwind, he said, "Now, Job, you have to shape up! Life is not fair." That's my interpretation and that's the way the book ended for hundreds of years. I agree with those of the opinion that the happy ending was spliced on many years later. If you read it, you'll note that the meter changes. People couldn't live with the original message. Here was a good man who came to unexplained grief, and the Lord told him: "That's the way it is. Don't challenge me. This is my world and you either live in it as I designed it or get out."

This was a great comfort to me in prison. It answered the question, "Why me?" It cast aside any thoughts of being punished for past actions. Sometimes I shared the message with fellow prisoners as I tapped through the walls to them, but I learned to be selective. It's a strong message which upsets some people.

Rhinelander also passed on to me another piece of classical information which I found of great value. On the day of our last session together he said, "You're a military man, let me give you a book to remember me by. It's a book of military ethics." He handed it to me, and I bade him goodbye with great emotion. I took the book home and that night started to read it. It was the *Enchiridion* of the philosopher Epictetus, his "manual" for the Roman field soldier.

As I began to read, I thought to myself in disbelief, "Does Rhinelander think I'm going to draw lessons for my life from this thing? I'm a fighter pilot. I'm a technical man. I'm a test pilot. I know how to get people to do technical work. I play golf; I drink martinis. I know how to get ahead in my profession. And what does he hand me? A book that says in part, 'It's better to die in hunger, exempt from guilt and fear, than to live in affluence and with perturbation.'" I remembered this later in prison because perturbation was what I was living with. When I ejected from the airplane on that September morn in 1965, I had left the land of technology. I had entered the world of Epictetus, and it's a world that few of us, whether we know it or not, are ever far away from.

In Palo Alto, I had read this book, not with contentment, but with annoyance. Statement after statement: "Men are disturbed not by things, but by the view that they take of them." "Do not be concerned with things which are beyond your power." "Demand not that events should happen as you wish, but wish them to happen as they do happen and you will go on well." This is stoicism. It's not the last word, but it's a viewpoint that comes in handy in many circumstances, and it surely did for me. Particularly this line: "Lameness is an impediment to the body but not to the will." That was significant for me because I wasn't able to stand up and support myself on my badly broken leg for the first couple of years I was in solitary confinement.

Other statements of Epictetus took on added meaning in the light of extortions which often began with our captors' callous pleas: "If you are just reasonable with us we will compensate you. You get your meals, you get to sleep, you won't be pestered, you might even get a cellmate." The catch was that by being "reasonable with us" our enemies meant being their informers, their propagandists. The old stoic had said, "If I can get the things I need with the preservation of my honor and fidelity and self-respect, show me the way and I will get them. But, if you require me to lose my own proper good, that you may gain what is no good, consider how unreasonable and foolish you are." To love our fellow prisoners was within our power. To betray, to propagandize, to disillusion conscientious and patriotic shipmates and destroy their morale so that they in turn would be destroyed was to lose one's proper good.

What attributes serve you well in the extortion environment? We learned there, above all else, that the best defense is to keep your conscience clean. When we did something we were ashamed of, and our captors realized we were ashamed of it, we were in trouble. A little white lie is where extortion and ultimately blackmail start. In 1965, I was crippled and I was alone. I realized that they had all the power. I couldn't see how I was ever going to get out with my honor and self-respect. The one thing I came to realize was that if you don't lose integrity you can't be had and you can't be hurt. Compromises multiply and build up when you're working against a skilled extortionist or a good manipulator. You can't be had if you don't take that first shortcut, or "meet them halfway," as they say, or look for that tacit "deal," or make that first compromise.

Bob North, a political science professor at Stanford, taught me a course called Comparative Marxist Thought. This was not an anticommunist course. It was the study of dogma and thought patterns. We read no criticism of Marxism, only primary sources. All year we read the works of Marx and Lenin. In Hanoi, I understood more about Marxist theory than my interrogator did. I was able to say to that interrogator, "That's not what Lenin said; you're a deviationist."

One of the things North talked about was brainwashing. A psychologist who studied the Korean prisoner situation, which somewhat paralleled ours, concluded that three categories of prisoners were involved there. The first was the redneck Marine sergeant from Tennessee who had an eighth-grade education. He would get in that interrogation room and they would say that the Spanish-American War was started by the bomb within the *Maine*, which might be true, and he would answer, "B.S." They would show him something about racial unrest in Detroit. "B.S." There was no way they could get to him; his mind was made up. He was a straight guy, red, white, and blue, and everything else was B.S.! He didn't give it a second thought. Not much of a historian, perhaps, but a good security risk.

In the next category were the sophisticates. They were the fellows who could be told these same things about the horrors of American history and our social problems, but had heard it all before, knew both sides of every story, and thought we were on the right track. They weren't ashamed that we had robber barons at a certain time of our history; they were aware of the skeletons in most civilizations' closets. They could not be emotionally involved and so they were good security risks.

The ones who were in trouble were the high school graduates who had enough sense to pick up the innuendo, and yet not enough education to accommodate it properly. Not many of them fell, but most of the men that got entangled started from that background.

The psychologist's point is possibly oversimplistic, but I think his message has some validity. A little knowledge is a dangerous thing.

Generally speaking, I think education is a tremendous defense; the broader, the better. After I was shot down my wife, Sybil, found a clipping glued in the front of my collegiate dictionary: "Education is an ornament in prosperity and a refuge in adversity." She certainly agrees with me on that. Most of us prisoners found that the so-called practical academic exercises in how to do things, which I'm told are proliferating, were useless. I'm not saying that we should base education on training people to be in prison, but I am saying that in stress situations, the fundamentals, the hardcore classical subjects, are what serve best.

Theatrics also helped sustain me. My mother had been a drama coach when I was young and I was in many of her plays. In prison I learned how to manufacture a personality and live it, crawl into it, and hold that role without deviation. During interrogations, I'd check the responses I got to different kinds of behavior. They'd get worried when I did things irrationally. And so, every so often, I would play that "irrational" role and come completely unglued. When I could tell that pressure to make a public exhibition of me was building, I'd stand up, tip the table over, attempt to throw the chair through the window, and say, "No way. Goddamit! I'm not doing that! Now, come over here and fight!" This was a risky ploy, because if they thought you were acting, they would slam you into the ropes and make you scream in pain like a baby. You could watch their faces and read their minds. They had expected me to behave like a stoic. But a man would be a fool to make their job easy by being conventional and predictable. I could feel the tide turn in my favor at that magic moment when their anger turned to pleading: "Calm down, now calm down." The payoff would come when they decided that the risk of my going haywire in front of some touring American professor on a "fact-finding" mission was too great. More important, they had reason to believe that I would tell the truth—namely, that I had been in solitary confinement for four years and tortured fifteen times—without fear of future consequences. So theatrical training proved helpful to me.

Can you educate for leadership? I think you can, but the communists would probably say no. One day in an argument with an interrogator, I said, "You are so proud of being a party member, what are the criteria?" He said in a flurry of anger, "There are only four: you have to be seventeen years old, you have to be selfless, you have to be smart enough to understand the theory, and you've got to be a person who innately influences others." He stressed that fourth one. I think psychologists would say that leadership is innate, and there is truth in that. But, I also think you can learn some leadership traits that naturally accrue from a good education: compassion is a necessity for leaders, as are spontaneity, bravery, self-discipline, honesty, and above all, integrity.

I remember being disappointed about a month after I was back when one of my young friends, a prison mate, came running up after a reunion at the Naval

Academy. He said with glee, "This is really great, you won't believe how this country has advanced. They've practically done away with plebe year at the Academy, and they've got computers in the basement of Bancroft Hall." I thought, "My God, if there was anything that helped us get through those eight years, it was plebe year, and if anything screwed up that war, it was computers!"

4.2

Technology and Responsibility: Reflections on the New Tasks of Ethics

HANS JONAS

All previous ethics—whether in the form of issuing direct enjoinders to do or not to do certain things, or in the form of defining principles for such enjoinders, or in the form of establishing the ground of obligation for obeying such principles—had these interconnected tacit premises in common: that the human condition, determined by the nature of man and the nature of things, was given once for all; that the human good on that basis was readily determinable; and that the range of human action and therefore responsibility was narrowly circumscribed. It will be the burden of my argument to show that these premises no longer hold, and to reflect on the meaning of this fact for our moral condition. More specifically, it will be my contention that with certain developments of our powers the *nature of human action* has changed, and since ethics is concerned with action, it should follow that the changed nature of human action calls for a change in ethics as well: this not merely in the sense that new objects of action have added to the case material on which received rules of conduct are to be applied, but in the more radical sense that the qualitatively novel nature of certain of our actions has opened up a whole new dimension of ethical revelance for which there is no precedent in the standards and canons of traditional ethics.

I

The novel powers I have in mind are, of course, those of modern *technology*. My first point, accordingly, is to ask how this technology affects the nature of our acting, in what ways it makes acting under its dominion *different* from what it has been through the ages. Since throughout those ages man was never without technology, the question involves the human difference of *modern* from previous technology. Let us start with an ancient voice on man's powers and deeds which in an archetypal sense itself strikes, as it were, a technological note—the famous Chorus from Sophocles' *Antigone.*

Many the wonders but nothing more wondrous than man.
This thing crosses the sea in the winter's storm,
making his path through the roaring waves.
And she, the greatest of gods, the Earth—
deathless she is, and unwearied—he wears her away
as the ploughs go up and down from year to year
and his mules turn up the soil.

The tribes of the lighthearted birds he ensnares, and the races
of all the wild beasts and the salty brood of the sea,
with the twisted mesh of his nets, he leads captive, this clever man.
He controls with craft the beasts of the open air,
who roam the hills. The horse with his shaggy mane
he holds and harnesses, yoked about the neck,
and the strong bull of the mountain.

Speech and thought like the wind
and the feelings that make the town,
he has taught himself, and shelter against the cold,
refuge from rain. Ever resourceful is he.
He faces no future helpless. Only against death
shall he call for aid in vain. But from baffling maladies
has he contrived escape.

Clever beyond all dreams
the inventive craft that he has
which may drive him one time or another to well or ill.
When he honors the laws of the land and the gods' sworn right
high indeed is his city; but stateless the man
who dares to do what is shameful.

(lines 335–370)

This awestruck homage to man's powers tells of his violent and violating irruption into the cosmic order, the self-assertive invasion of nature's various domains by his restless cleverness; but also of his building—through the self-taught powers of speech and thought and social sentiment—the home for his very humanity, the artifact of the city. The raping of nature and the civilizing of himself go hand in hand. Both are in defiance of the elements, the one by venturing into them and overpowering their creatures, the other by securing an enclave against them in the shelter of the city and its laws. Man is the maker of his life *qua* human, bending circumstances to his will and needs, and except against death he is never helpless.

Yet there is a subdued and even anxious quality about this appraisal of the marvel that is man, and nobody can mistake it for immodest bragging. With all his boundless resourcefulness, man is still small by the measure of the elements: precisely this makes his sallies into them so daring and allows those elements to tolerate his forwardness. Making free with the denizens of land and sea and air, he yet leaves the encompassing nature of those elements unchanged, and their generative powers undiminished. Them he cannot harm by carving out his little dominion from theirs. They last, while his schemes have their shortlived way. Much as he harries Earth, the greatest of gods, year after year with his plough—she is ageless and unwearied; her enduring patience he must and can trust, and to her cycle he must conform. And just as ageless is the sea. With all his netting of the salty brood, the spawning ocean is inexhaustible. Nor is it hurt by the plying of ships, nor sullied by what is jettisoned into its deeps. And no matter how many illnesses he contrives to cure, mortality does not bow to his cunning.

All this holds because man's inroads into nature, as seen by himself, were essentially superficial, and powerless to upset its appointed balance. Nor is there a hint, in the *Antigone* chorus or anywhere else, that this is only a beginning and that greater things of artifice and power are yet to come—that man is embarked on an endless course of conquest. He had gone thus far in reducing necessity, had learned by his wits to wrest that much from it for the humanity of his life, and there he could stop. The room he had thus made was filled by the city of men—meant to enclose, and not to expand—and thereby a new balance was struck within the larger balance of the whole. All the well or ill to which man's inventive craft may drive him one time or another is inside the human enclave and does not touch the nature of things.

The immunity of the whole, untroubled in its depth by the importunities of man, that is, the essential immutability of Nature as the cosmic order, was indeed the backdrop to all of mortal man's enterprises, including his intrusions into that order itself. Man's life was played out between the abiding and the changing: the abiding was Nature, the changing his own works. The greatest of these works was the city, and on it he could confer some measure of abidingness by the laws he made for it and undertook to honor. But no long-range certainty pertained to this contrived abidingness. As a precarious artifact, it can lapse or go astray. Not even within its artificial space, with all the freedom it gives to man's determination of self, can the arbitrary ever supersede the basic terms of his being. The very inconstancy of human fortunes assures the constancy of the human condition. Chance and luck and folly, the great equalizers in human affairs, act like an entropy of sorts and make all definite designs in the long run revert to the perennial norm. Cities rise and fall, rules come and go, families prosper and decline; no change is there to stay, and in the end, with all the temporary deflections balancing each other out, the state of man is as it always was. So here too, in his very own artifact, man's control is small and his abiding nature prevails.

Still, in this citadel of his own making, clearly set off from the rest of things and entrusted to him, was the whole and sole domain of man's responsible action. Nature was not an object of human responsibility—she taking care of herself and, with some coaxing and worrying, also of man: not ethics, only cleverness applied to her. But in the city, where men deal with men, cleverness must be wedded to morality, for this is the soul of its being. In this intra-human frame dwells all traditional ethics and matches the nature of action delimited by this frame.

II

Let us extract from the preceding those characteristics of human action which are relevant for a comparison with the state of things today.

1. All dealing with the non-human world, i.e., the whole realm of *techne* (with the exception of medicine), was ethically neutral—in respect both of the object and the subject of such action: in respect of the object, because it impinged but little on the self-sustaining nature of things and thus raised no question of permanent injury to the integrity of its object, the natural order as a whole; and in respect of the agent subject it was ethically neutral because *techne* as an activity conceived itself as a determinate tribute to necessity and not as an indefinite, self-validating advance to mankind's major goal, claiming in its pursuit man's ultimate effort and concern. The real vocation of man lay elsewhere. In brief, action on non-human things did not constitute a sphere of authentic ethical significance.

2. Ethical significance belonged to the direct dealing of man with man, including the dealing with himself: all traditional ethics is *anthropocentric*.

3. For action in this domain, the entity "man" and his basic condition was considered constant in essence and not itself an object of reshaping *techne*.

4. The good and evil about which action had to care lay close to the act, either in the praxis itself or in its immediate reach, and were not a matter for remote planning. This proximity of ends pertained to time as well as space. The effective range of action was small, the time-span of foresight, goal-setting and accountability was short, control of circumstances limited. Proper conduct had its immediate criteria and almost immediate consummation. The long run of consequences beyond was left to chance, fate or providence. Ethics accordingly was of the here and now, of occasions as they arise between men, of the recurrent, typical situations of private and public life. The good man was he who met these contingencies with virtue and wisdom, cultivating these powers in himself, and for the rest resigning himself to the unknown.

All enjoinders and maxims of traditional ethics, materially different as they may be, show this confinement to the immediate setting of the action. "Love thy neighbor as thyself"; "Do unto others as you would wish them to do unto you"; "Instruct your child in the way of truth"; "Strive for excellence by developing and actualizing the best potentialities of your being *qua* man"; "Subordinate your individual good to the common good"; "Never treat your fellow man as a means only but always *also* as an end in himself"—and so on. Note that in all these maxims the agent and the "other" of his action are sharers of a common present. It is those alive now and in some commerce with me that have a claim on my conduct as it affects them by deed or omission. The ethical universe is composed of contemporaries, and its horizon to the future is confined by the foreseeable span of their lives. Similarly confined is its horizon of place, within which the agent and the other meet as neighbor, friend or foe, as superior and subordinate, weaker and stronger, and in all the other roles in which humans interact with one another. To this proximate range of action all morality was geared.

III

It follows that the *knowledge* that is required—besides the moral will—to assure the morality of action, fitted these limited terms: it was not the knowledge of the scientist or the expert, but knowledge of a kind readily available to all men of good will. Kant went so far as to say that "human reason can, in matters of morality, be easily brought to a high degree of accuracy and completeness even in the most ordinary intelligence";[1] that "there is no need of science or philosophy for knowing what man has to do in order to be honest and good, and indeed to be wise and virtuous. . . . [Ordinary Intelligence] can have as good hope of hitting the mark as any philosopher can promise himself";[2] and again: "I need no elaborate acuteness to find out what I have to do so that my willing be morally good. Inexperienced regarding the course of the world, unable to anticipate all the contingencies that happen in it," I can yet know how to act in accordance with the moral law.[3]

Not every thinker in ethics, it is true, went so far in discounting the cognitive side of moral action. But even when it received much greater emphasis, as in Aristotle, where the discernment of the situation and what is fitting for it makes considerable demands on experience and judgment, such knowledge has nothing to do with the science of things. It implies, of course, a general conception of the human good as such, a conception predicated on the presumed invariables of man's nature and condition, which may or may not find expression in a theory of its own. But its translation into practice requires a knowledge of the here and now, and this is entirely non-theoretical. This "knowledge" proper to virtue (of the "where, when, to whom, and how") stays with the immediate issue, in whose defined context the action *as the agent's own* takes its course and within which it terminates. The good or bad of the action is wholly decided within that short-term context. Its moral quality shines forth from it, visible to its witnesses. No one was held responsible for the unintended later effects of his well-intentioned, well-considered, and well-performed act. The short arm of human power did not call for a long arm of predictive knowledge; the shortness of the one is as little culpable as that of the other. Precisely because the human good, known in its generality, is the same for all time, its realization or violation takes place at each time, and its complete locus is always the present.

---

[1] Immanuel Kant, *Groundwork of the Metaphysic of Morals*, preface.
[2] *Op. cit.*, chapter 1.
[3] *Ibid.* (I have followed H. J. Paton's translation with some changes.)

IV

All this has decisively changed. Modern technology has introduced actions of such novel scale, objects, and consequences that the framework of former ethics can no longer contain them. The *Antigone* chorus on the *deinotes*, the wondrous power, of man would have to read differently now; and its admonition to the individual to honor the laws of the land would no longer be enough. To be sure, the old prescriptions of the "neighbor" ethics—of justice, charity, honesty, and so on—still hold in their intimate immediacy for the nearest, day by day sphere of human interaction. But this sphere is overshadowed by a growing realm of collective action where doer, deed, and effect are no longer the same as they were in the proximate sphere, and which by the enormity of its powers forces upon ethics a new dimension of responsibility never dreamt of before.

Take, for instance, as the first major change in the inherited picture, the critical *vulnerability* of nature to man's technological intervention—unsuspected before it began to show itself in damage already done. This discovery, whose shock led to the concept and nascent science of ecology, alters the very concept of ourselves as a causal agency in the larger scheme of things. It brings to light, through the effects, that the nature of human action has *de facto* changed, and that an object of an entirely new order—no less than the whole biosphere of the planet—has been added to what we must be responsible for because of our power over it. And of what surpassing importance an object, dwarfing all previous objects of active man! Nature as a human responsibility is surely a *novum* to be pondered in ethical theory. What kind of obligation is operative in it? Is it more than a utilitarian concern? Is it just prudence that bids us not to kill the goose that lays the golden eggs, or saw off the branch on which we sit? But the "we" that here sits and may fall into the abyss is all future mankind, and the survival of the species is more than a prudential duty of its present members. Insofar as it is the fate of *man*, as affected by the condition of nature, which makes us care about the preservation of nature, such care admittedly still retains the anthropocentric focus of all classical ethics. Even so, the difference is great. The containment of nearness and contemporaneity is gone, swept away by the spatial spread and time-span of the cause-effect trains which technological practice sets afoot, even when undertaken for proximate ends. Their irreversibility conjoined to their aggregate magnitude injects another novel factor into the moral equation. To this take their cumulative character: their effects add themselves to one another, and the situation for later acting and being becomes increasingly different from what it was for the initial agent. The cumulative self-propagation of the technological change of the world thus constantly overtakes the conditions of its contributing acts and moves through none but unprecedented situations, for which the lessons of experience are powerless. And not even content with changing its beginning to the point of unrecognizability, the cumulation as such may consume the basis of the whole series, the very condition of itself. All this would have to be co-intended in the will of the single action if this is to be a morally responsible one. Ignorance no longer provides it with an alibi.

Knowledge, under these circumstances, becomes a prime duty beyond anything claimed for it heretofore, and the knowledge must be commensurate with the causal scale of our action. The fact that it cannot really be thus commensurate, i.e., that the predictive knowledge falls behind the technical knowledge which nourishes our power to act, itself assumes ethical importance. Recognition of ignorance becomes the obverse of the duty to know and thus part of the ethics which must govern the ever more necessary self-policing of our out-sized might. No previous ethics had to consider the global condition of human life and the far-off future, even existence, of the race. Their now being an issue demands, in brief, a new conception of duties and rights, for which previous ethics and metaphysics provide not even the principles, let alone a ready doctrine.

And what if the new kind of human action would mean that more than the interest of man alone is to be considered—that our duty extends farther and the anthropocentric confinement of former ethics no longer holds? It is at least not senseless anymore to ask whether the condition of extra-human nature, the biosphere as a whole and in its parts, now subject to our power, has become a human trust and has something of a moral claim on us not only for our ulterior sake but for its own and in its own right. If this were the case it would require quite some rethinking in basic priciples of ethics. It would mean to seek not only the human good, but also the good of things extra-human, that is, to extend the recognition of "ends in themselves" beyond the sphere of man and make the human good include the care for them. For such a role of stewardship no previous ethics has prepared us—and the dominant, scientific view of *Nature* even less. Indeed, the latter emphatically denies us all conceptual means to think of Nature as something to be honored, having reduced it to the indifference of necessity and accident, and divested it of any dignity of ends. But still, a silent plea for sparing its integrity seems to issue from the threatened plenitude of the living world. Should we heed this plea, should we grant its claim as sanctioned by the nature of things, or dismiss it as a mere sentiment on our part, which we may indulge as far as we wish and can afford to do? If the former, it would (if taken seriously in its theoretical implications) push the necessary rethinking beyond the doctrine of action, i.e., ethics, into the doctrine of being, i.e., metaphysics, in which all ethics must ultimately be grounded. On this speculative subject I will here say no more than that we should keep ourselves open to the thought that natural science may not tell the whole story about Nature.

V

Returning to strictly intra-human considerations, there is another ethical aspect to the growth of *techne* as a pursuit beyond the pragmatically limited terms of former times. Then, so we found, *techne* was a measured tribute to necessity, not the road to mankind's chosen goal—a means with a finite measure of adequacy to well-defined proximate ends. Now, *techne* in the form of modern technology has turned into an infinite forward-thrust of the race, its most significant enterprise, in whose permanent, self-transcending advance to ever greater things the vocation of man tends to be seen, and whose success of maximal control over things and himself appears as the consummation of his destiny. Thus the triumph of *homo faber* over his external object means also his triumph in the internal constitution of *homo sapiens*, of whom he used to be a subsidiary part. In other words, technology, apart from its objective works, assumes ethical significance by the central place it now occupies in human purpose. Its cumulative creation, the expanding artificial environment, continuously reinforces the particular powers in man that created it, by compelling their unceasing inventive employment in its management and further advance, and by rewarding them with additional success—which only adds to the relentless claim. This positive feedback of functional necessity and reward—in whose dynamics pride of achievement must not be forgotten—assures the growing ascendancy of one side of man's nature over all the others, and inevitably at their expense. If nothing succeeds like success, nothing also entraps like success. Outshining in prestige and starving in resources whatever else belongs to the fullness of man, the expansion of his power is accompanied by a contraction of his self-conception and being. In the image he entertains of himself—the potent self-formula which determines his actual being as much as it reflects it—man now is evermore the maker of what he has made and the doer of what he can do, and most of all the preparer of what he will be able to do next. But not you or I: it is the aggregate, not the individual doer or deed that matters here; and the indefinite future, rather than the contemporary context of the action, constitutes the relevant horizon of responsibility. This requires imperatives of a new sort. If the realm of making has invaded the space of essential action, then morality must invade the realm of making, from which it had formerly stayed aloof, and must do so in the form of public policy. With issues of such inclusiveness and such lengths of anticipation public policy has never had to deal before. In fact, the changed nature of human action changes the very nature of politics.

For the boundary between "city" and "nature" has been obliterated: the city of men, once an enclave in the non-human world, spreads over the whole of terrestrial nature and usurps its place. The difference between the artificial and the natural has vanished, the natural is swallowed up in the sphere of the artificial, and at the same time the total artifact, the works of man working on and through himself, generates a "nature" of its own, i.e., a necessity with which human freedom has to cope in an entirely new sense. Once it could be said *Fiat justitia, pereat mundus,* "Let justice be done, and may the world perish"—where "world," of course, meant the renewable enclave in the imperishable whole. Not even rhetorically can the like be said anymore when the perishing of the whole through the doings of man—be they just or unjust—has become a real possibility. Issues never legislated on come into the purview of the laws which the total city must give itself so that there will be a world for the generations of man to come.

That there *ought* to be through all future time such a world fit for human habitation, and that it ought in all future time to be inhabited by a mankind worthy of the human name, will be readily affirmed as a general axiom or a persuasive desirability of speculative imagination (as persuasive and as undemonstrable as the proposition that there being a world at all is "better" than there being none): but as a *moral* proposition, namely, a practical *obligation* toward the posterity of a distant future, and a principle of decision in present action, it is quite different from the imperatives of the previous ethics of contemporaneity; and it has entered the moral scene only with our novel powers and range of prescience.

The *presence of man in the world* had been a first and unquestionable given, from which all idea of obligation in human conduct started out. Now it has itself become an *object* of obligation—the obligation namely to ensure the very premise of all obligation, i.e., the *foothold* for a moral universe in the physical world—the existence of mere *candidates* for a moral order. The difference this makes for ethics may be illustrated in one example.

VI

Kant's categorical imperative said: "Act so that you *can* will that the maxim of your action be made the principle of a universal law." The "can" here invoked is that of reason and its consistency with itself: *Given* the existence of a community of human agents (acting rational beings), the action must be such that it can without self-contradiction be imagined as a general practice of that community. Mark that the basic reflection of morals here is not itself a moral but a logical one: The "I *can* will" or "I *cannot* will" expresses logical compatibility or incompatibility, not moral approbation or revulsion. But there is no self-contradiction in the thought that humanity would once come to an end, therefore also none in the thought that the happiness of present and proximate generations would be bought with the unhappiness or even non-existence of later ones—as little as, after all, in the inverse thought that the existence or happiness of later generations would be bought with the unhappiness or even partial extinction of present ones. The sacrifice of the future for the present is *logically* no more open to attack than the sacrifice of the present for the future. The difference is only that in the one case the series goes on, and in the other it does not. But that it *ought to go on*, regardless of the distribution of happiness or unhappiness, even with a persistent preponderance of unhappiness over happiness, nay, even of immorality over morality[4]—this cannot be derived from the rule of self-consistency *within* the series, long or short as it happens to be: it is a commandment of a very different kind, lying outside and "prior" to the series as a whole, and its ultimate grounding can only be metaphysical.

An imperative responding to the new type of human action and addressed to the new type of agency that operates it might run thus: "Act so that the effects of your action are compatible with the permanence of genuine human life"; or expressed negatively: "Act so that the effects of your action are not destructive to the future possibility of such life"; or simply: "Do not compromise the conditions for an indefinite continuation of humanity on earth"; or most generally: "In your present choices, include the future wholeness of Man among the objects of your will."

It is immediately obvious that no rational contradiction is involved in the violation of this kind of imperative. I *can* will the present good with sacrifice of the future good. It is also evident that the new imperative address itself to public policy rather than private conduct, which is not in the causal dimension to which that imperative applies. Kant's categorical imperative was addressed to the individual, and its criterion was instantaneous. It enjoined each of us to consider what would happen *if* the *maxim* of my present action were made, or at this moment already were, the principle of a universal legislation; the self-consistency or inconsistency of such a *hypothetical* universalization is made the test for my *private* choice. But it was no part of the reasoning that there is any probability of my private choice *in fact* becoming universal law, or that it might contribute to its

---

[4] On this last point, the biblical God changed his mind to an all-encompassing "yes" after the Flood.

becoming that. The universalization is a thought-experiment by the private agent to test the immanent morality of his action. Indeed, real consequences are not considered at all, and the principle is not one of objective responsibility but of the subjective quality of my self-determination. The new imperative invokes a different consistency: not that of the act with itself, but that of its eventual *effects* with the continuance of human agency in times to come. And the "universalization" it contemplates is by no means hypothetical—i.e., a purely logical transference from the individual "me" to an imaginary, causally unrelated "all" ("*if* everybody acted like that"); on the contrary, the actions subject to the new imperative—actions of the collective whole—have their universal reference in their actual scope of efficacy: they "totalize" themselves in the progress of their momentum and thus are bound to terminate in shaping the universal dispensation of things. This adds a *time* horizon to the moral calculus which is entirely absent from the instantaneous logical operation of the Kantian imperative: whereas the latter extrapolates into an everpresent order of abstract compatibility, our imperative extrapolates into a predictable real *future* as the open-ended dimension of our responsibility.

VII

Similar comparisons could be made with all the other historical forms of the ethics of contemporaneity and immediacy. The new order of human action requires a commensurate ethics of foresight and responsibility, which is as new as are the issues with which it has to deal. We have seen that these are the issues posed by the works of *homo faber* in the age of technology. But among those novel works we haven't mentioned yet the potentially most ominous class. We have considered *techne* only as applied to the non-human realm. But man himself has been added to the objects of technology. *Homo faber* is turning upon himself and gets ready to make over the maker of all the rest. This consummation of his power, which may well portend the overpowering of man, this final imposition of art on nature, calls upon the utter resources of ethical thought, which never before has been faced with elective alternatives to what were considered the definite terms of the human condition.

a.   Take, for instance, the most basic of these "givens," man's mortality. Who ever before had to make up his mind on its desirable and *eligible* measure? There was nothing to choose about the upper limit, the "threescore years and ten, or by reason of strength fourscore." Its inexorable rule was the subject of lament, submission, or vain (not to say foolish) wish-dreams about possible exceptions—strangely enough, almost never of affirmation. The intellectual imagination of a George Bernard Shaw and a Jonathan Swift speculated on the privilege of not having to die, or the curse of not being able to die. (Swift with the latter was the more perspicacious of the two.) Myth and legend toyed with such themes against the acknowledged background of the unalterable, which made the earnest man rather pray "teach us to number our days that we may get a heart of wisdom" (Psalm 90). Nothing of this was in the realm of doing and effective decision. The question was only how to relate to the stubborn fact.

But lately, the dark cloud of inevitability seems to lift. A practical hope is held out by certain advances in cell biology to prolong, perhaps indefinitely extend the span of life by counteracting biochemical processes of aging. Death no longer appears as a necessity belonging to the nature of life, but as an avoidable, at least in principle tractable and long-delayable, organic malfunction. A perennial yearning of mortal man seems to come nearer fulfillment. And for the first time we have in earnest to ask the question "How desirable is this? How desirable for the individual, and how for the species?" These questions involve the very meaning of our finitude, the attitude toward death, and the general biological significance of the balance of death and procreation. Even prior to such ultimate questions are the more pragmatic ones of who should be eligible for the boon: persons of particular quality and merit? of social eminence? those that can pay for it? everybody? The last would seem the only just course. But it would have to be paid for at the opposite end, at the source. For clearly, on a population-wide scale, the price of extended age must be a proportional slowing of replacement, i.e., a diminished access of new life. The result would be a decreasing proportion of youth in an increasingly aged population. How good or bad would that be for the general condition of man? Would the species gain or lose? And how *right* would it be to preempt the place of youth? Having to die is bound up with having been born: mortality is but the other side of the perennial spring of "natality" (to use Hannah Arendt's term). This had always been ordained; now its meaning has to be pondered in the sphere of decision.

To take the extreme (not that it will ever be obtained): if we abolish death, we must abolish procreation as well, for the latter is life's answer to the former, and so we would have a world of old age with no youth, and of known individuals with no surprises of such that had never been before. But this perhaps is precisely the wisdom in the harsh dispensation of our mortality: that it grants us the eternally renewed promise of the freshness, immediacy and eagerness of youth, together with the supply of otherness as such. There is no substitute for this in the greater accumulation of prolonged experience: it can never recapture the unique privilege of seeing the world for the first time and with new eyes, never relive the wonder which, according to Plato, is the beginning of philosophy, never the curiosity of the child, which rarely enough lives on as thirst for knowledge in the adult, until it wanes there too. This ever renewed beginning, which is only to be had at the price of ever repeated ending, may well be mankind's hope, its safeguard against lapsing into boredom and routine, its chance of retaining the spontaneity of life. Also, the role of the *memento mori* in the individual's life must be considered, and what its attenuation to indefiniteness may do to it. Perhaps a non-negotiable limit to our expected time is necessary for each of us as the incentive to number our days and make them count.

So it could be that what by intent is a philanthropic gift of science to man, the partial granting of his oldest wish—to escape the curse of mortality—turns out to be to the detriment of man. I am not indulging in prediction and, in spite of my noticeable bias, not even in valuation. My point is that already the promised gift raises questions that had never to be asked before in terms of practical choice, and that no principle of former ethics, which took the human constants for granted, is competent to deal with them. And yet they must be dealt with ethically and by principle and not merely by the pressure of interests.

b.   It is similar with all the other, quasi-utopian powers about to be made available by the advances of biomedical science as they are translated into technology. Of these, *behavior control* is much nearer to practical readiness than the still hypothetical prospect I have just been discussing, and the ethical questions it raises are less profound but have a more direct bearing on the moral conception of man. Here again, the new kind of intervention exceeds the old ethical categories. They have not equipped us to rule, for example, on mental control by chemical means or by direct electrical action on the brain via implanted electrodes—undertaken, let us assume, for defensible and even laudable ends. The mixture of beneficial and dangerous potentials is obvious, but the lines are not easy to draw. Relief of mental patients from distressing and disabling symptoms seems unequivocally beneficial. But from the relief of the *patient*, a goal entirely in the tradition of the medical art, there is an easy passage to the relief of *society* from the inconvenience of difficult individual behavior among its members: that is, the passage from medical to social application; and this opens up an indefinite field with grave potentials. The troublesome problems of rule and unruliness in modern mass society make the extension of such control methods to non-medical categories extremely tempting for social management. Numerous questions of human rights and dignity arise. The difficult question of preempting versus enabling care insists on concrete answers. Shall we induce learning attitudes in school children by the mass administration of drugs, circumventing the appeal to autonomous motivation? Shall we overcome aggression by electronic pacification of brain areas? Shall we generate sensations of happiness or pleasure or at least contentment through independent stimulation (or tranquilizing) of the appropriate centers—independent, that is, of the objects of happiness, pleasure, or content and their attainment in personal living and achieving? Candidacies could be multiplied. Business firms might become interested in some of these techniques for performance-increase among their employees.

Regardless of the question of compulsion or consent, and regardless also of the question of undesirable side-effects, each time we thus bypass the human way of dealing with human problems, short-circuiting it by an impersonal mechanism, we have taken away something from the dignity of personal selfhood and advanced a further step on the road from responsible subjects to programmed behavior systems. Social functionalism, important as it is, is only one side of the question. Decisive is the question of what kind of individuals the society is composed of—to make its existence valuable as a whole. Somewhere along the line of increasing social manageability at the price of individual autonomy, the question of the worthwhileness of the whole human enterprise must pose itself. Answering it involves the image of man we entertain. We must think it anew in light of the things we can do to it now and could never do before.

c.   This holds even more with respect to the last object of a technology applied on man himself—the genetic control of future men. This is too wide a subject for cursory treatment. Here I merely point to this most ambitious dream of *homo faber*, summed up in the phrase that man will take his own evolution in hand, with the aim of not just preserving the integrity of the species but of modifying it by improvements of his own design. Whether we have the right to do it, whether we are qualified for that creative role, is the most serious question that can be posed to man finding himself suddenly in possession of such fateful powers. Who will be the image-makers, by what standards, and on the basis of what knowledge? Also, the question of the moral right to experiment on future human beings must be asked. These and similar questions, which demand an answer before we embark on a journey into the unknown, show most vividly how far our powers to act are pushing us beyond the terms of all former ethics.

VIII

The ethically relevant common feature in all the examples adduced is what I like to call the inherently "utopian" drift of our actions under the conditions of modern technology, whether it works on non-human or on human nature, and whether the "utopia" at the end of the road be planned or unplanned. By the kind and size of its snowballing effects, technological power propels us into goals of a type that was formerly the preserve of Utopias. To put it differently, technological power has turned what used and ought to be tentative, perhaps enlightening plays of speculative reason into competing blueprints for projects, and in choosing between them we have to choose between extremes of remote effects. The one thing we can really know of them is their extremism as such—that they concern the total condition of nature on our globe and the very kind of creatures that shall, or shall not, populate it. In consequence of the inevitably "utopian" scale of modern technology, the salutary gap between everyday and ultimate issues, between occasions for common prudence and occasions for illuminated wisdom, is steadily closing. Living now constantly in the shadow of unwanted, built-in, automatic utopianism, we are constantly confronted with issues whose positive choice requires supreme wisdom—an impossible situation for man in general, because he does not possess that wisdom, and in particular for contemporary man, who denies the very existence of its object: viz., objective value and truth. We need wisdom most when we believe in it least.

If the new nature of our acting then calls for a new ethics of long-range responsibility, coextensive with the range of our power, it calls in the name of that very responsibility also for a new kind of humility—a humility not like former humility, i.e., owing to the littleness, but owing to the excessive magnitude of our power, which is the excess of our power to act over our power to foresee and our power to evaluate and to judge. In the face of the quasi-eschatological potentials of our technological processes, ignorance of the ultimate implications becomes itself a reason for responsible restraint—as the second best to the possession of wisdom itself.

One other aspect of the required new ethics of responsibility for and to a distant future is worth mentioning: the insufficiency of representative government to meet the new demands on its normal principles and by its normal mechanics. For according to these, only *present* interests make themselves heard and felt and enforce their consideration. It is to them that public agencies are accountable, and this is the way in which concretely the respecting of rights comes about (as distinct from their abstract acknowledgment). But the *future* is not represented, it is not a force that can throw its weight into the scales. The nonexistent has no lobby, and the unborn are powerless. Thus accountability to them has no political reality behind it yet in present decision-making, and when they can make their complaint, then we, the culprits, will no longer be there.

This raises to an ultimate pitch the old question of the power of the wise, or the force of ideas not allied to self-interest, in the body politic. What *force* shall represent the future in the present? However, before *this* question can become earnest in practical terms, the new ethics must find its theory, on which do's and don'ts can be based. That is: before the question of what *force*, comes the question of what *insight* or value-knowledge shall represent the future in the present.

IX

And here is where I get stuck, and where we all get stuck. For the very same movement which put us in possession of the powers that have now to be regulated by norms—the movement of modern knowledge called science—has by a necessary complementarity eroded the foundations from which norms could be derived; it has destroyed the very idea of norm as such. Not, fortunately, the feeling for norm and even for particular norms. But this feeling becomes uncertain of itself when contradicted by alleged knowledge or at least denied all sanction by it. Anyway and always does it have a difficult enough time against the loud clamors of greed and fear. Now it must in addition blush before the frown of superior knowledge, as unfounded and incapable of foundation. First, Nature had been "neutralized" with respect to value, then man himself. Now we shiver in the nakedness of a nihilism in which near-omnipotence is paired with near-emptiness, greatest capacity with knowing least what for. With the apocalyptic pregnancy of our actions, that very knowledge which we lack has become more urgently needed than at any other stage in the adventure of mankind. Alas, urgency is no promise of success. On the contrary, it must be avowed that to seek for wisdom today requires a good measure of unwisdom. The very nature of the age which cries out for an ethical theory makes it suspiciously look like a fool's errand. Yet we have no choice in the matter but to try.

It is a question whether without restoring the category of the sacred, the category most thoroughly destroyed by the scientific enlightment, we can have an ethics able to cope with the extreme powers which we possess today and constantly increase and are almost compelled to use. Regarding those consequences imminent enough to still hit ourselves, fear can do the job—so often the best substitute for genuine virtue and wisdom. But this means fails us towards the more distant prospects, which here matter the most, especially as the beginnings seem most innocent in their smallness. Only awe of the sacred with its unqualified veto is independent of the computations of mundane fear and the solace of uncertainty about distant consequences. But religion as a soul-determining force is no longer there to be summoned to the aid of ethics. The latter must stand on its own worldly feet—that is, on reason and its fitness for philosophy. And while of faith it can be said that it either is there or is not, of ethics it holds that it must be there.

It must be there because men act, and ethics is for the ordering of actions and for regulating the power to act. It must be there all the more, then, the greater the powers of acting that are to be regulated; and with their size, the ordering principle must also fit their kind. Thus, novel powers to act require novel ethical rules and perhaps even a new ethics.

"Thou shalt not kill" was enunciated because man has the power to kill and often the occasion and even inclination for it—in short, because killing is actually done. It is only under the *pressure* of real habits of action, and generally of the fact that always action already takes place, without *this* having to be commanded first, that ethics as the ruling of such acting under the standard of the good or the permitted enters the stage. Such a *pressure* emanates from the novel technological powers of man, whose exercise is given with their existence. *If* they really are as novel in kind as here contended, and if by the kind of their potential consequences they really have abolished the moral neutrality which the technical commerce with matter hitherto enjoyed—then their pressure bids to seek for new prescriptions in ethics which are competent to assume their guidance, but which first of all can hold their own theoretically against that very pressure. To the demonstration of those premises this paper was devoted. If they are accepted, then we who make thinking our business have a task to last us for our time. We must do it in time, for since we act anyway we shall have some ethic or other in any case, and without a supreme effort to determine the right one, we may be left with a wrong one by default.

A Philosophical/Historical Perspective on Contemporary Concerns and Trends in the Area of Science and Values

ROBERT BAUM

William Blanpied's "Subjective Impressions" in Newsletter 8 [*Science, Technology, & Human Values*, Harvard University] is certainly a much more important essay than its modest title suggests. It is, in my opinion, certainly a quite accurate analysis of the present state of affairs within the scientific/technological community. In so far as its primary intent was to initiate new discussions of the problems in this area, there is little that needs to be added and even less that needs qualification or change. The present essay is intended to complement, rather than supplement, Dr. Blanpied's essay by outlining the historical and philosophical context of the present-day situation, as perceived by a professional philosopher, rather than a scientist. In particular, an effort will be made to illuminate the factors underlying the recent surge of interest in the study of the inter-relations of science and values, and to indicate the directions in which this work should move to escape the problems of the past and present.

HISTORICAL PERSPECTIVE TO 1900

The history of ethics and value theory is essentially co-extensive with the history of culture. For the present purposes it is necessary to restrict the discussion to the history of the *intellectual* traditions in *Western* cultures. Even with these restrictions, the topic is so broad that entire volumes have been devoted to it. The reader who is interested in pursuing the subject in depth is encouraged to refer to any of several general studies of the field.[1]

A study of any of these histories will reveal that numerous diverse theories of ethics and values have been formulated, promulgated, advocated and debated in the 2500 years of the recorded history of the Western intellectual tradition. The significant (and perhaps, to some, surprising) feature of this near-plethora of theories of ethics and values is that with few exceptions (such as the Sophists and Skeptics) Western intellectuals have shared what must be considered a fundamental assumption of this tradition—the assumption that knowledge of ethical truths (about "the Right") and of values ("the Good") provides the paradigm of all true knowledge. The only other subject-matter which even approached the status of ethics in Western thought prior to the twentieth century is mathematics, and the challenges to its absoluteness were generally more numerous and penetrating than the attacks on the objectivity and absoluteness of ethics.[2]

As late as the end of the eighteenth century, Immanuel Kant took as the starting point of his major works (*The Critique of Pure Reason* and *The Critique of Practical Reason*) the question "Given that mathematics and morals are examples of universal truths which can be known with certainty, how is such knowledge possible?" And although the "hard-headed" nineteenth-century British Empiricist John Stuart Mill had reservations about attributing absolute certainty to any knowledge at all, including elementary arithmetic, he asserted that ethical knowledge was at least as objectively grounded as any other kind.

In contrast to the situation in mathematics (at least prior to the nineteenth century) the strong consensus of opinion in support of the objectivity of ethical knowledge did not reflect a consensus on a particular moral system. In point of fact, despite the generally shared cultural base of the Judeo-Greco-Christian tradition of most of the major Western intellectuals, *every* theoretically possible basic type of ethical theory and theory of value was formulated and advocated by at least one philosopher prior to the twentieth century. This point perhaps requires a bit of amplification.

Much is often heard today about the need for a "new" ethic, without much clarification as to what might be "new" about it. Philosophers have formulated a typology which allows for the categorization of every logically possible type of ethical theory. Thus, all normative ethical theories are such that they are either teleological or deontological (where "deontological" is essentially defined as "non-teleological"). In turn, all teleological theories are either act or rule theories; all act teleological theories are egoistic, utilitarian or altruistic, etc. Since theories have already been propounded which fit each of the categories at the third or fourth level (e.g., Mill was probably a normative teleological act utilitarian and Kant a normative deontological single rule formalist) any "new" theory could differ from the "traditional" ones only at an even more specific level of differentiation.[3]

It should also be noted that the history of the 2500 years prior to 1900 is a record of the struggle for respectability on the part of the physical sciences in the Western intellectual temples. Those who ranked the laws of the physical universe in the same lofty class as the laws of ethics and mathematics, did so generally in terms of their being knowable (or at least provable) independently of observation and sense experience. The ultimate prestige of the experimental sciences derived not so much from intellectual justifications as from the confirmation in practice of Francis Bacon's dictum that "Knowledge is Power."

THE TRIUMPH OF THE POSITIVIST ETHIC

The status of ethics in Western cultures was clearly displayed during the era of colonial exploitation of the rest of the world. There was no question in the minds of the Europeans that the natives of these countries were "uncivilized" and in need of education. There is no reason to question the sincerity of those who said farewell to family and friends to endure the hardships of the missionary life to bring the "truth" to these "ignorant" people. Even cultures with technologies that approached the European in sophistication were considered uncivilized—in part, at least because of their lack of a "true" ethic. It was only at the beginning of the twentieth century that anthropologists using "the scientific method" discovered the "fact" of ethical relativism—that the ethical systems of these other cultures are simply different from the ethical system(s) of the West.[4] They "discovered" that there was no way to provide an objective "scientific" proof of the superiority of one ethical system over another. All that the scientist could do was to provide a description of the various systems, this description itself being "objective" and "value-neutral." The failures of the missionaries to recognize their own biases and their arrogance in attempting to force their own subjective preferences on peoples whose ethical systems were now seen to be just as valid as those of the missionaries were soundly condemned by those who had achieved the scientific "enlightenment." Unfortunately, these scientific critics were placing themselves in a position where they would ultimately become subject to similar criticisms. But this is getting too far ahead in our story.

The general position associated with the "discovery" of the relativity of ethical systems is sometimes referred to as "scientism," but I shall refer to it in this discussion as "Scientistic Positivism" or simply "Positivism" with a capital "P." I am using this term because the key principles of this position are quite similar to the basic tenets of the philosophical school of Logical Positivism,[5] although of course they have never been as explicitly or elaborately formalized. And whereas Logical Positivism was a position shared by a relatively small group of philosophers over a relatively brief span of years. Scientistic Positivism is a widely held view which has now been passed on through several generations of practicing scientists. Whether the rise of Scientistic Positivism was a product of the "discovery" of ethical relativism or the "discovery" of ethical relativism the result of the rise of Positivism is a chicken-and-egg question which cannot be pursued here. The logical connection between Scientistic Positivism and ethical relativism can be seen in an analysis of four of the basic axioms of Positivism.

The entire *Weltanschauung* of Scientistic Positivism is perhaps contained in the four basic assumptions which are directly relevant to a consideration of ethical and human value questions. These four "axioms" are:

1. The physical sciences provide the paradigm of objective knowledge.
2. Science is value-neutral; that is, the scientist *qua* scientist makes no value judgements.
3. Ethical and other normative "judgements" are actually only expressions of emotions, attempts to evoke emotions in others, and/or commands.

4. Human behavior is entirely determined by antecedent events and states of
    affairs in such a way that they can theoretically be completely described or
    predicted by basic natural laws, as established in accordance with the
    scientific method.

Although there certainly is a growing segment of the scientific community that
would reject one or more of the above axioms, the Positivist metaphysic is so very
deeply entrenched in the scientific way of thinking that few discussions of
questions of ethics by professional scientists are free of them to any significant
degree. Even when they are explicitly rejected in one part of an essay, they sneak
back in at a crucial point in the argument. Two brief quotations from essays on
ethics by prominent scientists bring out this point.

> Science rests upon a strictly *objective* approach to the analysis and interpretation of
> the universe, including Man himself and human societies. Science ignores, and must
> ignore, value judgments. Knowledge yet discloses and inevitably suggests new
> possibilities of action. But to *decide* upon a course of action is to step out of the realm
> of objectivity into that of values which, by essence, are non-objective and therefore
> can't be derived from objective knowledge. There is strictly no way of *objectively*
> proving that it is BAD to make war, or to kill a man . . . [6]

> Ethical statements are not statements of objective fact, as are the statements of
> science, nor are they derivable from such statements of fact. Neither are ethical
> statements analytically true, as are the statements of mathematical theorems. Rather,
> ethical statements are expressions of people's individual feelings. [7]

In point of fact, the science-education process today is totally permeated with
iterations of these axioms in both implicit and explicit form. By the time a student
has completed a training program in any of the sciences, he or she is more often
than not a confirmed Positivist, able to recite these axioms by rote, although he or
she is probably not conscious of this fact, or if conscious of it, not really aware of
any serious alternatives to it. Such a person has few if any tools for dealing with
ethical and human value questions when they arise in professional contexts, other
than the traditional "ostrich defense," without becoming entangled in a web of
serious inconsistencies. And this is all too painfully displayed in an examination of
much of what is being written today on the topics of ethics and values by scientists
and engineers. It is therefore worthwhile to point out some of the problems
generated by the four tenets of Positivism listed above, in hopes of helping to avoid
them in the future. As with most metaphysical systems, the most telling criticisms
of Positivism are grounded on the principles of Positivism themselves. We will
begin with axiom 4 and work backwards.

A POSITIVIST CRITIQUE OF POSITIVISM

Anyone with even a passing acquaintance with the history of Western philosophy
should be aware that point 4—the matter of determinism vs. "free-will"—is

anything but a new problem. It has been approached from almost every conceivable angle during the last two thousand years, and it appears as though the debates over it could very well continue for at least that many years into the future (assuming, of course, that the human species survives that long). The Positivists have contributed little, it anything, new to this ongoing debate, other than perhaps clouding the issues a little in their amplifications and variations of William James' attempted distinction between "hard" and "soft" determinism, which many critics consider to be a distinction without a difference.

The importance of this component of the Positivist philosophy is therefore not so much theoretical as pragmatic. In so far as one takes seriously the *theory* of determinism, one is able to construct a theoretical justification for ignoring ethical questions entirely. It is this aspect of the Positivist thesis which has been taken up unquestioningly by Skinner and other behaviorists,[8] but which is not restricted to their domain. The basic consequence of the determinist thesis is that it becomes meaningless to hold a person morally responsible for his or her actions since these actions are "caused" by factors which are by definition beyond the control of the individual agent. Praise and blame become simply tools for influencing the behavior of others, rather than concepts of evaluation. Thus, for example, in the "Summerlin/Sloan-Kettering Affair" of the past year,[9] the individual involved was not really held responsible in so far as his actions were considered to be the product of environmental factors (the structure of the research/grants system) or medical factors (he was suffering from an "emotional disturbance"). "Punishment" can only be justified on the grounds that it will act as a deterrent for similar acts in the future. It is tempting to perceive such an approach as being much more humane than the unjust administration of retributive punishment as meted out in the past. But, as Thomas Szasz[10] and others have pointed out, the "treatment" of "mental illness" is often indistinguishable in practice from what in the past was considered punishment, with the exception that it can be carried out without the protection of due process of law. Thus, there is little in the determinist/rehabilitationist's position which makes it stand out as more humane than the traditional retributionist position. The justice is to be found more in the administration of the system than in its theoretical foundations.

A fundamental problem of the deterministic thesis is that it would seem to eliminate the notion of truth from science, at least in so far as these concepts are understood by the positivist. The only "reason" that a scientist has for believing one thing rather than another is that he or she was *caused* to believe it by certain environmental and/or psychological factors. Thus, if axiom 4 is true, then the Positivists cannot accept it *because* it is true. Rather, if it is true, they accept it because they were *caused* to accept it. This is just one instance in which Positivists all too frequently fail to recognize the full implications of their basic assumptions. This inconsistency or paradox (whichever it may be) can be removed by appeal to an appropriately modified theory of truth, but this would in essence destroy the basic Positivist metaphysic and epistemology.

The real problem of the determinist thesis, however, is not its potential for misapplication, nor even internal inconsistencies. Rather, it is its sterility with regard to the most basic concern of ethics—that of personal decision making. Once a person has made a decision, the determinist can (theoretically, at least) provide a description of the factors which caused that person to make that decision. Likewise, the determinist claims to be able (theoretically, again) to tell us how to cause another person to make a specified decision. But the determinist can tell us nothing about how to go about making the *right* decision, or how to determine what the right decision might be. And as Dr. Blanpied so clearly pointed out (p. 144), if modern science and technology have done anything, they have created increasing numbers of increasingly difficult decisions which must be made. And it would seem to be one of the basic tenets of modern science that decisions should be made *rationally*—i.e., "scientifically," in so far as point 1 places science as the paradigm of objectivity.

That "rational" decisions are impossible on the Positivist thesis is brought out most explicitly in axiom 3, which we shall look at more closely now. The main thrust of this axiom is that it is completely improper to attempt to talk of the truth or falsity of expressions such as "X is right" or "Y is bad," just as it is improper to discuss the truth or falsity of utterances such as "Wow!!" or "Stop!" Also, such expressions cannot be logically related to any of the objectively true descriptive statements of science itself. At most, ethical discourse is an attempt to control the feelings, attitudes, or behavior of other persons. At worst, it is the pitiful parading of one's personal feelings. In any case, for the Positivist, it is not subject to rational debate as to its truth or falsity.

And it is precisely at this point where the Positivist begins to become entrapped in his own system. Being confronted with real decisions, many of which are generated by new developments in science and technology, the Positivist can find little comfort and absolutely no assistance in the principle that all human behavior is determined. But even worse than this, it is impossible to discuss these choices or to seek advice, in so far as question "What is the right thing to do?" or "What ought I to do in this situation?" cannot be rationally answered. The answer which most of us desire to such a question is not an exclamation or an expletive, or even a command. What most of us desire and expect is a reasoned response such as "You *ought* to do such and such *because . . .*" But the Positivist can at most interpret such an utterance as an attempt to manipulate his behavior.

Axiom 2 also provides a number of difficulties for the Scientistic Positivist. The assertion of the value-neutrality of science is beginning to smack more and more of the naivete and arrogance which the Positivists themselves attributed to the missionaries fifty years earlier. As universal compulsory education has become a reality in many countries, the scientific priesthood has enlightened the masses with what it considers to be the objective truth about the real world. But with the increasing political power of other elements of these societies, there has been an increasing resistance to the imposition of this "truth" on the "uneducated" segments of the societies similar to the resistance of the "uncivilized" societies to the "truth" of the missionaries. The recent decisions concerning the teaching of evolution in California and Texas provide just one example of this growing resistance. And the Positivist is in at least as frustrating a position as the missionary, for his only response is "But I have the *Truth* as established by the value-neutral methods of objective science, whereas my opponents merely have myth and superstition." There is no way that the Positivist can begin a meaningful dialogue with the opposition from such a position.

A WAY OUT?

There are a variety of routes that the Positivist can take in response to problems such as those outlined above. The simplest, and unfortunately the one still most frequently taken, is the ostrich defense—making the problems disappear by refusing to recognize their existence. (Of course, one of the basic tenets of science and all other rational enterprises is that if there is no problem there is no need to seek a solution). Another route is to recognize the problems but to simply accept them fatalistically as being irresolvable in so far as the scientist has the Truth and nothing can be done to change it. Another route is to go to the extremes of irrationality and take the position that ethics is irrational and subjective, but science is also. None of these routes appeals to me, nor do any of them seem practical in the long run in so far as each would ultimately lead to the demise of science or the destruction of society, or both. But the situation is far from hopeless and there is at least one potential escape route which appears most promising to me.

It is my belief that trends have already been established in recent years which if pursued vigorously will enable science (and society as a whole) to escape the problems of Positivism (although as with any new endeavor, there is no guarantee that new problems will not be generated). One important trend is in ethics, the other in science—but it is perhaps most fruitful to view them as two sides of the same coin.

The new promising trend which is of such great importance in modern science is the recognition of *the ultimate inconsistency of Scientistic Positivism—the failure of the Positivist to subject science itself to a scientific analysis.* A prime example of this is in the area of the study of perception—both physiological and psychological optics. Considerable evidence has been acquired in the last several decades to indicate that different persons can literally *see* different things when subjected to what are presumed to be identical stimuli.[11] The implications of this for the traditional assumptions of the role of observation as the foundation for the objectivity of the sciences are potentially immense, but most scientists have little if any awareness of these scientific results and even less awareness of their ramifications for their own work.[12]

To take only one additional example of the potential importance of the scientific study of science: The very people who brought us the "discovery" of the relativity of ethics—that is, the social scientists—have finally gotten around to initiating similar studies of science. The results are still scattered, and the conclusions still preliminary and tentative, but there is good reason to believe that now that someone has taken the effort to look, significant similarities are going to be uncovered between scientific systems and value systems—in so far as (surprise of surprises and horror of horrors to the Positivist) science is itself a complex value system!!

The complementary trend to the scientific study of science is the return to a more reasonable approach to ethics. The Positivist account of ethics was clearly an over-reaction, however natural, which was made when the opportunity arose for science to dethrone the long-standing paradigm of intellectual respectability. If they have discovered anything, cautious social scientists today recognize that at most the empirical evidence indicates that it is *difficult* to resolve disagreements between individuals who are products of different cultures—but this is as true of disagreements about evolution as it is about ethical matters. No adequate evidence exists to prove that it is *impossible* to resolve such disagreements in science or ethics. Thus the study of ethics is taking on new life, this time from a more reasonable perspective than that which it had before being so rudely dethroned. The ethicist has finally learned, and the scientist is beginning to painfully relearn, that in matters of this type humility is a virtue and arrogance is an evil.

AN APOLOGY AND CLOSING HOMILY

The preceding discussion is excessively terse and condensed, simplistic and dogmatic, and (hopefully) raises more questions than it answers. An attempt might be made to excuse such a performance in terms of space limitations, the nature of the subject matter or the diverse interests of the intended audience. But all of these would be intellectually unfair, since this essay itself must be seen as an example of some of the very problems with which it attempts to deal. In so far as there is a disagreement (and I assume there will be many) with a specific point (or with the entire essay for that matter), it is hoped that this will be viewed as a failure on the part of the reader as much as on the part of the author. It is only with such a *moral* foundation that communications channels can be kept open among individuals—particularly among individuals from different disciplines or cultures.

One final reiteration, in case the communications process has broken down completely in some cases and the reader may not have been able to make heads or tails of anything that has been said up to this point. What I have identified as being the main failure of both ethicists and scientists in the past is the *lack of self-awareness*—particularly of one's own limitations. The arrogance and lack of self-awareness of the traditional ethicist who considered ethics to be the ultimate in knowledge was matched (albeit for a briefer span of time) by the Scientistic Positivist. The trend which I have suggested holds the greatest promise of providing a solution to many of the problems of science and ethics can be essentially described as the trend towards increased *self*-awareness on the part of both scientists and ethicists.[13]

The frustrating thing at this point is that this "trend" can be traced back at least as far as Socrates, who 2500 years ago emphasized to his students that it is important above all else to "Know thyself." Have we finally gotten his message?

Notes

1. One such study is Alasdair MacIntyre. *A Short History of Ethics*, The Macmillan Company (New York: 1966).

2. This point is more fully developed and documented in Robert J. Baum, *Philosophy and Mathematics: From Plato to the Present*, Freeman, Cooper & Company (San Francisco: 1973).

3. One of the best presentations of the typology of formal ethics and value theory is in Richard Garner and Bernard Rosen, *Moral Philosophy*, The Macmillan Company (New York: 1965).

4. See, for example, Ruth Benedict, *Patterns of Culture*, Houghton Mifflin (New York: 1934).

5. One of the best historical treatments of the philosophical school of logical positivism is in J. O. Urmson, *British Philosophy between the World Wars*, The Clarendon Press (Oxford: 1959).

6. Jacques Monod, "Knowledge and Values," in Watson Fuller (ed), *The Biological Revolution*, Anchor Books (Garden City: 1971), p. 12.

7. Gerald Feinberg, "Survival? Yes. But in What Form?" *Bulletin of the Atomic Scientists* (May, 1971).

8. Of course, Skinner's *Beyond Freedom and Dignity* is the not-so-revised standard version of this doctrine.

9. As reported in *Science*, 14 June 1974, Vol. 185, p. 1154.

10. For example, see Thomas Szasz, *The Manufacture of Madness*, Harper and Row (New York: 1970).

11. The scientific results are described in works such as R. L. Gregory, *Eye and Brain*, McGraw-Hill (New York: 1966).

12. The implications of this for scientific research itself are drawn out in works such as N. R. Hanson, *Perception and Discovery*, Freeman, Cooper & Company (San Francisco: 1969).
An example of such self-awareness in the scientific literature is Charles Tart, "States of Consciousness and State-Specific Sciences," *Science*, 16 June 1972, vol. 176, pp. 1203–1210.

13. Dorothy Zinberg, "A Strategy for Science Education in the 1970's," *Science*, vol. 179, p. 118.

4.4

Autonomy and Obligation: Is There an Engineering Ethics?

K. R. PAVLOVIC

INTRODUCTION

There are those willing to argue that the medical and legal professions expose their practitioners to unique ethical issues and that it is therefore appropriate to speak of "*medical* ethics" and "*legal* ethics" in a strict sense. There is a temptation to follow this line with regard to the engineering profession, and it is likely that many occurrences of the phrase "engineering ethics" represent a yielding to this temptation. This is how I view, for example, discussions of ethics and engineering that make reference to engineering's being a profession and to the alleged duty of a professional to serve the public above all others. The truth of the matter is that concern with professional ethics is in most cases the result of a growing dissatisfaction on the part of the lay public with the conduct of so-called professionals. Whether the lay public is justified or not in this, it is clear that the dissatisfaction as an attack upon engineers and engineering is rooted in certain conceptions of a profession, its obligations, and the obligations membership imposes upon individuals. It is equally clear upon closer examination that the way to meet this attack is *not necessarily* to meet it on its own grounds. That much of the discussion remains centered around the issue of professionalism reflects the fact that both the professions and the public believe that they have found in the concept of professionalism a means of manipulating the other party into a congenial position and mode of conduct. But this kind of jockeying and manipulation is politics, and politics and ethics are not necessarily the same thing. Individuals who are concerned with the conduct of professions, of individual members, and of the public over and against the professions ought not to allow the politics of the situation to dictate the parameters of their thinking on this issue.

This article is based on a paper of the same title delivered at the Annual Meeting of the American Society of Mechanical Engineers in New York, December 1979.

Samuel Florman has argued against teaching ethics to engineering students. He argues that engineers are not less ethical than anyone else, that they have little trouble telling right from wrong, and that, in any event, what engineers *are* is naive and unsophisticated in their understanding of the world at large. I maintain, and presume Florman would agree, that there is no "*engineering* ethics" in the strict sense. At the same time I wish to argue that (1) there is an ethical dimension to the practice of engineering, (2) that this dimension has no essential connection with engineering's being a profession, (3) that professionalism in engineering is rather an issue parasitic on this ethical dimension, and, finally, (4) that all this notwithstanding there are factors in the practice of engineering that make a discussion of ethics in the context of engineering appropriate.

2500 years ago the Greek philosopher Protagoras, in response to Socrates' asking who amongst the Greeks were the teachers of virtue, replied that *all* were, that each within the limits of his competence was an expert, those who could not meet the minimum standards having been either put to death or banished. According to the tradition Protagoras was a sophist, i.e., a dealer in specious arguments, so I will offer you a modern reason for acceptance of the conclusion. We live in such a society. Ethics concerns *all* human activity, *all* human knowledge, and in this society that is not a circumscribable area in which one could either become an expert in the usual sense or have recourse to experts in the usual sense.

In the following I am going to confine the context of my remarks and observations to the individual in 20th Century American society. The perspective of the individual in a given society faced with the question of right and wrong and what he ought to do, not in general, but in a specific situation has not been seriously dealt with in the recent past by ethics, ethics having given that perspective over to the social scientists, to the people who study decision-making, and to the therapists.

I want to call your attention to two of this society's more interesting aspects from the standpoint of the individual—"autonomy" and "obligation."

AUTONOMY AND OBLIGATION

Our concept of autonomy is a compound one composed of three simpler concepts. Two of these refer to external conditions: one of these conditions is the concept of a relative lack of restrictions on action, the other the concept of a relative absence of coercion. The third concept refers to an internal condition and is that of the actor as *the initiator of action* rather than simply its medium. This third concept presents problems because, when examined more closely, it appears at some times to be an illusion and to disperse itself into the first two concepts. It is sufficient, however, for our purposes to say that, if an individual feels that he is the initiator of an action, he is for practical purposes the initiator. Now, within this compound concept we can and do distinguish between a variety of forms or species of autonomy, depending on what kinds of restrictions and coercion are or are not present and on the basis of the manner in which we initiate whatever action is taken. I want to call your attention to one form of autonomy that is of particular interest. I will call this "informed autonomy." What I have in mind is the situation where I understand and articulate the restrictions and coercions inherent in my situation and then, within the range of actions left open by these restrictions and coercions, choose and initiate one course of action on the basis of further considerations that I consciously introduce into the situation. Let me make this clearer by means of an observation and illustration.

While in a given case the nature and sources of our restrictions are usually fairly obvious, the nature and sources of our autonomy are not so obvious. This has led some people to deny that there is such a thing as autonomy. They suggest that autonomy is an illusion born of ignorance of the true state of things and that as we learn more and more about why people act as they do we see more and more the restrictions on their behavior until, extrapolating, we will someday in total knowledge see that our actions are totally determined and thus hardly autonomous. I suggest that our *experience* is exactly the opposite of this. While it is true that daily the human sciences offer us more information about all the factors that condition human behavior and thus apparently teach us that we are less and less free, our experience as individual actors is that knowledge offers us ever more alternatives under given circumstances. Our perception of our problem is often that we have too many choices.

Now, if you will reflect a little further you will see that actually the situation is more complicated than this and that the observation above needs to be amended. Actually we find here a double movement. Let me illustrate this with an example. I, a lay person, wish to erect a structure to attain a given end. Within the state of my knowledge I judge there to be only one way of doing this, and for some reason or other I judge this way to be impossible. I go to you, an engineer, and explain my problem. You inform me that I am correct in judging the potential solution I know of to be impossible, but that I am incorrect in judging this to be the only solution— in fact, there is a variety of ways in which to accomplish what I have in mind. I have now gone from apparently no option to an embarrassment of riches. This is the first movement—the one referred to above. But, now, you continue by pointing out that, while a variety of solutions are all possible, they are not all equal. This one is more economical, that one more durable, this one requires very sophisticated construction techniques, that one not, and so on. Now we have begun the second movement. I re-examine my problem and the resources available to me and find in further consultation with you that, taking all this into account, one of the solutions you initially proposed stands out from the rest. I have now gone back to a relative scarcity of solutions—one. We can make this even more pointed by changing the example to a case where I initially see one and only one solution, but for a variety of reasons do not wish to use it. All these reasons I see as restrictions on my autonomy. At the end of the process above I have returned to a single solution, but now I do not view the fact of my having only one solution as a restriction (it is a mis-applied metaphor to speak of the factors that led to the choice of the final solution as "forcing" it upon me), rather I feel that I have exercised my autonomy—I have chosen. It is this curious double movement turning on the acquisition of further knowledge that characterizes informed autonomy.

This then is what I mean by "informed autonomy." We have a situation where I understand and articulate the restrictions and coercions inherent in the situation (viz., my project of a structure to achieve a given end and your knowledge as an engineer of the factors excluding certain structures), and within the range of actions left open by these restrictions and coercions (viz., the possible structures) I choose and initiate one course of action on the basis of further considerations (viz., questions of cost, durability, ease of construction, etc.). Please note that it is not necessary that two or more people engage in this. Were I an engineer I would have gone through essentially the same process in a dialogue with myself.

To bring this back to twentieth century America, I would like to point out to you that, not only does this society recognize this sort of autonomy, it actively encourages it. We live in a knowledge society with twin emphases on the acquisition and utilization of knowledge. That this is so is certainly diminished but not denied by the fact that there are groups within the society who as groups are actively encouraged not to acquire and utilize knowledge and that each of us is discouraged from acquiring certain kinds of knowledge. The fact that it is not universal is one of the things that makes this subject of interest. I am not going to recite to you all the myriad ways in which this society recognizes, rewards, punishes and restricts autonomy in this sense. With what I have given you so far you can with a little reflection come up with as many examples as you have the endurance to reflect and examine. I want to call your attention to some general features of one way in which informed autonomy is restricted in this society—obligation.

We speak of being obliged to do various things and distinguish between a variety of different ways in which we can be so obliged. Consequently there are a variety of kinds of obligations that can be distinguished. That obligations exist in our society can hardly be doubted. Nonetheless some people apparently do and so I would like to straighten out that issue at this point. Tables exist, they do not have to exist and there have been times and places where they did not. Where certain materials have been brought together into certain configurations, there we have tables. Similarly, we might be able to construct a society where the factors that constitute obligations did not obtain. Then we would be free of obligations. Such a society might even be better than the one in which we live. We live manifestly, however, in a society where obligations do exist. We have obligations that are imposed by law, we have obligations that are sanctioned by public pressure. We feel certain obligations towards certain people, and we create obligations when we perform certain actions, e. g., when we make a promise to someone. Here again I am not going to go into a detailed recital of all the forms of obligation that can occur within this society. With a little reflection you can do a lot of this work yourselves, and when you have come to the end of your abilities there is no lack of philosophers, writers, and other thinkers who have analyzed this area in greater or lesser detail. I do, however, want to call your attention to one form of obligation that is particularly germane in this context.

This is not an obligation that one necessarily feels, nor is it an obligation that is forced upon one, although once incurred it may be enforced by either the legal machinery or public pressure. It is the obligation that one accepts on the basis of knowledge of factors involved and sometimes, but not always, clearly perceived self-interest. As you can see this is not unrelated to what I have called "informed autonomy"—it is the obligation that is incurred by an act of informed autonomy. Let me illustrate this with another example. Let me again desire to erect a certain structure. Of this structure I know nothing except that I want it to accomplish certain specifiable functions and to be erectable given specifiable financial, material, and labor resources. I go to you, an engineer, describe my problem and some time later, having accepted the task, you supply me with the plans for a structure that meets these specifications. Now allow me to make several observations here. First, in your acceptance of this task you have incurred an obligation to supply me with the best possible solution given the specifications I laid down (I include here as a possible solution your judgment that within the parameters laid down there is no solution). Voiced or unvoiced, this is clearly my expectation in coming to you. Second, in this I am surrendering my autonomy—"informed," if I have the knowledge required to do this job myself, "uninformed" if not. Third, I have not, however, totally surrendered my autonomy—in fact, I have exercised it in choosing you rather than someone else—again, "informed," if I have reason to believe you to be the best engineer available, "uninformed," if I have chosen you at random. Fourth, I am in all probability acting to discharge some other obligation that I have incurred, e.g., to my employer, my family, a friend, etc. Fifth, in accepting this obligation you are in all probability similarly acting to discharge some other obligation previously incurred.

First and foremost I want to draw your attention to the inter-locking nature of the instances of autonomy and obligation. Notice (1) that autonomy is surrendered in an act of autonomy, e.g., my autonomy in choosing a solution is surrendered in the autonomous act of choosing you, (2) that obligations are incurred through acts of autonomy, e.g. your choosing to accept the job, (3) that obligations can be incurred in order to discharge other obligations, and (4) that one can incur an obligation to exercise one's autonomy, e.g., autonomously accepting the job you incur the obligation to exercise your professional autonomy in finding a solution.

Second, let me also point out that, although I have chosen the case of a so-called "professional" encounter to illustrate these phenomena, they are not confined to that sphere and in fact are widespread throughout this society. Even in our most informal relations with others we exercise our autonomy and incur obligations in this way. Part of this may be due to the fact, suggested earlier, that we live in a knowledge society, where we are constantly relying on formal and informal knowledge supplied by others. A great deal of emphasis has been placed upon the fact that our society has grown very complex and life within it heavily fragmented into roles. I would suggest that the nature of autonomy and obligation have not gone unaffected. Attempts have been made to deal with this complexity and fragmentation by gathering up areas of autonomy and obligation into individual bags that seemingly can be assigned to individuals in an attempt to simplify the situation, e.g., professionalism. I would further suggest, however, that the concept of e.g., engineering professionalism is too gross a category to deal effectively with the variety of autonomy and obligation the individual encounters.

Finally, you will have undoubtedly noticed that I have given in this discussion a privileged position to what I have called informed autonomy and obligation, only alluding to other forms that admittedly exist. My reason for this is simple. I maintain that in this society these are the privileged forms from the point of view of the individual. It is through acts of informed autonomy and the informed incurrence of obligations that we operate with and within whatever other forms of autonomy are presented to us and discharge whatever other kinds of obligation we incur. One might go so far as to say that this society obliges us in many instances to do this precisely because it is *not* ordered in such a way that our every action is mandated and because we do not live our lives, not metaphorically but literally, as simple cogs in a vast machine.

My reason for drawing your attention to these two forms of autonomy and obligation is that, where they function, they work precisely against attempts to compartmentalize and simplify the ethical dimension of human existence. Whereas it may be possible to take all other forms of obligation and divide them up into neat groups, e.g., legal obligation, social obligation, professional obligation, etc., keeping them nicely separated and dealing with each group as a whole, "informed autonomy" works precisely to tear down these neat structures, building thousands of bridges of obligation across any dividing lines and tying them all together into a vast web.

This vast web of autonomy and obligation—through which we navigate with the aid of informed autonomy and obligation and which we are constantly re-structuring through our actions—constitutes the greater part of the ethical dimension of life in this society. I say the greater part for I have not touched upon, and will do so here only in passing, another aspect of what we call ethics—values. On the view I am propounding values enter in as restrictions on autonomy, they also play a part in the setting of goals. I am playing down this aspect because I believe that in the past the attention paid to it has obscured other aspects. I am also playing it down because it seems to me that from the individual's viewpoint values play a subordinate role. In this society the typical ethical dilemma for an individual concerns, not so much values per se (most people have some and agreement is, I believe, more widespread than is usually supposed), as it does values embodied in obligations, projects and restrictions. The typical ethical question is "What should I do?" not "What do I value?" or "What should I value?" "What should I do?" means "How should I exercise my autonomy in order to satisfy the obligations I have and not incur further obligations that will pose similar problems." Ethical dilemmas occur for the individual when he is not sure which obligations he has in fact incurred, or when he is not sure how his autonomy should be exercised with regard to existing or potential obligations, or when he perceives a conflict (real or apparent) between obligations. When one focuses too heavily on values it is natural to assume that ethical dilemmas are at root problems of values. From there one moves easily to pondering justifications for values, concentrating on consistent sets of values, all the while retreating farther and farther from the case of the concrete ethical dilemma.

As developed thus far this analysis and argument applies to all members of this society. I want now to look at the special case of the engineer. The remarks apply to a greater or less extent to all professional, knowledge workers, but the engineer represents in my view the most extreme case.

ENGINEERS AND ETHICS

There are two factors that contribute to the engineer's unusual position with respect to autonomy and obligation and thus to ethics. *First*, not only do we live in a knowledge society whose shape and character are largely determined by the interaction between various spheres of knowledge and their practitioners, but we live in a technological society, thus making the engineers privileged knowledge workers. The engineer finds himself at the heart of the web of obligation constituted by the exercise of his autonomy as a knowledge worker and the exercise of others' autonomy with respect to the knowledge he possesses. *Second*—and this has been remarked with obsessive and monotonous regularity in recent years—in this web of obligation the engineer is far removed from the "ultimate consumer" of the results of his autonomous activity as an engineer. The engineer (whether he be in private practice, industry, or government) deals with (i.e., incurs obligations on the basis of knowledge supplied by) intermediaries—his immediate boss, government or industry representatives, public officials, etc. When something goes wrong, as it occasionally must, because of failures of communication anywhere along the line in what is in reality a long chain of intermediaries, the engineer as the knowledge worker who supplied the technical expertise will be found at the end of the search for responsibility. In this technological society what the engineer does has far-reaching and widespread effect. When the engineer supplies the right solution to the wrong problem (wrong because he was given the wrong parameters) his action will in all probability have far-reaching and widespread *bad* effects. The engineer with (in all probability) good faith accepts the parameters, supplies the solution and carries on in blissful ignorance until one day someone calls him to account for his action as an engineer, or he in one way or another comes to doubt the accuracy of relevance of the parameters he is receiving or someone comes to him with a proposition that smells faintly underneath the perfume of arguments he does not know how to evaluate. He is not equipped, i.e., does not have the knowledge, to deal with this situation—he is Florman's babe in the woods. To which knowledge worker does he turn to exercise his autonomy by surrendering his autonomy in order to discharge his obligation to himself to successfully deal with this situation?

Thus, it seems clear to me that there is an ethical dimension to engineering and that it makes sense to discuss and investigate that dimension. Insofar as engineering is a discrete entity with fairly sharp boundaries, we might well expect to find certain general principles that will help to sort out the confusion in specific cases of ethical dilemmas involving engineering. In this sense it makes sense to talk of "engineering ethics." It is equally clear to me that it makes very little sense to speak of an "engineering ethics" in a strict sense, as if there were some set of values, obligations or ethical principles peculiar to engineering. Every candidate for such a value, obligation or ethical principle can be found to a greater or lesser degree in other areas of human activity. But this means that these are not truly uniquely distinguishing characteristics—we have a distinction of degree rather than of kind.

I submit to you that talk of engineering ethics in a strict sense represents a desire to have ethical problems be more like the stereotypical engineering problem—to have the lines within the web of autonomy and obligation more clearly drawn and circumscribed—a desire to be able to effectively deal with ethical problems by, for example, formulating and promulgating codes of ethics and teaching courses in engineering ethics to prospective engineers. If this were a harmless yearning I would not burden you with these reflections. Unfortunately it is not harmless.

First, it is contrary to fact, as I have tried to indicate. One simply cannot in this society draw a line around a sub-set of values, obligations and principles and deal with them in a way that will be relevant to individual cases. That is to say that, because it is contrary to fact, measures based on it will be ineffectual. If a man is in a quandary over whether to give a bribe in order to obtain work for his company in order to keep it afloat, it will not be very helpful to tell him that it is unethical for an engineer to give a bribe. Presumably, he already believes that—otherwise he would not be troubled by the situation. What he needs to know is that "everyone does it" is not a germane argument—and it is amazing how many people have heard that and yet neither know why it is not nor even believe on faith that it is not. He needs to be able to trace out and evaluate all the obligations or supposed obligations that come together to create this situation—obligations to his family, his employees, other engineers, the community, etc. He needs to know what his options are and how exercising those will effect his obligations. If an engineer is troubled because he believes that the product that he is working on for his company poses a significant threat to the public health, safety and welfare, he does not need to be told that he is to hold paramount the public health, safety and welfare. Again, he presumably believes something like that already. He needs to be able to trace out the web of obligations, to know where to go and how to get the information he needs, what options are open and how to evaluate them. In both cases, doing this will quickly take the individual far beyond the boundaries of engineering and even of business. What they both need is knowledge, in Florman's phrase, "sophistication" and the tools to deal with that knowledge. Before talking about where they can get these I want to talk about one place and one way in which they cannot get them.

If you think you can nicely circumscribe a problem area, the usual next step is either to bring in or manufacture the "experts," "professionals" who will generate the specialized knowledge and then stand ready to supply professional solutions. While this approach makes sense in many areas it makes no sense in ethics. Ethics is not a subject matter in the usual sense. Most areas of human knowledge concern circumscribable areas of human activity, but ethics concerns all natural and human activity insofar as it may enter into the web of autonomy and obligations. Any and all knowledge can become the basis of or for the incurrence of an obligation—any and all activity can bear in one way or another on one's discharging or failing to discharge an obligation. The hard and unpleasant fact of the matter is that there are and can be no experts in this sense. In matters of ethics you cannot surrender your autonomy to an expert in the way that you apparently can to other experts.

CONCLUSION

Thus I agree with Florman. The problem with engineers—and not just engineers but all of us (even philosophers and ethicists) is that we are naive and un-sophisticated outside our chosen areas of expertise. Because when the ethical dimension of our lives takes us outside our own area of expertise, there is no ethical expert to whom we can turn for customized, sophisticated solutions to problems, we have no choice but to become our own experts—if only minimally competent ones. The question then is what do you need in order to become such an expert.

First, you need certain critical and analytical tools in order to initially sort out your situation. This is where philosophers and ethicists can be of help. In 2500 years a large number of ethical theories and analyses have been produced. None of these is satisfactory as a definitive theory of ethics. On the other hand most of them can contribute to an understanding of a concrete ethical dilemma. Theories of value can help you get clear on what values are or ought to be operant in your situations. Theories of duty, autonomy, justice, obligation, etc., can all help in an analysis and articulation of your situation and often reveal possibilities that would not otherwise occur to you. Insofar as consequences and the utility of actions and possible actions are concerned, consequentialist and utilitarian ethics can give you very valuable tools of analysis. Insofar as certain conceptual and linguistic fallacies seem to be involved in our ordinary ethical thought and discourse a great deal of 20th Century work in ethics can be of use. Lest you think I am talking about some high flown and abstruse course of study, let me give you an example. A great many of the case studies that are used in courses in engineering ethics appear to offer a large number of possible actions, and one is bewildered by all these possibilities. An exercise as silly and simple as writing down these possibilities and then assessing the possible consequences for each often reveals that many come to the same conclusion, thus diminishing the bewildering complexity. This is basically a consequentialist analysis and there is nothing difficult about it except overcoming the aversion to doing it. Most of the really technical and abstruse points in ethics come up when the theories are arguing amongst themselves as to which is true (in some sense or another), but I am talking, not about using these theories as algorithms for the mechanical solution of problems or as pictures of reality, but rather about using them as tools for the analysis of a concrete situation.

The second thing needed is what I would call imagination. I have in mind here a rather pedestrian use of imagination. I mean the ability to visualize or verbalize to oneself *what is not* or *what is not yet*—to imagine carrying out various courses of action and their consequences—to imagine the motivations and actions of other individuals—to reconstruct one's recent history. Clearly these are elements in the process of assessing one's situation and evaluating possible courses of action. This is shaky ground given the sorts of controversies that can and have in the past arisen around the issue of imagination. My own experience with students and others is that this kind of imagination can be aided and developed through literature, history and the social sciences. If nothing else they supply it with raw material. If it should turn out, as some insist, that this kind of imagination cannot be developed— one simply has it or not—then the unhappy conclusion here is that one who does not have it is at a disadvantage in this kind of society.

Finally, one needs knowledge and information outside one's area of professional competence. In analyzing and assessing one's situation it may be necessary to draw upon knowledge and expertise of which one is not the master. You have to know at least enough about these areas to choose a competent source and to assess the relevance and applicability of the knowledge and expertise received. One needs to know enough that one's surrender of autonomy in these areas is an informed act of autonomy. At the interface one must know enough of what lies on the other side to be able to monitor transactions across the interface. I have tried to show you why in this society competence in the ethical dimension demands this. Studies have supposedly shown that a large number of engineers become engineers because of their perception of engineering as a clean, clear discipline without fuzzy areas and a high degree of certainty—as a discipline where one works with things and their marvelous constancy, rather than with people and their frustrating change-ableness. I will leave it to you to decide whether engineering as a technical science lives up to this. I am on firm ground when I assert that engineering as a profession only remotely resembles it.

These three constitute the skills and knowledge necessary to deal with the ethical dimension of human existence. In this constellation philosophy and ethics occupy a privileged position in that they supply the critical and analytical tools. These tools are, however, only one element in an area that calls for the exercise of individual judgment. These tools cannot be transformed into mechanical procedures that in the hands of either the layman or the ethicist can mechanically generate solutions to concrete ethical dilemmas. What ethics has to offer is an expertise that must be exercised *by* the individual—not *for* the individual.

I have tried to show the broad general facts concerning the ethical dimension of this society from the viewpoint of an individual with an actual or potential ethical dilemma. I have tried to show what the individual must do to function competently within this area and given those facts. If you ignore the facts, for whatever reason, the facts will, as one person put it, take their revenge.

4.5

I Gave Up Ethics—To Eat!

Here is the most candid and shocking article ever published on professional ethics. For the first time an engineer tells how he has been forced to deal with commission agents and dabble in competitive bidding in order to be permitted to engage in the practice of his profession. His story may not be true for all branches of private practice nor in all parts of the country, but his story is authentic for his field of practice and also his state. We checked it.

Editors, *Consulting Engineer*

I was not among those brave but foolhardy young engineers who saved or inherited a few thousand dollars and then set themselves up in private practice with nothing to support their decision but an overwhelming self-confidence and a diploma on which the ink was still fresh. Quite the contrary, I have always been a conservative, and my background was such that even today I place more faith that most in the old maxims. I believe that it takes intelligence, experience, and hard work to become a qualified engineer, and I am of the opinion that engineering is a great calling that deserves in its ranks only those who abide by its ethics.

But I also know that what should be is not always what is.

When I first entered private practice, I was well qualified. I was graduated, with high honors, from a distinguished (perhaps the most distinguished) engineering school. My teachers not only had given me a good technical education, but even had provided me with a grounding in professionalism and engineering ethics. After I received my degree, I served a long apprenticeship with an old, established engineering firm. Only after I had directed many responsible jobs and had headed important projects successfully was I convinced that I could step out on my own.

Without question I was qualified technically, and I properly considered myself something of an authority on professional ethics. My special field of experience was in public works—highways, bridges, docks, sewage plants, and water works—those fields in which the client generally is some government body such as a state, county, or city. (This is important to my story, for had I gone into a branch of

233

private practice in which architects or private owners were the clients, I might have no story to tell.)

I thought that with my background and my willingness to work, I would, before long, be making a good living from my practice.

Instead, I soon was starving.

I did not understand why, but when I went to the state and municipal agencies to see about handling some of their projects, I never seemed to be able to find the right person. They were polite, but each would explain that the final choice was not up to him. I was on a treadmill leading nowhere. For a full two years I had no work, and this was not in the depression but immediately after the War.

Then a friend of mine, an engineer, who worked for the state, tipped me off. Nobody, he explained, gets engineering work without practicing the fine art of "political engineering."

I got the idea, but I did not like it. I also did not like the idea of hunger in a land of plenty or the pile of bills that kept pyramiding on my desk.

It was then that I turned to myself and began to rationalize. I took the position that as an engineer dealing with governments, I must wear two hats and cloak myself in two separate codes of ethics. One code applied to the engineering aspects of my work, another applied to my business behavior. By the development of this schizophrenic professional personality, I could secure work under one code of ethics and then perform it under another. I hungrily adopted this philosophy.

My lawyer introduced me to a "public relations counsel" who was supposed to be a close friend of the mayor in a small city that was planning to build a multimillion dollar housing project.

At this point, my counsel and I did not have a dollar between us after we purchased our train tickets for the trip to see the mayor. On arrival the first act required of us was to take the mayor's campaign manager to lunch. We squeezed through by drinking coffee while he dined on a hamburger.

We got to see the mayor.

"Hi, Joe. Remember me? I helped carry the first ward for you and made a few campaign contributions," my counsel said to the mayor.

"Sure, glad to take care of you. What can I do?" the mayor replied to his old friend.

I learned later that my counsel had never seen the mayor before in his life, but we got the job.

I was in business. I had my first project as a consulting engineer. My counsel received the customary 10 percent of my fee.

When getting jobs this way, I learned, you also are supposed to "do something" for the mayor. I have no way of knowing what my commission agent did. I made it a policy to give him a straight commission, and how he managed his end of the business was entirely up to him. He probably had to give the mayor most of his commission on that job.

My counsel and I had a satisfactory working relationship for a number of years.

Then there was a small misunderstanding. On one project, I was asked at the last minute to work with another consulting engineer. Naturally, my fee was only half of what I had expected, but the agent wanted 10 percent of what I would have received if I had been the only consultant on the job. He sued me, but the case was thrown out of court.

His contacts were a little limited anyhow.

I MEET "THE REVEREND"

Fortunately, he had introduced me to another man (I always called him "The Reverend") who was a former member of the legislature, a former mayor of a sizable city, and currently was a lobbyist (representing, so far as I could determine, his own interests). He also considered himself a public relations counsel.

The Reverend was in pretty good with the public works department and other state agencies. On a 10 percent commission, he made about $40,000 a year from me. I even made him a vice president in my firm. The title did things for his ego although he received no salary and had nothing to say about how my practice was managed.

The Reverend had his own Code of Ethics, and he abided by it. He was so crooked he could have worn a corkscrew for a tie-clasp, but when he worked for me, I received his undivided loyalty. His loyalty was well rewarded. I never accepted a project from another agent. I always worked through him even on a few jobs that I could have obtained by more conventional methods.

I never asked The Reverend how he made or kept his political connections. It is a criminal offense to offer a state official a bribe in order to be granted a project. I always assumed The Reverend had more sense than to do anything like that. But there are ways of showing your appreciation to the proper people without giving them money.

Sometimes I could hear him in operation on the telephone in his office. The conversation might go like this. "This is The Reverend. Has the word come in from the fat one? . . . No! . . . Has The Rabbit been up to the Capitol yet? . . . When he does, make sure he mentions the party from Hillsdale . . . Yeah, that should be enough. Call me back tonight at the Bellmont number."

The Reverend had a fascinating habit of calling all of his acquaintances by some descriptive label rather than by name. I never met any of these interesting personalities of the political world.

"SOCIAL ACTIVITIES"

I know one engineer who charters planes and takes all of his "social" acquaintances to out of town boxing matches. His social circle seems to include a preponderance of state officials, especially if a large contract is in the offing. It always amuses me to see these little excursions mentioned in the social columns of our paper.

An architect of my acquaintance has a "public relations counsel" who is currently an active member of the legislature. He draws a five figure salary from the architect whether he brings in any business or not.

Many of this counsel's fellow legislative members are influential back home. For example, let's say Senator Jones has a $3-million school project back home. The counsel informs the Senator that his firm of architects is the best in the country and then reminds the Senator that he might need a few legislative votes in the future. He agrees that the Senator can have his vote on three issues that he ordinarily would have no reason to favor—that is, if the correct architects are chosen for the school.

The architects seem to be even more adept and better organized than the engineers in the fine points of "political practice."

Other people have their own formulas for paying off the officials without being so uncouth as to flash money around.

The cousin of the governor was given office space in one consultant's office—even was moved here from another state. This engineer has been getting a surprising amount of work from the state since he got his new tenant. There are other and less obvious ways that can be used. Suppose our counsel hears it on good authority that a certain man could give us some business. Maybe the counsel can use his influence to get the state official a promotion, or an increase in salary—for which the official would be less than human if he did not reciprocate when the opportunity arose.

Or maybe we are doing a highway project. The Reverend knows about this, and as a result the commissioner's brother-in-law just happens to buy available land where the highway is going to be located. He could make a nice profit this way.

CAMPAIGN CONTRIBUTIONS

Campaign contributions also play an important part in getting the more lucrative state jobs.

I always make a large contribution to both parties. Last year, I gave $10,000 to the party with the best odds, plus an insurance contribution of $2500 to the other party. All parties keep excellent records on who donates and how much.

Immediately after the elections, a list of acceptable engineers is compiled from the campaign contribution records. At one time this was a list of only five or six firms. Now all of the engineers are catching on, and the list looks like the classified section of the telephone book.

One more pointer—you have to make the contributions before elections. This seems to count more in your favor than later tokens of appreciation to the winner. This business of campaign contributions is a delicate affair.

One year we gave a close relative of the governor $20,000 for the campaign. Several contracts came up, but we were not considered for any of them. Finally, The Reverend went to the governor and asked what was wrong. He hesitantly mentioned that we neglected him during the campaign.

When The Reverend explained that we had given $20,000 to his relative, the governor refused to believe him and wanted to know if we were accusing his family of stealing. Fortunately, the relative "remembered" the $20,000 when the governor asked him about it.

From then on, we received our share of the work. Meanwhile, we had missed several large contracts.

As far as I can find out, nobody gets a government job in this state without a go-between. I have discussed this frequently with other engineers, who readily admit (off the record) that they have commission agents.

In some districts, Federal work is just as hard to come by. When you hear some engineer casually say "the Army called us in," you might find that he has a retired Army officer on his staff to insure that he receives the "call."

The Founder Societies pretend that such things as commission agents and competitive bidding are rare or do not exist. If it makes them happy to believe this officially, let them. If they acknowledged the truth, they might feel they had to do something about it.

COMPETITIVE BIDDING

Once I deliberately got mixed up in competitive bidding just to see how it worked. We were invited to bid on a sewer system. The city described the job roughly, and we were to bid on a percentage design fee plus per diem rates for a survey party.

I thought sure I was the low bidder, but a large, and "reputable" firm of engineers got the bid. There must have been some political angle that I missed.

Shortly afterwards, I received another invitation to bid on a project. This time, I forwarded the letter to my Founder Society, and was informed, by letter, that complying with this invitation to bid would violate our Code of Ethics. The Society instructed me to send the mayor a letter explaining that I would be happy to discuss the project with him, if he wanted to hold the discussions to a professional instead of a price basis.

One of the leading engineering firms in the East got this job. I sent the announcement telling who was "low bidder" (clipped from a newspaper) to my Society headquarters. That ended our correspondence. I never even got an answer.

I know that this firm got the job on a bid basis, because the mayor told me what their bid was.

Competitive bidding seems to be permissible if your firm is big and reputable enough. No embarrassing questions seem to be asked by the Founder Societies in these instances.

However, I have since had nothing to do with competitive bidding. I stick to my political knitting.

TECHNICAL VS. BUSINESS ETHICS

I did not want to go into this business of using a commission agent. I was forced into it by my "ethical" contemporaries. And if political engineering is necessary, then I am going to do a thorough job of it. I also can state with some pride that no matter how I get my work, I always have given my client the very best engineering I know how to give. Perhaps my low business ethics makes me compensate when I turn to my technical tasks. I have never so much as taken a lunch from a manufacturer or a cigar from a contractor. My designs and specifications are my own, and I see to it that they are met. I have saved the taxpayers more money than they will ever know, and I have many projects of which I am, as an engineer, rightly proud.

I am not proud of either my or my competitors' dealings with commission agents, but this problem can be solved only if we acknowledge that it exists. I will be glad to give up one of my two hats when the others do.

But I don't like starving. I tried it.

4.6

The Story Behind the Recent National Scandals Involving Engineers

BRIAN J. LEWIS,[1] F. ASCE

THE TIME—1955: THE CITY—BALTIMORE, MARYLAND

Lester Matz and John Childs, two ambitious engineers, resigned from their employment in the City of Baltimore to form the consulting engineering firm of Matz-Childs & Associates. Moving into the exploding Baltimore County suburbs, doing both public and private work, Matz-Childs started specializing in development work such as sewers, roads, storm drains, sidewalks and subdivisions.

One of their friends was H. Bob Hammerman, a real estate developer. Another was Jerome B. Wolff, an engineer in private practice. The three of them established a business relationship. However, they were not able to receive any of Baltimore County's public work projects despite their repeated attempts to break into the circle of firms that did.

One day Matz learned that he had lost a job to someone who was paying off. A colleague suggested he report the bribe to the engineering society, but Matz spurned the idea. He told his partner that henceforth *no one* would ever take a job away from him.

1960. Matz and Childs became acquainted with the chairman of the Baltimore County Zoning Board of Appeals, a Republican attorney, Spiro T. Agnew. They became involved in certain transactions with Agnew and a man subsequently referred to in the press as "the close associate" (later identified as J. Walter Jones, an Annapolis realtor and banker).

1962. In what appeared to be a doomed cause, Matz and Childs donated $500 to Agnew in his campaign for Baltimore County Executive. Agnew was elected.

---

[1]Group Vice Pres., Home Office Group, Weston Environmental Consultants-Designers, West Chester, Pa.

Later, the close associate advised Matz that the two of them would be able to make a lot of money. Decoded, this message meant that Matz, working through the close associate, would receive county engineering contracts in return for certain kickbacks. Matz was asked to prepare a suggested scale of kickbacks. He recommended 5% on engineering contracts and 2½% on surveying contracts. His suggestion was followed.

The system was very simple. When Matz learned of contracts to be let, he would advise the close associate of those he wanted. He usually delivered the kickback to the close associate in his office in a plain envelope containing cash, making installment payments as he received payment for the work performed. As the volume of Matz-Childs work increased, a cash bind arose due to the volume of cash needed for the kickbacks. A scheme was devised for employees to kickback "bonuses." The "bonuses" were converted to cash.

1964. Matz complained that he was not getting enough county work. In a meeting at Agnew's home, the County Executive promised to contact county officials and direct them to increase the flow of contracts to Matz-Childs.

Matz also generated cash by having very good clients who were close friends pay for work performed by his firm in cash. In this manner, between 1967 and 1972, he generated $80,000 in cash.

Matz formed a phony public relations agreement with Joel Kline, under which scheme he mailed Kline a monthly check for $2,400 with Kline returning his "fee" in cash, thus acting as a "laundry." (Mr. Kline later served time in jail for obstructing justice.) The "Joel Kline System" netted Matz an additional $50,000 in illegal money for job payoffs.

The employees who received bonus checks and kicked them back understood they were donating for political contributions. They did not know where the money was going. In this manner, a further $50,000 in cash was generated. As proof there is no honor among thieves, Matz put some of this money in his own pocket.

1966. County Executive Agnew announced as a candidate for Governor of Maryland. While a candidate, he participated with Lester Matz and others in the purchase of a 107-acre tract on the western approaches to the Chesapeake Bay Bridge. Later, as Governor of Maryland, Agnew authorized the design and construction of a parallel bay bridge across Chesapeake Bay by the J. F. Greiner Company of Baltimore.

1967. Agnew appointed Jerome B. Wolff Chairman of the State Roads Commission. Wolff sold his engineering firm to his employees who formed two different firms, James A. Petrica & Associates and Whiteford-Faulk.

Matz-Childs, with the cooperation of Jerome Wolff, began to enjoy a steady flow of state highway contracts. The payoff scheme remained the same; however Wolff, as Chairman of the State Road Commission, became a third partner, receiving 25% of the kickbacks paid for contracts received. Other firms received work in the same manner for the same consideration.

After some period of operation in this manner, Matz suspected the close associate of skimming money off the top of his cash payments. He went to Annapolis for a face-to-face talk with Governor Agnew, and in the Governor's office, made arrangements to deal directly with him, handing him cash, an arrangement that was to continue in the future.

1967. Agnew's replacement as County Executive was Dale Anderson, a Democrat. Matz rapidly became his friend, and his firm continued to receive sizeable contracts from the County under the Democratic executive. The modus operandi remained the same as with his Republican predecessor.

1967. Governor Agnew asked Matz for a $5,000 donation to Nelson Rockefeller's Presidential campaign. Matz reply, "Cash or check?" The response, "A check will be fine." Later when Governor Rockefeller withdrew from the Presidential race, the check was returned.

July 1968. Matz's payments to Agnew already totaled $20,000, but with fees earned but not yet received, he anticipated payment of a further $30,000 cash. A cash flow crisis prevailed. He turned to a former client who generated large sums of cash in his business and arranged a "loan." Matz-Childs & Associates loaned this former client $30,000, making payments by corporate check. In turn, the client delivered $30,000 cash to Matz. The client indicated the loan on his book as being repaid in installments of $1,700. When each loan installment of $1,700 was received, it was transmitted to Agnew in the State House in Annapolis.

At the Republican convention in Miami, Spiro T. Agnew was chosen as Richard Nixon's running mate.

November 1968. Spiro T. Agnew was elected Vice President of the United States of America.

January 1969. Spiro T. Agnew took the Oath of Office as the Vice President of the United States.

Agnew appointed Jerome Wolff Vice Presidential Assistant for Science and Technology.

Shortly thereafter, in his office in the White House, Lester Matz paid the Vice President $10,000 in cash, the sum owing from his contracts received in the State of Maryland. He reviewed his calculations with the Vice President to show how this sum had been derived and advised him that he might owe him even more, and when his fees were received, he would call again when he had "more information."

Matz returned to his office in Baltimore and told his partner he was a "shaken man." He had just paid off the Vice President of the United States in the White House.

May 5, 1970. After 15 years in business, Matz and Childs sold their firm to the Walter Kidde Company, a conglomerate, for 85,000 shares of Kidde Common

Stock with a market value of $2,539,375—not bad for funny business. They also received a five-year consulting agreement with the firm. Subsequently, they earned a total of 124,000 shares of Kidde stock. Kidde's scrutiny of Matz-Childs' books apparently failed to reveal the large amounts of illegal cash flow being used for payoffs, or, if it did, they didn't say.

1971.   Matz made a payment of $2,500 to Vice President Agnew in return for a Federal contract awarded to a subsidiary of Matz-Childs.

January 1973.   The United States Attorney for Maryland initiated an investigation into the affairs of Dale Anderson, Baltimore County Executive and called as witnesses before the Grand Jury, Lester Matz and Jerome Wolff, who by this time was President of Greiner Environmental Systems.

William E. Fornoff, who served as Administrative Officer of Baltimore County under four executives, including Spiro Agnew and Dale Anderson, pleaded guilty to a minor tax law violation prior to cooperating with the Grand Jury. He admitted being a conduit in the kickback scheme for engineering contracts in Baltimore County.

January 15, 1973.   Lester Matz when questioned by Joseph H. Kaplan, Assistant United States Attorney in Baltimore, in regard to the Dale Anderson investigation, admitted to Kaplan that he had been paying off the Vice President of the United States.

January 20, 1973.   Spiro T. Agnew took the oath of office for a second term. The U.S. Attorney's investigation continued.

August 7, 1973.   Vice President Agnew issued a terse two-sentence statement, "I have been informed that I am under investigation for possible violations of the criminal statute. I will make no further comment until the investigation has been completed other than to say that I am innocent of any wrongdoing, that I have confidence in the criminal justice system of the United States, and that I am equally confident that my innocence will be affirmed."

August 11, 1973.   The Maryland State Attorney announced an investigation of work done in Anne Arundel County, Maryland, where Matz-Childs had had 81 agreements for $2,500,000 dating back to October, 1965. The same news story indicated that Greiner Environmental Systems had more than $120,000 in business with Anne Arundel County during fiscal 1972 and the firm of Whiteford-Faulk and a subsidiary of the Greiner Corporation were also listed for contracts totaling $161,000. Zollman Associates had entered 15 contracts since December, 1968, with a total value of $212, 406.

The head of the county purchasing office stated to reporters that he always awarded noncompetitive contracts on the basis of recommendations from the County Department Chief, who in turn conferred with the County Executive, Joseph M. Alton, Jr.

In a separate statement, Alton confirmed that he concealed the names of campaign contributors by "the fifty-dollar loophole" in the campaign law, by which he indicated ticket purchases of under $50 for campaign dinners were listed only as ticket sales without the names of the donors.

October 10, 1973.   The following letter was received by the U.S. Secretary of State, "I have today resigned the office of Vice President of the United States." Signed Spiro T. Agnew.

October 27, 1973.   Lester Matz and John Childs were fired from Matz-Childs Associates by the Walter Kidde Corporation, owners of the company. A spokesman for Kidde indicated they were unaware of the payoffs made by Matz-Childs.

December 19, 1973.   William A. Hasfurther, former Chief Engineer for Anne Arundel County, complained to the Maryland State Society of Professional Engineers that his recommendations for consultant contracts were consistently ignored in favor of Matz-Childs, Green Associates, and Jerome Wolff of Environmental Systems. Several times he had sent up three other firms for projects, and their names were crossed out in pencil by Alton, the County Executive.

February 1, 1974.   At the Federal Court corruption trial of Dale Anderson, County Executive, Lester Matz testified that he had paid $175,000 in illegal kickbacks to public officials between 1962 and 1970, including some $15,000 to William E. Fornoff, Anderson's former Administrative Officer. John C. Childs also testified that he had made kickback payments to Fornoff for contracts.

March 3, 1974.   During the Anderson trial, Jerome Wolff and Lester Matz said that they had paid off officials in jurisdictions outside of Baltimore County. At that time Matz-Childs had branch offices in Rockville, Md., Manassas, Va., and Newark, Del., all rapidly growing suburban communities. Later, in 1976, the County Executive of New Castle County, Delaware, was found guilty of receiving illegal payments.

March 15, 1974.   At the Anderson trial, Paul L. Gaudreau, President of Gaudreau, Inc., a Maryland architectural firm, testified that he had paid approximately $31,000 in kickbacks to William Fornoff between April, 1968, and the spring of 1972 in return for his firm receiving a $700,000 contract for designing the county's new $10,000,000 court house. He made cash payments of 5% of the fees he received.

March 20, 1974.   Dale B. Anderson, Baltimore County Executive, was convicted of conspiracy and 27 counts of extorting $38,000 from architects and engineers.

May 1, 1974.   Anderson was sentenced to five years in prison for his conviction of extortion and tax evasion charges. He was convicted of evading almost $60,000 in income taxes from 1969–1972. William Fornoff testified during the Anderson trial that he had a list of 14 firms of architects and engineers who had made illegal kickback payments to Anderson through him.

June 1974.   Matz, Childs, and Green were expelled from Membership in the American Society of Civil Engineers.

June 3, 1974.   John Hochader, the 57-year-old Vice President of George W. Stephens, Jr. and Associates, a Towson engineering firm, testified before the Grand Jury that he had made cash kickback payments to former County Executive Dale Anderson for engineering assignments. Other engineers who allegedly paid kickbacks to Anderson through his former assistant, William Fornoff, were Eugene Hsi of Hsi, Brynner & Day, Blair Overton of Baker-Wiberly, Maurice Bender of Knoerle-Bender-Stone, William Muth of Hurst-Roche and W. W. Ewell of Bomhardt Associates.

August 13, 1974.   The Maryland Society of Professional Engineers expelled eight engineers, including Jerome Wolff, Lester Matz, John Childs and James A. Petrica.

October 19, 1974.   A Federal judge sentenced Hochader for making illegal payoffs to William Fornoff and Dale Anderson.

November 9, 1974.   The United States Attorney charged Allan I. Green of Green Associates of making payoffs. Reportedly Green told prosecutors he paid Agnew about $22,000 between 1967 and 1969 and received 10 state contracts with fees totaling between $3,000,000 and $4,000,000. He continued the payoffs to Agnew while Vice President, making them to him in his office in the executive office building or at his apartment at the Sheraton Park Hotel. He indicated he paid Agnew a total of $50,000 in cash between 1966 and 1972.

November 25, 1974.   Allan I. Green was sentenced to 12 months in prison and a $5,000 fine. At the time of his sentencing, Green said "I admit my criminal and moral guilt." He indicated he felt at the time the payoffs were needed to expand his business. During his argument, his attorney told the judge, "I personally know of significant corruption at the Federal level that is the same as that listed in Maryland."

November 27, 1974.   Joseph W. Alton, Jr., Executive of Anne Arundel County was charged with conspiring to extort kickbacks from consultants on county projects.

December 19, 1974.   Maurice E. Bender, founder of Knoerle-Bender-Stone Associates, admitted to falsifying the engineering firm's 1971 corporate tax returns with the assistance of his partner, Joseph Knoerle. They made false entries in deductions from a cash fund to cover up kickbacks paid to William Fornoff and Dale Anderson.

March 24, 1975.   Joseph W. Alton, Jr., former Anne Arundel County Executive, pleaded guilty to conspiring to extort kickbacks and was sentenced to eight months in prison without trial.

EPILOGUE

This is a true story. None of the names were changed to protect the innocent. Those found guilty are real engineers and real public officials. This was not the first time public officials in Maryland have been found guilty of corruption in office.

Maryland is not alone. Officials in other states have long participated in similar kickback schemes. The president of the engineering firm currently working on the Philadelphia Airport has admitted to the press that he has frequently made payoffs to get design assignments.

Only those in the engineering profession who are prepared to stand up and be counted when they have first knowledge of such kickback schemes can, in fact, help purify the engineering profession. Tacit acquiescence, or turning one's head in the other direction, is tantamount to being an active accomplice.

Stronger professional registration laws must be enacted and enforced. Certainly one is disgraced when dismissed from ASCE, The National Society of Professional Engineers or any other professional organization. However, membership in such associations is not necessary to practice as an engineer or architect. The doorway to continued practice is guarded by the professional registration board. We need to encourage them to act quickly and summarily.

Above all, we as professional civil engineers should stand in shame that our profession has become thus corrupted and vow each to do whatever is necessary for us to regain the respect of the society of which we are so vital a part and in which we have and can play a distinguished and valuable role.

4.7

Three Leave GE...

CHRIS BARNETT

It was a helluva Monday morning for Greg Minor. Sixteen years with General Electric and a secure $35,000-a-year middle management job were about to be snuffed out by the resignation letter tucked inside his coat pocket.

It wasn't your normal "I regret I am leaving the company, but I have accepted another position more in line with my personal and professional objectives" type of resignation letter. Minor's letter was cold and to the point: "My reason for leaving is a deep conviction that nuclear reactors and nuclear weapons now present a serious danger to the future of all life on this planet. I am convinced that the reactors, the nuclear fuel cycle and waste storage systems are not safe."

Minor, 38, manager of advanced control and instrumentation for GE's nuclear energy division in San Jose, California, planned to give the letter directly to his boss, Harry Hindon, GE's manager of control and instrumentation engineering. But he was out, so Minor gave it to his boss's supervisor and called Hindon with his decision. At noon he met with his eight direct subordinates at the Laundry Works, a restaurant popular with the San Jose engineering community, not mentioning the resignation.

That evening Minor disconnected his home telephone after some three to four dozen people called. They congratulated, questioned, or scorned him for his action. More than a few were shocked and felt betrayed—puzzled that someone they'd known so long could do "this." Many of his colleagues, thunderstruck at the notion that any engineer would pull out of one job without another offer, repeatedly asked, "How could you give up this well paying job?"

It's amazing," says Minor today, "how many people put that ahead of the issue that caused me to make this move."

Minor's resignation alone probably would have caused little stir. But on that sunny California morning, two other GE managing engineers—Dale G. Bridenbaugh, 44, manager of performance evaluation and improvement, and Richard B. Hubbard, 38, manager of quality assurance in the control and instrumentation department—also quit. Each wrote a two-page letter to his supervisor citing what he considered to be the "immorality" of reactor building and the immense risks to mankind that can result from technological mistakes.

Wrote Bridenbaugh, whose 23 years with GE included ten months on a special assignment evaluating containment safety and adequacy at 25 U.S. nuclear powerplants: "It is hard for the mind to comprehend the immensity of the power contained in the relatively small reactor core and the risk associated with its control. In the past we have been able to learn from our technological mistakes. With nuclear power, we cannot afford the luxury."

The trio's simultaneous defection from General Electric on February 2 made global headlines. The three had never worked for any other company. All were married to their childhood or college sweethearts. Each has three school-age children, and none of their wives works. They lived quiet, comfortable, upper middle class, suburban lives. And all three were not only quitting, they were going to work for the opposition—donating their time to Project Survival, a group formed to put an antinuclear initiative on the state's June 8 ballot.

Proposition 15, the Nuclear Safeguards Initiative, would ban the construction of 28 nuclear powerplants planned in the state unless they met stringent new safety requirements. It would require that the state legislature approve construction of any new plant by a two-thirds vote. In addition, "Prop. 15" could shut down California's three existing nuclear plants unless the federal government boosts its liability limits in case of an accident to something higher than the current \$560 million ceiling. Lastly, the initiative, if passed, would cut power output from present plants by 10 percent a year unless two-thirds of the state legislators endorsed tougher safety requirements and waste disposal methods.

Before the three engineers tendered their resignations, only a comparative handful of the state's residents were even aware of the antinuclear initiative, let alone were for or against the measure. Outside of plant site communities and "nuclear cities" like San Jose (so called because of the GE offices), few people were excited about the issue. But since February 2, a media blitz has blanketed the state and much of the nation calling attention to the threat of nuclear power. Minor, Hubbard, and Bridenbaugh have been to Washington and to Sacramento testifying repeatedly that "nuclear power is a technological monster that threatens all future generations."

Articulate, credible, unemotional, and seemingly talking from the heart and gut, the three patiently recite their reasons for leaving GE in straightforward sentences.

In recent months, they've been running at a furious pace. With a savvy "media team" at Project Survival's Palo Alto headquarters scheduling as many as five interviews a day, the trio and their wives have fielded questions from a horde of local and national newspaper and magazine writers, radio and television interviewers.

The press attention has produced some negative fallout. Not everyone thinks the three acted solely on their consciences. Critics have pounced on the fact that they are all longtime members of the Creative Initiative Foundation, a controversial religious group that seeks to reconcile science and religion. Detractors claim the resignations were a strategically timed move to garner interest and support for Proposition 15. Indeed, the head of Project Survival, Jim Burch, a former vice president and creative director in the San Francisco office of Batten, Barton, Durstine & Osborne, the big Manhattan-based advertising agency, used to be president of the 2,000 member Creative Initiative group. Project Survival claims 8,000 members (600 from Creative Initiative) and 600 active workers.

Burch admits that he knew Minor, Hubbard, and Bridenbaugh from Creative Initiative. But in no way, he maintains, did he "arrange" their defection, nor did other CIF members pressure the three to quit. The three strenuously deny they were cajoled or enticed to make the move. There were no job promises, nor is there any interim financial support. Only their travel expenses are reimbursed, they say.

What is it like to have your lifestyle change so abruptly, to be thrust into the public eye, stripped naked of the personal and professional armor most engineers cloak themselves in? How do men who are technically trained to work in scientific environments fare on the political treadmill, stumping for a cause that even campaign-hardened politicians are avoiding?

New Engineer took a firsthand glimpse at their new life on a warm March day in Palo Alto, a picturesque college town adjacent to Stanford University. The cramped Project Survival offices were stuffed with some 25 volunteers, mostly bright, highly verbal, married women in their mid-thirties, well educated and well groomed. No one had his own desk; rather they jumped from phone to phone handling calls, setting up interviews, revising schedules among the volunteers— teachers, interior designers, professional people, housewives.

A scheduled 11 A.M. interview was running late. Minor and his wife Pat, and Hubbard and his wife Rachael were being briefed on an upcoming television appearance. Bridenbaugh and his wife Charlotte were unavailable because they had another interview scheduled that day, according to the keeper of the two "Master Appointment Charts" labeled "XGE's" and "XGE's Wives."

Half an hour later, the interview began in a cramped office hastily vacated by some volunteer typists. Hubbard recalled how he broke the news to his boss that morning. He had interrupted him in the middle of a budget presentation and handed him the letter. "I was sentimental about it," he remembered. "I had worked with these people, had trained many of them, and I was not thinking logically. The logic told me [that the resignation itself] was right."

Hubbard, who had 150 people under him, took his top two managers to lunch to explain his actions. Later he telephoned his other managers, expressed his appreciation for their loyalty, and offered to answer any questions. He claims he didn't try to sway them to his viewpoint. Socially, he told them, they would have to take the initiative if they wished to maintain a relationship. "I told them I'd feel uncomfortable calling, but if they called, I would like to see them."

So far no one has called.

Some of the broken friendships have hurt him deeply. A college roommate at the University of Arizona, a man Hubbard had known since 1955, told him a friend had never threatened his job before. "My first reaction was to take it personally, but then I said it 'wasn't anything I was doing that was threatening your job. What is threatening it is the truth.'"

Both Minor and Hubbard claim they have no animosity toward General Electric. They received all severance benefits due them and are not trying to embarrass the company or their coworkers. As far as they know, GE has engineered no personal reprisals against them. (GE, in a public statement, said the resignations had come as a "complete surprise," adding, however, that the reasons given for quitting simply repeated "the emotional claims of Project Survival," which had devised "publicity plans designed to exploit" the men's actions. Privately, GE appears to be trying to discredit the trio as religious zealots.)

Both men and their wives had long talks with their children about their decisions. Hubbard, moustached and goateed, who looks a little more professorial than the others, says his oldest son at 14 "became very defensive and defended what his dad did." Hubbard claims he presented both sides of the story to him.

Until a year ago, Minor didn't really discuss his job with the family, either Pat or his children. "But the Nuclear Safeguards Initiative really brought the issue into our lives for the entire year."

He maintains his children were also "supportive of me," although uncertain over the issues. Just before he handed in his resignation letter, he broke the news to his youngsters. Son Mark, 16, broke into a wide smile, Minor recalls, and said, "Dad, that's great. Now you're just like me. We're both looking for a job."

Minor's wife, Pat, until February 2 a stay-at-home housewife who never got involved in much of anything, admits the resignation unharnessed her insecurities. Before Proposition 15 surfaced as an issue, she had never really discussed nuclear energy with her husband despite the fact he was making a comfortable living helping to create it. "I always said, 'Everything will be fine.' It was something about wanting to protect my home, my children. My security was important to me."

Rachael Hubbard has similar recollections. Her "first inkling" there was another side to the nuclear question was the ballot initiative. She, also, claims she had never questioned just what it was her husband did to pay the mortgage and the bills. "I thought Dick had a good job, and he was providing what I wanted, and I liked it that way," she says.

"So it was pretty hard for me when I was hit with the reality that my husband was working in an area that was a cause of death and a very insidious, indirect, and invisible cause. The whole issue of radiation and what it does to human cells—the genetic mutations—that's what turned me against it."

Both couples stoutly maintain they made no snap decisions. Rather, the decisions were "evolutionary" over the period of a year. Although the men actually made the final decisions, in each case it was a husband and wife collaboration.

The tidal wave of publicity was a "psychic shock" at first, says Pat, who concedes it's "not normal" to pick up a newspaper or flick on the television and see yourself staring back. She insists, however, the insights gained on how government works at various levels have "strengthened our conviction that it's only going to happen if individuals speak out. That feels good."

Nevertheless, Hubbard, after dozens and dozens of interviews, still can't feel comfortable with television reporters. "All of a sudden the lights go on, somebody sticks a microphone in front of you and tells you to 'say it all' in 30 seconds.

"As a scientist you're not used to summarizing quite that quickly, and you're also not used to making absolute statements. We're taught to qualify everything we say with 'based on the data we have' or other phrases like that. Now we have to be definitive."

They also have to be ready to pack at a moment's notice. When Minor volunteered to work for Proposition 15, he thought he would have to fly to Los Angeles for one press conference and not much more. Now, the three frequently traverse the state and have crisscrossed the country. Antinuclear groups in London, Canada, Japan, and Australia have invited them to speak.

Despite the drain on savings, neither man says he's moonlighting during the Proposition 15 campaign. "I just don't have the time or energy," maintains Hubbard. "Now that I'm unemployed, I've never worked harder in my life. The hours are long, and the work is tougher, and the pay is terrible."

The future? Rachael Hubbard says she isn't worried. "This has been the most difficult part because you don't know what the next day will bring, let alone the next month." She admits her frantic new lifestyle bothers her, but so far it has gone well. "You just go with your convictions, and you think everything will work out."

No one at this point wants to hazard a guess if they'll ever get accepted back into the "establishment"—industry. Pat Minor is doubtful, but Dick Hubbard is optimistic. "We've had hundreds of letters from people all over the world who support us, and I'm sure that somewhere in the business and professional community, there are people who will come forward when the election is over and offer us new positions."

And if an offer isn't forthcoming? Hubbard finds the academic world appealing. He once taught night school at General Electric and enjoyed it.

Minor, meantime, wants to get into something where "I feel I am part of the solution to the problem." Both agree this was part of the motive for leaving General Electric in the first place. As Hubbard puts it, "We all got into the nuclear field because we thought we could do something good for mankind. We chose not to go into aerospace, and we picked GE. We liked GE, stayed with GE, and thought we were doing something good. And when the controversy came, we had always thought of ourselves as the good guys who were providing the solutions. Suddenly we were the ones who were creating the problem."

Hubbard, who holds a master's in business administration in addition to a B.S.E.E., chimes in that he has only been taught how to solve problems, never how to identify which problems were worth solving.

Why didn't their departure set off an exodus of other intelligent, questioning colleagues? Hubbard thinks some are leaving, "but they're going quietly." In the nuclear energy field it's not uncommon for an engineer to feel his job is perfectly all right, and if there is a problem it's outside his area, he explains. "We were working on light water reactors, and we had another group in Sunnyvale working on breeders. We all used to say, 'Those breeder guys, that's really dangerous.' Everyone would try to get the cause of any accident blamed on someone else so that it didn't count on his tally."

Both men feel that anyone questioning his job would probably move sideways to another position as a safe means of escape short of quitting. They're also convinced younger engineers will be making such heavy decisions much quicker than they did.

"The kids coming out of school," says Hubbard, "are asking better, harder questions. They're taking much less for granted than we did. When I came to General Electric, I had a clear vision. I wanted to be a manager, and I was willing to work 16 hours a day to get ahead. Students today have fewer illusions than I did."

For Minor, the whole question was once summed up by an engineering student who sat in on a two-day nuclear symposium attended by industry representatives, suppliers, and researchers. Most of the discussion centered on how rosy the industry's outlook was, he recalls, with the light water reactor people gushing about how great the breeder reactor folks were.

"Basically everybody was just building up each other's morale when all of a sudden one lone student got up in the back of the room and said, "I have some serious questions about nuclear power. I would like to know how you can continue to build nuclear powerplants without having tested the safety systems, and how you can meet the seismic criteria today with your existing designs?"

"A pall fell over the whole auditorium," says Minor. "Everyone was incensed that someone from the outside world would come into that hallowed group and question whether it was right. Our only real question at that time was whether we were going fast enough or far enough and who had the best product."

4.8

How Would *You* Apply Engineering Ethics to Environmental Problems?

HERBERT POPPER
ROY V. HUGHSON

Particularly during the past year, a great deal of invective and self-examination has centered on the question of who has the ultimate responsibility for preventing pollution. Does this responsibility rest with the:

- Engineers who design and operate industrial plants and municipal treatment facilities?
- Managers who can reward, punish and direct the engineers, and who can either accept or reject the engineers' recommendations?
- Governmental officials who can frame codes and thus create constraints for engineers and managers?

In regard to the first of these groups, engineers are now being told by some industry executives, as well as by educators and sociologists, that they have social as well as technical responsibilities in the environmental area—that they should take a broad view of these responsibilities and not merely limit themselves to finding least-cost ways of solving technical problems and of meeting existing codes.

If this viewpoint is valid in principle, translating it into practice would add a whole new area to engineering ethics, one that would involve a wide range of "how far to go" and "what course to take" decisions.

In 1963, when *Chemical Engineering* published a pioneering series on ethics, the emphasis was on proprietary information, consulting practices, and vendor relations; but in the decade ahead, the major emphasis may well fall on some of the gray areas in the environmental field—areas that either are not covered by existing canons of ethics (such as the AIChE code), or that leave doubt as to how a canon should be interpreted or implemented.

To illuminate some of these gray areas, we have set up four hypothetical case studies and are asking you, the engineer or technical manager, to judge the desirability or practicality of various courses of action, and to fill in the ballot. The results will be tabulated and published early in 1971.

Furthermore, your judgments on these case studies will be considered by an "environmental ethics" panel that will be asked to voice agreement or disagreement, to explore some of the issues and to make constructive suggestions. To bring a variety of environmental viewpoints to this panel, we were fortunate in obtaining the services of:

LOUIS N. CARMOUCHE
Manager, Functional Products
and Systems
Dow Chemical Co.

NEIL C. ELPHICK
Director, Environmental
Planning Dept.
FMC Corp.

IRWIN B. MARGILOFF
Manager of Environmental
Technology
Scientific Design Co.

ROBERT N. RICKLES
Commissioner, Dept. of
Air Resources
The City of New York

AARON J. TELLER
President
Teller Environmental Systems, Inc.

We believe that the survey of readers, the panel discussion and related editorial projects will prove helpful by pointing out the conflicts that exist, and influential by proposing some solutions to these conflicts. Your input is needed to achieve these goals, and we strongly urge you to participate by evaluating the case studies and sending in your ballot.

CASE 1:   HOW TO HANDLE THIS WATER PROBLEM?   The Marginal Chemical Corp. is a small outfit by Wall Street's standards, but it is one of the biggest employers and taxpayers in the little town in which it has its one and only plant. The company has an erratic earnings record, but production has been trending up at an average of 6% a year—and along with it, so has the pollution from the plant's effluents into the large stream that flows by the plant. This stream feeds a lake that has become unfit for bathing or fishing.

The number of complaints from town residents has been rising about this situation, and you, as a resident of the community and the plant's senior engineer, also have become increasingly concerned. Although the lake is a gathering place for the youth of the town, the City Fathers have applied only token pressure on the plant to clean up. Your boss, the plant manager, has other worries because the plant has been caught in a cost/price squeeze, and is barely breaking even.

After a careful study, you propose to your boss that, to have an effective pollution-abatement system, the company must make a capital investment of $1 million. This system will cost another $100,000 per year in operating expenses (e.g., for treatment chemicals, utilities, labor, laboratory support). The Boss's reaction is:

"It's out of the question. As you know, we don't have an extra million around gathering dust—we'd have to borrow it at 10% interest per year and, what with the direct operating expenses, that means it would actually cost us $200,000 a year to go through with your idea. The way things have been going, we'll be lucky if this plant *clears* $200,000 this year, and we can't raise prices. Even if we had the million bucks handy, I'd prefer to use it to expand production of our new pigment; that way, it would give us a better jump on the big boys and on overseas competition. You can create a lot of new production—and new jobs—for a million bucks. And this town needs new jobs more than it needs crystal-clear lakes, unless you want people to fish for a living. Besides, even if we didn't put anything in the lake, this one still wouldn't be crystal clear—there would still be all sorts of garbage in it."

During further discussion, the only concessions you can get from your boss is that you can spend $10,000 so that one highly visible (but otherwise insignificant) pollutant won't be discharged into the stream, and that if you can come up with an overall pollution-control scheme that will pay for itself via product recovery, your boss will take a hard look at it. You feel that the latter concession does not offer much hope, because not enough products with a ready market appear to be recoverable.

If you were this engineer, what do you think you *should* do? Consider the alternates below:

A. Report the firm to your state and other governmental authorities as being a polluter, and complain about the laxness of city officials (even though the possible outcome might be your dismissal, or the company deciding to close up shop).

B. Go above your boss's head (i.e., to the president of the company). If he fails to overrule your boss, quit your job, and then take Step **A**.

C. Go along with your boss on an interim basis, and try to improve the plant's competitive position via a rigorous cost-reduction program so that a little more money can be spent on pollution control in a year or two. In the meantime, do more studies of product-recovery systems, and keep your boss aware of your continued concern with pollution control.

D. Relax, and let your boss tell you when to take the next antipollution step. After all, he has managerial responsibility for the plant. You have not only explained the problem to him, but have suggested a solution, so you have done your part.

Assuming the Marginal Chemical Co. treated its engineers very well in regard to salary and working conditions, and that the firm was a responsible influence in the community in nonenvironmental matters, do you think most of its engineers would actually take Step **A, B, C,** or **D** if they became involved in a pollution problem of this type today?

CASE 2:   INVOLVEMENT IN SOLIDS DISPOSAL   You are the division manager of Sellwell

Co.—a firm that has developed an inexpensive chemical specialty that you hope will find a huge market as a household product. You want to package this product in 1-gal. and ½-gal. sizes. A number of container materials would appear to be practical—glass, aluminum, treated paper, steel, and various types of plastic. A young engineer whom you hired recently and assigned to the packaging department has done a container-disposal study that shows that the disposal cost for 1-gal. containers can vary by a factor of three—depending on the weight of the container, whether it can be recycled, whether it is easy to incinerate, whether it has good landfill characteristics, etc.

Your company's marketing expert believes that the container material with the highest consumer appeal is the one that happens to present the biggest disposal problem and cost to communities. He estimates that the sales potential would be at least 10% less if the easiest-to-dispose-of, salvageable, container were used, because this container would be somewhat less distinctive and attractive.

Assuming that the actual costs of the containers were about the same, to what extent would you let the disposal problem influence your choice? Would you:

A. Choose the container strictly on its marketing appeal, on the premise that disposal is the community's problem, not yours (and also that some communities may not be ready to use the recycling approach yet, regardless of which container-material you select).

B. Choose the easiest-to-dispose-of container, and either accept the sales penalty, or try to overcome it by stressing the "good citizenship" angle (even though the marketing department is skeptical about whether this will work).

C. Take the middle road, by accepting a 5% sales penalty to come up with a container that is midway on the disposability scale.

Do you think the young engineer who made the container-disposal study (but who is not a marketing expert) has any moral obligation to make strong recommendations as to which container to use?

A. Yes. He should spare no effort in campaigning for what he believes to be socially desirable.

B. No. He should merely point out the disposal-cost differential, and not try to inject himself into decisions that involve marketing considerations about which he may be naive.

CASE 3:   ACTING ON A "MAYBE" AIR POLLUTANT.   Stan Smith, a young engineer with two years of experience, has been hired to assist a senior engineer in the evaluation of air and water pollution problems at a large plant—one that is considering a major expansion that would involve a new product. Local civic groups and labor unions favor this expansion, but conservation groups are opposed to it.

Smith's specific assignment is to evaluate control techniques for the effluents in accordance with state and federal standards. He concludes that the expanded plant will be able to meet these standards. However, he is not completely happy, because the aerial discharge will include an unusual byproduct whose effects are not well known, and whose control is not considered by state and federal officials in the setting of standards.

In doing further research, he comes across a study that tends to connect respiratory diseases with this type of emission in one of the few instances where such an emission took place over an extended time period. An area downwind of the responsible plant experienced a 15% increase in respiratory diseases. The study also tends to confirm that the pollutant is difficult to control by any known means.

When Smith reports these new findings to his engineering supervisor, he is told that by now the expansion project is well along, the equipment has been purchased, and it would be very expensive and embarrassing for the company to suddenly halt or change its plans.

Furthermore, the supervisor points out that the respiratory-disease study involved a different part of the country and, hence, different climatic conditions, and also that apparently only transitory diseases were increased, rather than really serious ones. This increase might have been caused by some unique combination of contaminants, rather than just the one in question, and might not have occurred at all if the other contaminants had been controlled as closely as they will be in the new facility.

If Smith still feels that there is a reasonable possibility (but not necessarily certainty) that the aerial discharge would lead to an increase in some types of ailments in the downwind area, should he:

- **A.** Go above his superior, to an officer of the company (at the risk of his previously good relationship with his superior).

- **B.** Take it upon himself to talk to the appropriate control officials and to pass their opinions along to his superior (which entails the same risk).

- **C.** Talk to the conservation groups and (in confidence) give them the type of ammunition they are looking for to halt the expansion.

- **D.** Accept his superior's reasoning (keeping a copy of pertinent correspondence so as to fix responsibility if trouble develops).

CASE 4: JONES VS. THE PHOSPHATE FOES.   Jerry Jones is a chemical engineer working for a large diversified company on the East Coast. For the past two years, he has been a member—the only technically trained member—of a citizens' pollution-control group working in his city.

As a chemical engineer, Jones has been able to advise the group about what can reasonably be done about abating various kinds of pollution, and he has even helped some smaller companies design and buy control equipment. (His own plant has air and water pollution under good control.) As a result of Jones' activity, he built himself considerable prestige on the pollution-control committee.

Recently, some other committee members started a drive to pressure the city administration into banning the sale of phosphate-containing detergents. They have been impressed by reports in their newspapers and magazines on the harmfulness of phosphates.

Jones feels that banning phosphates would be misdirected effort. He tries to explain that although phosphates have been attacked in regard to the eutrophication of the Great Lakes, his city's sewage flows from the sewage-treatment plant directly into the ocean. And he feels that nobody has shown any detrimental effect of phosphate on the ocean. Also, he is aware that there are conflicting theories on the effect of phosphates, even on the Great Lakes. (E.g., some theories put the blame on nitrogen or carbon rather than phosphates, and suggest that some phosphate substitutes may do more harm than good.)

To top it all, he points out that the major quantity of phosphate in the city's sewage comes from human wastes rather than detergent.

Somehow, all this makes no impression on the backers of the "ban phosphates" measure. During an increasingly emotional meeting, some of the committee men even accuse Jones of using stalling tactics in order to protect his employer who, they point out, has a subsidiary that makes detergent chemicals.

Jones is in a dilemma. He feels that his viewpoint makes sense, and has nothing to do with his employer's involvement with detergents (which is relatively small, anyway, and does not involve Jones' plant). Which step should he now take?

A. Go along with the "ban phosphates" clique on the grounds that the ban won't do any harm, even if it doesn't do much good. Besides, by giving the group at least passive support, Jones can preserve his influence for future items that really matter more.

B. Fight the phosphate foes to the end, on the grounds that their attitude is unscientific and unfair, and that lending it his support would be unethical. (Possible outcomes: his ouster from the committee, or its breakup as an effective body.)

C. Resign from the committee, giving his side of the story to the local press.

4.9

Environmental-Ethics Panel Offers Views and Guidelines

ROY V. HUGHSON
HERBERT POPPER

After most of the reader response to the four case studies in our Nov. 20, 1970, issue had been tabulated, a panel of five experts was called in to offer an analysis, and to see whether some guidelines could be proposed.

The panel was chosen to represent a broad cross-section of the type of people who have to make environmental decisions in operating companies, design-engineering firms and governmental agencies. All members of the panel have degrees in chemical engineering or chemistry, all have some managerial as well as technical responsibility—but otherwise, their backgrounds and functions differ, as described in the boxes on the ensuing pages. For now, we will just mention their names and affiliations: LOUIS N. CARMOUCHE (Dow Chemical Co.); NEIL C. ELPHICK (FMC Corp.); EDWARD T. FERRAND (The City of New York's Dept. of Air Resources)°; IRWIN B. MARGILOFF (Scientific Design Co.); and AARON J. TELLER (Teller Environmental Systems, Inc.).

In addition, four members of the *Chemical Engineering* staff (including the co-authors of this series) took part in the discussion. For simplicity, these editors will all be referred to as "*CE*."

The organization of the discussion was simple. First, *CE* presented a summary of the problem, together with data on how our 1,050 respondents voted. (Since all this material appeared on pp. 106 to 112 of the Feb. 23, 1971 issue, it will not be repeated here—except for a short summary of each problem.) Then, the panelists were encouraged to present their views.

The panel discussion lasted about three hours, and the transcript covered over 100 typescript pages. Consequently, we are able to print only a small portion

---

°Dr. Ferrand substituted, at short notice, for Robert N. Rickles, Commissioner of the Dept. of Air Resources, who was hoping to attend but had to be in Washington that day.

of the remarks (some of which were quite far-ranging and will provide valuable input for future editorial projects).

The words that follow are those of the panelists; we have tried to quote rather than to condense or reword in order to preserve the informal discussion flavor.

CASE 1. THE WATER-PROBLEM DILEMMA. *Summary of Problem.* Marginal Chemical Co., a profit-squeezed firm that is the biggest employer in a small community, is polluting a local lake. The lake has become unfit for bathing or fishing, though partly due to pollution by others. As senior engineer of Marginal's only plant, you estimate that effective control of your plant's effluents will call for $1 million in capital investment, plus $100,000 yr. in operating costs. Your boss, the plant manager, turns this down, because the company can't possibly afford so high an expenditure. He is willing to spend $10,000 to get rid of a highly visible but otherwise insignificant pollutant, but won't even consider a rigorous control system unless it can be made to pay for itself via product recovery—and this does not appear feasible. What *should* you do?

Carmouche: I didn't find any of the proposed courses of action as satisfactory as I'd like. I didn't have enough alternatives. The senior engineer in the case study would have many more options open to him than were described.

There are opportunities inside the company to make its officers aware of the problem, and its effect on the community, without being a blatantly overt destroyer of either your boss's position or your own.

The problem of pollution in the river and the lake is apparently broader than just Marginal Chemical's. Concerted activity by the community *can* solve the larger problem, and alleviate some of the cost burden for Marginal, if the corporation really is important enough to the community. The municipality can sell bonds, build a facility, lease back to the corporation for much lower interest rates (and at no immediate capital cost for Marginal), or do any combination of these things.

Even the million-dollar solution that the engineer proposes can be implemented in stages—it doesn't all have to be done immediately. Product recovery, which the study comments upon, may not pay all the costs, but it sure would reduce the net cost.

So I see this fellow being able to operate inside his company, and in the community, in a much more versatile manner.

And I think he should do that. I don't think he should give up his job at this point—he hasn't even tested the water. I don't think he should go over his boss's head, but I think he should darn well see that he goes with his boss to inform more people.

Margiloff: The case study implies that in Marginal's community, as in many communities, there is a big pollution input simply from the local population. And it isn't at all clear that a comprehensive program taken by an industrial firm can hope to solve the entire problem. Now, this pushes the town into facing the very thorny questions of how much these jobs are worth, and to what extent it is willing to commit its own resources, fiscal and moral, to improving the environment so this plant can continue to do business and contribute the jobs.

It's perfectly conceivable that ultimately there may be no solution—the plant may have to shut down. But many matters ought to be explored, particularly the joint treatment of wastes. I'm familiar with one case where there was joint treatment of industrial and municipal wastes at very little cost to the community.

Marginal's town has been more tolerant of its own wastes than it ought to be. A strong effort is ultimately going to have to be made—for instance, to get both the municipality and the principal industry together in some kind of a special tax district, or to launch some kind of a special construction program to handle these wastes jointly.

Teller: If we take the facts in the case literally, none of the alternatives on the ballot offers a real solution to Marginal's problems. If the company continues to pollute, the chances are that the city will close it down. The other alternative is to control via the conventional system that has been proposed. Which means that they're out of business either way—the company could not exist with a million-dollar abatement system.

The real question is, has the engineer applied sufficient engineering and innovative capability to come out with recycled components so that the system will not cost as much to install or operate?

The engineer should go to management, establish what the alternatives are, and set up a program so that a system of less loss, less waste, can be instituted at the plant. If innovative engineering can be applied, at least on a pilot-scale unit, to determine whether this can be done at a reasonable cost, it should be recommended.

To return to the original problem, I think *CE* put in a restrictive boundary. This plant manager represents the kind of view of a plant manager that I had when I was a junior engineer. We caricaturized him—he's embittered, stubborn, vindictive. When I became a plant manager, I saw the plant manager as pleasant, receptive to ideas, broadminded.

So, I think you've placed boundaries that are not necessarily applicable in a real situation. The fact is that it's up to this young engineer to help the company survive. It's going to survive only if it solves its pollution problems—because that's what the public has decided. And it's going to survive only if it's done in an economical manner.

The young engineer is going to have to leave convention and go to innovation. That's *his* alternative. He can't sic the public on the company; you just don't do that. His job as an engineer is to see that this company prospers.

CE: Dr. Ferrand, instead of putting yourself in the engineer's position on this problem, perhaps you might want to make some comment from the government enforcement viewpoint. Do you feel that there are special cases where enforcement ought to be slowed down to adapt to the company's ability to make the change?

Ferrand: Society has pretty well accepted that pollution has got to go. Firms that can't solve serious problems will have to shut down, just like a number of companies have shut down in New England for other economic reasons—just shut down and walked out of town.

I don't think that you can get away from the fact that something has to be done. A company like this is just not going to be allowed to pollute. It either has to come up with a solution, or just go out of business.

And if it means that some people are going to go out of business, I think that this is going to have to be accepted as part of the free-enterprise system. This is the way it works. You've got your costs, and you have your profits, and this is part of the whole game. They just can't operate if they're going to continue to pollute.

Elphick: If I were the engineer in this problem, instead of taking the initial reaction I got from my boss, I think I might go back two or three times, which is what happens in real life in business, I think.

I'd go back and I'd try to sell it some other way, some other day, with some other approach.

There's one other thing I wonder about. This is the plant's senior engineer. And he's come up with a study that results in a capital investment of a million dollars. This must have taken considerable engineering and thought and development on his part, if not his subordinates. And I wonder why his boss just found out about it. Perhaps the engineer would have gotten a better reception if he had involved his boss in the study at an earlier stage, instead of bringing him a great big surprise package.

CASE 2. INVOLVEMENT IN SOLIDS DISPOSAL. *Summary of Problem.* Research on a container for a new household product shows that municipal disposal costs for the various possible types can vary by a factor of three. The company marketing expert thinks the hardest-to-dispose-of container has the most sales appeal. He estimates a 10% sales penalty if the most easily disposable container is chosen, because it would be less distinctive and attractive. If you were the division manager with overall responsibility for this product, and assuming the actual costs of the containers were about the same, which would you choose?

Margiloff: I think most of us would recognize that the whole field of marketing and of package design is subjective. It's subject to persuasion. It isn't something founded in science as reactor design is, for instance. And this is probably an area where the will to change the recyclability or disposability of a package can have a very great effect on the nature of the eventual business decision.

One probably ought to choose the easy-to-dispose-of container, and get the best talent involved in creating an attractive physical form.

Throw out the consultants who say you can't sell it, and hire ones who say you can and who tell you how. Because as someone used to say, "If you don't like what one lawyer tells you, fire him and hire a different one." If it's true in law, it ought to be true in marketing.

Ferrand: One of the things that has to be considered is the attitude of the governmental agencies that can wash out all these considerations very quickly. For example, if somebody comes to us and says we're going to bottle something in, say, polyvinyl chloride and sell it in New York City, my advice to him would be, "Don't bother." It very likely will be prohibited.

If it looks like it's going to have a significant effect on the environment, we tell them to go slow, and evaluate it very carefully before they invest any kind of money in it.

Carmouche: It seems to me that's unfortunate. The technology to burn halogenated plastics already exists and can be used. But because a community doesn't choose to go that route, then it bans the product.

Ferrand: Well, if we had the disposal facilities that could handle it, existing right now . . .

Margiloff: We can build such facilities.

Ferrand: It's easier said than done.

Margiloff: It's easier said than paid for.

Teller: There seems to be a commitment, almost, that we will assume that the existing technology is all we have, to cope with the environment. No mention seems to be developing of the creation of a technology that is unique in itself that *can* cope with environmental hazards.

I'm a little incredulous, because we are the engineers. But the bounds here establish an assumption of an existing technology. *I* assume we're going to build *better* incinerators. We may be able to take advantage of pollution for the benefit of society if we do this.

Isn't there a forward-looking approach, rather than a hindsight approach, that we should be using?

Carmouche: I'm concerned that we're going to *ban* things instead of learning how to dispose of them, or use them, or recycle them. But, we ban them first . . .

Teller: The easy way.

Carmouche: Yes, but it's not the long-range beneficial way for society.

CE: Can you be a little more specific about approaches to the container-disposal problem?

Teller: Well, for example, if you talk incineration, I assume that the concern is with the disposal of a glass, PVC, polypropylene or paper container. While your paper, PVC and polypropylene can be burned (of course, you can't "burn" glass), most dry systems for control of pollution from incinerators are utter failures for the level of achievement that we want.

Halogen control systems? We've built them for many companies, and for much higher chlorine concentrations. These systems have been disposing of chlorinated materials for years. They are far more efficient for particulate removal than anything that we've got on incinerators in the municipal enterprises. And they're less costly. Halogens really don't constitute a technological problem in combustion.

We ban halogens because we don't have the control equipment on our municipal incinerators—but really because we have an obsolete piece of equipment that doesn't work anyway!

Another factor is that we are a full-belly society. We're accustomed to a certain degree of convenience. We have alternatives. We can go back to the inconvenient society, but there's going to be a slight case of revolution among the people before they accept it.

Margiloff: I don't think it's quite out of the question to say that some of our habits ought to be unlearned, or ought to be modified. Our habits are very much the result of the particular way our economic system has operated.

Teller: In everything that's happened to date, there's been an evolutionary form. Admittedly, there is a wastefulness in the American society. A fellow I've been talking to said it made him proud to be an American because "we have more garbage than anybody else."

The point is, are you going to make the changes in an evolutionary manner, or are you going to upset the society, and hide behind ancient technology?

Ferrand: We don't want any kind of progress that's going to make our problems worse. Now I think progress, so-called, has got us into a fix up to now, and any future "progress" is going to be looked at with a jaundiced eye. Yes, I think it's a sad state of affairs when we have to impose restrictions on the free enterprise system, but there are times when it has to be done.

We have made a major improvement in the sulfur dioxide content in the air—at the expense of the pocketbooks of the citizens of New York. And I think that any other improvements we make are going to be at the expense of the people's pocketbooks.

Teller: New York has reduced the sulfur content of its air by using a low-sulfur fuel, and now we're having a low-sulfur-fuel shortage.

Elphick: And we're in a sulfur glut.

Teller: Right! But just take your solid-waste problem. We are still hauling garbage (at \$27/ton in New York) down to the central incinerator. We are still burning it, and we are still not controlling the emission.

Yet the garbage in New York is worth 1,600 megawatts! The mayor and Con Edison have battled over new power plants, but 1,600 megawatts, the quantity we need in New York, resides untapped in our garbage. The only reason we haven't tapped this is that we've looked at garbage disposal as a necessary evil, and not in terms of what might be recycled in the process, or sold.

CE: We have time for just a few more words on Case 2. Any last thoughts?

Carmouche: The marketing estimates quoted in this container-selection case study really show no discernable difference; 10% is within the limit of error. I think that if the engineer presented his case right, he'd be heard and would cause the container-of-choice to be the disposable one.

Ferrand: How about if the difference were 50%?

Carmouche: He's got a problem with that. (*Laughter*) But he should still present the thing in a way that will get heard the best.

CASE 3.   ACTING ON A "MAYBE" AIR POLLUTANT.   *Summary of Problem.*   A young engineer is involved in a study to see if a planned new facility to make a novel product will meet all pollution codes. He finds that it will. However, one of the byproducts, not covered specifically in any code, has been implicated in one earlier study as a possible cause of minor respiratory disease. Emission of this byproduct cannot easily be controlled. The earlier study had remained obscure, and its existence was apparently not known to a local conservation group that had voiced opposition to the new plant for less specific reasons. The young engineer's supervisor feels that the data in the old study are too vague to necessitate a halt or costly delay in the plant-building project, particularly inasmuch as the plant will be meeting all existing codes. What course of action should the young engineer now choose?

Elphick: I don't see that the company in the third case study has a great deal of choice. There's sufficient doubt, so something's got to be done. There's no positive assurance that this particular effluent is not going to damage the environment—is not going to harm people.

Someone has to recommend a crash research program in this particular area, with whatever agencies and whatever experts can be mustered to the cause, and get an answer as quickly as possible. Is this effluent really damaging? And, if it is, how can it be captured?

If the answer is that everything's all right, you've still got a live project. If it turns out that the effluent is damaging, and no control method is available, then you may have to wash the whole project down the drain. There are companies that would.

Carmouche: I look at this whole thing as a communications problem within the company. Before we can suggest a course of action to the young engineer, we've got to know some things: Who besides this engineering supervisor is receiving the data about Compound X? And what is the general policy in regard to that type of problem?

If a communications network is at all normal, there are likely to be several people aware of the potential hazard. These people are likely to represent disciplines not under the control of the senior engineer. And this group would serve as a forum for discussion of the problem and the working out of an acceptable procedure to deal with it.

On the kind of data presented, it's difficult to gain much real thrust to stop the project. There's plenty of reason, however, for the company to immediately initiate tests in a recognized biochemical lab, to find out if Compound X is a threat to public health.

It's inconceivable that they couldn't control it if they found that they needed to. It would cost them more money, but they've got to take that risk and go ahead with the project.

Elphick: Right!

Margiloff: I think the question is how you handle uncertainty. The uncertainty is how to prove that an emission is harmful. Proof consists of long-term inferences, threshold studies, the elimination of background interferences, and the study of synergisms.

All of these things make it very difficult to prove beyond reasonable doubts that a particular emission is pollution and not a harmless emission—if there is any such thing.

We handle uncertainty in different ways. A development man generally takes an optimistic view. That is, if we're on a course of discovery, we have to assume that there's something there to discover. If we are in design, we generally handle uncertainty by overdesign, by conservatism.

In the pollution area, we generally handle uncertainties by conservatism or pessimism. In this case study, we have come upon the central aspect of dealing with pollution. There are very great uncertainties. Honest differences of opinion can occur and can continue to occur for decades, even when all parties are faced with the same evidence.

CE: We gave a talk in which we discussed these problems with junior and senior chemical engineering students at a local college, Cooper Union. One of the students seemed distressed because we hadn't given him all the data he needed. You just don't know everything about this sort of thing, and . . .

Carmouche: You never will.

CE: Yes, but he didn't realize that. We told him that sometimes you have to make decisions even when you don't know everything. But he felt that, scientifically, he had to have all the data. Otherwise he couldn't make a decision.

Carmouche: Your young engineer is going to find out, when he gets out in industry, that people that have to have all the data before they can make a decision never get paid very much.

Ferrand: This is true in all fields. In my area I have a lot of scientists working for me. And one of my major problems in managing is to get these fellows to act, because they want to turn everything into a detailed three-year study when I need the answer tomorrow. Or, maybe a week from now. But I need some answer—the best answer they can give me.

CASE 4.   JONES VS. THE PHOSPHATE FOES.   *Summary of Problem.*   Jones, a chemical engineer, is the only technically trained member of a citizen's pollution-control committee working in his city. Some members of the committee have started a drive to ban phosphates from detergents sold in that city. Jones opposes such a ban because he believes that there is still a question as to the effect of phosphates (e.g., compared with phosphate substitutes), and also because detergents furnish only a minor part of the phosphates in the city's sewage, and this sewage flows into the ocean where phosphates may not cause much harm anyway. But Jones cannot convince the other committee members that the ban would be ill-advised or premature. A crisis is reached when they accuse him of stalling in order to protect his employer, who has an interest (though minor) in the detergent-chemical industry. If you were Jones and held his views, what would you do now?

Margiloff: I think that the alternatives aren't politically realistic in terms of the way our society operates. You can't expect unanimity on every issue.

It isn't reasonable to be a fanatic and to say, as the engineering individual here was given the opportunity to say, that "I've got to fight this to the very end." What we really have (and I've served on a number of quasi-public and informal committees) is a system where even if occasionally you may find some infallible wisdom as an input to the process of making decisions, it is almost never found in the output.

What Jones should do in my opinion is to vote "no," state his reasons, and say to the members of his committee, "I'm not with you on this—you haven't persuaded me. I'll be with you on other things, as you will with me, but it isn't clear to me that banning phosphates is going to do a bit of good in this town. I don't think it is necessary or wise for me to resign, and I hope you don't think so either."

Carmouche: If that engineer resigns from the committee and gives his great story to the press, it may make the papers for one weekend, but from there on out his voice is still—unless he gets another committee started.

However, if he stays in there and battles, he's news all the way. In the first place, he can prove that the course of action the committee is taking is ineffective—even if you ban all phosphates from detergents, you don't do the job, not even a third of it. Secondly, he can show engineering results from a number of installations around the country that have attacked this whole problem in a much better fashion by taking all the phosphates out of sewage effluent. These systems are commercial, and they allow you to put as much phosphates as you want into a detergent.

Elphick: Speaking in more-general terms, Jones' ethical responsibility is to try and bring to his community a rational and scientific understanding of its environmental problems. And if, because of his position, this committee gets him thrown off, and it breaks up as a result thereof, I say, good. That committee didn't deserve to exist in the first place, and will not become a proper force in the community.

Ferrand: Well, if you play by the rules of the game we have set up—if you ask yourself what you should do if you were Jones and shared his viewpoint—I agree; stick to your guns. I think committees have come a long way, and I doubt whether a committee would throw its only technical expert overboard over one issue if he has been effective on others.

THE GENERAL QUESTION OF ENVIRONMENTAL ETHICS

CE: There are a number of reader comments that we might talk about.

One of them involves the theory that although a professional engineer is spontaneously ethical, in practice he needs a little help through appropriate coercion. Legal qualification and registration of all engineers would be the first step.

Somebody else suggested that to give the engineer support where he had a dispute with a management of the Marginal Chemical Co. type, the AIChE or some other organization should have an Environmental Ethics Review Board that would apply codes of conduct as fairly and uniformly as possible.

We just wondered if you have any thoughts on these suggestions—or if you think such steps are unnecessary because the engineer may have more influence than he realizes, or because they wouldn't do any good anyway.

Carmouche: Or because even if "Ethics Boards" try to be effective, they bog down. I've never known one to be very successful in trying to legislate or committee-ize ethics. I don't think you can do that. I think it's an inherent responsibility in the individual to police his own ethic.

Elphick: You're legislating honesty, and that's just not possible.

Margiloff: It seems to me that the area in which the professional association can do the most good is not in adopting legislation passing on the ethics of individuals, but in serving as a common meeting ground for people in certain professional or technical areas where one can shed the supervisor-subordinate relationship, and discuss matters on a scientific or technological plane, divorced from competitive pressures.

Carmouche: Engineers, it seems to me, have one responsibility as a group. The public in general is not aware of facts. When the facts are presented, they're usually so damned dull that nobody ever reads them and, secondly, nobody ever seems to give an honest interpretation of what such-and-such a happening really means to the guy in the street.

So I think that engineers generally have a responsibility to communicate facts, and interpret them as objectively as they can.

CE: You may be interested in one thing *Chemical Engineering* has been doing. Since last spring, every month we've sent a small four-page newsletter to members of Congress. We call it "Technology Alert from Chemical Engineering," and it's a brief rundown of technical developments in our magazine that might have some impact on the national life in one way or other. But let's return to what the engineer can do in the communications area.

Margiloff: The question is whether, in the public's mind, anyone employed in industry is necessarily an "industry spokesman."

That's really the core of the problem with respect to so-called industry representatives on state boards of pollution control. Are they representing their employers and industries, or are they professionals who represent the public and who just happen to derive their income from industry?

There is a taint attached, that anyone who derives his income from an industrial employer automatically expresses the viewpoint of industry, rather than his own personal convictions.

Carmouche: If there's any taint attached to the group that works for industry, that taint is there only because it's assumed by the public. What has not been brought to the public's attention is that everybody—but everybody—is a member of some kind of group. Each group has its own biases, and they may be just as bad, or just as good, as industry's.

Margiloff: I think you can say that quality will out eventually. If your point of view is based on sound technical assessment, and especially if the others are not, then eventually you're going to be vindicated.

Carmouche: Engineers could do industry and their employers a service if they'd take the trouble to do a little of what I call preventive maintenance.

In pollution, industry has a reasonably good story to tell on many occasions, but it seldom tells this to the public until after something's happened that has put it behind the eight-ball. Engineers as a group could start helping to educate their employers—and I mean that literally—to the advantages of going to the press and to the city governments and pointing out the things they're doing, and how they're doing them, and the programs they have going. This would offset some of the blind reaction that comes when some little thing happens.

We've done this for years in Midland, Mich., because we have been a big chemical plant in a little city, on a little river, and the townspeople all know what's going on at Dow.

There may have been a fish kill once in a while by a leak or a pipe breaking, or some other source, but they also know that we'll shut those off, and there are now big holding ponds for automatic diversion of all that stuff. They know these programs, and they just know the river isn't going to get fouled up. And it doesn't.

But if we had never told anybody what we were doing, and then killed a thousand fish in the river, we'd have the whole state water-resources commission on our neck.

As it is, when somebody from the community calls up the water resources commission and says, "Dow did so and so," the commission says, "Yes, we know it, they called us yesterday and told us about it. We know their program to take care of it, and we also know what their long-term program is."

Elphick: I think engineers should get involved inside their own organizations; and chemical engineers in particular, because they represent the industry that has the greatest potential for causing trouble.

But an engineer, by his very background and training, can bring a scientific approach to environmental problems. And we are going to solve our problems with science, not with emotion.

Carmouche: Well, the fact is, as has been said many times by many people, that pollution control is not a technical problem; it's social, it's political and it's economic, but it's not technical. Most of the technical solutions already exist.

Elphick: Dr. Teller isn't necessarily satisfied with all the solutions—I'll say that in his behalf.°

Carmouche: I am not satisfied with all the solutions either. There *are* better ways!

EDITORIAL POSTSCRIPTS

At this stage, we (Popper and Hughson) take the liberty of summarizing our own feelings and making a few brief final observations about some aspects of our "environmental-ethics" project:

1.   A very large number of engineers and technical managers *in all age groups* are concerned about pollution problems. Based on previous experience, 1,100 ballots represent a very large return for this type of survey (i.e., one that requires considerable thought and yet offers no premium—not even a return envelope). Of course, this still does not necessarily make the answers statistically valid for our entire readership, but the pattern of answers and job titles does indicate across-the-board concern.

2.   The majority of engineers feel that the professional man should work from within in order to improve his company's responsiveness to a pollution-control problem; but a large number did feel that he should contact outsiders (e.g., control agencies) once he has exhausted all internal channels. And in this connection, the ballots—and particularly the panel discussions—pointed up that "all internal channels" means more than just talking to the boss; it means working with company committees, taskforces, staff specialists, and so on.

3.   In using the above channels effectively, a documented, carefully thought-out presentation—as well as the knack of being able to "disagree agreeably"—can be a real asset. It is a mistake to suppose that any pollution-control idea that you consider promising will automatically sell itself—you have to demonstrate why it is more promising than other approaches, and better for the company's long-range interests than the status quo. †

°Dr. Teller had to leave after the first two hours of the discussion, and was no longer present at this point.

†For some pointers on how to sell ideas, see: "Good Ideas Need Selling!," included in "Modern Technical Management Techniques," Popper, H. (ed.), McGraw-Hill, 1971.

4. In this connection, some companies' top managements may also have some communicating to do to demonstrate to lower-level managers that they are solidly and sincerely behind environmental protection. And this involves not only making speeches but also backing them up by making pollution-control success or failure an important factor in promotions and salary increases for the production and engineering staff. Although industry has come a long way, there are plant managers around (like the hypothetical one from Marginal Chemical Co.) who feel that their chances of getting a promotion are based solely on the profitability of their plants—that while good environmental control may be a nice "plus," especially if you can make it pay for itself via product recovery, it's the short-range profitability comparisons that hold the key to personal advancement.

5. Most readers and panelists felt—and we agree—that the preferred road to a better environment lies in improved technology and more thorough application of existing technology rather than in the abrupt banning of products such as phosphates or nondeposit bottles (or, for that matter, internal combustion engines). Granted, as Dr. Ferrand pointed out, there are times when a municipality or state must ban something that it lacks the technology or funds to dispose of safely, or something whose disposal costs are out of proportion to its value to begin with. But, in general, we feel that the environmental emphasis of the 1970's should be on better or more-broadly-applied control technology, and on improved processes that produce fewer harmful emissions to begin with, rather than on the wholesale banning of products. The "banning" approach, applied indiscriminately, runs the risk of eventually losing public support for environmental improvement—e.g., because of the jobs that are banned along with the products, or because hastily conceived substitute products may work a hardship on the public without enough offsetting benefits.

6. There is a need for better communication in the whole environmental area—not only vertically, between the engineer and management, but also horizontally, between the engineer and his professional societies and his technical magazines; *Chemical Engineering* will continue to do what it can to further this horizontal communication process, and to furnish the engineer and manager not only with technical ideas to use on pollution problems but also with the latest thinking in the "environmental ethics" area. Of course, it now appears that the latter area will also be getting "outside" attention—note, for instance, consumer-advocate Ralph Nader's article on "A Code for Professional Integrity" in the Jan. 15, 1971 issue of *The New York Times*.°

°In this piece, (which appeared too late to be discussed by our panel), Nader proposes three basic changes to end what he feels is the reluctance of engineers and scientists to dissent openly from their employer-organizations' policies; these changes would involve Congress, professional societies and professionals themselves, and would be aimed at assuring that the competent professional employee who "exercises his right of dissent" is not made the victim of arbitrary treatment.

Nader has just followed up on this proposal by setting up a new organization to encourage what he calls "responsible whistle-blowing" by engineers, scientists and other professional employees of industry and government. The organization, called the Clearinghouse for Professional Responsibility, will operate as part of Nader's public-interest group (which consists of ten young lawyer-lobbyists), and will accept information in confidence.

POLICY IN THE SEVENTIES

One more postscript about the environmental/ethical/economic interface during the 1970's:

It is generally recognized that a decade of large pollution-control expenditures by the chemical process industries would improve their "sociological posture," but it is less generally recognized that profitability might also improve.

Let us say that during the seventies we stay on the capital-expenditure curve envisioned in the sixties, but spend more than envisioned on pollution control, and make offsetting cuts in new-capacity types of expenditures. Some companies are already embarked on such a course; if they stay with it, and others follow, the results could well include:

- A great increase in environmental technology that has worked successfully in the CPI, and that the CPI can then apply profitably to other industries and municipalities.
- An increase in product yields, and in general processing efficiency.
- A sounder balance between capacity and demand, and hence a fairer price structure.

In connection with this latter point, the overcapacity that characterized some of the CPI's products during a good part of the past decade has seldom brought big windfalls to the consumer—e.g., even substantial, overcapacity-induced cuts in the price of synthetic fibers have had rather unsubstantial effects on the price of a suit or dress. A rechanneling of some capital funds from expansion into pollution control would thus seem to hold more advantages than disadvantages for the general public, as well as for the CPI.

Of course there are two requirements for making this "rechanneling" policy work out fairly: (1) sensible, nonconflicting codes that spell out minimum requirements in such a way that there is no temptation to steal a march on competitors by spending less on pollution control than *they* are spending, and (2) trade policies or standards that prevent unfair competition from goods produced in those countries that do not require industry to pay for a reasonable degree of pollution control.

Let us close on a hopeful note. Perhaps the present concern for the environment will have a bonus for the CPI—one that will enable them to reestablish their status as all-around growth industries where profits grow in line with physical volume and in line with the "better living" supplied to the general public. As the tangible nature of this bonus becomes more apparent to more people in more companies, it could well be that the type of ethical problems we have examined in this series will become less frequent, because the objectives of the engineer, of his employer and of society will be in close accord.

In the meantime, we hope this series will prove helpful, and we thank all who contributed their thoughts.

Selected Additional Readings

Adkins, A. W. H., *Merit and Responsibility*, University of Chicago Press, Chicago, 1975.

Aristotle, *Nicomachean Ethics*, Bobbs-Merrill, Indianapolis, 1962.

Baum, R., *Ethical Arguments for Analysis*, Holt, Rinehart, Winston, New York, 1976.

Bazelon, D. L., "Risk and Responsibility," *Science*, July 1979, pp. 277–280.

Bok, D. C., "Can Ethics Be Taught?" *Change*, October 1976, pp. 26–30.

Bok, S., *Lying: Moral Choice in Public and Private Life*, Pantheon, 1978.

Callahan, D., Hunt, P. B., MacIntyre, A., Barker, B., Gray, B., and Cooke, R. A., "Ethics and Regulations," a series of articles, Hastings Center Report, February 1980, pp. 25–41.

Cebik, L. B., "Ethical Trilemmas," *Technology and Society*, Committee on Social Implications of Technology, IEEE, December 1979, pp. 4–6.

Chalk, Rosemary, "Ethical Dilemmas in Modern Engineering," *Technology and Society*, Vol. 9, March 1981, pp. 1, 8–12.

Cohan, R. M., and Witcover, J., "The Investigation and Resignation of Vice President Spiro T. Agnew," *A Heartbeat Away*, pp. 91–96, Viking Press, New York, 1974.

Durniak, A., "—And Three Leave G.E.," *New Engineer*, May 1976, pp. 35–36, 39–40.

Epictetus, *Enchiridion*, Bobbs-Merrill, Indianapolis, 1963.

Garner, R. T., and Rosen, B., *Moral Philosophy*, The Macmillan Company, New York, 1967.

Gert, Bernard, *The Moral Rules*, Harper & Row, New York, 1966.

Goodpasture, K., and Sayre, K., "An Ethical Analysis of Power Company Decision-Making," *Values in the Electric Power Industry*, edited by K. Sayre, pp. 238–288, University of Notre Dame Press, Notre Dame, IN, 1977.

Houston, C. W., "Experiences of a Responsible Engineer," *Proceedings*, Conference on Engineering Ethics, American Society of Civil Engineers, May 1975, pp. 25–30.

Hughson, R. V. and Kohn, P. M., "Ethics, Ethics, Ethics. . . ," *Chemical Engineering*, September 22, 1980, pp. 132–147.

Kant, I., *Foundation of the Metaphysics of Morals*, Bobbs-Merrill, Indianapolis, 1969.

Kohn, P. M. and Hughson, R. V., "Perplexing Problems in Engineering Ethics," *Chemical Engineering*, May 5, 1980, pp. 96–107.

Leiman, S. K., "The Ethics of Lottery," *Quarterly Report*, The Kennedy Institute, Summer 1978, pp. 8–12.

Lockhart, T. W., "Safety Engineering and the Value of Life," *Technology and Society*, Vol. 9, March 1981, pp. 3–5.

Machiavelli, N., *Discourses*, Modern Library, New York, 1950.

Machiavelli, N., *The Prince*, Modern Library, New York, 1950.

MacIntyre, A., *A Short History of Ethics*, The Macmillan Company, New York, 1966.

MacIntyre, A., *After Virtue*, University of Notre Dame Press, Notre Dame, IN, 1981.

Mantell, M. I., *Ethics and Professionalism in Engineering*, The Macmillan Company, New York, 1964.

Marcus Aurelius, *Meditations*, Bobbs-Merrill, Indianapolis, 1963.

Mavrodes, George I., "The Morality of Chances," *The Reformed Journal*, March 1980, pp. 12–15.

Mayer, R. B., "I'm Not That Hungry—Yet!", *Consulting Engineer*, December 1975, pp. 42–44.

McCuen, R., "The Ethical Dimension of Professionalism," *Issues in Engineering*, American Society of Civil Engineers, April 1979, pp. 89–105.

Mill, J. S., *Utilitarianism*, Hackett, 1979.

Miller, S. H., "The Tangle of Ethics," *Harvard Business Review*, January–February 1960, pp. 51–64.

Nietzsche, F., *On the Genealogy of Morals*, Random House, New York, 1967.

Plato, *Protagoras*, Bobbs-Merrill, Indianapolis, 1956.

Popper, H., and Hughson, R. V., "Engineering Ethics and the Environment: The Vote Is in!," *Chemical Engineering*, February 22, 1971, pp. 106–112.

Price, D. K., "The Ethical Principles of Scientific Institutions," *Science, Technology and Human Values*, Winter 1979, pp. 46–60.

Reed, E., "The Engineer Faces Unethical Practices," *Mechanical Engineering*, March 1977, p. 83.

Stanley, C. M., "A Jekyll and Hyde Approval," *Consulting Engineer*, December 1975, pp. 45–48.

Turley, Paul W., "Professional Codes of Ethics: A Handbook of Conduct or an Insulation from Competition," CSEP Occasional Papers, No. 3, December 1980, Center for the Study of Ethics in the Professions, Illinois Institute of Technology, Chicago.

Professionalism: In Whose Interest?

Characteristically, the professions have sought to distinguish themselves from other commercial endeavors by pointing to their public service components, the specialized training they generally require, the close relationship between client and professional. Yet none of these elements is absent from the services provided by auto mechanics, insurance salesmen, or butchers. The level of education demanded may be higher, the scarcity value created greater—but the fundamental ecomomic nature of professional transactions is no different from that of other consumer-producer confrontations.

John H. Shenefield
Antitrust Division,
Department of Justice

The question is usually put: Do the professions serve their own interests or those of the public? Thus, for example, several Supreme Court decisions have forbidden professional practices that, according to the court, constituted the exercise of monopoly power and thus could serve only the interests of the profession and harm the interests of the public. In many ways, however, this particular question is simplistic. The question seems to presuppose that a profession and the public are each a kind of superindividual, and we tend then to think of them and to act toward them as if they were individuals with interests, etc. In truth, both the professions and public are very complex groups composed of individuals. The real individuals who make up these groups do have interests—a wide variety of often contradictory interests. Frequently these interests come into conflict, for the individual professional is also a member of the public.

"Profession vs. public" is, therefore, a misleading description of a complex situation. The simple question should be replaced by a series of questions. What interests do real individuals in our society have? How are their interests served or harmed? Which interests need to be protected, and what are the best ways to protect them? This permits us to see that the professions and the public are not the sorts of things that have interests. They are concepts constructed for convenience that (1) presuppose that certain interests need to be protected and (2) are themselves solutions to that need, although not necessarily the very best solutions.

The organization of members of society into *The Public* as a group having certain rights, obligations, and duties is such a solution. The Public is designed to protect certain interests that real individuals have—the Declaration of Independence and the Constitution make this clear. The Public has come by custom and law to expect a certain level of performance from *The Professions* and has granted The Professions certain privileges in return. The Professions are likewise solutions to the need to protect certain interests of certain individuals, for example, the reputation, economic well-being, and personal development of members of the profession—and also to satisfy the obligation to protect the health, safety, and welfare of The Public. Each of the articles in this chapter addresses a part of this issue and some of the questions posed above.

Ralph Nader in "The Engineer's Professional Role" looks at the interests of individuals who use the technology that engineers produce and suggests ways that engineers through their professional organizations could better protect these interests. In "Can Professionalism be Attained within the Corporate Structure?" Martin Goland addresses the apparent predicament of the engineer employed within a corporation. Taking the position that "the virtues of professionalism are independent thought, the free expression of opinion, the creative dedication to the dual and compatible goals of advancing the employers' interests while serving the just needs of the society," Goland argues that it is the responsibility of engineers to find the solution to the questions and problems inherent in the circumstances of their employment.

Robert Baum's article, "The Limits of Professional Responsibility," presents the dilemma of the engineer working on a project of which he or she does not approve. Baum points out that the profession, primarily through the instrument of the codes of ethics, fails to protect such an engineer's interests. Robert Whitelaw is of the opinion that engineers in the corporate setting have certain interests that are protected neither by the law nor by the professional organizations. In the article "The Professional Status of the American Engineer," he identifies those interests and proposes protecting them through a "Bill of Rights."

Michael Bayles in "Against Professional Autonomy" returns to the interests of nonprofessionals and argues that the professions as organized today cannot protect these interests. Inasmuch, he says, as the professions have been given a wide area of autonomy so that they could protect those interests, this autonomy should be withdrawn. Milton Lunch outlines the recent court decisions that dealt with the professions' right to set standards of conduct and practice and analyzes the reasoning apparent in the decisions. He argues that, whether the courts, the public, or the professions realize it or not, these decisions are dismantling an apparatus that was intended to protect a variety of interests and are replacing it with an economic apparatus that protects only a short-term interest we have in marketplace competiton. In a short editorial comment that appeared originally in the *Washington Star* and later in *Professional Engineer*, Edwin Yoder reflects on the contemporary penchant for interpreting all relationships in economic and commercial terms and the effects of this on the professions.

Discussions of these and other issues of concern to the professions and professional organizations often take place without a clear concept of what constitutes a profession. Not uncommonly, debate degenerates into argument over definitions of "profession," "professional," etc. Morris Cogan discusses the difficulties involved in defining a profession, discusses the characteristics of a profession, and suggests ways to overcome some of the difficulties inherent in the problem of definition in a society that uses the term loosely.

The final two articles in this chapter discuss more specific issues within engineering as a profession. In "Professional Licensing and Public Policy," Robert Reich analyzes the practice of professional licensing. He argues that licensing is a question of balancing our interest in marketplace competition with the interest in protection of the consumer and outlines ways in which this can be done. Carl Nelson and Susan Peterson analyze a pattern of systematic avoidance of confrontation that they find in engineers. They suggest this avoidance of confrontation by engineers is a form of ethical "drift"; that is, the engineer considers his technological solutions to be value free and the ethical questions that may arise with the use of those solutions are someone else's problem. Nelson and Peterson conclude that the common practice of avoidance of confrontation of ethical questions by engineers serves to protect neither the interests of engineers nor the public.

The Engineer's Professional Role: Universities, Corporations, and Professional Societies

RALPH NADER

The theme of this Middle Atlantic Section meeting of the American Society for Engineering Education is "Engineering Education in a Changing Society." Perhaps a more realistic title would be, "The Engineering Crisis in a Changing Society"—a crisis which comprises university, industry, and technical society circles. Visible to all of us are the consequences of this crisis—the endemic pollution of air and waterways, the contamination of soils and food products, the traumatic epidemic on the highways, the old and new forms of industrial dangers to workers, chemical and radiation hazards, and many others which are accumulating into a massive assault on the human biosphere. Not so visible are the attitudes, conditions and disincentives which have prevented a comprehensive development of remedial engineering to reduce markedly the social costs of private enterprise and ameliorate the painful by-products of great engineering accomplishments.

TECHNOLOGY OUT OF CONTROL

At the outset, it is useful to frame our inquiry in concrete terms. People are being killed, injured, and inconvenienced by the products and processes of technology. Many of these casualties are overtly traumatic and therefore promptly observable. Others, and this is the trend of the affliction, are much more insidious in their impact and increasing debilitation—to wit, cigarettes. Insofar as technology does things to us that we do not want to endure, to that extent can it be called out of control. As long as there is undue and parochial attention paid to the short range economic utility of product and process at the same time that the short and long range biological consequences are treated with indifference or contempt, our society is going to plunge into deeper collective cruelties.

What is now out of control may soon be running amuck in an arena of macabre anarchy so enmeshed in giant bureaucratic structures whirling in furious activities over *means*, that the accountabilities for the *ends* of human welfare are blurring more and more. And with the jet-paced growth of new technologies, full of potential for both ease and unease, the lag between a framework of responsibility for the safety of the manmade environment and the increasingly far-reaching impact of corporate decision-makers threatens to render the future significantly more challenging to our humane values than the past.

All this may sound inexorable, as did the dreary projection of traffic death and injury statistics until very recently. After decades of inaction, the Nation has begun to establish a more rigorous framework of responsibility for reducing highway casualties. Fundamental to the emergence of this national policy on motor vehicle safety is the recognition of *a value* and *a capability*. The value was the right of individuals not to have their physical integrity violated by hazardous vehicles— whether by product design or construction. The capability is an engineering one— the capability to invent the technological future once we decide that we want the benefits of such a future.

The coupling of deeply felt values with graspable remedies represents a very dynamic impulse to reform. Yet, why did auto safety reform occur so late—in 1966—and not two or three or four decades ago? Why have there been so few programs—and so late—to counteract other environmental and consumer hazards developed by the hand of man? Why do we continue to wait for disasters or near disasters before some action commences? Why have the "bodily rights" of people against the incursions of old and new technology been so inadequately articulated and protected?

WHERE ARE THE ENGINEERS?

One summary of these questions is another question: *Where are the engineers?* This audience should be the last to think that such a question attributes to engineers a mission of unattainable grandeur. But it certainly is a grandeur unattained. The role of the engineer, as befits professionalism, should be a dual one—to meet empirical human needs and to insure that the human costs of meeting these needs are kept to the barest minimum. The contribution of his expertise is to foresee and forestall the risks of his innovations, both individually and, as in the case of pollution, cumulatively.

But what is the practice? On one public safety frontier after another, the engineer is not in the vanguard. Indeed, it is the rare occasion when he can be found reluctantly bringing up the rear. Reporters, public servants, lawyers, physicians, artists, poets, even scientists precede him. But the peculiar role which he has rendered a vacuum cannot be so easily usurped. The engineer is not expendable. He is crucial. And his lack of action to pave the way for the genius of his techniques is consequential in terms of death, injury, torment, waste, sorrow, consequential in terms of decades of delay before the adoption of remedies.

I believe the time has come when we can no longer ignore the plight of the engineer as minion to corporate management or other allegiances to which he is chattel. For this plight is fast becoming a serious affront to the public. In the near future, it will be the subject of rising indignation as the realization spreads of what the engineer *can* do in contrast to what he *does* do to diminish the hazards to life and limb and to accelerate the adoption of more efficient, cheaper and aesthetic technologies, be it in transportation, building, communications, or many other areas.

An instructive inquiry to increase our understanding of what is holding back engineering statesmanship and involvement in the social problems of technology centers around the institutional framework through which engineers are engineered. This framework includes universities, corporations and technical societies.

PROBLEMS OF UNIVERSITY ENGINEERING EDUCATION

Universities afford the young engineering student his first exposure to what is to be expected of him when he goes to work after graduation. Thus, he is grilled in the general tools of the trade but, by and large, not in the demands of the profession. Numerous academic commentators have deplored the narrowness of the curriculum and have urged a little more liberal education. The important reason for such urgings in my judgment, is that presumably, studies in the arts and social sciences can enable the engineer to know and appreciate a wider array of values so that he can bring them to bear, along with the technical values, on his search for and choice of engineering alternatives to a given problem. These are the values that are most jeopardized by the "social costs" of a narrowly conceived technology.

Although there are interesting experiments at work, such as at Princeton University, the broadening of the conventional engineering education has not been done well. It has been formalistic, adding a course option or two and letting the student "get a little culture." Tack-on courses cannot be expected to do more. Fusing these varied inputs around a social problem of great engineering content might be a more productive approach. Studying mass transit systems and air pollution control systems offers the kind of very real, concrete attachment that demands the broadest, professional engineering embrace. These are problems where engineering choices and priorities become very different under one set of values, as expressed by a particular industry, or another set of values insisted upon by virtue of professional stamina. Further, under joint, interdepartmental ventures, the student engineer, as well as the student economist, psychologist, sociologist, etc., can learn how whole a man must be if he is to be a skilled practitioner of his discipline, and how related skills and institutions must become for problems to come nearer to solutions.

LACK OF AUTOMOTIVE ENGINEERING IN THE CURRICULUM

The problems of the engineering curriculum are certainly not foreign to this distinguished gathering. I should like, however, to note my astonishment at engineering schools (regarding the subject of motor vehicles) which began during my student days at Harvard Law School in the mid-fifties. After realizing that there were no physical laws which required people to be killed or severely injured in automobile collisions, I went about the Cambridge environs to study what was available about automotive safety. Not at M.I.T. or Harvard's engineering departments, but at the Harvard School of Public Health did I find relevant material and people interested in a more humane automotive technology, a more forgiving automobile.

Consider the situation years later, during which nearly 500,000 people perished and some 40 million were injured in motor vehicle collisions. Not a single university or college in this country gives an undergraduate or graduate degree in automotive engineering. A professor recently tried to interest book publishers in a proposed text on automotive engineering and received the reply that there was no market for such a volume. In an era of specialized engineering degrees, it is jolting to know that the biggest technology in this country—the motor vehicle transport system which yearly takes 1 out of every 5 retail dollars—does not derive the nourishment of engineering research and policy leadership from our institutions of higher education.

As a leading institution, the Massachusetts Institute of Technology cannot be a poor sampling of this wasteland. What has this great center of engineering learning contributed to automotive safety? Its technical output can be generously described as trivial. Its journal, *Technology Review*, though consistently containing automotive industry advertisement, has not devoted articles to this great problem confronting engineering. Its posture has been one of aloofness to the kind of challenge posed in 1961 by a graduate of M.I.T., Dr. William Haddon, who told the Society of Automotive Engineers: " ... The success of your profession in the present decade will largely be weighed in terms of its success in handling this overwhelming problem." Only in the past year, spurred by a four-year, million dollar grant from General Motors, hastily announced on the eve of the first Senate hearing, did that institution begin to awaken to its potential. A one-million-dollar grant, spaced over four years, which GM said was for a "long-range, in-depth, quantitative analysis of all facets of the safety problem—the car, the road, the driver, and their various interactions," does not have much stretch. But it did begin to interest some of the gentlemen on the banks of the Charles in M.I.T.'s relevance to the vehicular safety of millions of people on the roads.

CONSEQUENCES OF AUTOMOTIVE ENGINEERING INACTIVITY

It is not difficult to point to some of the consequences of this national inactivity on the engineering campus. The technical literature of quality in automotive engineering has been pitifully small. No independent centers of basic and applied research exist, with the exception of a few sporadic and spotty projects. The understanding, for example, of the closed loop car-driver servomechanism under varied highway conditions is primitive. There is shocking theoretical ignorance of vehicle dynamics. Participation by faculty members in the decisional work of automobile technical societies is little more than nominal.

Contributions to public policy by the country's engineering faculties can be measured by their total absence during the voluminous testimony on auto safety before Senate and House committees in the past year. They can be measured by a governmental contribution to safety work in this field during 1965 of less than $2 million. They can be measured by the utter lack of critical review of the automobile industry's legendary avoidance of safety research and development. They can be measured by the obsessive but successfully preserved secrecy of the auto companies and the non-existence of a professional society for research, communication, and criticism dealing with traffic safety problems and their solutions. Finally, they can be measured by the stagnating technology represented in contemporary automobiles and expressed ultimately in avoidable accidents and injuries on the highways.

The automobile industry has not displayed concern over this state of affairs at university engineering departments. The policy of automobile companies over the years has been to take the engineer out of college or technical institute and teach him "the industry way." This is a process of acculturation to the status quo that is to be expected from a management policy that has relegated the engineering function to a state of conditioned response. Quite basically, the issue here can be viewed on two levels: (1) The degree to which engineers feel free to dissent, innovate, and advocate innovation through open channels of communication to top management, and be naysayers to ruthless cost analysts, whimsical stylists and remote executives, in a genuine attempt to improve the overall corporate performance; and (2) The degree to which engineers can live professional lives of their own in technical societies and public forums apart from their employee status and without fear of overt or covert retaliation whether in the form of dismissal, demotion, or the freezing of promotion.

HOW FREE IS AN ENGINEER?

In essence, how free is an engineer within a large corporate government, whose primary mission is profit-maximization via all possible shortcuts, and whose bureaucratic structures pose real problems for individual expression and initiative both in matters of skill and conscience? A study of postwar automobiles indicates that he has been neither very free nor effective. Change has come about primarily as a result of external, not internal, pressures on the companies, chiefly by the threat or passage of legislation. The engineer's subordination to the stylist over the years in such areas as external vehicle design, or design of instrument panel, door, roof, and seat structures—where the stylist operates with full approval of management as quack engineer—is another syndrome of his degraded professional status. At the bottom of the hierarchy are the safety research engineers (apparently now more in demand), whose frustrations and powerlessness constituted their occupational hazards. They have had facilities and funds to work with up to the level of their primary utility—impressing visiting legislative and other delegations with a few crashes of rejected automobiles on company lots.

It was not surprising then to hear Ford Motor Company's Vice President, Donald Frey, tell an engineering gathering not long ago: "I believe that the amount of product innovation successfully introduced into the automobile is smaller today than in previous times and is still falling. The automatic transmission was the last major innovation of the industry." (The automatic transmission was first adopted on a mass-production basis in 1938–39.)

Mr. Frey can be a refreshingly candid executive at times. Addressing a national industrial research conference at Purdue University January 10, 1966, he stated:

> It's a sad commentary, but some of the most reactionary people in industry are engineers. Fresh new departures that require creative thinking and innovation can wind up in the file marked NIH—Not Invented Here. It is up to management to prevent this waste by creating engineering organizations that are mentally attuned to trying the new.

Then, as a former professor of engineering, Mr. Frey uttered the words so long awaited from a top auto industry executive.

> Invention can be predicted with a fair degree of accuracy and it can be scheduled. In the automobile industry, our technology has advanced to the stage that our engineers can invent practically on demand. Almost any device we can dream up, the engineers can make.

Perhaps in the future other auto industry executives will tell us candidly about management's responsibility to encourage with full backing the engineer's contribution to rapid cost reduction of these innovations for prompt mass production. They may also explore the proposition that many of the "reactionary" traits of engineers in corporations are induced by a management policy of disinterest, contempt, and obstinancy which is filtered downward in authoritarian patterns.

Even in the most elemental freedom of communication between company and outside researchers on non-trade-secret matters—a cardinal canon of scientific and technological progress—the strictures are unrelenting. Complaints to me by a number of academic researchers over the intense difficulty in communication over safety principles illustrate the comprising situation in which a professional scientist or engineer finds himself in the industry.

Anyone who reads, for example, GM's "corporate procedure for approval of technical publicity" will understand how such incommunicados are enforced. The printed output of auto company research in operating and crashworthy safety of the vehicle is so small as to leave even the cynical person incredulous. One can read all the public papers written by GM researchers in the past 15 years on vehicle crashworthiness in a short afternoon, even allowing for their redundancy.

PROFESSIONAL COMMITMENT VS. CORPORATE ALLEGIANCE

The conflict between professional commitment and blanket corporate allegiance raises the more encompassing question of corporate constitutionalism, particularly the rights and recourses of professional employees within the private government of the corporation. It is encouraging to see some attention to this problem given by the *American Engineer* for October 1966 in an article by Robert T. Howard, entitled, "A Bill of Professional Rights for Employed Engineers?"

The malaise of extreme corporate discipline over engineers who take their professional dictates seriously infects the performance and very structure of the so-called professional engineering societies. These societies, such as the Society of Automotive Engineers (SAE) and the American Society of Mechanical Engineers (ASME), behave like manufacturers' associations.

Society of Automotive Engineers

In the motor vehicle area, the SAE is a technical society dominated by the auto industry. SAE committee members are mostly engineers on company missions and company time; they utilize company facilities for SAE standards work; they are expected to advance company interests, not to further their own profession's directions when they conflict with company imperatives. Vehicle safety standards are issued by SAE without any published technical justification. SAE members-at-large, outside the particular committee issuing the standard, are not permitted to see proposed standards so that they can comment on them before promulgation. Where it infrequently exists, consumer representation—broadly meaning non-industry representation—is little more than window dressing without the remotest potential of influencing the outcome of the standard, much less having the votes to defeat it. Auto company representatives largely decide what topics and papers are selected at the technical meetings and largely determine which are published.

The auto industry's SAE representatives refused, for example, to publish the aforementioned address by Dr. William Haddon, presently the National Highway Safety Agency's administrator. Obviously, the absence of any technical paper critical of specific auto company products and policies is entirely consistent with the industry's dominance in SAE's programming. So is the absence of any written code of ethics for the SAE. So was the recent ban on any recording of presentations or discussions at last year's Stapp car crash conference, conducted by the SAE—a remarkable restriction on an open scientific and engineering conference.

American Society of Mechanical Engineers

The American Society of Mechanical Engineers has higher pretensions than SAE. It possesses a code of ethics, yet it has denied long-time, distinguished members access to minutes of standards committee meetings. Its constitution and bylaws tread ground familiar to other engineering organizations, by assuring a self-perpetuating control to a small number of entrenched officials. There are members of these engineering societies who know that if they do the right things, they are quite likely to become presidents of their respective organizations at such and such a year. The lines of succession are reasonably clear for anyone who starts up the hierarchy and singlemindedly persists in the practice of "getting along by going along." The path of ascension is more in the form of an escalator than a ladder. A ladder requires some individual effort to climb, while an escalator demands only that you get on it and up you go.

It is sad enough to observe how little due process there is for the dissenting member to rely upon in trying to secure his rights within these organizations. More important is that the defects and defaults of these organizations have ceased to be private vagaries and have become public menaces. This is becoming increasingly the situation, owing to the recent determined effort by industry to obtain a governmental imprimatur on the safety and other standards set by these technical societies and rubber-stamped by the American Standards Association. ASA was renamed and reorganized in 1966 as the United States of America Standards Institute (USASI), but remains convincingly dominated by these technical societies.

United States of America Standards Institute

The saga of USASI and its portentous consequences for public safety, innovation, small business, competition, foreign trade and other key issues are beginning to unfold. What is at stake is whether there is to be an officially recognized institution of private government that decisively controls public functions.

The control of a standards-setting process carries with it significant power. When this control is in the hands of industry-dominated technical societies, with no real consumer representation, problems arise. The fact that these standards are designated to be voluntary is of little comfort, for the principle of "consensus" by which they are adopted insures a great deal of economic power behind their acceptance. Since the principle of consensus is a principle of economic power, the political repercussions are there as well. Note how frequently the SAE standards have found their way into legislation.

However serious this concentration of private standards power has been up until now, at least there was the potential of government moving in to correct abuses and fill needs that were being clearly neglected. Such a movement reached a climax September 1966 with the enactment of motor vehicle safety legislation requiring the government to set safety standards. Pending is a bill in Congress to have the Federal Power Commission set safety standards for natural gas pipelines. In both these areas the voluntary industry standards have failed to afford adequate protection to the public.

USASI STRATEGY

Consider the latest strategy by the industry-dominated USASI. There is every indication that USASI will ask Congress to grant it a Congressional Charter—giving it a quasi-official status and recognition, together with possibly an explicit congressional endorsement of USASI to represent the United States in international standards negotiations and activities. Additional legislation has already been offered, and is presently pending, by the Secretary of Commerce which, among other objectives, would permit the government to make grants to, or contracts with, USASI and would place Congress on record as recognizing the system of voluntary standardization based on the consensus principle as expressive of the public interest.

The restructuring of USASI, from the old American Standards Association, is a formal maneuver to create the image that it has meaningfully broadened its representation and *modus operandi* to include consumer and government representatives on its councils. A reading of USASI's constitution and bylaws will dispell any illusions. The same industry dominance via the technical societies who are members of USASI and set the standards for USASI to give its *pro forma* label of approval to are in charge. The consumer council on USASI serves only in an advisory capacity and, by the bylaws, could be overwhelmingly dominated, in numerical terms, by industry and commerce. To presumptously designate the old ASA standards and new ones forthcoming under USASI as U.S.A. Standards amounts to a deceptive trade label.

To permit the USASI proposal before Congress to prevail would constitute a radical assumption of governmental functions by private monopolies and wither away the potential of public options for mandatory government standards. Above all, this little known but exceedingly significant drive, if successful, will further violate a cardinal principle of the integrity of democratic government; namely, that business or any other special interest group be prevented from differential access to the governmental process and policymaking councils under official cognizance.

At the very least, the decision for or against the USASI proposal in Congress should not be made without a complete airing of the facts and issues so that the public can understand what the situation imports. Engineers can play an important role in clarifying the issues and presenting the judgments, so that USASI's greatest asset—the esoteric nature of the subject matter as far as the public is concerned—is rendered ineffective.

I wish to thank the members of the Middle Atlantic Section of ASEE for this invitation to candor.

In Aristophanes' play, *The Birds*, an unbound Prometheus is shown to believe that he can escape the attention of Zeus by staying under an umbrella. Engineers can no longer operate under a comparable impression that their activities will be shielded from the lay publics severe scrutiny. This is partly so, because engineers have shown this public in recent decades how very much they can contribute to man's progress. But progress frequently brings with it terrible costs. And the public is more and more realizing something rather new—that the engineering remedy to eliminate or diminish these costs can be as impressive as the engineering achievement that developed the technology.

It is the recognition of this gap between promise and performance that is producing the pressures which will continue to mount on the engineering profession and demand that it assert itself toward its most magnificent aspirations—for so much of our future is in your trust.

Can Professionalism Be Attained within the Corporate Structure?

MARTIN GOLAND

The members of a profession are, first, the custodians of a specialized branch of knowledge which is beneficial for the human cause. Second, the professional is obligated to use his specialized skills and knowledge only to advance the welfare of society. The custodianship function is not merely a passive one; it requires that the professional contribute toward the further advancement of his art, and that he prevent charlatans and incompetents from abusing the public under the guise of professional practice.

Finally, and of the utmost importance, he must practice his profession independently, as an individual answerable to his own conscience. To use his professional talents in a cause he believes to be unworthy is a violation of a sacred social trust.

In reviewing this list it is at once clear that professional ethics are in every sense in consonance with an honest employer-engineer relationship. Why, then, are so many engineers frustrated in their corporate roles? Why do they consider that they are merely dealt with as "hired hands"?

Modern, enlightened management is keenly aware that its greatest resource lies in the abilities and motivations of its staff members, and this is particularly true of the creative elements of its organization in which engineering is included. It is for this reason that today's progressive corporations expend considerable resources toward programs which enable the ambitious employee to improve his skills and further his education.

Admittedly, not all corporation managements recognize their full responsibilities along these lines. The cooperating engineering societies are to be commended for their drafting and subsequent adoption of a Guideline for Employers which includes provisions relating to professional advancement. In time the Guidelines will become an effective force for improvement of the conditions of professional employment.

The need for the Guidelines program highlights the most sensitive area in any analysis of engineering professionalism—the fact that the large majority of engineers cannot escape the employee category. As such, even as professionals, they are subject to the tides of management decisions, to layoffs during times of economic stress, and to the constraints of the flow of authority which is necessarily present in large and complex organizations. Under these circumstances, can the engineer remain at all times true to himself and to his professional conscience? Can he remain an *individual* voice, a quality indispensible to professionalism?

Further complicating this issue is the team nature of corporate enterprise. Unlike the doctor and the lawyer, whose services are delivered directly and personally to his client, the engineer is usually one member of a larger team. His specialized contribution is but one of many needed to realize the end product of his work, and in turn he is dependent on many others to make his own activities meaningful in the marketplace. Team efforts necessarily tend to subjugate the individual personality to the group median. Here again, the question arises, can the professional engineer cope with this environment?

The answers to these questions, it seems to me, rests primarily in the hands of the engineer himself and not in the corporate employer. Engineers will be respected as professionals to the extent that they, as a group, perform as professionals. The virtues of professionalism are in independent thought, the creative dedication to the dual and compatible goals of advancing the employer's interests while serving the just needs of the society. When these attributes are made part of the daily credo of engineering, the practitioners of the art will be accorded the respect and consideration due them.

The sad fact is that too many graduates of engineering schools do not earn the mantle of professionalism. They remain content with career assignments which are little more than the exercise of sophisticated technician skills. They are inactive in professional affairs; the great majority of so-called engineers are not members of a professional society. Their participation in community activities is limited, even in matters directly concerned with engineering and the issues of technology.

A diploma from an engineering school does not constitute an automatic entry into a profession. The privileges of a professional are earned; they are accorded those who, in the interests of public and professional service, give more of themselves than the circumscribed demands of job and livelihood.

The issue of engineering professionalism is at another crossroad. The interest and concern generated in recent years can be permitted to gradually subside, so that history is repeated with another period of much talk but little concrete action. The alternative is to convert our good intentions into effective action.

5.3

The Limits of Professional Responsibility

ROBERT BAUM

Most academic discussions of the ethical responsibilities of engineers (and other professionals as well) are at a level of generality and abstractness such that they are of minimal value to individuals confronted with real-life moral decisions. Even the codes of ethics of the various engineering societies, although formulated and adopted by practicing members of the profession, are in many respects so abstract and general as to be essentially useless in dealing with specific cases of certain kinds. Let us consider two hypothetical (but not unrealistic) cases in order to better understand the fundamental problems confronting professional engineers who want to perform in a morally responsible way.

1. An experienced civil engineer, employed for ten years by a medium-sized company, personally believes quite strongly that the construction of a proposed new public works project would in the long run be to the detriment of the residents of the local community. After a lengthy public debate, in which the engineer chooses not to participate, a bond issue for the project is approved by 82% of the voters. (86% of the registered voters voted.) Bids, including price estimates, for the engineering work are solicited by the government purchasing office. The engineer, who also believes that competitive price-bidding is unethical and not in the best interest of the general public, is assigned by her firm to prepare their bid. She also believes that her firm is the most qualified for the job, and that she has the most experience of anyone in the company on projects of this kind. What should she do?

2. A young engineer's first job after receiving his B.S. involves working on a major construction project. Shortly after starting work he encounters what he believes to be a serious potential hazard which could result in harm and even death for some of the workers employed on the project. He discusses

the situation with his supervisor, who tries hard to explain why in his judgment nothing should be done to lessen the risks to the workers, emphasizing that he has had a great deal of experience on projects of this kind. The young engineer still believes that something more should be done to protect the workers. What should he do?

What guidance do the codes of ethics of some of the major engineering societies give to engineers in situations such as these? Because of the essential nature of codes as providing statements of general ideals, it is not necessarily surprising to find that none of them provide specific guidelines as to how engineers should proceed in such concrete cases. It *is* somewhat surprising that none of the published interpretations of the codes by official groups (such as the NSPE's Board of Ethical Review) include considerations of any cases similar to these. The specific cases analyzed by the engineering groups are concerned primarily with matters of protocol, violations of civil laws and governmental regulations, unfair competitive practices, and even trivial matters such as the determination of whether the use of the name of an engineering firm on Little League baseball uniforms is ethical (it is, according to NSPE Opinion 74-1). There are numerous analyses of cases dealing with various aspects of the problem of competitive price-bidding, and some codes even have sections dealing explicitly with this issue, but most of these discussions are now in need of reconsideration in light of the recent Supreme Court ruling and actions of several regulatory agencies.

Although none of the specific case analyses published to date by the engineering societies deal with issuess such as those brought out in the above cases, these discussions do provide some useful insights into the essential nature of the codes of the various groups. The most striking fact revealed by these discussions is that most of the more specific items in the engineering society codes are ultimately grounded on or limited by a general principle requiring consideration of the welfare and safety of the public. For example, the restrictions on advertising and price-bidding have always been justified primarily on arguments to the effect that such restrictions are necessary to protect the public from harm. Even principles dealing with honesty and the limitation of practice to areas of individual engineers' competence, which could stand on their own, are sometimes discussed in terms of contributing in one way or another to the general welfare. Other principles, such as those concerning loyalty to employers usually include qualifications to the effect that concern for the general welfare must be given greater priority.

Not only are most of the specific principles of engineering codes grounded in or limited by tacit requirements to consider the health and welfare of the general public, but most codes contain an explicit principle to the effect that such considerations should be of paramount concern to professional engineers. It is important to note here that there are some significant differences in the nature of the responsibilities being assigned in the various codes. In some instances, the engineers' responsibilities to society are characterized in the essentially passive sense as the duties to *not* do anything that would cause harm or decrease the general welfare. Thus, section 1-b of the ECPD Guidelines states that "Engineers shall not approve nor seal plans and/or specifications that are not of a design safe to the public health and welfare and in conformity with accepted engineering standards." Other sections of this and other codes stipulate a more positive responsibility to take steps to *prevent* harm to the public. For example, section 1-c-3 of the ECPD Guidelines requires that "Should engineers observe conditions which they believe will endanger public safety or health, they shall inform the proper authority of the situation." Similarly, the IEEE Code of Ethics specifies that "Engineers shall, in fulfilling their responsibilities to the community, protect the safety, health and welfare of the public and speak out against abuses in these areas affecting the public interest" (Article IV-1).

The codes of some societies, e.g., IEEE, do not assign any more positive responsibilities to engineers for the general welfare, but a number of major codes do make the much stronger requirement that engineers should do that which *benefits* the public, in addition to merely preventing harm. Thus, the first of the four "Fundamental Principles" of the ECPD Code of Ethics states that "Engineers uphold and advance the integrity, honor and dignity of the engineering profession by using their knowledge and skill for the enhancement of human welfare." And the Preamble of the NSPE Code requires engineers to use their "knowledge and skill for the advancement of human welfare."

These basic principles in the engineering codes are certainly not original or unique, especially in traditional Western cultures and philosophical systems. They all fall under the general category of "teleological" principles, that is, principles which require a consideration of the consequences of an action in determining its moral rightness or wrongness. This general type of principle and most of the specific instances of it have been extensively, if not exhaustively, discussed and analyzed by philosophers and others in the Western intellectual tradition for centuries, and most of the strengths and weaknesses of the various formulations are today widely recognized. However, there is one issue that has received minimal, if any, attention in these discussions and which is of particular relevance to the question of the social responsibilities of engineers. The remainder of this paper will focus essentially on this one problem, rather than attempt a survey of all of the numerous other strengths and weaknesses of teleological theories.

Because the problem which is to be discussed applies to all of the different kinds of teleological principles, it will be sufficient to limit the discussion to only one of them. And to make this argument as strong as possible, it will focus on the weakest of the various teleological principles, namely that of doing no harm. It should be noted that this principle is so central to most cultures that it perhaps should not even be considered a *moral* principle. It might instead be included as one of the criteria for being a normal person, insofar as a person who intentionally causes harm to another person for no other reason than the causing of pain and suffering itself is commonly considered to be "crazy," "possessed" by outside powers, or in some other way "less than human," rather than immoral. Even egoistic theories, including those expostulated in popular books such as the novels of Ayn Rand and Robert Ringer's *Looking Out For #1*, explicitly assert that everyone should try to get as much as they can for themselves as long as it doesn't directly harm others. (It is interesting to note that most of these egoistic theories themselves are ultimately grounded on some more altruistic principle insofar as their advocates assert that if everyone seeks their own best interest this will in fact result in the greatest possible good for everyone.) Although the specifically moral status of the principle of doing no harm may be open to some question, we will continue to assume that it is a moral principle for the purposes of this discussion, since the criticisms to be made of it are for the most part applicable to all other teleological principles. It is also the case that in many "real life" situations, particularly those encountered by engineers in their professional practice, it is impossible to distinguish between failing to do good and doing harm. That is, violation of a principle to do good necessarily involves violating the principle of doing no harm. But this point need not be further developed here.

The problematic feature of the basic principle of doing no harm (and a feature which is shared with almost all of the more positive teleological principles) is that it is almost always formulated in terms of actual harm, or more accurately, in terms of the *knowledge* that harm will result from a specific action. For example, as we noted earlier, the ECPD Guidelines tell engineers that they should not approve plans that are not of a design safe to the public welfare. But this formulation does not take into consideration at all the fact that in many situations it is possible for rational persons, even intelligent and competent engineers, to hold conflicting beliefs in good faith as to what is and is not safe. It is even possible for two professionals to hold conflicting beliefs as to what the exact consequences of an action will be, independent of any judgments about the goodness or badness of those consequences. And it is precisely in those situations where there are conflicts of belief about such matters, as illustrated in the two hypothetical cases above, where engineers are in greatest need of assistance and guidance in making morally correct decisions. When some persons believe that an action will (or will probably) cause harm to the general public and other persons believe that the same action will not cause any harm, none of the principles in any of the engineering codes provide any useful guidance, and in fact none of the major ethical theories generated in the Western philosophical tradition are any more helpful.

A commonly offered "solution" to the problem of how to apply teleological principles to cases where there are conflicting (reasonable) beliefs about the consequences of a specific action is to simply formulate the principle into something like the following statement: "It is morally wrong to do anything which you believe will cause harm." This strategy is totally inadequate in that it entails a form of relativism where the same action, if done by two different persons with different beliefs about the nature of the consequences of the action, can be both morally right and morally wrong. Insofar as ethical judgments are about actions independent of the identity of the agents, and insofar as logical consistency is a desirable feature of an ethical (or any other kind of) theory, a theory which permits the same action to be both right and wrong (i.e., not right) is seriously flawed. If it is wrong for engineer X to work on a project, then it must be wrong for all other engineers to do the same work on the same project (assuming they all have the same skills and competence, etc.). In a situation like that in case 1, this version of the no-harm principle could even lead to the conclusion that it is wrong for the engineer in that case to bid on the job *and* that it is right for a less competent engineer who believes (sincerely) that the project would do no harm to bid on it. In case 2, this principle could lead to a confusing situation, if not open conflict, since both parties are working for the same company, and they would both be equally justified in following exactly opposite courses of action. Although the problems with this attempt to make the no-harm principle applicable to cases involving conflicts of belief could be elaborated on in considerably more detail, this seems to be unnecessary in the present context.

A second way of dealing with the problem of conflicting beliefs is to modify the principle to specify that in situations of this kind, it should always be assumed that the employer's or client's beliefs are the correct ones. Although such a strategy may appear on the surface to be clearly absurd, it is not an uncommon one, and it must be analyzed more carefully. One of the more persuasive arguments for this version of the principle is given by Samuel Florman in his book *The Existential Pleasures of Engineering* and more recently in an article in Harper's (October, 1978). The essence of his argument is as follows:

> Engineers can (and should) contribute to public policy as citizens, but this is very different from filtering their everyday work through a sieve of ethical sensitivity. As a class, engineers have neither the power nor the right to plan social change.... If each person is entitled to medical care and legal representation, is it not equally important that each legitimate business entity, government agency and citizens' group should have access to engineering advice? If so, then it follows that engineers (within the limits of conscience) will sometimes labor in behalf of causes in which they do not believe.

Although there are some serious flaws in this argument, several significant insights into this problem can be gained by a careful analysis of it.

The right to medical care and legal representation referred to in the premise of Florman's argument is generally understood to be a right to access to such services regardless of one's economic status. The only consequence of this right is the justification of medicare and legal aid programs and pro bono service on the part of professionals. It does *not* imply that a specific professional must take a specific client under any and every circumstance.

It is not at all clear in what sense a physician or lawyer would be obligated or even justified in laboring "in behalf of causes in which they do not believe." It is true that a lawyer has an obligation to defend a client in certain cases even when the lawyer personally believes that the client is guilty. But in these cases, the lawyer must be acting in what he or she believes is in the best interest of the client. Moreover, this obligation is grounded in the basic principle that every person has a right to be assumed innocent until proven and judged guilty by a jury of peers. And presumably all lawyers do believe in the validity and priority of this principle, and when they take on a case it is in the "cause" of supporting this principle.

With regard to other situations, lawyers not only do not have an obligation, they do not even have a right to do many things for clients that they believe to be wrong or harmful to the public. Thus, just because someone asks a lawyer to assist in working out a scheme to defraud members of a pension fund or to trick an elderly person into changing a will, even if this can be done within the letter of the law and would be in the best interests of the client, there is no obligation, and probably not even a moral justification for the lawyer to perform such tasks.

There is no generally accepted principle governing the medical profession that requires or even permits a physician to treat a patient in a way that the physician does not believe to be proper. If a patient comes in and asks the physician for a prescription for penicillin or barbiturates, the physician has no moral or legal right (as recognized by either the profession or society at large) to give the prescription if he/she does not believe it to be medically justified. And certainly from the point of view of medical ethics, a physician employed by a corporation or other organization is never morally justified in laboring "in behalf of causes in which they do not believe," if this results in harm to employees of the organization or to the general public. In this regard, the code of ethics of the medical profession is, if anything, more explicit and forceful than the engineering codes. However, like the engineering codes, the medical codes do not address the issue of the uncertainty and the problems of disagreements of beliefs as to the consequences of specific courses of treatment.

Since neither the codes nor practice of law and medicine provide any justification for a member of either profession to "labor in behalf of causes in which they do not believe," in any sense that is analogous to cases of the kind encountered by engineers, Florman's attempted argument by analogy fails to support his conclusion.

Although his argument does not support it, Florman's conclusion is *partially* correct. Engineers are indeed *sometimes* justified in working on projects, even if they personally believe that these projects will result in harm to the general public or to specific subgroups. The critical factor in justifying this principle is the providing of criteria that will adequately identify the specific situations in which the engineer is justified in working on such a project. And the necessary criterion can be found in an aspect of medical practice which Florman has not considered.

As previously indicated, the physician has no moral responsibility, and perhaps not even the right, to provide a treatment that he/she does not believe to be in the best interest of his/her patient. However, there is a second limitation on the physician to the effect that he/she cannot perform any treatment on a patient to which the patient does not freely consent. Thus, even if the physician believes that a specific treatment is in the best interest of the patient, the patient must still consent to this course of action before the physician can act. (In cases where the patient is deemed to be incompetent to make such a decision, precedures have been established for a relative or court to give—or withhold—a proxy consent.) The principles underlying this procedure are essentially three. One is that the potential harm or benefit resulting from the treatment of a patient is to be assumed to fall entirely on the patient. Thus, it is unnecessary from a moral point of view to consider the impacts of the treatment on relatives, friends, employers/employees or the general public. (This principle is certainly challengeable, but it is in fact part of the currently operational procedure of informed consent.) The second principle is that every individual has a right to make all decisions having to do primarily with their own health and welfare. The third principle is that the physician has a responsibility to provide to the patient complete and accurate information relevant to the making of a decision in an understandable way.

In relating the procedure of informed consent to the engineering context, it is not hard to recognize that the first underlying principle simply does not hold. It is clear from the earlier discussion and the sample cases that there is no basis in fact or theory that the effects of engineering activity will apply only to the client. It is also the case that the "client" in most cases is not a single individual. These facts do not invalidate the second principle, but they do indicate the need for a modification of it to the effect that every person has a right to participate in making decisions to a degree proportional to the extent to which their own health and welfare is affected in comparison with other persons. Thus, insofar as a person may be affected by the results of an engineering decision, the individual has a right to participate in some degree in the making of that decision. The third principle remains applicable without change, imposing on the engineer an obligation to provide complete and accurate information essential to the making of a decision by all affected parties.

The operational implications of this for professional engineers are much more specific than Florman's somewhat vague conclusion. The most significant implication is that the codes of the engineering societies must be rewritten to play down the significance of (or perhaps even delete) the various teleological requirements to hold paramount the public's health and safety. These principles should then be replaced with new statements of the responsibilities of engineers to:

1. Recognize the right of each individual potentially affected by a project to participate to an appropriate degree in the making of decisions concerning that project.
2. Do everything in their power to provide complete, accurate and understandable information to all potentially affected parties; and
3. To carry out to the best of their ability assignments which have been appropriately approved by the potentially affected parties, even if the engineers believe that they may (or even will) have harmful consequences.

It should be recognized immediately that under these principles engineers can justifiably believe that they are doing the morally right thing even if they also believe that it will have harmful consequences as long as the affected parties have given uncoerced "informed consent." Thus, in our first hypothetical case, the engineer would not be morally wrong to bid on the project, insofar as the public has approved the project itself and has also approved the process of competitive bidding (through both the legislative and judicial process) even though the engineer personally believes that both the project and the process (of competitive bidding) are potentially harmful. It is especially important to recognize that doing something that one believes is harmful is not morally wrong, if the parties most affected want it done. (It should also be noted that it is even possible that the affected parties believe that a project will be harmful to them and still choose to approve the project.)

Several points do have to be considered about the first case. First, a judgment should be made as to whether the persons voting on the project, especially those approving it, are those who may be harmed by it. This is especially important for the protection of minority groups, although it must also be recognized as being difficult to judge. And second, the engineer should consider whether the public have available to them all of what she considered to be the relevant facts. If they did not have all the facts, then she probably should have participated in the public debates to supply essential information in an understandable form.

With regard to the second case, where the no harm principle (and all of the other teleological principles) failed to give any guidance, the principle of the right of potentially affected individuals to participate in the making of the decision would clearly require that both the young engineer and his supervisor present their beliefs to the workers who should then be allowed to participate in making the decision about whether additional precautions need to be taken. The only question that is left open in this case under the informed consent principle is that of determining the proper degree of participation of the workers in comparison to other parties (i.e., other persons who might be affected by the decision) in the decision-making process.

There are a number of obvious objections which will inevitably be raised against the preceding discussion. Many of them can be best answered by pointing out that this has at best been a *sketch* of what a complete analysis of this topic should look like. There are definitely many elements which require elaboration and/or clarification, and there is a multitude of specific cases—particularly "borderline" cases—which need to be discussed and used as tests of the applicability and adequacy of the proposed alternative to the traditional teleological principles that comprise the foundation of most engineering codes. Although most of these criticisms can't be dealt with any further in the context of this paper, there is one other basic objection which can and must be touched on briefly in closing.

It is obvious that the hard facts about the "real" world are such that it is simply impossible to apply the informed consent procedure and the related principles to engineering practice today. There are very real expectations on the part of many people that engineers and other professionals are looking out for, or at least are responsible for, their best interests, and they will not readily accept immediate pronouncements from the profession telling them that they are responsible for watching out for their own interests. There are also engineers who will perceive the loss of this supposed responsiblity as a loss of their autonomy and a threat to their professional status. Both perceptions are mistaken, but it will require a lengthy dialogue and discussion within the profession and between the profession and the general public over a period of years to provide an adequate understanding and appreciation of the realities of the situation. During this period substantial changes will have to be made not only in the codes of the various societies, but also in the structures and policies of governmental agencies, corporations, and other elements of the societal fabric. This will be difficult, but it *is* feasible. We have been trying to make infallible and omniscient deities out of mortal and finite professional engineers for more than fifty years to no avail. It is time now to try to make real progress by developing realistic and reasonable codes and guidelines that engineers can follow and the public can accept and respect.

5.4

The Professional Status of the American Engineer: A Bill of Rights

ROBERT L. WHITELAW, P.E.

Society accords professional status to those who have acquired proficiency in a learned art and are free to practice it in a spirit of public service, their only fetters of conduct being personal conviction and professional ethics. The operative words are "free," "learned," and "public service;" the art must require disciplined learning as well as skill; the practice must be free of all coercion from others within the bounds of professional ethics; and the principal motivation to practice (in the public mind at least) should be public service, i.e., the protection and enhancement of human well-being.

It follows that an individual who is not free to practice a learned art within the bounds of professional ethics and personal conviction of what is in the public interest, is not entitled to professional status. No matter how great his or her proficiency, or how many credentials can be displayed, or how many "state boards" have been passed, or licenses acquired, so long as the individual is looked upon as an employee rather than as a free artisan, to that extent there is no professional status.

In our society, those who enjoy true professional status today, both in fact and in public opinion, are the following: almost all those in the medical and legal professions, the architect, the public accountant, a certain fraction of clergymen, teachers, authors, and artists, and that very tiny fraction of engineers who are consulting engineers. But the engineer employees of the consulting engineer are not included. Nor is this picture about to change, either in the public view or in the graduate engineer's view or in the engineering teacher's view until certain essential ingredients come to be recognized as an inalienable part of engineering employment.

No matter how many orations are made about professional status at engineering conventions by leading engineers (who are usually corporate managers first and engineers second), no matter how many professional engineering societies, sections and chapters are formed or journals published, no matter how many legal tomes are filled with "Acts for Licensing Professional Engineers," or licensing boards set up, or examinations held, and no matter how many threats are made of fine or imprisonment against misuse of the word "engineer," or fine-spun definitions of what is or is not an engineer—all this will be of no avail. The vast majority of American graduate engineers will never acquire professional status, either in the public mind or even in their own, until they enjoy certain minimum rights as a recognized part of their vocation, rights which mark the line between true professional status and mere pretense.

Let us then consider nine rights essential to true professional status—rights that should not be waived by an engineer nor denied by those he serves.

I. The Right of Private Practice, Regardless of Employment. A professional engineer, just as a doctor or lawyer, must be looked upon as self-employed, at all times retaining the right of private practice compatible with the ethics of the profession, i.e., scrupulous adherence to morality and law, never divulging to one client the confidential information gained from another, nor letting a commitment to one be impaired by services to another. Even where the principal (or even sole) client is a large corporation or a government agency, the engineer should never enter into a relationship of captive employee nor should any client be allowed to look upon the engineer as an employee but rather as a professional whose services are to be engaged by contract without preempting his or her right to serve others in the same capacity.

II. The Right to Contractual and Private Terms of Engagement. Just as the doctor, lawyer, and other professionals, are engaged for the services they are skilled to perform, so should be the professional engineer. Never should he or she be thought of as a mere unit on a payroll acquired and controlled by a personnel department. Rather, the engineer should be viewed as a specialized and expensive service acquired by negotiation and contract through the purchasing department. The contract should specify the services to be supplied, the time when they are needed, and the terms of remuneration mutually agreed upon.

This right automatically includes the right to privacy; the terms of engagement should not be published or disclosed, except as required of public contracts. Above all, these terms should never be bargained collectively by some non-professional agency, as if the engineer were just a pair of hands to be purchased like a bushel of potatoes. Nor should these services be reckoned in terms of working hours; or controlled by punching a clock; or manipulated and measured by any of the operations research techniques presently in vogue for large bodies of faceless employees, to be treated by accountants and computers as merely the "labor" cost component of a refrigerator, or a box of soap.

III. The Right to Refuse Unethical Activity without Predjudice or Loss of Contract. No employed engineer, whether of corporation or government, truly enjoys this right today, and many submit in silence to doing that which is not in the public interest and sometimes even morally wrong for the sake of salary or advancement. Only a strong professional society, free from domination by corporate management or government officialdom can secure this right for professional engineers and bring them the dignity that goes with it.

Can one imagine the censure that would fall upon a hospital, both from the public and the medical profession, if it attempted to "fire" or otherwise punish a doctor who refused to perform an operation he or she thought to be wrong? The doctor enjoys the status of a professional. So also should the professional engineer, not as a privilege but as a matter of right.

IV. The Right of Total Remuneration for Professional Services Rendered. It is inconceivable that when paying the bill from a doctor, lawyer, or public auditor, one would first deduct from the bill the doctor's income tax, social security, personal retirement plan or other so-called "fringe benefits," on the assumption that the doctor has neither the honesty, intelligence nor foresight to take care of these matters independently. What is more, if legislation were introduced to require doctors' and lawyers' clients to make such deductions it would be resisted by their powerful professional societies as a thing intolerable to their professional status and would probably also be laughed to scorn by the public.

Even the plumber, electrician, gardener, and newspaper deliverer enjoy this prerogative when they submit their separate bills for services rendered, whether to a private home, a corporation or even a government agency. The entire bill is due and payble according to the contract for the service performed.

But not so for the vast majority of professional engineers wherever they are engaged as employees of a company, a university, or other government agency. They are mere ciphers in an employee payroll, to be coddled with various corporately planned and managed fringe benefits and to have these costs plus their taxes withheld from payment for their professional services because they may not be trusted to save their earnings and pay for their own just debts and taxes. The corporate managers who dominate our engineering societies show no awareness of this absurdity, nor inclination to remedy it, no matter how much they trumpet in favor of engineering registration.

All remuneration for services rendered by a registered professional engineer, employee or not, should be based on a contract or invoice with no deductions of any kind except by mutual agreement—neither income tax, social security, health care, or anything else. Until this is so, engineering "professionalism" is an empty word, and registration is of little value.

V. The Right to Lifetime Registration, Barring Proven Misbehavior or Incompetency. The titles "doctor," "barrister," "judge," "governor," and "general" are granted for life and are so used by their recipients, except when they are publicly revoked for due cause such as fraud in acquiring them or gross incompetence or misbehavior in their use. Their use is not based upon payment of an annual fee, membership in a certain society, or residence within a certain state. No Ph.D. calls himself mister simply because he has moved to another state or dropped his membership in the society of his discipline.

So should it be with the professional engineer. His registration should give him the right to advertise his services and enjoy their fruits for life, except on a showing of incompetence or fraud or misbehavior to the same degree that would disqualify a doctor, dentist, or lawyer from continuing practice.

VI. The Right to Freedom from Discrimination by Reason of Mere Age.   We live in a time when many legislatures (and particularly Congress) are loaded with septuagenarians, each claiming greater ability to "serve the people" with every passing election; when the courts throughout the land take pride in their venerable judges; and when aged doctors and lawyers are held in public esteem for no other reason than the wealth of experience they have gained. It should, therefore, be the foregone conclusion in every corporate management, in every university, and in every government agency, that the value of the practicing professional engineer must increase with age, short of proven senility. Most certainly, the idea of discriminating against professional ability by reason of age alone—or suddenly reducing it to zero value on a certain calendar day—should be repugnant both to common sense and to common fairness. In any case, such a continuing inequity should be just as vigorously resisted by professional engineering societies as it would be resisted by the medical or legal profession if any management dared to impose such an inequity on them.

Yet we find the experience of almost all practicing engineers entirely otherwise. Whether employed in a company, a university, or a government agency, they are told that retirement policy or labor policy requires that their employment be severed at an arbitary point in time determined by birthdate; "retirement policy" being dictated by insurance actuaries who think of people as computer digits and "labor policy" being dictated by labor union economists who think of jobs as pairs of hands.

In short, the engineer is told that no matter what remarkable talents he or she may have for creating new products (which create new jobs), no matter what unique experience may have been gained to pass on to the next generation, unless the individual is a doctor, lawyer, judge, corporate director or politician, value as a productive employee must be reduced to zero on a certain calendar day, simply because a computer or some half-baked labor policy so dictates.

No one would engage a bricklayer to build a garden wall on the condition that the trowel must be layed down the day this person reaches 65 on the ground of sudden incompetence or dismiss a piano teacher for one's children on the ground that suddenly there is nothing to teach. Yet thousands of engineers every year are subjected to this absurdity, while not a voice in our 55 state licensing boards is raised to point out the folly of it. Nor do we hear a voice from our powerful engineering societies recognizing that such policy long-continued may well be a form of technological suicide by slow degree.

Never before has the world seen a society so dependent on technology as ours; never before has the general public—including learned individuals in other professions and disciplines—been so ignorant of the various black boxes which we all take for granted in everyday living. Whether it be a carburetor, a telephone, a computer, or even a three-way light switch, these are mysteries to the average person today. The survival of such a society, not to speak of its progress, is more than ever dependent on the years-of-transfer available from one generation to the next, both to avoid the mistakes of the past and to understand the new black boxes of the present. The need is even greater as we see the engineering world ruled more than ever by acronym-labeled computer programs (SPLOSH, PERT, STRUDL).

Can anyone seriously imagine a corporate contract with a medical clinic or a legal firm which stipulates that the contract become null and void when all the doctors or all the lawyers reach age 65, or that the services of any one doctor or lawyer be barred when that person has reached a certain birthday? Yet this is the kind of nonsense with which engineers suffer silently every day while Congress and our state legislatures ferret out "unfair discrimination." Surely it is time to banish this, along with other relics of 19th century.

VII. The Right to National Registration and Freedom to Practice. A man in Bristol, Virginia, calls his lawyer in Bristol, Tennessee, three blocks down the street to represent him in a court in Virginia and gets the response, "Oh, but I'm sorry, sir, I'm not licensed to practice in Virginia." A farmer in Southern Pennsylvania with a desperately sick child calls his friend, a doctor, who lives across his back fence in Maryland. "Oh, I'm sorry, Jack, but I let my Pennsylvania license expire last week and can only practice in Maryland. I can only come over and give her first aid."

We smile at these absurdities. But the public guardians of these professions bristle indignantly when they are pointed out. "Surely you don't propose that doctors or lawyers be free to practice wherever the occasion for their services arises. How could we maintain our standards? Such confusion." In the case of medicine and the law, plausible arguments may thus be advanced to maintain local control and restrict competition to a certain area. But no such arguments can be made for licensing professional engineers.

The fact that doctors and lawyers are licensed to practice in a given state merely reflects the local character of these professions plus the continuance of an early tradition in American and Canadian licensing that is rarely found elsewhere. It should certainly not be followed in the licensing of professional engineers. Of all professions, engineering is least geographical in character—either in relation to clients or final applications. With respect to public safety, ships, planes, automobiles, electrical appliances, and a host of other hazardous products can never be geographically isolated. Buildings and bridges are the exception rather than the rule. Hence an engineer should be qualified by national examination, registered for life without further fees of any kind, and free to offer his services anywhere.

Those who pose as guardians of public safety in the engineering world should thus concern themselves with increasing both the opportunity and the competitiveness of the engineering profession on a nationwide basis.

VIII. The Right to Freedom from Surveillance, Psychological Manipulation, and Other Job Evaluation Techniques. These kinds of things are familiar today as the products of the new-think by which industrial engineering and personnel management have become operations research. Every employee (including professional engineers) must now be treated as a statistic in an elaborate computer program to be measured, weighed, evaluated, grouped, motivated, sensitivity-trained—everything except treated as a unique, responsible, and self-reliant individual.

The professional engineer is entitled to the same assumptions of competence and ability to manage contractual obligations and ethical responsibility with which one approaches a lawyer, doctor, or even an insurance agent or plumber. The PE should be thought of as a unique individual, with creative and analytical skills to be judged solely by the merits and consistency of his or her products and who is willing and able to go hungry for a while whenever the product does not come up to that of the competition or meet with customer's requirements. Under no circumstances should the engineer or the profession tolerate the current practice in monster corporations of keeping a "stable" of faceless engineer lackeys on call to be shuffled around, assigned, hired, fired, and retired on the basis of some computerized management or evaluation program, some upper-level labor deal between big business and union boss, or some actuarial hocus-pocus by the corporation's pension and retirement planning board.

IX. The Right to Practice Regardless of Health or Physical Fitness (Where They Do Not Obviously Impair Ability to Perform Professional Service).

Such an outrageous intrusion upon personal privacy as the mandatory physical exam is totally irrelevant to the ability to perform the skills expected and solely the result of the spillover into industry of the military mentality toward the draftee, plus the corporate concern to protect the funding of its employee health-care plan; such nonsense is the annual experience of thousands of professional engineers, accepted without question by most because of the carrot of a company-paid health care plan. And nobody in the upper ranks of the professional societies or the state boards cries out against such indignities to remind management that engineers are minds, not bodies, and have a right to be treated as such.

Imagine a corporation engaging learned counsel in a court case, or a certified accountant to look over its books, or a management consultant to plan a reorganization and saying, "Now before you commence work we must have our company doctor give you a physical examination to see if you have any missing parts or deformities or other possible health problems that might impair your work or jeopardize our company health insurance fund." Just as any self-respecting professional in those fields would tell such a company to jump in the lake, it is high time that professional engineers do likewise instead of submitting to such impertinence for the sake of fringe benefits and juicy retirement plans.

Yet the most creative minds in engineering history have more often than not been found in bodies afflicted with suffering, or deformity, and well past the population notion of prime of life. One only needs to consider Edison or Tesla or Steinmetz or Charles Kettering of General Motors, few of whom would have passed the physical exam of many "modern" companies.

These then are the nine rights to which every engineer is entitled if he or she is to be called a "professional" and to be accorded the dignity and responsibility which that title implies in American society. Until the heads of our professional societies and the officials who sit on our 55 state boards go out and campaign for and win these rights for every practicing engineer who has passed the PE exam and see to it that they are protected in these rights against management, government, and organized labor, all the reams of annual print and speechifying about professional status for the American engineer will continue to be so much sounding brass and tinkling cymbals.

5.5

Against Professional Autonomy

MICHAEL D. BAYLES

Talcott Parsons has written that "the professional complex, though obviously still incomplete in its development, has already become the most important single component in the structure of modern societies."[1] As the professions become dominant in society, they threaten democratic values. Their interests and those of the public are not always congruent. By restricting membership, for example, many professions keep the price of their services artificially high. Yet they have traditionally claimed the right to be autonomous and self-regulating. Increasing professional dominance plus autonomy may erode democracy and eventually lead to an elitist society.

Some inroads on professional autonomy have been made recently. The Supreme Court has stricken down minimum fee schedules and prohibitions on advertising.[2] I shall examine three general arguments for autonomy in law, medicine, and college teaching. If the arguments are unsound as applied to these professions, they are not likely to be sound for other professions.

First, I wish to distinguish two different areas of professional autonomy. One area concerns establishing or enforcing standards of professional membership and conduct; the other is the management of professional activity, either in the general operations of a professional organization or in particular cases of clients with respect to the choice of ends to be sought in helping a client or the means used to achieve them.

The three traditional arguments for professional autonomy rest upon the following important premises: only professionals have the expertise required to dispatch particular cases; professional judgment should be free of extraneous influences to assure that the client's interests are looked after; and professional autonomy is necessary to insure confidentiality in the professional-client relationship. The argument from expertise is usually considered to be the strongest. It begins with the assumption that the most important distinguishing feature of a profession is the possession of a body of knowledge attained through an extended period of education. Professional practice, the exercise of expertise, involves organizing particular facts into patterns and applying general principles and rules to them, and this requires the exercise of judgment in the absence of certainty. Choosing a drug for a patient or wording a contract for a client requires expertise involving both knowledge and judgment. Such judgments cannot be evaluated solely on the basis of evidence about the percentage of successful outcomes in a series of cases. Evaluating a particular case requires skill developed through experience, since each case is unique. Thus, lay persons are not competent to judge or direct the work of professionals, since they cannot possibly have the requisite skill.[3]

Despite its attractiveness this argument is, at best, too broad and, at worst, simply unsound. First, it does not apply to the area of establishing professional standards. Many professional standards are simply restatements of general, ethical principles which prohibit bribery, deceit, theft, etc. The organization of professional activity concerns the types of services to be rendered, the most efficient way to provide them, and the reduction of their costs. Intelligent lay persons may judge and direct this kind of activity as well as professionals.

Second, the argument falsely assumes that each case is unique and requires special judgment, ignoring the more frequent, routine, professional work, such as, treating flu, conveying real estate, and teaching introductory courses. Professional expertise is not needed to identify and judge gross mistakes—failures to perform simple, routine tasks adequately. For example, I once complained to a group of first-year law students that my previous lawyer had not done a good job preparing my will. Their immediate response was to ask how I, a lay person, was competent to evaluate the lawyer's work. The reply that he had failed to date the will silenced the argument from expertise. Similarly, a surgeon who leaves a sponge in a patient after an operation makes a gross mistake.

Third, the argument from expertise assumes that professional judgments are either value-free or rest upon indisputable value-judgments. If disputable values are involved, then professional judgments may conflict with those of clients or the public. It is hard for professionals not to be systematically biased. College teachers tend to believe that study of their disciplines is essential for everyone; physicians tend to promote the prolongation of life in all circumstances; and lawyers have a penchant for legal forms and written agreements. While it is understandable and perhaps valuable for professionals to have these built-in biases, it is *because* they have them that professional autonomy in establishing standards may result in standards contrary to the public interest. The American Bar Association's *Code of Professional Responsibility* does not even clearly require a lawyer to reveal his client's intention to commit a crime.[4] Lay persons might plausibly think it contrary to the public interest to permit lawyers to be accessories before the fact. It should also be noted that if the choice of goals in particular cases is for clients, then the argument from expertise has weight only for choices of permissible means to permissible client ends. As lay persons can judge the basic morality and competence of professional activity, direct its general organization, and as clients choose its ends, the argument from expertise does not support a general claim of professional autonomy.

Finally, in some areas professionals have little or no expertise. While college professors argue that people outside their discipline are not competent to evaluate their work, many of them also lack the expertise to evaluate teaching. Few college professors have seriously studied education; training graduate students to teach is often frowned upon. In one case, when a history department wanted to offer a graduate course in teaching, the dean of that graduate school refused to grant credit for the course. Although most professors learn how to teach through trial and error, they are no better prepared to evaluate colleagues whose classes they have never attended than are the students in the classes.

The second major argument for professional autonomy is that from independence, which takes slightly different forms among the professions. For college professors it centers on academic freedom. For lawyers and physicians it focuses on independence of judgment on behalf of their clients. (See ABA, *Code*, Cannon 5 and DR5-107 (C); AMA, *Principles of Medical Ethics*, Section 6.) To protect their independence, lawyers forbid non-lawyers owning interests in, or serving as officers of, legal firms.

The argument from independence applies primarily to autonomy in managing professional activity. It has little to do with establishing and enforcing professional standards. There are two basic arguments against independence in managing professional activity. First, complete independence assumes that professional ideals are practically absolute and that professionals fearlessly defend these "absolutes" against inimical, external influences. However, the values pursued by the professions are not absolute, not even the truth allegedly sought by academics. Other values, such as the prevention of suffering, sometimes outweigh the pursuit of truth, justice or health. Although by acculturation, professionals give great weight to the value represented by their profession, the health, legal status, or knowledge of a client may be less important than other values of the public or even the client. Nor are the professions always the best defenders of these values. In medical experimentation the well-being of individual clients is too often sacrificed for further medical knowledge. Lawyers sometimes sacrifice clients for precedent-setting (and reputation-gaining) appellate decisions, just as professors sacrifice their students' needs and interests in order to teach their current research topic.

Second, the strength of the argument from independence varies inversely with the generality of the subject. When decisions concern the general management of an organization, independence is not basic. A professional is not prevented from doing his best for a client if a lay board decides a legal services group should not take class action cases, a medical group will not provide cosmetic surgery for signs of old age, or a college will not offer a major in Asian studies. The key concern for those affected by decisions of lay boards is that people be able to obtain such services elsewhere. Such a concern is not one of professional independence.

With respect to the ends and means in specific cases, a professional's claim to the exercise of independent judgment becomes stronger. However, provided they are permissible and within the framework of public values, decisions about ends should be made by clients. A professional has a strong claim to independence of judgment only for the means used in a case, and few people would wish to interfere at that level. Interference is most likely to come from third-party payers, such as, insurance companies. A parent may wish to deny a dependent teenager an abortion and make her marry, but this concerns the autonomy of the teenager, not the physician.

The third major argument for professional autonomy is that its abridgement will intrude upon the confidentiality of the professional-client relationship. College professors have not pressed this argument, although it obviously pertains to recommending students and reporting their classroom comments. Discovering, as I once did, that an undercover policeman has been attending a class has a chilling effect on freedom of discussion. Physicians and lawyers consider confidentiality essential, for without it they claim they cannot obtain the information necessary to handle properly their clients' cases.

Without denying the importance of confidentiality in professional-client communications, one may seriously challenge the relevance and validity of the argument from confidentiality to professional autonomy. First, the argument is of limited scope. It does not apply to the establishment of standards or to the general management of an organization because particular clients are not involved. Moreover, confidentiality is not wrongfully breached if clients consent to disclosure of information. Thus, the argument pertains only to disclosures of information without client consent in enforcing standards and managing particular cases.

Second, lay employees of professionals have access to this information but preserve confidentiality. Other nonprofessionals can also keep information confidential. Only in the enforcement of standards might confidentiality be a serious consideration. Currently, most complaints against lawyers and professors are made by clients. A complaining client is usually willing to waive confidentiality, and a lawyer is entitled to disclose information to protect himself against charges of wrongful conduct [ABA, *Code*, DR4-101 (c) (4)]. More aggressive enforcement of standards of professional conduct might involve more cases not based on client complaints. As clients might be implicated in wrong-doing, they might refuse to waive confidentiality. The reason for professional-client confidentiality disappears then, because it is not a protection for criminal or unethical conspiracy. Hence, the argument premised on the necessity for confidentiality in the professional-client relationship does not adequately support the demands for professional autonomy as they are usually articulated.

If the traditional arguments for professional autonomy are strong only for the means used in particular cases, then the way is opened for innovative lay involvement with respect to professional standards and activity to keep them compatible with democratic values. I do not have a comprehensive theory of what such involvement should be. Determining appropriate lay involvement is a task for many persons—members of the lay public, of the professions, and of government. Nevertheless, I will suggest a few possible types of involvement and limits to it.

Some general limits are clear. I do not suggest a return to the nineteenth-century practice of unlicensed professions. Nor do I suggest that any lay person whatsoever should be involved—only reasonably informed and intelligent lay persons. Nor should the involvement primarily be that of governmental agencies, a federal professions commission.

With these caveats in mind, the following are some examples of lay involvement which may be worth exploration. With respect to standards of professional conduct, lay persons might serve on boards or panels to hear charges of misconduct. In several states, lay persons already serve on such boards for the legal profession[5] as well as ultimate medical licensing boards. Yet these are almost all appeals boards and not those which first hear complaints. To my knowledge, no college or university has outside lay persons on committees for hearing cases for dismissal of faculty, although the final decision usually rests with a board of trustees who are mostly nonacademics. Not one of the professions has lay persons involved in establishing the norms of professional conduct—drafting codes of professional ethics.

Potential lay involvement in professional activity could be even broader. One possibility might be to have lay members of governing boards of professional businesses to help direct the activities of professionals, such as, client or public control of hospitals, health maintenance organizations, and legal services groups. Some medical organization's already utilize lay members. Since the arguments for professional autonomy are stronger in particular cases, the plausible forms of lay involvement at that level are fewer. Yet lay persons might serve as members of hospital ethics committees to advise physicians in difficult cases.[6] They already serve on institutional review boards to evaluate the ethics of human experimentation projects. Colleges and even departments might have lay advisory committees about curricular matters. The legal profession could at least experiment with allowing legal firms to have lay officers and owners so that combined legal-real estate or legal-marriage counseling firms could be considered. Other possibilities may readily suggest themselves when the subject is further considered.

As the professions grow in importance in society, democratic values will be threatened unless public scrutiny of the professions increases. While the principle of professional autonomy is presented as a bar to lay involvement in the professions, the arguments from expertise, independence, and confidentiality do not support professional autonomy except in the narrow area of the choice of means to permissible client ends. The professions have perhaps constituted an American aristocracy whose theoretical foundations are no sounder than those of previous aristocracies. Hopefully, this American aristocracy will surrender its unjustified privileges more gracefully than the Ancien Régime did. By doing so, it may preserve its ability to continue to make valuable contributions to society.

Notes

1. "Professions." *International Encyclopedia of the Social Sciences,* (1968).

2. *Goldfarb* v. *Virginia State Bar,* 421 U.S. 773 (1975); *Bates* v. *State Bar of Arizona,* 97 S.Ct. 2691 (1977); *Virginia Pharmacy Board* v. *Virginia Consumer Council,* 425 U.S. 748 (1976).

3. "The profession bases its claim for its position on the possession of a skill so esoteric or complex that non-members of the profession ... cannot even evaluate the work properly." Eliot Freidson, *Profession of Medicine* (New York: Harper & Row, 1970), p. 45.

4. Disciplinary Rule 4-101 (C) (3) only permits a lawyer to do so. ABA *Opinion* 314 (1965) does require revelation, if the lawyer knows *beyond a reasonable doubt* that his client will commit a crime.

5. "A Grievance Committee at Work: The Michigan Story," *Juris Doctor 3* (October 1973): 22, 24, 30.

6. Robert M. Veatch, "Hospital Ethics Committees: Is there a Role?" *Hastings Center Report 7* (June 1977): 22–25.

5.6

Do the Professions Have a Future?

MILTON F. LUNCH

Only a few years ago the question as stated would have been considered either a "gimmick" to attract attention, or a foolish one. Who could have thought that those who have been recognized for their expertise in callings of high learning in areas of vital importance to the public since "the time that the memory of man runneth not to the contrary" would ever face such a question. While there has never been, and probably never will be, an absolute all-encompassing definition or standard to determine which callings are "professions," occupations represented in MAP are professions, entitled as such to recognition for their achievements in the public interest as vital forces in the building of modern society, and before.

But in recent years the professions have come under attack from agencies of the Federal Government, principally the Antitrust Division of the Department of Justice and the Federal Trade Commission, followed now by state attorneys general in some states. And they have as a consequence been subjected to public suspicion as self-serving groups with objectives alien to the public welfare. We have recently seen such newspaper headlines as:

"Rebellion Seen Brewing Against the Professionals" (Washington Post, Feb. 23, 1979); "F.T.C. Chief Criticizes Professionals as Cartels That Keep Fees High" (New York Times, Feb. 23, 1979); "Antitrust Officials Probe Licensing of Occupations" (Washington Star, Feb. 21, 1979).

PROFESSIONS UNDER ATTACK

Perhaps the most revealing comments reflecting the ultimate problem faced by the professions were made recently by the head of the Antitrust Division, John H. Shenefield, who said: "Characteristically, the professions have sought to distinguish themselves from other commercial endeavors by pointing to their public service components, the specialized training they generally require, the close relationship between client and professional. Yet, none of these elements is absent from the services provided by auto mechanics, insurance salesmen, or butchers.

The level of education demanded may be higher, the scarcity value created greater—but the fundamental economic nature of professional transactions is no different from that of other consumer-producer confrontations."

And in the same vein, the Chairman of the Federal Trade Commission, Michael Pertschuk, recently attacked all occupational licensing, including the state licensing of the professions, putting them into the same basket as well diggers, home improvement contractors, pet groomers, sex therapists and TV repairmen. "Licensing becomes a vehicle for the dominance and exclusive authority of a profession, legitimizing its monopoly on discerning and remedying needs, and enforcing its mystique by limiting access to special knowledge."

ANTITRUST HEADS ATTACK

How did the professions get into this kind of situation? Perhaps it was the cynicism generated by the Agnew Scandal, involving engineers and architects, possibly it was the fallout from Watergate, involving mainly lawyers, or was it charges of excessive medical fees paid under Medicare and Medicaid programs. While all of these may have been elements of recent history, it more likely was a deliberate decision by antitrust officials that they had an opportunity to move into a new area of activity by broadening their jurisdiction to the heretofore untouched group of society called the professions. The fact is that since the Sherman Act was enacted in 1890 it had been largely assumed that its restrictions on combinations and conspiracies "in restraint of trade" did not refer to the professions; they were not engaged in "trade or commerce."

But in the early 1970's the Justice Department made its move to test the waters in seeking to apply the Sherman Act to professions. It initially brought suits against the American Society of Civil Engineers, the American Institute of Architects and the American Institute of Certified Public Accountants, charging those societies in each case with a violation of the Sherman Act for having ethical provisions opposing competitive bidding as a basis for selecting a professional or a firm for the respective services. These cases were all settled in 1972 by consent decrees whereby the respective societies agreed to withdraw the ethical standard under challenge.

In December, 1972 the Department of Justice followed with a similar suit against the National Society of Professional Engineers. NSPE chose to resist and litigate the issue, claiming that its ethical rule was demonstrably a standard designed to protect the public from shoddy, inadequate and even dangerous results in a field vital to the public health and safety.

At that time, of course, it had not been decided whether the Sherman Act even applied to the professions. While the NSPE Case was in litigation the Supreme Court decided the Goldfarb Case in June, 1975, holding for the first time that the Sherman Act did not exempt at least the commercial aspects of professional practice from the antitrust law.

Interestingly, that case was not brought by the Government, but was a private action in which certain homeowners complained that a mandatory county bar fee schedule for title searches was an illegal conspiracy to fix prices. In upholding that charge the Supreme Court made several important observations.

FEE SCHEDULES ILLEGAL

First, in addition to resolving for the first time since the Sherman Act was adopted in 1890 that the professions do not have a blanket immunity from the antitrust laws, it held that a mandatory fee schedule for professional services is illegal. And, it held that the county bar fee schedule was not protected by the "state action" doctrine under the *Parker v. Brown* case of 1943, of which more later. The most significant declaration by the Supreme Court for the professions generally, however, was contained in a footnote, as follows:

> The fact that a restraint operates upon a profession as distinguished from a business is, of course, relevant in determining whether that particular restraint violates the Sherman Act. It would be unrealistic to view the practice of professions as interchangeable with other business activities, and automatically to apply to the professions antitrust concepts which originated in other areas. The public service aspect, and other features of the professions, may require that a particular practice, which could properly be viewed as a violation of the Sherman Act in another context, be treated differently. We intimate no view on any other situation than the one with which we are confronted today.

STATE AGENCY RELATIONSHIP

And of particular importance to state agency officials, in the body of the opinion the Court also stated:

> We recognize that the States have a compelling interest in the practice of professions within their boundaries, and that as part of their power to protect the public health, safety, and other valid interest they have broad power to establish standards for licensing practitioners and regulating the practice of professions. We also recognize that in some instances the State may decide that "forms of competition usual in the business world may be demoralizing to the ethical standards of a profession." *United States v. Oregon State Medical Society*, 343 U.S. 326, 336 (1952), see also *Semler v. Oregon State Board of Dental Examiners*, 294 U.S. 608, 611–613 (1935). The interest of the States in regulating lawyers is especially great since lawyers are essential to the primary governmental function of administering justice, and have historically been "officers of the courts." See *Sperry v. Florida*, 373 U.S. 379, 383 (1963); *Cohen v. Hurley*, 366 U.S. 117, 123–124 (1962); *Law Students Research Council v. Wadness*, 401 U.S. 154, 157 (1971). In holding that certain anticompetitive conduct by lawyers is within the reach of the Sherman Act we intend no diminution of the authority of the State to regulate its professions.

PRICE ADVERTISING

Soon thereafter in a case, involving the question whether a state law prohibiting price advertising of prepackaged drugs violated the First Amendment right of free speech, the Court held the restriction to be invalid, but pointedly noted:

We stress that we have considered in this case the regulation of commercial advertising by pharmacists. Although we express no opinion as to other professions, the distinctions, historical and functional, between professions, may require consideration of quite different factors. Physicians and lawyers, for example, do not dispense standardized products; they render professional services of almost infinite variety and nature, with the consequent enhanced possibility for confusion and deception if they were to undertake certain kinds of advertising.

On the basis of these cases the distinctions made by the Supreme Court between standardized commercial products and the nonstandardized services of professionals hopefully indicated that the Court would uphold professional ethical standards which were shown to be for the protection of the public from shoddy, misleading and fraudulent procedures.

NSPE CASE

But these hopes were dashed when the Supreme Court decided the NSPE Case on April 25, 1978. Despite what it had said in the earlier cases, the Court held that the rule against competitive bidding, was a per se violation, admitting of no showing of justification by the extensive record which had been made in the District Court. In fact, the Court said:

> It may be as petitioner argues, that competition tends to force prices down and that an inexpensive item may be inferior to one that is more costly. There is some risk, therefore, that competition will cause some suppliers to market a defective product. Similarly, competitive bidding for engineering projects may be inherently imprecise and incapable of taking into account all the variables which will be involved in the actual performance of the project. Based on these considerations, a purchaser might conclude that his interest in quality—which may embrace the safety of the end product—outweighs the advantages of achieving cost savings by pitting one competitor against another. Or an individual vendor might independently refrain from price negotiations until he has satisfied himself that he fully understands the scope of his customers' needs. These decisions might be reasonable; indeed, petitioner has provided ample documentation for that thesis. But these are not the reasons that satisfy the Rule of Reason . . .

The decision went further and laid down the principle that the "rule of reason" can be invoked only if the ethical point at issue " . . . is one that promotes competition . . . " or stated another way, " . . . the statutory policy precludes inquiry into the question whether competition is good or bad."

Thus, the professions are now confronted with the need to evaluate every ethical standard against the single measure of whether that standard promotes or inhibits competition.

Other than the limited ray of sunshine in *Ohralik v. Ohio State Bar Association*, 98 S. Ct. 1912 (1978) upholding a disciplinary action against a lawyer for "ambulance chasing," the outlook for meaningful professional ethics is dubious, at best. Added to the superior economic power of the Antitrust Division and Federal

Trade Commission to compel professional organizations into consent decrees over ethical standards which have not been declared invalid by the courts, there is the added burden of a more vigorous supporting action at the state level under the incentive of Federal financial aid to the states for antitrust enforcement (much of which is being directed at the professions) under the Hart-Scott-Rodino Antitrust Improvements Act (Public Law No. 94–435, 90 Stat 1383 (1976). A recent article on the status of the Federal-aid program notes that since the inception of the program state antitrust staffs have more than doubled. (See, Phillip F. Zeidman, "Parens Patriae: Another Victim of 'Illinois Brick,' " Legal Times of Washington, November 27, 1978.

PROFESSIONS' BEST CHANCE

The best chance for the professions to maintain ethical standards probably lies through the willingness of state professional licensing boards to write those ethical standards into the requirements for maintaining a license under the state licensing laws, hopefully protected from antitrust attack by Parker. But the Federal Trade Commission has moved to block that avenue of relief through its disputed claim that its trade rule regulation authority under 203 of the Federal Trade Commission Improvement Act authorizes it to preempt applicable state laws to the contrary. The FTC argues at length that its preemption authority is not foreclosed by Parker. (For the full text of the FTC's initial foray in this direction, see the trade regulation rule on Advertising of Opthalmic Goods and Services, *Federal Register*, Vol. 43, No. 107, June 2, 1978.)

Several organizations have filed suits in Federal courts challenging the FTC claim. More recently an FTC official has hinted that the agency may be backing off to some extent on its preemption claim over state regulation of the professions. Alfred H. Kramer, Director of the FTC's Bureau of Consumer Protection, reportedly explained to the antitrust committee of the National Association of Attorneys General that "We are not trying to preempt state authority to decide who can be licensed in the professions and occupations. We are concerned solely with the commercial aspects. We would not affect anything having to do with standards." (Quoted in 47 LW 2373, December 12, 1978.) How the FTC would distinguish between professional standards and commercial aspects of professional practice can only be a matter for wild speculation unless and until the state police power to control the professions is replaced with the "higher wisdom" of Washington.

The best line of defense for the professions to maintain adequate professional ethical standards lies in the Parker Doctrine, alluded to above. That doctrine refers to a 1943 Supreme Court decision, known as *Parker v. Brown*, which held that the Sherman Act was not intended to apply against certain state action; that the State "as sovereign, imposed the restraint as an act of government which the Sherman Act did not undertake to prohibit."

For over thirty years the Parker Doctrine was unchallenged, but in the past several years it has been subjected to all sorts of claims of qualification, restriction and narrowing, leading one to believe that the goal of the Department of Justice and the Federal Trade Commission is to eventually repudiate it in favor of the doctrine of Federal preemption. A brief recital of a few of the recent cases pertinent to the viability of Parker may help clear the air of some recent statements about its current status:

Goldfarb v. Virginia State Bar, 421 U.S. 723 (1975).   The Court rejected the *Parker* defense offered by the local bar association, holding that while the Virginia legislature had authorized the state Supreme Court to regulate the practice of law, the state Supreme Court had not adopted any rule or requirement that lawyers subject to its jurisdiction adhere to any form of fixed fee schedule. The test under Parker, the Court said, " . . . is whether the activity is required by the State acting as sovereign." It was not enough that the state Supreme Court's ethical code mentioned advisory fee schedules to qualify as "state action" under Parker. Likewise, it is not enough that the anticompetitive conduct is "prompted" by state action; the anticompetitive activities must be compelled by direction of the State acting as sovereign; thus, the county bar price schedule was private anti-competitive activity.

Bates v. State Bar of Arizona, 433 U.S. 350 (1977).   Best known for its rejection of a state supreme court ban on advertising by lawyers on First Amendment grounds, the decision is particularly significant on the Parker issue. The case came up also on the claim that the state ban on advertising was a violation of the Sherman Act. But the Supreme Court unanimously rejected that claim, ruling that the Parker exemption applied by virtue of the "affirmative command" of the Arizona Supreme Court as "the ultimate body wielding the State's power over the practice of law."

United States v. Texas State Board of Public Accountancy, F. Supp. (May 5, 1978).   This is a pending case now on appeal to the Fifth Circuit Court of Appeals in which the District Court, Western District of Texas, Austin Division, found against the licensing board and denied the Parker exemption for a board rule against competitive bidding by CPAs. The case is unique, however, in that the state licensing law, while authorizing the board to promulgate rules of professional conduct, conditions those rules becoming effective only upon approval of each proposed rule by a majority of permit holders voting in a referendum. The District Court held that the CPA board rule is thus a "conspiracy to fix prices since neither the Board nor the permit holders could put the rule into effect acting alone; each acted knowing the assent of the other was required." Thus, it is arguable that the board did not act as the agent of the sovereign and did not act as the independent voice of the state. Parenthetically, it is interesting that the District Court categorized the CPA board rule against competitive bidding as "price fixing," whereas the Supreme Court in the *NSPE Case* explicitly held that a similar rule of a private organization was not price fixing, even though otherwise offensive under the Sherman Act. The District Court also held that the Parker exemption applies only if the state mandates the anticompetitive conduct, or if the policy is dictated by the state, citing Lafayette for that proposition. This is in direct conflict with the Third Circuit's analysis in Princeton. If the Fifth Circuit upholds the District

Court's approach is may well set the stage for a definitive decision by the Supreme Court in light of the conflict of circuits.

This review of the recent cases would appear to leave no doubt that anti-competitive rules or procedures enacted by the state legislature are immune from the Sherman Act. If the critics of Parker have their way the result may well be that the state legislature itself must adopt each specific rule, regulation or procedure to withstand Sherman attack, thus negating the concept that a legislative body may elect to rely upon the expertise of its administrative agencies in formulating standards which implement the general regulatory scheme of the legislature.

At bottom, this review indicates that aside from the narrow legalisms which have formed the basis for the professional ethics cases, and those yet to come, the basic issue is one of social policy for the future.

Conceding some merit to Barzun's contention, does it follow that society is so disenchanged with the professions that it would discard the warning of Chief Justice Hughes that "Society would come to grief without Ethics, which is unenforceable in the Courts, and cannot be made part of Law..." (from an address at the Louis Marshall Award Dinner of the Jewish Theological Seminary of America, New York City, November 12, 1962). Or are we ready to abandon the principle enunciated by the Supreme Court that "... the community is concerned in providing safeguards not only against deception, but against practices which would tend to demoralize the professions by forcing its members into an unseemly rivalry which would enlarge the opportunities of the least scrupulous." (*Semler v. Oregon State Board of Dental Examiners*, 294 U.S. 608 1935). Admittedly, the Semler decision, upholding a state law banning the advertising of professional services, has been overturned by Bates, but the social principle expressed has yet to be totally repudiated.

The ultimate concern is not to protect predatory professional practices (such as price-fixing), but rather to look at the question in the broad range of attack on professional standards with the end not yet in sight.

The ultimate danger has been stated briefly by one of the leading scholars engaged in a study of the place of the professions in society:

> I detect a strong trend in our society to strip the professions of their autonomous and self-regulatory features, which are coming to be regarded as monopolistic and, hence, contrary to the public welfare. The trend is to invest the regulatory power in the government on the grounds that the professions can no more be trusted than can business to regulate themselves. The defect in this doctrine is that, if carried to its ultimate, it would accelerate the development of an all powerful state and weaken the system of voluntary associations which is one of the chief strengths of a democratic society.°

°Extracted from a letter to the speaker from Ernest Greenwood, Professor Emeritus, University of California, Berkeley, dated May 23, 1978.

5.7

Addendum to "Do the Professions Have a Future?"

MILTON F. LUNCH

Since the above-noted paper was prepared, there have been further developments in the overall area of attacks on professional ethics. Some of the major developments have involved:

AUTHORITY OF FTC

The initial attempt by the Federal Trade Commission to gain control of state licensing laws, through its claim of jurisdiction to preempt state laws through FTC trade rule regulations, has been thwarted, at least for the present. While the alleged authority remains on the books, the U.S. Court of Appeals for the District of Columbia has set aside the FTC's initial foray in that direction in a matter involving optometry. The Court remanded the FTC rule for further consideration in light of the Supreme Court decision in the *Bates Case* allowing advertising of professional services with reasonable limitations. The Court said that it should be assumed that the states would amend their laws restricting advertising of optometric services to conform to the Supreme Court decision, and that the FTC rule with regard to such advertising is not justified at this time. Also, more significant to the overall preemption question, the Court indicated serious doubt that the FTC has the authority it claims to override state licensing laws.

At the same time, the FTC has been under severe attack in Congress for abuse of its authority. As of April, 1980, Congress has declined to pass continued authorization legislation for FTC operations, and is likely to pass some form of legislative review and veto authority over future FTC regulatory controls.

TEXAS CPA CASE

Subsequent to the District Court decision holding the Texas CPA licensing board rule against competitive bidding to violate the Sherman Act, and denying the *Parker* exemption, the U.S. Court of Appeals, Fifth Circuit, upheld the District Court decision without discussing the merits. An appeal was filed with the U.S. Supreme Court, which declined to hear the case, without giving reasons, as usual. However, it is likely that the refusal of the Supreme Court to hear the case was based, at least in part, on the fact that the Texas legislature had previously amended the CPA licensing law to prohibit the board from adopting rules restricting advertising or competitive bidding by licensees unless it was demonstrated that such rules were necessary to prevent deception, fraud, or undue influence.

THE SUPPLANTING RULE

Most professional societies in the engineering and architectural field had ethical rules to the effect that a member may not seek to displace another member who had an ongoing contract for a particular project, or is engaged in definite steps toward his employment by a client. In a private antitrust action under this type of rule of the American Institute of Architects, Judge Sirica (of Watergate fame), on behalf of the U.S. District Court for the District of Columbia, held the AIA rule to be a violation of the Sherman Act, basing his decision to a large degree on the Supreme Court decision in the NSPE Case. AIA has the option of appealing, but it is indicated that further proceedings now in progress will center only on the amount of damages awarded to the member who was disciplined by AIA for violating the supplanting rule. AIA and the major engineering societies with codes of ethics have subsequently rescinded similar supplanting rules.

RULES ON FREE ENGINEERING,
CONTINGENT FEES AND DESIGN COMPETITIONS

The Antitrust Division, Department of Justice, has recently served a Civil Investigative Demand on the American Consulting Engineers Council, requiring production of information and documents on the above-cited rules of the organization. While the serving of a CID does not necessarily mean the Department of Justice will file a formal complaint alleging that such rules violate the antitrust laws, such action might be anticipated. If so, it is not known at this time if ACEC will resist or agree to rescind or modify its present rules.

STATUS OF PARKER DOCTRINE

There is continuing evidence that the Department of Justice intends to "break" the *Parker Doctrine* to the effect that the Federal antitrust laws do not apply to the states or to state action. Most recently the DOJ has met with the attorney general of Mississippi to urge that he require the state engineering registration board to rescind its rule of professional conduct barring competitive bidding for engineering services. The attorney general's office has the matter under study. If the DOJ pressure succeeds in Mississippi, it is anticipated that the DOJ will extend its pressure to other state boards which have similar rules of conduct. If the Mississippi attorney general and the registration board refuse to accede to the DOJ demand it could set the stage for a definite ruling on the application of the *Parker Doctrine* to state agencies.

5.8

PE Guest Comment

EDWIN M. YODER, JR.

Consult any classic storyteller, from Aristophanes to Dickens, and you find the ancient professions in caricature—predatory lawyers, quack doctors, tyrannical pedagogues—stock figures of ancient lineage, figures we love to hate.

Yet new trades always aspire to professional status, eager for the prestige and money that come with it even if those blessings are mixed with contempt.

"Our culture," said Chairman Michael Pertschuk of the Federal Trade Commission in a thoughtful speech, "is dominated by professionals. . . . Morticians advise us of what is required for a 'decent' burial. Sex therapists counsel us about what is required for intimacy. Professional hypertrichologists diagnose and treat our 'excessive and unsightly' facial hair. . . . The increasing ubiquity of professionals has had an insidious, intimidating effect. . . . A rebellion is brewing against [their] privileged position."

Mr. Pertschuk's speech is part of the trend he describes—another sign of public unrest about the professional "mystique." President Carter has scolded doctors and lawyers; the Supreme Court now permits lawyers to advertise; an FTC judge recently ruled that physicians may advertise "routine services," whatever they are.

When practitioners of auto repair, sex therapy, burial, "cosmetology" (and journalism, for that matter) affect professional trappings, the notion of professionalism depreciates. Even the "learned" professions suffer a damaging ruboff effect. Services, as Mr. Pertschuk observes, are increasingly spoken of as "commodities"; patients and clients seen as "consumers." The language of trade creeps in; the language of service fades.

What's so special anyway, we now ask, about the "learned" professions? Is there really much other than their price lists to distinguish a surgeon from an auto mechanic? I may seem to jest. But even medicine, the most sophisticated and arcane of professions, has its energetic gadflies nowadays: Ivan Illich, for instance. "To what end, [Illich] asks, do we develop all our medical technology but to delay grotesquely deaths which are inevitable?" (Mr. Illich's views are thus summarized, in part, by sociologist Nathan Glazer, in a recent *Commentary* article.)

316

According to Mr. Glazer, many of the more raucous attacks on professionalism take the Marxist view that professions are forms of "monopoly," enclaves of class conspiracy and privilege—although as he observes professionalism in America has been fluid and meritocratic, not caste-ridden.

Mr. Pertschuk thinks it healthy that American society has become keenly aware of—even obsessed by—the seamier side of professionalism; that the "commercial underpinnings" of the professions are increasingly visible. It may be healthy enough, up to a point. But few "transvaluations" of traditional values and customs come cheap, and a few caveats occur to me.

"Mystiques," including professional ones, have their uses. When useful mystiques dissolve, competition will probably supplant ethical self-restraint as society's safeguard against professional fraud, and the social tone will change too. A community of self-policing people offering reciprocal services is seemlier than the atomized, dog-eat-dog marketplace. That, at least, has been the civilized prejudice for centuries. Ethical codes can't entirely protect the public against fraud, of course. But it's gratuitous to assume that such ideals, imperfectly followed, are hypocritical and ready for discard.

Professionalism has relied traditionally on the principle of self-regulation. One must always ask what the alternative is. And the alternative frequently urged by current reformers is intrusive government regulation, or vigilante action by consumerist zealots, or both.

Mr. Pertschuk says the FTC is wary of "the extension of federal authority into areas which historically have been the province of the states." But what of the extension of political authority, federal or state, into provinces historically private? Tocqueville considered "voluntary association" part of the essential genius of American democracy, and professional self-policing is an aspect of voluntary association.

If we come to assume that every profession is predatory and its ethical code a sham, that we can be held from one another's throats only by official policemen, will we be better served—legally, medically, educationally, or even funereally? I doubt it. But there does seem to be a remarkable nostalgia, these days, for the state of nature.

5.9

The Problem of Defining a Profession

MORRIS L. COGAN

To define "profession" is to invite controversy. C. L. Stevenson has said, "To choose a definition is to plead a cause, so long as the word defined is strongly emotive."[1] It is quite apparent that "profession" is just such a word. When definitions of it are proposed they are rarely subjected to rational consideration. Reactions tend to be polarized toward an enthusiastic and uncritical acceptance or toward a rancorous and defensive rejection. So many advantages have accrued to profession, so many claims to it are made by so many people, that the cutting edge of a definition—be it ever so blunt—is almost sure to draw cries of protest from many aspirants to the title. If the definer is prepared to accept these risks, then he must in addition be prepared to see his work result in failure or at best seemingly insignificant accomplishment, for the concept with which he is dealing is extremely difficult to identify and describe. The man who addresses himself to the problem of defining a profession should be able to say with equanimity: "It is better to light a little lantern than to curse the darkness."

THE LITERATURE OF PROFESSION

There is a rather extensive literature of definitions of profession. Research into these writings reveals no general agreement on any "authoritative" statement.[2] The meanings found in most dictionaries are similar, and although these are often cited, they appear to offer to most writers only a convenient point of departure or, perhaps more precisely, a primary theme upon which radical variations are composed. The legal definitions of profession are almost always so closely related to the conditions of a specific case that they are rarely susceptible of wider application, and the arbitrary definitions commonly adopted for purposes of research or descriptive statistics are of almost equally restricted utility.

---

[1] Charles L. Stevenson, *Ethics and Language* (New Haven: Yale University Press, 1944), p. 210.
[2] For a summary and analysis of definitions, see Morris L. Cogan, "Toward a Definition of Profession," *Harvard Educational Review*, Vol. 23 (Winter 1953), pp. 33–50.

Some of the writers most notable for their interest in the analysis of profession simply refuse to venture a definition. Outstanding representatives of these are Alexander M. Carr-Saunders and P. A. Wilson, who, after an extensive treatment of the characteristics of profession, are forced to conclude that certain vocations, possessing these characteristics in a greater or lesser degree, approach more or less closely to the condition of profession.[3] Oliver Garceau writes: "There is no accepted definition of a 'profession.' Interpretation of such a concept is a matter of personal temperament."[4] More recently Roy Lewis and Angus Maude have written that "a moral code is the basis of professionalism...."[5] Beyond this common thread of a special morality, their efforts at analysis collide head on with the difficulty of identifying and clearly delineating such elusive concepts as the associational standards of admission and conduct, the general and special education, the unified body of knowledge, that are requisite to profession.

Flexner's criteria

In brief, a survey of some of the literature on this topic reveals that the promulgation of a satisfactory definition has progressed but little beyond the six criteria proposed by Abraham Flexner in 1915: (1) intellectual operations coupled with large individual responsibilities, (2) raw materials drawn from science and learning, (3) practical application, (4) an educationally communicable technique, (5) tendency toward self-organization, and (6) increasingly altruistic motivation.[6]

THREE LEVELS OF DEFINITION

Some confusion of terms is evident in many recent discussions of profession. This perplexity appears to derive in large part from the difficulty of communicating ideas and resolving differences when a single term (profession) is used to designate disparate referents. This disparity appears to arise from the use of "profession" at three different levels of definition. If productive consideration of the problem of defining a profession is a desideratum, then an initial clarification of some of the denotations of the word may be useful.

Historical and lexicological definitions

The first type of definition is historical and lexicological. If a specific profession is to be defined, and if such definition is not to be purely arbitrary, that is, not to be divorced from the rich historical and traditional associations clustering about the term, then the definition must be related to the general concept of profession. This concordance with the "principal features or structure of [the] concept [is necessary], partly in order to make it definite, to delimit it from other concepts, and partly in order to make possible a systematic exploration of the subject matter

[3] Alexander M. Carr-Saunders and P. A. Wilson, *The Professions* (Oxford: Clarendon Press, 1933), p. 4.
[4] Oliver Garceau, *Some Aspects of Medical Politics* (Ph.D. thesis in the Department of Government, Harvard University, 1939), p. 4.
[5] *Professional People in England* (Cambridge, Mass.: Harvard University Press, 1953), p. 64.
[6] Abraham Flexner, "Is Social Work a Profession?" *School and Society*, Vol. 1 (June 26, 1915), p. 904.

with which it deals."[7] Once a lexicological and historical examination has been made as the initial step in the process, then it is possible to begin to isolate the essential features of profession in general, and thus to describe—even if only tentatively—the boundaries that set it off from other allied terms: vocation, trade, craft, art, semiprofession, and so on. Research of this kind may provide the framework within which the process of defining a specific vocation can proceed with some assurance that the vocation will tend to remain, in spite of its differentiae, essentially a species of the genus profession. The task is certainly difficult. It involves a recognition of the complexity of the concept and a sustained effort to cut through the nuances and the multiple secondary meanings clustered about the core meaning.

The standard etymological works constitute valuable sources of definitions at the first level. The research in such treatises must however be supplemented by examination of other writings. Among the most illuminating of these are the statements made by Abraham Flexner,[8] Alexander M. Carr-Saunders and P. A. Wilson,[9] Robert Ulich,[10] and Alfred North Whitehead.[11]

An objective and rigorous analysis of such writings may provide the basis for a defensible statement of the essentials of profession. The results of one such attempt are reproduced below for purposes of illustration.

> A profession is a vocation whose practice is founded upon an understanding of the theoretical structure of some department of learning or science, and upon the abilities accompanying such understanding. This understanding and these abilities are applied to the vital practical affairs of man. The practices of the profession are modified by knowledge of a generalized nature and by the accumulated wisdom and experience of mankind, which serve to correct the errors of specialism. The profession, serving the vital needs of man, considers its first ethical imperative to be altruistic service to the client.[12]

After such a first-level statement has been formulated, the problem of defining a *specific* profession then becomes the concern of the practitioners, who may at this point address themselves to the unique conditions of their area of specialization.

Persuasive Definitions

The second level of definition is called "persuasive." Professor Charles L. Stevenson has noted that such statements are used "consciously or unconsciously, in an effort to secure, by [the] interplay between emotive and descriptive meaning, a redirection of people's attitudes."[13] Why is it desirable to encourage definitions of this second type, definitions designed to redirect people's attitudes?

---

[7]Morris R. Cohen and Ernest Nagel, *An Introduction to Logic and Scientific Method* (New York: Harcourt, Brace and Company, 1934), pp. 231–32.

[8]*Op. cit. supra* (note 6), pp. 901–11.

[9]*Op. cit. supra* (note 3).

[10]*Crisis and Hope in American Education* (Boston: The Beacon Press, 1951), pp. 133–35, 158–59.

[11]*Adventures of Ideas* (New York: The Macmillan Company, 1933), pp. 72–79.

[12]Morris L. Cogan, *op. cit. supra* (note 3), pp. 48–49.

[13]*Loc. cit. supra* (note 1).

The answer lies in the nature of profession and in the behavior of man: profession demands arduous training and the practitioner's personal commitment to an exacting ethical code; men have commonly needed to be persuaded to undertake such training and to follow the dictates of such a code. Both attributes are required, and a failure in either is a failure of profession. It is interesting to note in this connection that the most persistent criticisms of professional groups are in essence allegations of such failure. The public school teachers, for example, are accused of inefficiency due to inadequate preservice education; doctors and lawyers are often criticized for violation of their ethical code.

If profession is to remain desired and desirable, then powerful persuasive definitions are a sine qua non. In recent history it has been all too easy to dismiss such statements as hortatory or messianic. The derogatory force of these adjectives derives from the idea that such appeals do not really influence behavior. It is part of the pessimist myth of our times that the seer and the visionary are without followers, that the appeal to ideals must fail, whereas the appeal to self-interest is universally invincible. Yet it is clear that in human affairs, yesterday's idealistic commandments often become today's laws and are tomorrow translated into action. One has only to observe the series of persuasive definitions in the history of antislavery, recently culminating in the desegregation ruling of the Supreme Court. That there are daily regressions in the course of man's behavior is not to the point; the force of the decision is toward the ideal behaviors, and the persuasive powers of the Supreme Court's decision will probably be of greater influence than the immediate legislative and administrative actions deriving from it. Where, as in profession, it becomes necessary to redirect men's attitudes, the power of the persuasive definition should not be underestimated. Morris R. Cohen has written:

> It would be a mistake to conclude that the issue [of definitions] is merely verbal and of no real significance . . . no question of this sort can be *merely* verbal, because words are most potent influences in determining thought as well as action.[14]

Two striking examples of the force of the persuasive definition are to be found in the degree to which the Hippocratic Oath and the persuasive definitions offered by Abraham Flexner have both been translated into the program and behavior of contemporary medical societies. To be deflected, then, from formulating persuasive definitions of profession is to be deflected from the extremely important functions of definition at the second level. A few essays of this kind deserve special mention: Morris L. Cooke's "Professional Ethics and Social Change,"[15] Robert D. Kohn's "The Significance of the Professional Ideal,"[16] and Roscoe Pound's "The Professions in the Society of Today."[17]

The insights provided by the first- and second-level definitions are especially valuable for the construction of theoretical models within which profession may be examined and from which important hypotheses may arise. Such definitions have,

---

[14]Morris R. Cohen, *Reason and Nature* (New York: Harcourt, Brace and Company, 1931), p. 389.
[15]*American Scholar*, Vol. 15 (1946), pp. 487–97.
[16]*The Annals*, Vol. 101 (May 1922), pp. 1–5.
[17]*New England Journal of Medicine*, Vol. 241 (September 1949), pp. 351–57.

however, uniformly collapsed when they have been inappropriately applied as the operational criteria for decisions as to whether *specific* phenomena are to be called professional, nonprofessional, or unprofessional. The failures following upon such misapplication have tended to be generalized, and the positive values of such statements have therefore often been discounted or overlooked entirely.

Operational Definitions

The third level of definition is termed operational, and is designed to furnish the basis upon which individuals and associations may make specific decisions as to the behavioral concomitants of profession. The demand for operational definitions stems from the scientific temper of the times. It is a demand for the observable and the measurable; it will not be satisfied by lexicological and persuasive statements.

Operational definitions are the guidelines for the practitioner as he faces the day-to-day decisions of his work. They are, for example, the rules for professional conduct; they mediate the practitioner's relations to his client, to his colleagues, to the public, to his association. They set forth the specific criteria of general and special education for the professional, the requirements for admission to practice, the standards for competent service.

An example of operational definitions in action may be seen in the requirements for preservice medical education. At present this prerequisite to actual practice may be stated as "graduation from an accredited medical school." But behind this apparently simple statement lies a long history of operational definitions promulgated and applied. It has been necessary to set up standards (definitions) for the accredited schools: their admission requirements, curriculum, qualifications of the instructional staff, facilities for instruction, and the like.

The complex problems of professional ethics have been rather effectively attacked by associations of architects. They have studied and publicized certain "cases"; they have prepared, illustrated, and circularized "ethical documents" specifically designed to offer guidance to the architect in the complicated situations he daily faces, in which he must have operational definitions that will clarify his often conflicting responsibilities to client, public, inspectors, subcontractors and colleagues.

To build up the body of decisions, illustrative cases, and precedents is a task that makes severe demands upon the time and energy of the practitioners. That this time and energy are well spent is evidenced by the findings of Benson Y. Landis. Dr. Landis has made a sociological analysis of professional codes and has concluded that the optimal control of ethical behavior is generally achieved by those professions that have set up the machinery for control based on "a code containing clear definitions of situations."[18]

The major advantage of the operational definitions is that they restrict the area of idiosyncratic professional behavior and tend to stabilize the boundaries between genuine professionalism, unprofessionalism, and nonprofessionalism. It should not be assumed, however, that the operational definitions are at present an aqua regia in which all the obdurate problems of the professions can be dissolved.

---

[18]Benson Y. Landis, *Professional Codes* (New York: Teachers College, Columbia University, 1927), p. 96.

Such definitions are relatively rare, since most professions have contented themselves with the generalities of the first- and second-level statements. Where they do exist, they are often rather primitive, lacking precision. What is needed is research, examination of behaviors, trial and inevitable error in developing the theory of profession and in investigating its behavioral accomplishments.

PROBLEM AREAS AND RESEARCH

Most of the important studies of profession have been done, almost fortuitously, by sociologists, educators, and students of government. It would seem to be up to the organized professions themselves now to encourage research and self-examination. For their own welfare the professions need to support extensive investigations of their own history, their behavior, and their problems. Otherwise they may have to resign themselves to vagueness about their own rights, privileges, and responsibilities; they may expect creeping encroachment by pseudo professions; they may witness the gradual dissolution of the very attributes that have contributed to their survival and growth.

Professionalism under the State

Some of the areas in which research may be valuable deserve mention here. Both the state and the large corporations are absorbing more and more professionals. It is a moot question whether it is possible to maintain professionalism under such conditions. For example, the fact that most teachers are employees of a municipality may have inhibited their professional growth, since the conditions of teaching tend to obscure the locus of responsibility to the client (Is it to each child? to the class? to the local community? to the whole society?) and also tend to limit associational solidarity and self-regulation. At the very least the professions should be examining the enormous readjustments taking place in the relations between practitioners and the state under the socialist regime in Britain. These are of major importance to professions in the United States. Such processes need to be subjected to close scrutiny. The insight gained from this research may contribute to the survival of professions.

American Medical Association

A closely related problem is at present plaguing the medical associations of the United States. It is clear, for example, that the American Medical Association has failed to define its proper role to the satisfaction of its members and the public. The pressures building up around the question of socialized medicine owe a portion of their intensity to ideological differences, it is true; but some of this intensity is also due to the unwillingness of the AMA to define its role in relation to government. Indeed, it is safe to say that most professional associations have failed to examine the assumptions underlying the maintenance of political lobbies, the functions of the association in the certification of practitioners, the delegation of regulatory powers to nongovernmental groups, and the association's part in the regulation of practice.

Other Trouble Spots

Very little attention has been given to problems of individual and associational professionalism. Is profession inevitably a group phenomenon or may it be achieved individually? Are there two kinds of profession—personal and associational? If there are, can these two roles come into conflict, that is, is it possible that the dictates of personal professional ethics may force a practitioner to resign from an association whose activities appear unethical to him?

Another trouble spot in organized profession is the frequent conflict between the associational functions of internal protection of members and protection of the client. At the verbal level there appears to be no conflict at all: practically every code gives priority to the welfare of the client. In actual practice, however, the issue is confused, and a great deal of research will probably be necessary before it is clarified. For example, is highly selective recruitment of practitioners a guarantee of competent service to the client, or is it rather a guarantee of economic protection to the successful candidate and at the same time an artificial restriction of vital public services?

Among schoolteachers this issue is even more clearly drawn. Is a teachers' strike a deprivation of services vital to the children or is it rather an ethical imperative for the protection of these very children? Easy answers to such questions are not possible, yet answers must be sought. Otherwise the professions will have to be content with decisions which are no more than vagaries and with actions whose justifications are no more than a defensive posture.

Perhaps it would be appropriate to add a word of caution as to the power of definition to solve the problem of the arrogation of the title "profession" by certain vocations. This process is viewed with a great deal of alarm by many persons, who therefore seek a sharp definition that will serve to cut these claimants off from the title. It should be obvious that definition alone will not accomplish this purpose. Perhaps the best way to make such distinctions clear is to ensure that the genuine professions demonstrate such a consistently high order of professional behavior that ultimately every inappropriate assumption of the title will appear inept, ludicrous; thus the unqualified claimants will either labor to achieve genuine profession or retire from the field.

SUMMARY

In summary, it appears that some of the problems of defining a profession can be resolved if the practitioners (1) seek to establish a general framework within a lexicological and historical context, (2) provide ideal incentives through the persuasive definitions, and (3) develop the behavioral and operational definitions that give direction within the unique conditions of a specific profession.

5.10

Professional Licensing
and Public Policy

ROBERT B. REICH

The public's growing skepticism about professional licensing is founded on two decades of economic research showing that licensing is often an inefficient way of protecting consumers, but a highly efficient way of managing a professional cartel. The central challenge for public policy therefore is to develop the means for deciding when licensing is necessary and what form it should take. That challenge has proceeded in several distinct phases over the last few years.

The first phase concerned professional advertising. In a series of judicial and administrative decisions grounded upon the First Amendment, the antitrust laws, and the Federal Trade Commission's authority to prevent "unfair . . . trade practices," state restrictions on professional advertising have been found to be unwarranted. Regardless of the legal basis, the analysis has been substantially similar: the costs to consumers of such restrictions outweigh their benefits. The FTC's recent decision to lift restrictions on price advertising by opticians, optometrists, and opthalmologists, for example, was premised on the assumption that absent such restrictions consumers would, on the average, pay less for eyeglasses with no commensurate loss in terms of product quality. The Commission rejected industry arguments that widespread deception would follow the lifting of the advertising ban. Since false and deceptive advertising techniques were already prohibited in every state, said the Commission, "[t]o prohibit opthalmic advertising totally because of the possibility that a few practitioners will engage in deceptive advertising constitutes a classic example of regulatory overkill."*

The barriers to professional advertising have fallen quickly. State licensing boards, often prompted by rulings of state attorneys general, have agreed to permit advertising. Codes of professional ethics have been modified to allow advertising. Attorneys and opticians in particular are now advertising their wares in increasing numbers, and prices have started to fall.

---

*Trade Regulation Rule, Advertising of Opthalmic Goods and Services, 43 Fed. Reg. 23992 (1978).

The only important question remaining with regard to professional advertising concerns the extent to which a state or a professional association may prohibit certain types or forms of advertising, such as testimonials, discount prices, or direct mail advertisements, to prevent deception. These are close questions often requiring an intricate balancing of competition and consumer protection. Several weeks ago the FTC determined that the American Medical Association had engaged in an "unfair method of competition" by prohibiting physicians from advertising. At the same time, however, the Commission recognized the role of the AMA in formulating and adopting restrictions on physician advertising designed to prevent deception. Moreover, the Supreme Court recently determined that the states could, consistent with the first amendment, bar optometrists from advertising under a trade name.[*] The Court reasoned that trade names may effectively mislead consumers as to the identity of the professionals with whom they deal as well as the standard of optometrical care provided. Oddly, the Court's majority appeared to disregard the strong economic incentive of the owners of such an enterprise in preserving the value of their trade name by ensuring consistent quality.

The second phase of public scrutiny of professional licensing concerns restrictions on commercial practice which arguably bear little or no relation to quality of service. For example, most state statutes or regulations now prohibit licensed professionals from being employed by unlicensed parties or by non-professional corporations. As a result, professionals who wish to expand their practice have limited access to equity or venture capital. Similarly, many states place limitations on locations where professionals can have offices, often barring them from commercial areas and restricting the number of branch offices they can open. These restrictions arguably inhibit the development of high volume, lower cost professional services. The FTC's Bureau of Economics is currently conducting a study examining the relationship between commercial restrictions and the cost and quality of professional services. While the study is not yet completed, preliminary data indicate that the cost of eye examinations and eyeglasses from optometrists may be significantly lower in jurisdictions which do not restrict these commercial practices. Finally, state residency requirements and refusals to recognize licensing by other states also may drive up costs with little or no consumer benefit. A recent study by Dr. Lawrence Shepherd of the University of California, comparing the prices of dental services in states with and without reciprocity arrangements, concludes that the price of dental services and mean dentist income is 12 to 15 percent higher in nonreciprocity jurisdictions, at an annual cost to consumers of approximately $700 million.[†]

Already large department stores in some states are offering inexpensive dentistry, legal advice, and eye examinations. Evidence indicates that many consumers who would not otherwise seek professional help, are taking advantage of these low cost services. There is reason to believe that some of these services may permit substantial economies when provided on a large scale, with little or no sacrifice in quality.

---

[*] *Friedman* v. *Rogers*, 99 S. Ct. 887 (1979).

[†] Shepherd, "Licensure Restrictions and the Cost of Dental Care" (1977)

The third phase of public scrutiny of professional licensing concerns the provision of services by paraprofessionals such as paralegals and dental auxiliaries, and alternative, independent providers such as nurse practitioners and midwives. Most states substantially restrict the circumstances in which these two groups can provide their services directly to the public. But recent studies suggest that many of these restrictions are drawn too tight. In one study, for example, comparing amalgam restorations and stainless steel crowns placed by dentists and dental nurses in Canada, researchers found that the nurses perform as well, if not better, than the dentists.° Other studies confirm that many functions traditionally reserved to licensed dentists can be safely delegated to an auxiliary at a considerable savings without sacrifice in the quality of care.

But measuring the quality of care is a tricky business, and it is unclear to what extent federal policymakers should attempt to balance quality of care against competitive prices in determining what responsibilities paraprofessionals and alternative providers can safely undertake. One area where the FTC can be helpful, however, is in the provision of information, designing and funding studies that compare the effects of alternative regulatory systems on price and quality, and serving as a clearinghouse for the findings of other studies.

The fourth and final phase of public scrutiny of professional licensing concerns the processes by which licensing decisions are made. The evolution of certain occupations from being collections of individual sellers to tradespersons to certified professionals and, ultimately, to licensed professionals is well known. Licensing is often a sign that an occupational group has come of political age. Once licensed, the profession is often in a position to determine how much competition it will tolerate. Licensing boards dominated by members of the profession may act like any other cartel—adjusting prices and entry standards to protect the incomes of established practitioners. Unless the process of licensing is itself rendered resistant to these pressures, piecemeal attempts to reform particular licensing provisions will be swamped by a rising tide of special interests seeking protection from competition.

One state now licenses 86 occupations, including circus exhibitors, bowling-alley operators, florists, and billiard parlor owners. Another licenses tattoo artists; a third, lightning-rod salespersons. Until two months ago it was a misdemeanor for someone without a license to repair a watch in yet another state. The penalty was a six-month jail sentence, the same as for practicing medicine without a license. Is there any doubt that these provisions were enacted because the states had no mechanisms in place to differentiate measures designed to protect businesses from those designed to protect consumers?

Political decisions about professional licensing fall into three logical steps, each of which requires some mechanism to check the momentum of the particular occupational group.

The first step occurs when the occupational group seeks to be licensed. Although the group has a constitutional right to seek such protective legislation,† it

---

°See Ambrose, "A Quality Evaluation of Specific Dental Services Provided by the Saskatchewan Dental Plan," Saskatchewan Dept. of Health (1976); Hord, "The Ontario Dental Association Demonstration Project on Dental Auxiliaries with Extended Duties," *Ontario Dentist* 14–18 (1978).
†See e.g., *Eastern Railroad President's Converence v. Noerr Motor Freight, Inc.*, 365 U.S. 127 (1961); *California Motor Transport Co. v. Trucking Unlimited* YOU U.S. 508 (1972).

has no right to a free ride. I understand that the New York State Assembly's Higher Education Committee recently adopted a series of eighteen questions which the sponsor of any new licensure bill must answer before the proposal may come up for a vote; the questions are designed to elicit from the sponsor accurate estimates of the real need for licensing, and its prospective costs. Due to these questions as well as the active role of the Department of Education, the efforts of many groups seeking licensure have been curtailed including athletic trainers, marriage and family counselors, naturopaths, sanitarians, professional recreators, and tree pruners. The State is to be congratulated on its tenacity.

Other states have experimented with slightly different methods for screening requests for licensure. Minnesota and California have removed the decision to license health-related occupations from the legislature to an administrative agency, which functions as a sort of buffer against political pressure by the group seeking licensure. In Virginia, all requests for licensing are first referred to a Board of Commerce, consisting of nine public members. After conducting hearings, the Board may recommend to the legislature changes in civil or criminal law, registration, certification, licensure, or no action.

A second step occurs after the profession is licensed, when a board is established to issue detailed regulations governing the profession. Licensing statutes which themselves are relatively harmless can give birth to extraordinarily anticompetitive regulatory progeny. Through licensing board regulations, years of schooling and periods of residency as a prerequisite to licensing may gradually expand. Innovative ways of providing services may be restricted. Higher percentages of applicants may be rejected during periods of economic downturn, when there is less demand for their services. Advertising, promotion, and forms of business may also be restricted.

In response to this danger of creeping entry barriers, some state legislatures have provided for legislative vetoes of licensing board regulations. But the legislative veto is a clumsy instrument, at best. It is difficult for a busy legislature to monitor any but a small fraction of board regulations, and even then the legislature is ill-equipped to determine which regulations deserve most careful scrutiny.

Other states have placed a majority (or large minority) of "public" members on each professional licensing board. Because they are not members of the profession and have no financial interest in creating anticompetitive regulations, the public members can serve as a check upon the professional members. Indeed, this is the principle behind a rule recently proposed by the staff of the FTC requiring that a majority of the members of the boards of Blue Shield and Blue Cross be non-physicians. Arguably, principles of due process also require that licensing board decisions be made by individuals who do not have a pecuniary interest in the outcome.[*]

Another means by which to increase public scrutiny of the board's rulemaking process is the creation of broad rights of public intervention into licensing board proceedings, and the provision of funds specifically earmarked to finance such interventions by groups which could not otherwise afford to do so. I am not aware of any states that have established public participation programs of this sort. But

---

[*]*Gibson* v. *Berryhill*, 411 U.S. 546 (1937); *Wall* v. *American Optometric Assoc. Inc.*, 379 F. Supp. 175 (N.D. Ga.), *aff'd mem.*, 419 U.S. 888, *rehearing denied*, 419 U.S. 1061 (1974).

our experience over the last four years, during which time various public representatives received funding from the Federal Trade Commission to participate in its proceedings, has confirmed the importance of this mechanism in helping to redress the balance between organized economic interests and relatively disorganized consumers.

A third step occurs after a number of years of licensing, by which time judgments can be made about whether its anticipated benefits have exceeded its costs. Several states have enacted "sunset" provisions which require that licensing statutes and accompanying regulations be reevaluated after a fixed time interval. But such provisions have been relatively unsuccessful so far, because once a professional group has obtained licensing it is difficult to muster the political will necessary to repeal the license. Perhaps more important is the fact that legislatures have not appropriated funds necessary to properly evaluate the impact of their licensing laws.

Ideally a "base-line" study should be undertaken before the profession is licensed, seeking to measure both price and quality; then, at the time of the sunset review, a follow-up study should remeasure price and quality, controlling for extraneous changes. A strong argument can be made that when enacting licensing laws state legislatures have a responsibility to commit themselves to undertaking such base-line and follow-up studies.

The control of professional licensing is no easy task. It requires the ability—and political will—to balance the need for consumer protection against the cost of reduced competition, and to find less costly ways of protecting consumers. It requires complex economic research. And it requires an institutional framework which insulates the process from special interest politics. The task cannot and should not be left to the federal courts, nor should the issues be reduced to simple legal theories. Fundamentally the responsibility, and the challenge, is yours. And given public skepticism about government and the professions, the challenge is one you must meet head on.

5.11

Ethical Decisions for Engineers: Systematic Avoidance and the Need for Confrontation

CARL NELSON
SUSAN PETERSON

INTRODUCTION

The engineer and the engineering profession are at a threshold of challenge to their continuing professional contribution to society. If unable to cope with moral issues and the moral dimensions of their work in the coming decade, the days of the engineer as a professional will indeed be numbered in the public community.

This stern ten year warning is directed at each of us; every engineering educator, student and practitioner. It is a plea to encourage taking on individual moral responsibility for professional activities and jobs. The public will undoubtedly view the coming change in the engineering community as one result of another professional crisis and scandal: the indictment and resignation of U.S. Vice President Agnew less than a decade ago. This connection is inevitable from the public point of view only partly because Mr. Agnew suffered political disgrace; another aspect of his case was the revelation of unethical procedures and activities that had become "common practice" in his professional community. Now each profession is held under stricter public surveillance and suspicion. We think confronting the ethical issues in each profession head on is the appropriate measure to take, and hope you find our efforts somewhat enlightening and worthwhile.

National Project on Philosophy and Engineering Ethics, November 1978.

BACKGROUND

The engineering profession in the United States, recognized by the public in the mid-nineteenth century, has been well documented for *technical* achievement to date. Thus technical competence is almost a synonym for "engineer" in the public view. Volumes of works exist declaring the myriad of contributions made by engineers to the improvement in the quality of life in town and village throughout the United States and indeed throughout the world. In fact, recent bestsellers such as *The Great Bridge* and *The Path Between the Seas* by David McCullough have told success stories of *technical* know-how and phenomenal technological skills of engineers. Each of us can probably name a half dozen similar books in the last decade which have provided us with satisfaction and pride to be part of the engineering profession.

However, mere technical know-how is an insufficient basis for true professionalism, and a person whose primary asset is his or her technical ability is somehow "dehumanized," almost like a well-trained robot. Public images of technicians therefore have a negative connotation at times; the human touch is more and more required in all the professions according to public opinion. With doctors, more emphasis is laid upon actual doctor–patient contact, rather than the growing tendency in the medical profession to rely upon intermediaries such as nurses, lay medical persons,, orderlies, etc. Also in the legal profession, more emphasis is laid upon taking into account a more responsive and direct relationship between legal procedures and the individual involved. With respect to engineering, unfortunately there has been news media coverage of the BART, DC-10, Pinto, Firestone, Hartford Civic Center and Teton Dam cases of engineering "malpractice," so to speak, leaving the public with some legitimate concerns about the professionalism of the engineer.

This then is the dilemma: unless the moral issues of engineering activities are confronted and solved by engineers themselves, in recognition of their obligation to the public,[1] engineering will become a technical vocation managed from without, rather than a significant profession with an internal integrity.

PROFESSIONALISM

The American Society of Civil Engineers defines "profession" as a pursuit of a learned art in a spirit of public service. ASCE also provides the following amplification:

> A profession is a calling in which special knowledge and skill are used in a distinctly intellectual plane in the service of mankind, and in which the successful expression of creative ability and application of professional knowledge are the primary rewards. There is implied the application of the highest standards of excellence in the educational fields prerequisite to the calling, in the performance of services, and in the *ethical conduct of its members.*

---

[1]Wisely, William H., "Public Obligation and the Ethics System," ASCE 3415, October 1978.

Also implied is the conscious recognition of the profession's obligation to society to advance its standards and *to prescribe the conduct of its members.*

ASCE also defines civil engineering as a profession in which a knowledge of the mathematical and physical sciences gained by study, experience and practice is applied with judgment to develop ways to economically utilize the materials and forces of nature for the progressive well-being of mankind. Engineers should create, improve and protect the environment and provide facilities for high quality community living, industry and transportation as well as provide structures for the general use of mankind.

Thus, professionalism as a concept includes not only a technical mastery of certain skills and a body of knowledge, but a firm commitment to make progress in a qualitative sense in human society. This is usually referred to as a *commitment to the public interest.* Our research has revealed to us thus far that, philosophically speaking, the public interest is sometimes difficult to determine. Of course, many engineers realize this in practical terms, and understand that satisfying the public interest is sometimes a lost cause or a zero sum game, since it is difficult to please everybody all the time. However, it seems to us that the problem goes deeper than this and is more theoretical than practical in nature. For instance, the concept of "interest" is itself problematic. Traditionally, moral philosophers have treated the "interest" of a person as that which either provides for that person's happiness or pleasure, or on the other hand, promotes the well-being of that person. One problem is obvious here: is the interest of a person determined subjectively by that person himself or herself, or by objective data gathered by others? This theoretical problem has been under discussion in philosophy since the days of ancient Greece, and yet is still relevant to our concerns here today. One reason this is so is that engineers tend to use their technical know-how and data to determine and satisfy objective quality control criteria, and therefore tend to discount subjective evaluations of a person's "interest." On the other hand, common sense tells us that one way to find out whether a person's better interest has been served is simply to *ask* that person. And a perennial American value, embodied in the Constitution, is to promote a person's life, liberty and happiness, as judged by each of us individually. So the dilemma runs deep and is not easily solved. It is complicated also by the fact that what exactly constitutes the "public" is not always clear either. Is the public that group of persons affected by a single engineering project? Is it everyone in the country, or the world, who may be affected in the distant future? Does it include the engineer himself? The government? These questions are not easily answered either. Both sets of professionals in this research study (i.e., philosophers and engineers) are familiar with these problems, although one approaches them from a more practical perspective than the other, and both agree that some effort must be made to solve them. Our primary concern is that, generally speaking, most of these problems have remained undercover, have not been discussed either internally or publicly, indeed, *they have not been directly confronted by most engineers.* There seems to be a *systematic avoidance* on the part of practitioners to begin to sort out the myriad of messy ethical problems which exist both socially and individually. Our particular plea, then, is to urge a confrontation of issues on behalf of the profession, by way of *acknowledging the moral dimensions of one's work* and professional activities.

Professional ethics can be seen then as a critical examination of the ethical issues created by and surrounding the exercise of abilities and skills on behalf of a particular professional. Ethics itself is a philosophical inquiry into the nature and grounds of moral judgments, moral standards and normative rules of conduct.[2] Thus we can see that professional ethics is a sub-class of ethics *per se*, i.e., that ethical problems occur in all aspects of our lives, but that there are peculiar ethical problems which occur because one has taken upon oneself the role of a professional, and thus confronts different types of moral problems than occur off the job, as an ordinary person. This particular rendition of ethics and professional ethics will provide the basic perspective of our inquiry, both here and in our research in general. We urge you to keep them in mind throughout.

ETHICAL DECISIONS FOR ENGINEERS

Individual engineers face a myriad of ethical decisions during their career. The type of ethical problems which occur will vary with the individual's background, position (e.g., as employer or employee), and experience (e.g., technical or non-technical), and environment (e.g., corporate, public or private sector). Also, one's age, family and community background, and ideological attitudes also affect which particular ethical problems may arise. Each engineer has a basic value system, and these value systems will vary as well as band together in a rough homogeneity. In other words, our individual value systems may vary to some extent, but our culture provides a roughly standard given set of values which are shared by all to some extent or other. Thus we each think stealing is wrong, lying is bad, etc. Most engineers think *competency is a very important value*, and very few will vary from that value commitment. Moreover, engineers, like other professionals, acknowledge a calling to *serve the public interest with honesty, integrity and honor*.

In a day-to-day situation, both personal values and professional values merge together to determine what action will be taken by an individual engineer. A moral issue may be raised with the next telephone call, letter, or luncheon business appointment. Thus the engineer must by force rely upon his or her own value system to cope with ethical problems on a daily level. Whether or not this is convenient or comfortable is another matter, but it must be acknowledged that such occurrences are inevitable, and therefore cannot be ignored or handled by passing the buck to one's superior or colleague. Moreover, many times there are conflicting moral demands presented to an individual engineer, making the moral decision-making process difficult to operate.

Given these complexities in the moral and professional life of an engineer, the question arises: To whom can the individual engineer turn to help solve these ethical problems? There is one source not yet discussed—the professional engineering society. These societies usually have codes of ethics which presumably embody the values of the profession held collectively by the members. This code provides the individual engineer with a series of fundamental principles, canons and guides to morality, as developed by members of the profession. Many of these codes include rules of profession or business conduct, standards of ethical

---

[2]Taylor, Paul, *Principles of Ethics*, Dickenson, 1975.

evaluation and a list of do's and don't's as well. The individual engineer is required by the professional society or association to commit himself or herself to the code as part of the membership in the profession. Thus, the individual engineer making ethical decisions is able to seek guidance and resource from three sources:

1. His or her own moral value system
2. Public morality
3. Professional societies or association codes of ethics

In resolving a particular ethical situation, the engineer looks for the norm or standard which will consistently and practically result in integrity, honesty and honor. There is a danger here, however. Many engineers unfortunately regard professional codes of ethics as all-knowing and absolutely universal standards. The problem then is trying to apply very general and all-encompassing standards to one's particular problem or situation. There is no handy guideline or set of guidelines for the engineer to refer to with respect to ethical issues found on the job.

SYSTEMATIC AVOIDANCE

Engineers who search for a solution of a particular ethical problem usually look to professional codes of ethics, with little satisfaction. Next, one can look to the law or to social values, but since there are inconsistencies and a lack of clarity to these, little satisfaction is gained here as well. One can then look to the employer's policies and procedures, with perhaps somewhat better satisfaction, depending upon the organization. Finally, one can call to one's colleagues and associates, something seemingly rarely done. Thus it seems that the engineer bypasses his or her own moral value system entirely, constantly referring outwardly for guidance. One must at least entertain the possibility that this tendency is purposeful. Without attaching any blame to the individual engineer, let us admit that many engineers feel a lack of confidence in his or individual values, perhaps because of the nature of engineering education, where moral values are under stressed, if not overlooked entirely. (This seems to be changing to some extent, and the progress is very much welcomed.)

The systematic avoidance of ethical decision making has become a major deterrent to professional prestige of engineers in the eyes of the public. Case after case of an engineer's inability to cope with the temptations of kickbacks, inability to speak out in favor of public interest, lack of involvement in moral issues, preoccupation with liability problems, etc., have made enough headlines to have the public community suspect of the engineer's claim to be professional serving the public interest with honesty, integrity and honor.

Compounding the dilemma is the engineer's role in the business community. Over 90% of United States engineers work for major corporations and therefore are considered employees of business. Many engineers in fact "graduate" from being engineers on the job to being managers. The public has had very little evidence of late to convince them that engineers do make decisions in favor of the public interest. The public then places engineers in the rank and file of big business, not a very popular place to be these days. Every "lemon" off an assembly line convinces John Q. Public that the engineer professional has gone with big business against the private consumer.

ONE TYPE OF AVOIDANCE

There is a typical type of ethical avoidance commonly found among engineers: the response that there is nothing particularly ethical about these problems or situations, but they are really matters of public relations or merely a matter of professsionalism itself. In other words, an engineer who has made a bad moral judgment is seen instead as someone who failed his or her professional duties. This is one way of saying that the primary method of systematic avoidance on the part of engineers is *to deny the ethical dimension of their professional activities altogether. By translating ethical problems into either technical or operational ones, the avoidance is complete.* The dilemma here is that, on the "professionalist" position (as opposed to the "moralist" one), all problems, ethical and otherwise, can be solved by merely doing one's best as a professional. "If one is the best possible professional engineer, these problems will be adequately handled." The philosophical question here is: Can a commitment to mere professionalism solve ethical problems? *It seems to us clear that the answer is No.*

THE PROFESSIONAL AS AN HONEST CROOK

If it is possible to distinguish responsibilities which professionals have to their profession from those they may have to the public, as is frequently done in professional codes of ethics, then a very difficult problem arises in distinguishing good professionals from bad ones, or good persons from bad ones. Professional duties are those relating to integrity *within* a profession having to do with the proper manner of carrying out the duties, jobs, and tasks involved in one's work. These operate *within* larger social parameters of the general public and *independently of* moral duties to protect the public interest. Thus it is possible for a professional to totally ignore the larger, moral dimensions of his or her professional scope of activities and yet be a paragon of virtue within the profession. A professional who fulfills his or her professional duties is virtuous and considered to be a good person for doing so. A doctor, lawyer or engineer who operates with such integrity is not given credit only for that, in other words, but this integrity is taken as evidence of good character in general. Rather than accuse those who make this inference of ignoring a crucial distinction, let us admit the moral relevance of integrity in one's profession to judgments of moral character. We do not want after all, to praise someone as having good character if they are a *bad* professional.

However, such professional conduct of integrity may amount to no more than a code of honor among thieves, i.e., being loyal to your accomplices, not betraying your partners in crime, keeping promises made to one's collaborators, not penetrating the territory of one's competitors unfairly, being reliable (i.e., don't be late with the getaway car), improving one's professional specialty and skills (safecracking, burglary, extortion techniques, etc.), not charging less than competitor colleagues (hired killers seem to charge roughly the same amount for a murder contract), etc. Gangsters have loyalties to each other and it would seem that in terms of loyalty, only the military values loyalty higher than do gangsters (and perhaps engineers!) If anything, the structural reinforcements to professionalism in this illegal context are more conducive to encouraging "professional" behavior than in legitimate professions. In the underworld, if your conduct is exemplary, the rewards are high financially and your reputation spreads quite rapidly (through the prison system) increasing your business, and if you begin to cheat and deceive colleagues, on the other hand, you face imminent physical danger and almost certain death.

Thus it seems mistaken to consider the professional merely within the context of his or her given area of expertise. "Professionalism" *must* mean more than this internal governing of accepted practice and rules or moral heroism will amount to no more than achieving technical perfection at any cost. Every professional, to be sure, can think of someone who far exceeded (at great personal sacrifice) performance expected of them by their profession, where absolutely no moral dimension was involved. This heroism is really pseudo-heroism. It is irresponsible to mistake one's professional duties for moral duties, and the existence of both sets of duties provides a built-in provision for moral conflict between commitment to each. Somehow we must incorporate into the concept of professionalism a *public spiritedness* to provide morality as a necessary ingredient in the purview of the professional.

THE NEED FOR CONFRONTATION

The time is now, for engineers, engineering educators and students to begin a crusade to place the ethical and moral issues in engineering in a proper focus. Unless the obligation to the public interest in engineering practice returns to the high priority it warrants, the profession of engineering will not survive. *Where to begin? Here are some immediate steps*:

1. Introduce ethics in *every* undergraduate engineering *design* course
2. Introduce ethical considerations in *every* graduate engineering course
3. Establish ethics advisory committees, composed of educators, students, practitioners, and philosophers, at *every* engineering college or university
4. Establish ethical programs as a criterion for ECPD accreditation of engineering curricula
5. Conduct *semi-annual* ethics programs for engineering faculty and administrators
6. Conduct semi-annual ethics programs for engineering practitioners via local professional society units, (perhaps at regional conferences and conventions)
7. Establish ethics questions as part of EIT examinations
8. Create a singular "code of morality" for all engineering disciplines

9. Conduct ethics programs in career "development and education" activities via engineering societies, student organizational units at engineering colleges and universities
10. Add an ethical dimension to public relations programs via engineering colleges, national societies and local organizational units.

CONCLUSIONS

We think this paper, the National Project on Philosophy and Engineering Ethics and related activities are a step forward in trying to re-establish the engineering profession as the real public spirited profession it once was and still can be. The problems and opportunities are not merely confined to public image, but related to the very foundations of professional integrity in engineering. The beginning steps we have suggested are only a fragment of the work that needs to be done, but we hope they are practical suggestions which will be of value to the engineering and public communities.

The primary and fundamental step to be taken is to arrest the common and systematic avoidance of ethical issues in the engineering profession. This is not as small a step as it may seem, for once the issues are confronted directly, their solution may not be far behind.

Selected Additional Readings

Ackerman, A., "Slow Death of a Free Profession," *IEEE Transactions on Aerospace and Electronics Systems*, Vol. AES-7, Number 3, May 1971, pp. 418–428.

Bledstein, B. J., *The Culture of Professionalism*, W. W. Norton & Company, Inc., New York, 1976.

Bush, V., "The Professional Spirit in Engineering," *Mechanical Engineering*, March 1939, pp. 195–198.

Davis, G., "Self-Interest and the Professional Society," *Business & Professional Ethics*, Spring 1980, pp. 3–4.

Dixon, M. D., "But Are You *Really* a Professional?" *Engineer*, January–February 1968, pp. 17–19.

Eler, J., "The Engineer's Professional Image," *Chemical Engineering*, October 1978, p. 135.

Ewing, D. W., *Freedom inside the Organization*, E. P. Dutton, New York, 1977.

Flores, A., "Engineers' Professional Rights," *Issues in Engineering*, American Society of Civil Engineers, October 1980, pp. 389–396.

Fromson, D., "The Engineer's Legal Responsibility and Liability—What Price Professionalism?—Part I," *Hydraulics and Pneumatics*, August 1976, pp. 82–84.

Goland, M., "Professionalism in Engineering," ASME Roy V. Wright Lecture, November 14, 1973, American Society of Mechanical Engineers.

Kline, R. R., "Professionalism and the Corporate Engineer: Charles P. Steinmetz and the American Institute of Electrical Engineers," *Transactions on Education*, Vol. E-23, No. 3, August 1980. Institute of Electrical and Electronics Engineers, pp. 144–150.

Layton, E. T., "Engineering Ethics and the Public Interest: A Historical View," preprint American Society of Mechanical Engineers, Annual Meeting, December 1976.

Larson, M. S., *The Rise of Professionalism*, University of California Press, Berkeley, 1977.

Lurie, W., "Engineering and the Learned Professions," *The American Engineer*, July 1961, p. 33.

Metzger, W. P., "What is a Profession?" Seminar Reports, Program of General and Continuing Education, Vol. 3, No. 1, 1975, Columbia University, New York, pp. 1–12.

Nader, R., "Personal Integrity," *New York Times*, editorial, January 15, 1971.

"Opinions of the Board of Ethical Review," Vol. I–Vol. III, National Society of Professional Engineers, 1965–1971.

Page, B. B., "Who 'Owns' the Professions?" The Hastings Center *Report*, Vol. 5, No. 5, October 1975, pp. 7–8.

Pavlovic, K. R., "A Common Interest," *Proceedings*, Frontiers in Education Conference, 1981, Institute of Electrical and Electronics Engineers, pp. 196–201.

Sabatini, J. N., "Professionalism and Philosophy," *The American Engineer*, February 1962, pp. 35, 59–60.

"The Troubled Professions," *Business Week*, August 16, 1976, pp. 126–138.

Unger, S. H., "Engineering Societies and the Responsible Engineer," *Annals of the New York Academy of Sciences*, Vol. 196, Article 10, 1973, pp. 433–437.

Wilensky, H. L., "The Professionalization of Everybody," *The American Journal of Sociology*, September 1964, pp. 137–158.

Whistle Blowing

If, to please the people, we offer what we ourselves disapprove, how can we afterward defend our work?

George Washington

Blowing the whistle or raising a hue and cry, or living up to the ethical standards that are already embodied in the various codes of conduct, is part of the antidote to the poisonous abuse of power that is infecting our society.

Senator William Proxmire

"Organizational disobedience" is the technical term for the act of violating organizational policy in order to correct some wrong within the organization. "Whistle blowing" has come to be both the technical and popular term for a specific kind of organizational disobedience, the act of going outside an organization (a corporation, government agency, etc.) to some other organization (government, the media, etc.) in order to correct some wrong within the first organization. Both whistle blowing and organizational disobedience are the most publicly visible issues in engineering practice today. They arouse strong emotions and offend deep-seated loyalties no matter how they are done. Most commonly those who will be displeased are close associates, while support and praise will come only from a distance.

Whistle blowing in particular epitomizes the contemporary dilemma of the individual engineer and of the engineering profession: the dilemma of the three-way conflict mentioned earlier. Engineers' duties frequently lead them into a position of conflict between obligations to society, to other engineers, and to the client or employer.

From his position as a proponent of whistle blowing, Ralph Nader sees as vital that the potential whistle blower fully understand the underlying assumptions and consequent ramifications of engaging in whistle blowing. His article, "An Anatomy of Whistle Blowing," admirably lays out these assumptions and ramifications.

Nader emphasizes above all the need to ask, "What do I know and, more importantly, what do I not know?" One of the most obvious features of cases of whistle blowing that have been reported is a reluctance on the part of all concerned to admit that they might not know something. Donald Christiansen in "Fact vs. Feeling" claims this reluctance is common among scientific workers. He hypothesizes that this reluctance stems from fear that credibility will be harmed and from the belief that science, if it is truly science, provides unequivocal "yes" or "no" answers.

In this chapter, articles dealing with four cases of potential whistle blowing are presented. In the first two cases (A–7D Brake case and the BART case) whistle blowing did in fact occur, while in the last two (the DC-10 case and the Browns Ferry case) it did not. In considering these cases, assessing responsibility and alotting blame and praise to individuals is not the major concern. These cases will probably never occur again so there is little point in hindsight judgment of the participants. The cases should be used to gain concrete insight into the dynamics of the act of whistle blowing and the situations that give rise to potential acts of whistle blowing. If there are useful indictments to be made here, they are indictments of attitudes and institutional structures—not of specific individuals and specific institutions.

The discussion of the A–7D case includes "Engineers, Ethics, and Economics" by K. Vandivier, one of the principal participants, and extracts from the testimony of B.F. Goodrich officials and Government Accounting Office investigators before the Proxmire Committee. One of the most important aspects of this case is the differing assumptions held by the participants concerning the nature of engineering work. Is engineering a rigorous science that allows for unambiguous prediction of future results? If it is not, what kinds of attitudes should be fostered in people involved in engineering work and what kinds of procedures should be followed that would take this aspect of engineering into account? This is perhaps the fundamental issue in the A7–D case.

The BART case is presented in "The Bay Area Rapid Transit (BART) Incident" by Anderson, Jr., Otten, and Schendel, members of a group at Purdue University that studied the case. To this account is added "Engineering Ethics: the Amicus Curiae Brief of the Institute of Electrical and Electronics Engineers in the BART Case," prepared by F. and J. Cummings. The BART case resulted in legal action, and IEEE submitted this brief arguing that the code of ethics of an engineering society to which an engineer belonged should be ruled by the court to be an implicit clause in that engineer's contract. This is a most significant facet of the BART case. Had the court ruled favorably on this issue, it would have in effect said that engineers have not only an ethical but also a legal obligation to blow the whistle under certain circumstances. As is often the case, however, the legal actions resulting from BART were eventually settled out of court so this issue remains unresolved.

"The Case of the DC-10 and Discussion" by Fay Sawyier and "The Browns Ferry Case" by Vivien Weil deal with two situations where no one acted to correct or report apparently known hazards. The primary concern of both authors is to understand why no action was taken. They concentrate on analyzing the organizational structures involved and the assumptions that influenced the actions of the participants.

Sissela Bok concludes this chapter with "Whistleblowing and Professional Responsibilities." Her article is of the nature of an "interim report" on whistle blowing, summarizing the issues involved, sketching the current status, and suggesting directions for the future.

Many other articles relating to the phenomenon and action of whistle blowing have been published recently. A number of these are included in the selected additional readings provided at the end of the chapter.

6.1

An Anatomy of Whistle Blowing

RALPH NADER

Americans believe that they have set for themselves and for the rest of the world a high example of individual freedom. That example inevitably refers to the struggle by a minority of aggrieved citizens against the royal tyranny of King George III. Out of the struggle that established this nation some chains were struck off and royal fiats abolished. Americans became a nation with the conviction that arbitrary government action should not restrict the freedom of individuals to follow their own consciences.

Today arbitrary treatment of citizens by powerful institutions has assumed a new form, no less insidious than that which prevailed in an earlier time. The "organization" has emerged and spread its invisible chains. Within the structure of the organization there has taken place an erosion of both human values and the broader value of human beings as the possibility of dissent within the hierarchy has become so restricted that common candor requires uncommon courage. The large organization is lord and manor, and most of its employees have been desensitized much as were medieval peasants who never knew they were serfs. It is true that often the immediate physical deprivations are far fewer, but the price of this fragile shield has been the dulling of the senses and perceptions of new perils and pressures of a far more embracing consequence.

Some of these perils may be glimpsed when it is realized that our society now has the numbing capacity to destroy itself inadvertently by continuing the domestic chemical and biological warfare against its citizens and their environments. Our political economy has also developed an inverted genius that can combine an increase in the gross national product with an increase in the gross national misery. Increasingly, larger organizations—public and private—possess a Medea-like intensity to paralyze conscience, initiative, and proper concern for people outside the organization.

Until recently, all hopes for change in corporate and government behavior have been focused on external pressures on the organization, such as regulation, competition, litigation, and exposure to public opinion. There was little attention given to the simple truth that the adequacy of these external stimuli is very significantly dependent on the internal freedom of those within the organization.

Corporate employees are among the first to know about industrial dumping of mercury or fluoride sludge into waterways, defectively designed automobiles, or undisclosed adverse effects of prescription drugs and pesticides. They are the first to grasp the technical capabilities to prevent existing product or pollution hazards. But they are very often the last to speak out, much less to refuse to be recruited for acts of corporate or governmental negligence or predation. Staying silent in the face of a professional duty has direct impact on the level of consumer and environmental hazards. But this awareness has done little to upset the slavish adherence to "following company orders."

Silence in the face of abuses may also be evaluated in terms of the toll it takes on the individuals who in doing so subvert their own consciences. For example, the twenty-year collusion by the domestic automobile companies against development and marketing of exhaust control systems is a tragedy, among other things, for engineers who, minion-like, programmed the technical artifices of the industry's defiance. Settling the antitrust case brought by the Justice Department against such collusion did nothing to confront the question of subverted engineering integrity.

The key question is, at what point should an employee resolve that allegiance to society (e.g., the public safety) must supersede allegiance to the organization's policies (e.g., the corporate profit), and then act on that resolve by informing outsiders or legal authorities? It is a question that involves basic issues of individual freedom, concentration of power, and information flow to the public. These issues in turn involve daily choices such as the following:

To report or not to report:

1. Defective vehicles in the process of being marketed to unsuspecting consumers;
2. Vast waste of government funds by private contractors;
3. The industrial dumping of mercury in waterways;
4. The connection between companies and campaign contributions;
5. A pattern of discrimination by age, race, or sex in a labor union or company;
6. Mishandling the operation of a workers' pension fund;
7. Willful deception in advertising a worthless or harmful product;
8. The sale of putrid or adulterated meats, chemically camouflaged in supermarkets;
9. The use of government power for private, corporate, or industry gain;
10. The knowing nonenforcement of laws being seriously violated, such as pesticide laws;
11. Rank corruption in an agency or company;
12. The suppression of serious occupational disease data.

It is clear that hundreds and often thousands of people are privy to such information but choose to remain silent within their organizations. Some are conscience-stricken in so doing and want guidance. Actually, the general responsibility is made clear for the professional by codes of ethics. These codes invariably etch the primary allegiance to the public interest, while the Code of Ethics for United States Government Service does the same: "Put loyalty to the highest moral principles and to country above loyalty to persons, party, or Government department." The difficulty rests in the judgment to be exercised by the individual and its implementation. Any potential whistle blower has to ask and try to answer a number of questions:

1. Is my knowledge of the matter complete and accurate?
2. What are the objectionable practices and what public interests do they harm?
3. How far should I and can I go inside the organization with my concern or objection?
4. Will I be violating any rules by contacting outside parties and, if so, is whistle blowing nevertheless justified?
5. Will I be violating any laws or ethical duties by *not* contacting external parties?
6. Once I have decided to act, what is the best way to blow the whistle—anonymously, overtly, by resignation prior to speaking out, or in some other way?
7. What will be likely responses from various sources—inside and outside the organization—to the whistle blowing action?
8. What is expected to be achieved by whistle blowing in the particular situation?

In Part IV of this book [*Whistle Blowing*] we have developed a series of possible strategies with these questions in mind and in light of all the experiences presented in the intervening pages. But the decision to act and the answers to all of these questions are unique for every situation and for every individual. Presently, certitudes are the exception.

There is a great need to develop an ethic of whistle blowing which can be practically applied in many contexts, especially within corporate and governmental bureaucracies. For this to occur, people must be permitted to cultivate their own form of allegiance to their fellow citizens and exercise it without having their professional careers or employment opportunities destroyed. This new ethic will develop if employees have the right to due process within their organizations and if they have at least some of the rights—such as the right to speak freely—that now protect them from state power. In the past, as the balance of this book documents, whistle blowing has illuminated dark corners of our society, saved lives, prevented injuries and disease, and stopped corruption, economic waste, and material exploitation. Conversely, the absence of such professional and individual responsibility has perpetuated these conditions. In this context whistle blowing, if carefully defined and protected by law, can become another of those adaptive, self-implementing mechanisms which mark the relative difference between a free society that relies on free institutions and a closed society that depends on authoritarian institutions.

Indeed, the basic status of a citizen in a democracy underscores the themes implicit in a form of professional and individual responsibility that places responsibility to society over that to an illegal or negligent or unjust organizational policy or activity. These themes touch the right of free speech, the right to information, the citizen's right to participate in important public decisions, and the individual's obligation to avoid complicity in harmful, fraudulent, or corrupt activities. Obviously, as in the exercise of constitutional rights, abuses may occur. But this has long been considered an acceptable risk of free speech within very broad limits. Throughout this book [*Whistle Blowing*], in the stories of whistle blowers and in the chapters advocating reform and spelling out strategies, we set out many of the risks and the necessary limitations on blowing the whistle in the public interest.

Still, the willingness and ability of insiders to blow the whistle is the last line of defense ordinary citizens have against the denial of their rights and the destruction of their interests by secretive and powerful institutions. As organizations penetrate deeper and deeper into the lives of people—from pollution to poverty to income erosion to privacy invasion—more of their rights and interests are adversely affected. This fact of contemporary life has generated an ever greater moral imperative for employees to be reasonably protected in upholding such rights regardless of their employers' policies. The corporation, the labor unions and professional societies to which its employees belong, the government in its capacity as employer, and the law must all change or be changed to make protection of the responsible whistle blower possible.

Each corporation should have a bill of rights for its employees and a system of internal appeals to guarantee these rights. As a condition of employment, workers at every level in the corporate hierarchy should have the right to express their reservations about the company's activities and policies, and their view should be accorded a fair hearing. They should have the right to "go public," and the corporation should expect them to do so when internal channels of communication are exhausted and the problem remains uncorrected.

Unions and professional societies should strengthen their ethical codes—and adopt such codes if they do not already have them. They should put teeth into mechanisms for implementing their codes and require that they be observed not only by members but also by organizations that employ their members. Unions should move beyond the traditional "bread and butter" issues, the societies should escape their preoccupation with abstract professionalism, and both should apply their significant potential power to protecting members who refuse to be automatons. Whistle blowers who belong to labor unions have fared only slightly better than their unorganized counterparts, except when public opinion and the whistle blower's fellow workers are sufficiently aroused. This is partly a result of the bureaucratized cooptation of many labor leaders by management and the suppression of rank and file dissent within the union or local.

Government employees should be treated like public servants if they are to be expected to behave like them. Today, civil service laws and regulations serve two primary functions, each the exact opposite of those intended by Congress. First, they tend to reward or at least shield incompetence and sloth. Second, they discourage creativity and diligence and undermine the professional and individual responsibility of those who serve. You might say that the speed of exit of a public servant is almost directly proportional to his commitment to serve the public. The Civil Service Commission itself is in need of major reform. Its clients are the personnel managers of the various agencies, not the individual employees. Like other regulatory agencies it has been captured by the very group whose conduct it was created to regulate. A new administrative court should be created and invested with all the employee protection functions now given to the commission. Civil servants should be guaranteed the right to bring agency dereliction to public attention as a last resort. And they should have the right to go to court to protect themselves from harassment and discharge for doing their duty. To reduce the high cost of pursuing their lawful remedies, employees who challenge agency action against them should continue to receive their pay until all of their appeals are exhausted, and they should be permitted to recover the costs of their appeals from the government if they ultimately win.

All areas of the law touching upon the employee-employer relationship should be reexamined with an eye to modifying susbstantially the old rule that an employer can discharge an employee for acts of conscience without regard to the damage done to the employee. Existing laws that regulate industry should be amended to include provisions protecting employees who cooperate with authorities. The concept of trade secrecy is now used by business and government alike to suppress information that the public has a substantial need to know. A sharp distinction must be drawn between individual privacy and corporate secrecy, and the law of trade secrecy is a good place to begin. The Freedom of Information Act, which purports to establish public access to all but the most sensitive information in the hands of the federal government can become a toothless perversion because civil servants who release information in the spirit of the act are punished while those who suppress it are rewarded.

Whistle blowing is encouraged actively by some laws and government administrators to assist in law enforcement. Under the recently rediscovered Refuse Act of 1899, for example, anyone who reports a polluter is entitled to one-half of any fine collected—even if the person making the report is an employee of the polluting company. And corporations constantly probe government agencies to locate whistle blowers on their behalf. Consumers need routine mechanisms to encourage the increased flow of information that deals with health, safety, environmental hazards, corruption, and waste inside corporate and governmental institutions. Whistle blowing can show the need for such systemic affirmations of the public's right to know.

The Clearinghouse for Professional Responsibility, P. O. Box 486, Washington, D.C. 20044, will assist in the endeavor to establish such mechanisms and will suggest alternative actions to sincere persons considering blowing the whistle. Organization dissenters on matters of important public interest should feel free to contact the Clearinghouse with any information they believe will help citizens protect themselves from the depredations of large organizations.

The rise in public consciousness among the young and among minority groups has generated a sharper concept of duty among many citizens, recalling Alfred North Whitehead's dictum, "Duty arises from our potential control over the course of events." But loyalties do not end at the boundaries of an organization. "Just following orders" was an attitude that the United States military tribunals rejected in judging others after World War II at Nuremberg. And for those who set their behavior by the ethics of the great religions, with their universal golden rule, the right to appeal to a higher authority is the holiest of rights.

The whistle blowing ethic is not new; it simply has to begin flowering responsibly in new fields where its harvests will benefit people as citizens and consumers. Once developed and defended as recommended by the other participants in the Conference on Professional Responsibility and by the final chapters of this book, a most powerful lever for organizational responsibility and accountability will be available. The realistic tendency of such an internal check within General Motors or the Department of the Interior will be to assist traditional external checks to work more effectively in their statutory or market-defined missions in the public interest.

This book [*Whistle Blowing*] presents a detailed excursion through the pathways of courage and anguish that attend the exercise of professional and personal responsibility. We hope by this method to provide a context of case studies from which a broader view of the overall phenomenon of internal dissent within larger organization can develop. The focus is the range of conditions for whistle blowing and the possibilities for understanding and defending whistle blowers. However, the exercise of ethical whistle blowing requires a broader, enabling environment for it to be effective. There must be those who listen and those whose potential or realized power can utilize the information for advancing justice. Thus, as with any democratic institutions, other links are necessary to secure the objective changes beyond the mere exposure of the abuses. The courts, professional and citizen groups, the media, the Congress, and honorable segments throughout our society are part of this enabling environment. They must comprehend that the tyranny of organizations, with their excessive security against accountability, must be prevented from trammeling a fortified conscience within their midst. Organizational power must be insecure to some degree if it is to be more responsible. A greater freedom of individual conviction within the organization can provide the needed deterrent—the creative insecurity which generates a more suitable climate of responsiveness to the public interest and public rights.

6.2

Fact vs. Feeling

DONALD CHRISTIANSEN

Lately one hears a great amount of criticism of the technical community because it so seldom expresses an opinion unequivocally. What the lawmakers need, we are told, is a clear cut yes or no from the experts on questions involving permitting the Concorde into the U.S., or proceeding at full speed in constructing nuclear power plants. Engineers (or scientists) must "speak with one voice," we are told.

A danger resides in this repetitious demand for unity of opinion: technologists may be lured into forcing a consensus where one does not exist. Yet, given the choice between expert and nonexpert "lobbying" in scientific matters, most of us would prefer to see the expert view prevail even if that view is not based 100 percent on fact. Perhaps the thought here is: as long as our sociotechnical future is to be determined in the political arena, let's see to it that the emotion-quotient necessary to the formation of public policy be provided by the scientists and engineers rather than less well-informed interests.

There may be some sense to this view. We are living in an age when policy decisions too often must be made before the "facts are in." The too-frequent consequence: retrofits or even abandonment of misguided policy. Nevertheless, premature judgments and ill-informed decisions are being made daily whether engineers and scientists like it or not. So why not, the argument seems to go, make mistakes out of knowledge rather than out of ignorance?

There are several reasons why not. One danger is the damage that can be done to the overall credibility, within and outside the profession, of scientific judgment. A second danger is the possibility that as the expert allows himself to be lured into not-quite-expert judgments, he may begin to find simplistic answers habit-forming and self-deluding.

Alvin Weinberg, director of the Institute for Energy Analysis, observed in a recent issue of *Science:*[1]

---

[1]Weinberg, Alvin M. "Science in the public forum: keeping it honest," *Science*, vol. 191, p. 341, Jan. 30, 1976.

> The debate on most matters at the intersection of science and society is largely conducted in the public, not the scientific, forum. When scientists express opinions on scientific matters in the public forum they are not subject to the sanctions that regulate opinions expressed in the usual channels of scientific communication. Because these traditional sanctions do not operate, the extra-scientific debate often tends to be irresponsible scientifically; lower standards of proof are demanded in the public than in the professional debate, and half-truths are perpetrated on the public by scientists.

In this regard, he noted recent pronuclear power and antinuclear power petitions signed by groups of scientists, and he suggested that the signers of both petitions were implying that they possessed adequate knowledge to make judgments on nuclear power. "How many of the signers of either petition had studied nuclear power sufficiently to have a responsible scientific opinion on this complex issue?" asked Weinberg.

Suggesting that most issues at the junction of science and policy cannot be answered unequivocally—either because science has not progressed sufficiently or because the issue is unresolvable in principle—Weinberg pleads for greater responsibility on the part of scientists when they engage in scientific debate in the public forum. When they engage in "informed guessing," let them at least clearly distinguish between judgments that are firmly based on scientific fact and those that are based on less than scientific fact. (Such an exercise, he suggests, might mean more "I don't know" answers from the experts—not an undesirable result.)

The implementation of such a recommendation may be the only feasible way to separate fact from feeling, thereby permitting the technical expert to retain his credibility with both the public at large and in the eyes of his colleagues, while, at the same time, adding his expertise to public debate.

Some scientists have made notable attempts to testify only with regard to the facts when the public interest was at stake, and literally to disguise their own judgments, or to withhold them in the interest of retaining their scientific credibility. For example, the Reactor Safety Study (*Spectrum*, January, p. 41), which projected probable hazards due to the operation of 100 commercial nuclear power plants in the U.S., clearly makes no attempt to judge the acceptability of nuclear risks, but rather leaves it to a broader segment of society than that involved in the study to make that judgment.

On the other hand, it is quite possible that members of the study team, viewing the identical data and projections, might reach different conclusions concerning the desirability of nuclear power. Should they not be able to speak out, to expose their own assessments of the impact of the studies, and to make their individual recommendations known to the public?

We think the answer should be yes, and that they can retain their credibility as objective technical experts, while disagreeing about the way in which society should respond in its own best interests.

6.3

Engineers, Ethics, and Economics

K. VANDIVIER

The engineering profession is based upon exact science, and one would conclude, therefore, that the ethics of the engineer would be equally exact—firmly established within well-defined perimeters. However, the economic climate of the past few decades, primarily those years since the end of World War II, has produced an economic atmosphere in which not only the science of the engineer, but his ethics as well, have been subverted in the interest, real or imagined, of economics.

A notable example occurred at the B. F. Goodrich Co. aircraft wheel and brake plant in Troy, Ohio, in 1967–68. The company was awarded a contract to supply wheels and brakes for the U.S. Air Force for the A7D aircraft and, during the manufacturing and preproduction testing of the brakes, virtually ignored common ethics, reducing engineering science to mootness and endangering human life in order to meet an economic goal.

The situation was not resolved until an engineer and an engineering technician protested their unwilling involvement to federal authorities and eventually succeeded in presenting the matter before the Economics-in-Government Subcommittee of the Joint House Senate Economics-in-Government Committee.

Their disclosures resulted in drastic procedural changes in U.S. Government inspection techniques, although both men were forced to resign their positions in order to maintain their own individual ethical standards.

In early 1966, the B. F. Goodrich Co. made a successful bid to supply wheels and brakes for the A7D light attack aircraft being built for the U.S. Air Force by the Ling-Temco-Vought Co. (LTV) of Dallas, Texas. The amount of money involved in the contract was almost negligible, but Goodrich was especially jubilant about its successful bid because it was the first time in ten years Goodrich had made a successful bid on an LTV proposal . . . A decade earlier, a landing gear assembly built by BFG for LTV was a spectacular failure, and since that time Goodrich had been unable to get back into the good graces of the Dallas company.

Almost from the very moment BFG received the A7D contract, the word began

350

circulating through the BFG offices that the brake must be successful. There could be no more failures, regardless of the cost. One of the factors which induced LTV to award the contract to Goodrich, in addition to the very low price, was the weight of the proposed Goodrich brake. Although no weight penalty was imposed on the brake supplier, Goodrich's proposal was for a four-rotor brake weighing only 105 pounds.

The A7D project was assigned to John Warren, a senior design engineer. A new member of the engineering staff, Searle Lawson, although totally inexperienced in aircraft brake design, was assigned to assist Warren. It would be one of Lawson's tasks to oversee laboratory tests on the brake. As a data analyst and technical writer assigned to BFG's test laboratory, I would be responsible for analyzing and correlating test results and later, upon qualification of the brake to military specifications, to prepare a report of the tests.

From the very beginning of laboratory dynamometer tests, it was apparent that the brake was a failure. Four lining carriers were insufficient, and temperatures within the brake often reached 2000F, resulting in rapid lining failure and excessive torque which caused the lining carriers to disintegrate. The brake designer, John Warren, refused to believe that the design was wrong, claiming that the problem lay in the material used for the linings. He was supported in this belief by Projects Manager Robert Sink. Lawson, however, conducted his own experiments with a five-rotor brake which were successful. He urged that the four-rotor brake design be scrapped in favor of a five-rotor brake.

He was ignored and for nearly a year test after test was conducted on the brake, all of them ending in failure despite extensive experimentation with various lining mixes and configurations. During this time, LTV was never informed of the difficulties being encountered with the brake. In fact, LTV was told that the brake was successful and that flight tests could be conducted as early as June 18, 1968. To substantiate the claims of success, what was purported to be the test log of a 50-stop test conducted on the brake was submitted. Actually, virtually every entry on the so-called log was fabricated.

Despite the obvious defects in the brake design, formal attempts were made to qualify the brake to the specifications of MIL-W-5013-E, as required by LTV. During these tests, virtually all concepts of good engineering practice and common ethics were ignored.

Although military specifications require that a brake must be successfully subjected to a 50-simulated-landings test at the design landing speed and weight of the aircraft without mechanical rework or modification of the brake, the test brake was often dismantled between simulated landings and machined to remove warpage and other disfigurations. Improper testing techniques were employed to reduce the excessive torques developed during the landings. Fans were improperly used to reduce the abnormal heats developed within the brake. And on one occasion the brake pressure recorded was ordered deliberately miscalibrated to indicate a braking pressure of 1000 psi when in fact the actual braking pressure was 1115 psi.

Dozens of experimental and preliminary tests were conducted, and fourteen formal qualification attempts. None was successful, and it was obvious to all concerned that the brake was a failure. After the thirteenth unsuccessful qualification attempt, word was sent to the lab that one more attempt to qualify the brake would be made and that "regardless of what the brake does on test, it's going to qualify."

The message here was clear. Flight tests were scheduled to begin in about six weeks, and Goodrich now had no choice in the matter. Engineering officials had backed themselves into a corner, and since they already told LTV the four-rotor brake would be qualified and delivered on schedule, they had left themselves no choice but to issue a false qualification report on the brake.

Since I was the person responsible for issuing such a report, I vigorously protested. My supervisor, Ralph Gretzinger, also protested. Searle Lawson, who would be required to assisst me in writing the report, protested. Our protests, continuing through various channels and over several days, were heard but ignored. I was told to mind my own business and to "write the damned report and shut up about it!"

Accordingly, on June 5, 1968, just a week before formal flight tests were to begin at Edwards Air Force Base in California, an official qualification report was issued by the Goodrich Co., stating that the brake "had met the requirements of MIL-W-5013-E and therefore was qualified."

This official qualification report contained nearly 200 pages of elaborate engineering curves and other graphic displays, nearly all of which I had falsified. Some narrative portions of the report were outright lies; others had been twisted to render them ambiguous.

Flight tests were conducted as scheduled but suddenly were called off when difficulties were experienced with the brake. On one landing, for example, the brakes welded together and the pilot barely succeeded in skidding the aircraft to a halt. The wheel had to be removed and the brake pried apart before the plane could be towed from the flight line.

Within a week of the aborted flight tests, I had told my story of the A7D brake to the Federal Bureau of Investigation. A few days later, Lawson confirmed my statements to the FBI. Military officials were contacted and called a halt to all flight tests.

Yet, for nearly three more months, Goodrich continued to represent to LTV that the four-rotor brake had been successfully qualified, and it was not until I resigned in October 1968 that Goodrich acknowledged that the four-rotor design was a failure. That acknowledgement came 48 hours after my resignation was accepted.

It was apparent, however, that military officials had done little more about the situation than cancel flight tests, and it was not until Senator William Proxmire (D-Wis.) became interested in the subject and ordered an investigation by the General Accounting Office that the A7D scandals became public knowledge.

One day after a hearing on the matter before the Economics-in-Government subcommittee of the Joint Economics Committee, the Department of Defense ordered sweeping changes in inspection and reporting procedures in the defense industry. A DOD spokesman later admitted that the changes were a direct result of the Goodrich affair.

The story of the A7D brake is now a matter of record. Questions about what happened were answered fully before the congressional committee. What is not a matter of record is "Why?" Why would a reputable firm engage in such devious and fraudulent practices, especially when the total amount of finances involved was slightly less than $70,000?

Why would anyone, for that matter, engage in something which was totally foreign to his own moral and ethical concepts? I can answer only for myself.

My own reason—and it is only a reason, not an excuse—is that my own economic situation was not such that I could readily refuse. With six children to support, the choice between ethical practice and practical economics is a difficult one indeed.

There are some, however, who refuse to believe that such a thing can happen. R. G. Jeter, then vice-president and general counsel for Goodrich, was a witness at the congressional hearings. Said Jeter: "... I say to you that it is incredible that (Goodrich engineers) would stand idly by and see reports changed or falsified ... I mean you just do not have to do that working for anybody ... Just nobody does that."

Russell Line, manager of technical services at Goodrich and the man to whom I complained about my part in the A7D affair, told me that he only did what he was told and advised me to do the same.

H. D. Sunderman, Goodrich's chief engineer, called me irresponsible, adding that "There's nothing wrong with anything we've done here. You aren't aware of all the things that have been going on here." My charges, he said, amounted to "disloyalty."

Said Robert Sink, "We're only exercising engineering license." He admitted to changing data, but told Senator Proxmire, "That is part of your engineering know-how," adding that the changes had been made "only to make them more consistent with the over-all picture ... "

Although the A7D scandal received unusual publicity, it is probably not an isolated incident. It is obvious that as long as there are engineers who believe in "engineering license," as long as there are engineers who believe in always doing exactly what they are told, as long as a protest against lying and cheating is considered "disloyalty," then there are, and always will be, other such scandals fermenting somewhere.

Until codes of ethics are changed from meaningless cant to established principles, then every engineer stands in danger of that time when he, too, will be called upon to exercise "engineering license."

6.4

Extracts from
Air Force A–7D Brake Problem
Hearing
Before the
Subcommittee on
Economy in Government
of the
Joint Economic Committee
Congress of the United States
Ninety-First Congress
First Session
August 13, 1969

Chairman Proxmire: Our next witnesses are from the General Accounting Office. We have here Richard W. Gutmann, Guy A. Best, Stanley R. Eibetz, and Jerome P. Pederson.

Mr. Gutmann, you have the honor of leading this delegation.

Mr. Gutmann: Yes I do.

Chairman Proxmire: You have a concise statement. It will only take 3 or 4 minutes to read it. Let me say, before you begin, that the GAO report, "Review of the Qualification Testing of Brakes for the A–7D Aircraft," will be included in the record at the conclusion of today's proceedings.

Go right ahead, Mr. Gutmann.

Mr. Gutmann: Mr. Chairman and members of the subcommittee, we are pleased to appear before the subcommittee today in response to your invitation of August 7, 1969.

The review of the brake qualification testing performed by the B.F. Goodrich Co. was performed in response to your request of May 13, 1969, for an inquiry into (1) the accuracy of the reported qualification test results; (2) the effect of defective brakes on the test pilot's safety; (3) the identification of additional costs, if any, incurred by the Government to obtain an acceptable brake; and (4) the responsibilities of the Government, including Air Force actions, in the qualification testing.

The A–7D aircraft was purchased for the Air Force by the Navy from LTV Aerospace Corp., Vought Aeronautics Division, Dallas, Tex. LTV awarded a subcontract to the B.F. Goodrich Co., Aerospace and Defense Products Division, Dallas, Tex., for the development and production of brakes for the A–7D aircraft. The subcontract was performed at Goodrich's plant in Troy, Ohio.

In performing this assignment we took the following steps: (1) we reviewed the qualification test procedures and compared the actual and reported qualification test results at Goodrich; (2) we discussed qualification test procedures with LTV officials and Air Force engineers; (3) we reviewed the applicable specifications for qualification testing; (4) we discussed potential harm to pilot and aircraft resulting from defective brakes with the Federal Aviation Administration, Air Force, Navy, and LTV officials, and LTV, Navy and Air Force test pilots who had flown the A–7D aircraft; (5) we reviewed prime contractor and military flight reports and flight discrepancy sheets; (6) we reviewed the prime contract, the subcontract, and other documents and correspondence relating to the pricing and/or configuration of the brakes; (7) we discussed the effect of brake problems on aircraft testing and delivery with an Air Force engineer and LTV officials; and (8) we reviewed documents regarding the prime contractor and Government responsibilities with personnel of LTV, the Air Force, and the Defense Contract Administration Services District Dayton, personnel.

The results of the review have been summarized in the digest of our report to you, Mr. Chairman, dated July 3, 1969, and in a subsequent letter dated July 11, 1969.

In summary our work shows that:

In some instances Goodrich's test procedures for the four-rotor brake did not appear to comply with specification requirements or normal industry practice.

Goodrich's qualification report on the results of testing the four-rotor brake contained some discrepancies that may be considered significant.

Chairman Proxmire: Would you repeat that statement again?

Mr. Gutmann: Goodrich's qualification report on the results of testing the four-rotor brake contained some discrepancies that may be considered significant.

In our opinion Goodrich should have accurately reported the test results, since in the absence of accurately reported test results it is difficult, if not impossible, to properly evaluate product performance.

Opinions differed as to the potential danger to the pilot and damage to an aircraft due to brake failure. However, no significant aircraft damage due to the use of the four-rotor brake had been reported.

Goodrich offered to, and did, replace the four-rotor brake with a new five-rotor brake without any apparent increase in cost to the prime contractor or the Government. We were advised that the change did not cause any delays in the delivery or testing of the aircraft.

The prime contractor's procedures and those of the Defense Contract Administration Service District were inadequate to protect the Government's interests in the qualification tests of the four-rotor brake.

The Department of the Air Force protected the Government's interest by withholding approval of the qualification report.

As you know, Mr. Chairman, in this case we did not follow our usual practice of obtaining written comments on the matters discussed in the report from the parties involved; that is, Goodrich, LTV, and the Department of Defense.

This concludes my statement Mr. Chairman. My colleagues and I would be pleased to try to answer any questions that the subcommittee may have.

Chairman Proxmire: Our final witnesses this morning are from the Goodrich Co.: Mr. R. G. Jeter, vice president and general counsel, and Mr. Robert L. Sink, the projects manager, aircraft wheel and brake design.

Mr. Jeter, you go right ahead, sir.

Mr. Jeter: Mr. Chairman, members of the committee, my name is R. G. Jeter and I am vice president, general counsel and secretary of the B.F. Goodrich Co.

I have with me on my left Mr. Robert Sink, who is a senior wheel and brake design engineer and projects manager.

I would like if I may to read my statement and interject a few remarks in response to statements which have been made here this morning.

The B.F. Goodrich Co., now in its 99th year, for many years has been a leading manufacturer of airplane wheels, brakes, tires, and other equipment. Many thousands of our airplane brakes are now in service on commercial airlines and military planes, both in this country and throughout the world.

I am sure that some people who have listened to this testimony this morning might wonder whether our company is capable of making a satisfactory airplane brake, and let me assure them that we are. We are, in fact, one of the very leading manufacturers in the world.

We manufacture the brakes for the Boeing 707, 720, 727, and the new, very large Lockheed L–1011. The Lockheed Jetstar, the Beech Kingair, and other commercial and private planes also use B.F. Goodrich brakes.

In the military field we have supplied the wheels, brakes, and tires for the giant Lockheed C–5A transport, the General Dynamics F–111 fighter interceptor, the Lockheed SR–71, the North American XB–70 supersonic bomber, the LTV A–7D under discussion here, as well as a variety of other fixed wing and helicopter type military aircraft now in service, and we expect to manufacture a great many hundreds more before we are finished.

In listening to the testimony of the first two witnesses, I just wanted to suggest that to me it seemed incredible that more than 30 engineers, professional men who work at this plant, our Troy plant, would continue to work for a company which would countenance any, any of the conduct described by Mr. Vandivier and Mr. Lawson. It appears that everybody at our Troy plant is out of step, or were out of step, except these two men. That is the substance of their testimony.

Now the Chairman has raised the question, as has Mr. Conable, what did the company stand to gain by this? Why would the company do this? Why would we deliberately falsify records? Why would we produce a defective brake? There is not any reason under God's creation why we would do it. The fact of the matter is we contracted with LTV to manufacture satisfactory, workable, efficient brakes for an aircraft.

This is our obligation under our contract. We contracted to do this for a fixed sum of money—and we are talking about I think $90,000—and this is our obligation.

Now why go and produce a brake and go through all the agony and tests and everything else of producing a defective brake so that we could set about then and manufacture another brake, to design a new brake and manufacture another brake which would work?

What on earth would be the point of this? It escapes my imagination or conception. I do not understand why such a thing could possibly happen.

We engineer and manufacture these airplane brakes and wheels at the plant in Troy, Ohio, which specializes in only these products.

This plant includes laboratory testing facilities for aircraft brakes which are second to none in the industry.

In short, B.F. Goodrich enjoys an excellent reputation within the aircraft industry as a supplier of these products.

This hearing is concerned with airplane brakes supplied by B.F. Goodrich as a subcontractor of Vought Aeronautics Division, of LTV (hereafter referred to as LTV) for the A7D light attack aircraft of the U.S. Air Force, as you gentlemen know.

On June 28, 1967, B.F. Goodrich was awarded a contract by LTV to supply a four-rotor brake for the A7D aircraft.

The four-rotor brake was designed and several were made for the indoor laboratory tests required by LTV in the contract. You understand the tests that were to be performed on this brake were not all of the tests that have ever been conceived by man.

The tests that were to be performed on this brake were the tests specified in the contract, and that is what I shall be talking about. And I say that because at the Chairman's questioning a couple of gentlemen here have disagreed with the statement that the knitting would not have arisen in the laboratory test if they had been most precisely and properly performed, and they have said that this is not so.

Well, the point of the matter is it is so, and I insist upon the statement, if you are talking about laboratory tests prescribed by our contract, so let us just keep that in mind.

Now this brake was designed, several were made, and then upon completion of these indoor tests in May 1968, it was the judgment of the responsible B.F. Goodrich aircraft brake engineers that the four-rotor brake satisfactorily performed the indoor tests and was ready for field testing on the A7D aircraft. Brakes of this design were then installed on a test airplane and given flight tests by LTV pilots at Edwards Air Force Base.

Between May 1968 and January 1969, LTV pilots made, as reported by the General Accounting Office, 229 test flights using the four-rotor brake.

Now let us just keep this in mind also. We are talking about a brake that was on an airplane, or more than one airplane, whichever the case, on which 229 flights were flown by test pilots who, of course, deliberately subjected the equipment, the entire equipment, to the most stringent use.

In all of these 229 test flights, the four-rotor brakes performed the braking function. No one has said anything to the contrary here. They brought the plane to a stop, as a matter of fact, in less than the required distance, and these are the records.

There were no brake-related safety incidents involving, and we do not need to use weasel words about this, these are the facts, and the Air Force and LTV's records will substantiate it, there were no safety incidents involving either the pilots or the planes, a fact confirmed at page 11 of the GAO report.

However, in these 229 flights, as noted by the GAO report, page 11, the pilot reported 12 flights and I would like to emphasize this now if I may, out of the 229 flights they reported 12 flights during which there were, and I quote "potential" problems with the brake system.

Now that is the heading of the chart in which they record this problem. There were 12 flights out of 229 in which there were "potential" brake problems.

It should be understood at this point—and I am not sure that it is understood—that a brake system on an airplane consists of three principal components, one of which we manufactured, namely the brake, the other an antiskid mechanism, and the third, the brake hydraulic system.

In other words, this is the brake system, and all three of these things are required if you are going to have a brake that functions on an airplane.

Of the three parts, we manufactured the brake. We did not manufacture the other two items. We did not in our laboratory tests of the four-rotor brake prior to flight tests have available to us, we did not test the brake system, the entire brake system, hydraulic and the antiskid mechanism, we did not manufacture.

Pursuing these flight tests of the four-rotor brake—and I say listening to this testimony one might suspect that this brake fell off the airplane and on the first taxi down the runway—LTV flew 229 test flights, and these 12 potential problems which I have mentioned.

Now of the 12 potential problems, only two, in only two of the flights in which the brake system problem was noted was there a problem related to the four-rotor brake—in two flights out of these 229.

This problem was that the brake linings—and our engineers have described this—knitted or fused slightly at low speeds, and this was mentioned a moment ago in connection with safety. This problem occurred at low speeds.

As I say, it has never been determined whether the knitting problem resulted from the brake design, or was because of an incompatibility between the brake and its associated parts of the braking system.

This was not determined because we set upon another course rather than try to blame the problem on some other part.

In the same period—I have mentioned 229 test flights by LTV test pilots—in this same period, military pilots flew 38 test flights with this four-rotor brake. These, of course, were in addition to the 229. The military test pilots reported no brake system problems in the 38 flights which they conducted. Again I say this was not a defective brake that fell off the plane the first time they started down the runway with it. They flew 229 plus 38 flights, test flights, and there were 12 potential problems regarding the brakes in that total number of test flights, and there were two problems which involved this knitting or sticking together of the parts, whatever they were.

So let us reduce this thing. We can be factual about it I hope, and let us reduce it to its facts. Let us see what the problem is, and this is the problem.

We had a fusing or knitting in this brake in two flights out of 229 plus 38, a total of 267 flights, and that is the only real brake problem they had on this airplane.

I make the statement—and the chairman has questioned a couple of witnesses about it—and I want to make the statement again, and stand by this statement.

I meant exactly what I said in the statement, notwithstanding what some of the witnesses have said. I say it is most significant that the four-rotor brake knitting problem referred to above could not have been predicted from it, from the indoor laboratory tests which the contract required, regardless of how precisely these tests may have been performed. And the knitting problem was the only significant field problem with the four-rotor brake.

Now I trust I make the point that the statement is based upon testing the brake pursuant to the specifications of our contract, which is what we did.

It was during the period from June 14 to July 5, 1968, in flight testing at Edwards Air Force Base that the knitting problem appeared. The company's engineers immediately initiated a two-phase program to deal with this problem without regard to whether the real cause was the brake itself, or the brake system.

The program was: (1) we set about to study other possible linings for the four-rotor brake which would have a higher fusion level; and (2) we set about to design a five-rotor brake in the event the lining development was unsuccessful.

During September 1968, we started testing a new five-rotor brake. This is a month before Mr. Vandivier left our company, but he apparently did not know anything about this.

Concurrently, B.F. Goodrich obtained from LTV an antiskid mechanism and tested it in combination with the four-rotor brake. These were our first laboratory tests of the entire brake system. The tests of the entire brake system were in no way provided for or involved in our contract, but since we had a problem, we obtained the system, the rest of the system, so that we could make a test with the entire system. And with the entire system our engineers designed a test, and note my words, we designed a test for the purpose of attempting to simulate or duplicate the knitting problem in the laboratory.

We did determine from this test, we concluded at least from this test that a brake lining change would probably not solve the problem, whereas a 5-rotor brake design probably would be satisfactory.

Our engineers reached that conclusion.

Now the chairman has stated that B.F. Goodrich did not "qualify" the five-rotor brake for the A7D until after GAO had made its investigation and recommended improved procedures.

The facts are that B.F. Goodrich satisfactorily completed evaluation tests on the five-rotor brake on October 17, 1968, and the results of these tests were reviewed with LTV on October 21, 1968. The five-rotor brake was formally recommended to LTV on October 29, 1968. The fact of the matter is that the GAO investigation was not begun until many months later, to be precise, on May 28, 1969.

Formal qualification testing of the five-rotor brake was completed in December 1968, and the first shipment of these new five-rotor brakes was delivered to LTV on January 12, 1969.

On February 13, 1969, the formal qualification report for the five-rotor brake was approved by LTV. Beginning January 27, 1969, and continuing to the present time, I am told, the five-rotor brake has been flight tested on an A–7D by both LTV and military test pilots. The test pilots have reported no significant brake problem, and I guess that has been agreed to by everyone here.

The essential facts of this matter I summarize as follows:

1. There have been no safety incidents—personal injury or property damage— resulting from the flight testing of either the four-rotor or the five-rotor brake, the GAO report on pages 10 and 11 are my authority for this statement.

    "Air Force, Navy, and LTV/VAD officials generally agreed that the brakes did not endanger the life or safety of the test pilots."

2. The total experience with both brakes was a typical one. And I understand the chairman is talking about the four-rotor brake, but I am talking about both brakes, because we had a contract to perform for the Government, well, for LTV and the Government, and I am telling the story of what happened in the performance of this contract including the four-rotor brake total experience with both brakes on the A–7D, a new airplane, was a typical one. And I am telling you we have thousands of brakes on airplanes flying in this country.

    Obviously, the reason a new airplane brake must be subjected to rigid flight testing is because indoor laboratory tests alone do not always provide a reliable guide to the performance of the brake on the aircraft, and I daresay that the Air Force will not disagree with that statement.

3. Now further, B.F. Goodrich engineers moved quickly, and I think my listing of dates substantiates that certainly, to provide a solution to the knitting problem with no delay in the aircraft program, and the GAO report supports this.

4. The substitution of the 5-rotor brake, and Mr. Chairman, I trust that you will not further question the conclusion of the General Accounting Office, which was that the 5-rotor brake was in fact provided under the contract with no additional costs to anybody except B.F. Goodrich, and this is the fact of it; and

5. Both the LTV and the Air Force have expressed their satisfaction with our 5-rotor brake.

GAO REPORT OF ITS INVESTIGATION

Now I want to refer to the General Accounting Office Report of its investigation of it and I would like to point out at this time that Mr. Vandivier has testified here that he has told his tale to the General Accounting Office. It had the benefit of this story of his for whatever it was worth in the course of their investigation.

On May 28, 1969, the GAO began its investigation of the B.F. Goodrich A7D aircraft brake program.

I might interject at this point to say it was a rather thorough investigation. They were at our plant 8 or 10 days, I am not sure of the exact count, but quite a few days. They were provided access to all test data which had been compiled in the indoor testing of the 4-rotor brake.

Engineers of B.F. Goodrich who were directly associated with this project and who had worked in the program cooperated fully in answering questions and explaining the test data.

We now know that early in July 1969 the GAO delivered its report to Senator William Proxmire. The GAO did not submit copies of the report to the Defense Department, LTV, or B.F. Goodrich for either review or comment. In fact, B.F. Goodrich first received the GAO Report on August 4, 1969, one day after Senator Proxmire had delivered his statement to the news media.

Further, the Senator did not contact B.F. Goodrich until after his public accusations.

There are only two "Findings and Conclusions" in the GAO Report which are in the least critical of B.F. Goodrich. It is significant that the news statement B.F. Goodrich issued on August 4, 1969, in response to Senator Proxmire's press release, incorporated all the "Findings and Conclusions" of the GAO Report.

FIRST—B.F. GOODRICH'S TEST PROCEDURES

Now as to the critical findings, the first regarded our test procedures. GAO stated that in some instances our laboratory test procedures for the 4-rotor brake "did not appear to comply with specification requirements or normal industry practice." Our answer:

a.   LTV approved our interpretation of the contract specifications to end the test runs in a rolling stop. The energy level over each test run was maintained as required, which is the principal purpose of the test runs. The precedent for the rolling stop used in this test was established at the Wright-Patterson Air Force Base testing laboratory a number of years ago when the Air Force was qualifying military aircraft brakes, and this technique is often used in testing other aircraft brakes.

You may be interested to know that the brakes of the XB70 supersonic bomber were qualified in test runs with a rolling stop.

b.   GAO questioned the sequence in which the normal energy stops and the overload energy stops were run in this test, but GAO admits (GAO Report, page 7) that test sequence requirements were not specified in the contract. The sequence of the normal and overload energy stops B.F. Goodrich used was discussed with and approved by LTV.

c. Finally, GAO states (report, page 7) that, for some runs of the test, stators were interchanged between the No. 1 and No. 3 positions within the brake. The report acknowledges B.F. Goodrich's explanation that this was a laboratory technique for studying special wear effects. All components were subjected to the full qualification test, however.

SECOND—THE ALLEGED DISCREPANCIES

Now the second finding that I refer to as criticism perhaps in the GAO Report relates to alleged discrepancies. The second and only other respect in which the GAO Report is critical is that it states that "Goodrich's qualification report on the results of testing the four-rotor brake contained some discrepancies that might be considered significant."

I want to say in this connection, Mr. Tremblay has testified here from the Air Force that Goodrich refused to furnish data to the Air Force. This is not a fact. We refused to transport this data to Wright Patterson field, and we offered to submit the data for review at our Troy plant.

Now those are all the facts on the submission of data to the Air Force. We do not take the data out of our plant, and this is our practice I am told for many years. But the offer was made for the Air Force to review this data at the Troy plant.

Chairman Proxmire: Let me just interrupt on this point, please.

Mr. Jeter: Yes, sir.

Chairman Proxmire: Did you at any time say that this data was proprietary?

Mr. Jeter: We did say it was proprietary.

Chairman Proxmire: You said it was proprietary.

Mr. Jeter: And we kept it in our plant.

Chairman Proxmire: But you did say it would be available if the Air Force would come to the plant?

Mr. Jeter: At our plant, yes, sir. We did say, sir, it was proprietary. We did say that.

Now responding to this claim by GAO:

Our answer: Appendix A of the report identifies 16 data items characterized as "discrepancies that may be considered significant." That is to say, there are 16 data items criticized out of more than 250 items the GAO examined. Therefore, we must conclude that 94 percent of the items were not questioned.

Of the 16 so-called "discrepancies," three of them are against our interests for the purpose of the test report; in other words, three discrepancies showed worse results than the actual data.

Next of the 16 so-called "discrepancies," seven of those items were not discussed or reviewed with our representatives. As I recall the GAO states in the report that it did not have time, or time did not permit or something to that effect, but they were not even discussed with us, so we do not know really anything about what the problem is on them.

Next, the GAO report does not identify the size or importance or relative importance of these alleged discrepancies. During the examination of the test data, we, in fact our engineers who were working with it did in fact observe many "deviations" being noted by GAO which were less than 1 percent.

Finally, the stop times of our qualification report for the five overload stops were taken from a digital tape record. The results of this test are simultaneously recorded by two different methods—one is a digital or computer produced record. The other is a record visually interpreted by the laboratory operator. Whenever these two records of this test are at variance, the computer produced record is accepted. This is our practice, and this is what we did in this instance.

Now there would be discrepancies between the two records, and in those cases we accepted the computer produced record.

In the opinion of our engineers, and I can assure you that in view of the charges made here this matter has been reviewed thoroughly, the data and the entire subject by our aircraft brake engineers at our Troy plant, and it is their opinion that none of the foregoing criticisms are relevant or significant as to whether the brake was or was not qualified.

ACCUSATION BY SENATOR PROXMIRE

Now as to the accusations of the Chairman in the statement which he released, I think in view of that statement it is most—well, the statement charged us with falsifying test reports to hide defects in brakes which we had made for this A7D attack plane.

We think it most significant that the GAO report, contrary to the Senator's statement, does not even suggest that test records were falsified.

Now I have read the report a number of times and I have not yet, Senator, found that accusation in it. I have found the section about discrepancies, which I have just finished discussing.

B.F. Goodrich emphatically denies that any test data relating to the A7D four-rotor brake were in any respect falsified.

The Air Force has mentioned here, and I am sorry I did not get all their names, so I will refer to the group, that some data in the report which was presented, did not meet some of the specifications. It must be obvious that these data were not changed so as to make it look like the brake was qualified.

We have presented data which the Air Force says did not meet some of the specifications.

As has already been shown, the field test problems encountered by the four-rotor brake were wholly unrelated to the question of whether the test data on the four-rotor brake were or were not changed.

This entire controversy in our opinion, if I may respectfully say so, completely overlooks the fact that judgments based upon years of professional training and experience are required in the interpretation and evaluation of test data. And I do not mean by this whether a thermometer reads 98 or 108, I am talking about an interpretation and an evaluation of this entire mass of test data, and in arriving at a judgment, a professional decision as to whether this brake qualifies under the test procedures prescribed by the contract. This is what I am talking about.

Judgments must be made to reach reasonable conclusions regarding the likely performance or the product being tested. Test data were not changed or falsified, but in interpreting and evaluating the data the project design engineer arrived at judgments as to the reasonableness and validity of the test results.

This is opinion, based not only upon data but also upon professional training and experience, and may result and in fact I am sure it did, in a rejection of some data in arriving at conclusions.

Obviously, we are talking about something very different from changing or falsifying data.

As has been said, the judgments of our engineers were correct, because tests we were told—were instructed—to run under a contract would not have produced this problem which arose and there was not any other problem with this brake.

In Senator Proxmire's press release of this past August 4, it was said to have been based upon statements made by a technical writer who formerly was employed by B.F. Goodrich and who left our employ in October 1968. At the outset it should be understood that the technical writer is a high school graduate with no professional training.

Now while we are on that subject, I was informed by our people that Mr. Vandivier was in fact a high school graduate, and he so testified here two or three times, but his application for employment written in his handwriting shows that he attended high school for 2 years, September whatever the date is 1941 to 1943, and under "graduated" the word "no."

For the record just to keep it straight I think I will leave a Xeroxed copy of that employment report.

This technical writer told the Senator that he had written in the report that A–7D four-rotor brake was not qualified, and that someone changed the statement to read that it was qualified.

If the Senator had inquired of us, which he did not, we would have told him that the sentence of the report referred to was changed as stated. The change was made by the senior project engineer who was responsible for the ultimate decision as to whether in the light of the test results, as prescribed by the contract, the specifications, and the views of LTV, the brake did or did not qualify.

This senior engineer was of course professionally trained, and through his discussions from day to day with the LTV engineers regarding test results, knew that they had accepted the tests.

Now this, of course, Mr. Vandivier knew nothing whatsoever about. He had nothing to do with contacting the LTV engineer who was on the job watching the day-to-day progress.

Those discussions were held between the senior project engineer and the LTV engineer.

Furthermore, the ultimate decision as to whether the brake was qualified obviously could not be made and should not be made by a technical writer with no professional training.

The conclusion of qualification of the brake was later formally approved by LTV.

Mr. Lawson, who testified here this morning, apparently disagreed with the conclusion that this brake qualified. I may say for your information and for whatever it is worth that Mr. Lawson is an engineer and was an engineer, but this was in fact, so I am told, the first airplane brake upon which he had ever worked.

He had come to us, to our employment just shortly before, and the decisions with respect to this brake were in fact made by the project engineer, Mr. Warren, who was directly in charge of this project.

The source of the accusations that the company's records were falsified is the technical writer who was a former employee of our company. It has not been easy to find someone who would give any credence to his bizarre tale of the falsification of records.

In the fall of 1968 he attempted unsuccessfully to peddle this story to his employer, the Troy Daily News. He next reviewed his accusations with a lawyer. From there he went to the FBI and then on to the General Accounting Office.

Failing to interest any of this impressive group, including two Federal agencies, he then broadcast his tale to the newspapers and magazines throughout the country according to a quote of him in a newspaper the other day.

Significantly, we think, the first public notice given to the story was in the statement of Senator Proxmire on August 4, 1969.

CONCLUSION

In conclusion I want to say that on the record it has been established that B.F. Goodrich at a reasonable cost, and without any delay in the Air Force schedule, produced a brake for the A–7D, the performance of which exceeded the aircraft requirements. Throughout the entire program there were no safety incidents related to either the 4- or 5-rotor brake.

Further the GAO after an intensive examination, and we think it was intensive, has found no evidence of any change or falsification of any data.

Upon that record, we emphatically urge that there be an end to this wholly unjustifiable pointless attack through releases in the national press and congressional hearings which can only be harmful to the fine reputation enjoyed by our company.

Now that concludes my statement. I have Mr. Sink with me. I would like to ask him if time permits, Mr. Chairman, to comment on three or four items very briefly, if I may do that.

Chairman Proxmire: Very good. Time is getting along but I think that is only fair. Mr. Sink, you go right ahead.

Mr. Jeter: Mr. Sink, you have been referred to in the testimony here this morning by I think both Mr. Vandivier and Mr. Lawson.

Did you tell either Mr. Vandivier or Mr. Lawson to make any changes in the report as they were writing the report?

Mr. Sink: During the time that this report was written, I was on the west coast supervising the certification tests on the 727–200 aircraft. That was during the months of April, May, June and up to the early part of July. I had reviewed some of this data prior to leaving for the desert, but it had not gotten into report form yet.

Mr. Jeter: Who, in fact, would have directly supervised the writing of this report during this period that you were away?

Mr. Sink: In my absence John Warren, the senior project engineer assumed this duty.

Mr. Jeter: Reference has been made, Mr. Sink to a photograph, and mention I think if I understood it of an extra pressure disk or something, whatever that statement was. Would you just comment on that briefly?

Mr. Sink: Yes, sir. In the Mil Spec, the purpose of the worn-brake RTO, as Mr. Tremblay has told you, is one for information only. Also, as has been stated, prior to putting in the pressure plate, we actually had a spacer prior to the 45 stop condition. This also is well known. The pressure plate was put in so that adjusters could be installed and the correct pressures run for this RTO stop.

As far as our ability to evaluate the heat stack for its capability to conduct an RTO, it did not change in one way the capability of the brake to demonstrate performance.

Mr. Jeter: Were adjustments made in the data to make—in the report I mean of the data, to make the brake look better than it really was?

Mr. Sink: Yes sir; but not to make it look better. If I may ask the Senator a question concerning—

Chairman Proxmire: This is unusual procedure, Mr. Jeter.

Mr. Jeter: Yes, you are getting a little bit out of bounds.

Chairman Proxmire: I will be happy to let you do it, and if you want to ask me a question I will be happy to reply.

Mr. Sink: Fine, thank you, sir.

Mr. Jeter: I did not suggest this, Senator, I assure you.

Chairman Proxmire: Perfectly all right.

Mr. Sink: The submission that I believe was made earlier contains several of these changes which were made, and which were reviewed by the GAO. One of these changes that was made, of course, is that on the overload stops, the pressure was run at approximately 850 p.s.i. In the report you will see that this has been corrected to show the correct pressure to compensate for the adjusters which were not present during the overload stops, and what this would show is what the correct presure would have been if the adjuster pressure were compensated for.

Another example is on the 45-stop condition stop time, and the 5-stop condition stop time. When the original copy of the report was submitted to engineering, Mr. Warren determined from the data presented that the stop times which were taken from the dynamometer log sheet and the torque traces, which were shown to represent the torque curve, did not coincide. There was a difference in stop time. Mr. Warren had a torque trace plot made from the digital data, and he determined from this what the actual stop time was, and informed LTV of these stop times, since they were longer than the specification stop time allowance.

It is our understanding that LTV in turn discussed these stop times with the Air Force, and that LTV told us that these stop times were acceptable, to present them as we got them off the digital data and they would be acceptable to LTV. They subsequently did approve the report.

Now we also found an example in this report where after going over the report in the July 27, 1968, session, to see how the raw data checked actually with the report, we found that at stop 33 of the tire change, the thermocouples for the fuse plug, bead seat, and tube well were interchanged on the slipring and what we detected up to stop 33 they had been reading in order: tube well, bead seat, fuse plug. We found after stop 33 they were reading in sequence bead seat, fuse plug, tube well.

Mr. Jeter: These two readings, I do not understand this myself.

Mr. Sink: In this situation what I am pointing out—

Mr. Jeter: The reading before and the reading after, were they near one another or quite different? This is what the Senator and the Congressman would like to know.

Mr. Sink: Yes, sir. We determined from this that here was another case where we should have changed the data from the raw data that was presented because it was perfectly obvious that when you have temperatures running in a sequence, during parts of the test they are not going to just arbitrarily change and run in a completely different sequence from one stop to the next. So there were areas where we should have changed the data in the report from the raw data, but we missed it due to the relative unimportance of these temperatures compared to other data that we did observe more closely.

Mr. Jeter: Does that answer your question? You might say a word about comparison, for example, of brake design program for an aircraft for a military fighter plane as compared with the design of an aircraft brake for—

Chairman Proxmire: Will this be the last?

Mr. Jeter: Yes, sir; it will, Mr. Chairman.

   For a commercial plane?

Mr. Sink: Brakes for commercial aircraft are designed like work horses, since these things perform many stops every day, day in day out service accumulating many stops in a very short period of time. There is a lot of beef in them. They are the—well, work horse is a good comparison.

   The fighter aircraft, by comparison to this, is more like a thoroughbred horse. It is skinned down. There is not an ounce of excess weight in it. It is very high performance, and as a result of this, the development program required to develop these fighter plane brakes is much longer than it is for commercial brakes.

Mr. Jeter: Let me say only one word, Mr. Chairman, if I may.

   I understand quite well, I think, that the Chairman and members of this committee are concerned and should be concerned that the Government gets and obtains good products when it is required to buy them, and that it buy these products at a reasonable cost. This is a very proper concern of this committee, and I understand that.

   I do want the committee to understand likewise that we at the B.F. Goodrich Co. have this same concern, identical concern. We are in this business. We are very big at it. We hope to continue in it for, well, more years than I can say, and we do not know of any way to continue in it and be successful and be one of the major producers and suppliers in the world than to produce quality products at reasonable prices. There is not any other answer, because this is a competitive field.

   I do want to thank you, Mr. Chairman, and Mr. Conable, for your attention to our statement.

Chairman Proxmire: I want to thank you, Mr. Jeter and Mr. Sink. It is a vigorous, forthright statement. There is no question where you stand on this and it is very helpful.

At the same time, just to take up what you have said, I think I would not be a responsible member of the Senate if two men coming to me as these two men did with what I thought was a sincere effort to express their concern about this very serious matter and told a story that sounded plausible, if I had not asked that an investigation be made. I did that.

The GAO made an investigation. You and I differ very much on what the GAO said.

Let me just start with that.

You say in your statement that, "The GAO report does not even suggest that test reports were falsified." Well, falsified is a word of intent which I am sure the GAO and everybody else is very careful about using. I did use it frankly in my release.

We did have testimony by the GAO engineer this morning and he told us that data did not show an honest picture.

Now—

Mr. Jeter: If I may interrupt.

Chairman Proxmire: Yes, sir.

Mr. Jeter: This is something he has added since the report was written.

Chairman Proxmire: Well, yes.

Mr. Jeter: OK.

Chairman Proxmire: As I say, when I asked the GAO to conduct an investigation of this, I did not anticipate that they would come in, and they never do come in, and make charges that somebody is lying or that kind of thing. They do their best to find out what the facts are with relation to the data, with the substantive data which is being considered whether it is accurate or inaccurate, so that when I asked them whether this was an honest picture or not, I think their answer satisfied me in that regard. But I think that you and I could talk, argue all day about whether this was a falsification or whether it was a discrepancy that cannot be so categorized.

Mr. Jeter: Let me say I am not to any degree whatever questioning the chairman's motives. I want to make that clear.

Chairman Proxmire: All right.

Now let me ask you this. Professional engineer opinion has been unanimous, designer Searle Lawson, your employee, Air Force engineer expert, Bruce Tremblay, and GAO engineer, Guy Best, have all testified this morning that the engineering practices at B.F. Goodrich in qualifying the four-rotor A7D brake was unacceptable. In fact, each of these gentlemen has testified this morning that Goodrich is still making questionable statements about the A7D four-rotor brake, this morning, for example, from your statement. You say, "It has never been determined whether the knitting problem resulted from brake design or was because of incompatibility between the brake and associated parts of the braking system."

As I recall, Mr. Best said that he was not sure on the basis of the data he had that he could make an assertion on that. The other two men were emphatic in saying that the knitting problem did result from the brake design.

Then in the second place they all testified where you say it is most significant "that the four-rotor brake knitting problem referred to above could not have been predicted from the indoor laboratory tests which the contract required regardless of how precisely these tests may have been performed." There again they disagree and their disagreement seems to be one of impressive unanimity.

Mr. Jeter: The record will show, Mr. Chairman, but I think they testified as a general proposition and without any reference to the tests prescribed by the contract.

Chairman Proxmire: And then that they also testified where you say, "It was the judgment of the responsible B.F. Goodrich aircraft brake engineers at that time four-rotor brake satisfactorily performed the indoor tests and was ready for field testing on the A7D," they disputed this.

Mr. Jeter: Yes, sir.

Chairman Proxmire: So that the experts that we have here disagree with you. Now let me ask you did B.F. Goodrich personnel add a spacer to the brake that was undergoing qualification tests? Did you testify that they did?

Mr. Jeter: I am sorry, I did not hear the question.

Chairman Proxmire: You testified on this picture.

Mr. Jeter: Yes sir.

Chairman Proxmire: On the basis of that picture and the other testimony I am asking you did B.F. Goodrich personnel add a spacer to the brake that was undergoing qualification tests?

Mr. Sink: Yes.

Chairman Proxmire: And that to assure that the brake would not run out of piston travel during the tests?

Mr. Sink: Yes, sir.

Chairman Proxmire: Didn't the other engineer say that that was unacceptable?

Mr. Sink: The other engineer was not in a position to know whether it was acceptable or not, Senator. In fact, I think if you will look again at the—

Chairman Proxmire: When you say the other engineer you are referring to Mr. Lawson?

Mr. Sink: Yes, sir.

Chairman Proxmire: Why wasn't he in a position to know? He was the employee of Goodrich. You gentlemen hired him and you gave him this assignment. It was your responsibility to put a competent man on it.

Mr. Sink: This is true, and he was a new engineer. We did try to give him guidance, but he preferred to have his own convictions.

Chairman Proxmire: You did try to give him guidance but he preferred to have his own convictions?

Mr. Sink: Yes, sir.

Chairman Proxmire: An interesting response.

Mr. Jeter: I think it is quite understandable if I may say so, Mr. Chairman. He is a new employee. This is the first brake he has worked on. He is directly supervised by Mr. Warren, who is a very experienced design engineer, who has been with our company many years and designed many, many brakes.

I cannot believe this young man knew everything there was to know about brakes, and I am not criticizing him.

Chairman Proxmire: Nevertheless, it was testified by the other independent experts that the insertion of this kind of a spacer was not acceptable. Do you feel that those men are not qualified?

Mr. Sink: As pointed out under item 2 under a spacer—

Mr. Jeter: You are referring to what?

Chairman Proxmire: The spacer you have in that picture in front of you.

Mr. Sink: I am referring to what I assume was introduced earlier. It says following the overall stops the stack deflections was eight hundred fifty-nine one-thousandths and the minimum travel allowable, the maximum to be more than this but the minimum with all the tolerances was eight hundred eighty one-thousandths which says had there not been a spacer in there the piston still would not have bottomed.

We had gone through the 45-stop test and this takes quite a while to do. We wanted to assure ourselves that the pistons would not bottom-out, and discontinued the tests on this heat sink that had the 45 stops so the spacer was put in as a precautionary measure.

Chairman Proxmire: Now that the Air Force has indicated this is not acceptable you continue to test your brakes this way?

Mr. Sink: If we had the same situation come up today I believe the thing we should probably have done was to go back to LTV and let them go to the Air Force to have the Air Force witness this test.

I believe under these conditions that this procedure would be approved if it were properly handled beforehand.

Chairman Proxmire: Was in fact the wheel allowed to coast to a stop as indicated by GAO and confirmed by Mr. Vandivier's testimony this morning?

Mr. Sink: Yes, sir.

Chairman Proxmire: It was?

Mr. Sink: Yes.

Chairman Proxmire: And is this, Mr. Sink, a standard test procedure in the B.F. Goodrich, Troy, Ohio field and brake plant?

Mr. Sink: This is a common procedure. I would not say it is a standard procedure. It is a common procedure, and it has precedence in that this procedure was used for the qualification of military brakes when Wright Field was qualifying military brakes in that facility.

Chairman Proxmire: Didn't the Air Force say that this is unacceptable as quoted in the GAO report?

Mr. Sink: I have heard that this has been quoted in that report.

Chairman Proxmire: Once again, here is the same kind of a problem. The Government experts who testified to us tell us that this is not an acceptable procedure to give valid results.

Mr. Sink: But our contracting agency, sir, is LTV, and we discussed this procedure with LTV engineering and they approved this procedure prior to our conducting these tests.

Chairman Proxmire: You have a responsibility, I am sure both of you gentlemen are very honorable gentlemen, you recognize you have a responsibility to the Federal Government, too.

Mr. Sink: Yes, sir.

Chairman Proxmire: You are producing a plane, a product ultimately to be used as a U.S. plane, and in view of the fact that the policies adopted by the Air Force, which is the ultimate customer, seem to contradict your position, it would seem to me that that would govern rather than any negotiation or any agreement you might make with the contractor.

Mr. Sink: It is always the function of the contracting agency to set up the requirements for the design and qualification of a brake, and where the contractor has set up these requirements, these take precedence in many or most cases over the military specification.

Chairman Proxmire: Mr. Vandivier has testified before this committee that there are more than 80 falsifications of data in the Report 26031. He and others that are or were employed by B.F. Goodrich wheel and brake plant in Troy contend that they were coerced by their superiors into writing a distorted qualification report. Would you agree that there is misinformation in this report?

Mr. Sink: I would say that the information that is in the report represents a fair analysis of the performance of the 4-rotor brake during the qualification testing. There have been changes made in the data as we have noted before, but only to make them more consistent with the overall picture of the data that is available.

Chairman Proxmire: You stand alone in making that assertion. No other witness has indicated that. Certainly the GAO did not indicate it, the Air Force did not indicate it, neither did of course Mr. Lawson who is the engineer directly working on this project. Why do you need a 5-rotor brake if this is the case?

If this 4-rotor brake is qualified, why did you have to go to the trouble and expense to yourself of the 5-rotor brake?

Mr. Jeter: Let me answer, Mr. Chairman.

When you say this brake, the 4-rotor, 5-rotor, or whatever, is qualified, when we say this in our report, which we did, to the prime contractor, we are not saying this brake will fly on an airplane. We have not flown it on an airplane. We are saying that this brake is qualified under the specifications and tests prescribed by the contract, period, the indoor tests prescribed by the contract. We cannot state that the brake will pass the flight test.

Let me say just this about this whole subject. If this science were such an exact science that we could design and produce a brake with or without, say with laboratory tests, and this brake will perform on an airplane exactly as it is supposed to, then there would be an absolute waste of money to conduct these flight tests, and they conducted almost 300 of them here, an absolute waste of money to conduct the tests. You just put them on a plane and fly it.

Chairman Proxmire: With all due respect—

Mr. Jeter: I mean why do you have the flight tests?

Chairman Proxmire: Why did you change the design? Why did you go from a 4-rotor brake to a 5-rotor brake if the 4-rotor brake met the qualifications?

Mr. Jeter: Well, the problem developed in flight. What I am trying to say to you is that we said the brake qualified on the basis of the laboratory test prescribed for us by the contract, and we do not certify to another thing. We cannot. We have not even had the brake working in conjunction with the rest of the braking system. We have not even seen that happen.

All we can certify to, and I think this ought to be clear, is that on the basis of the tests prescribed by our contract, the laboratory tests prescribed by our contract, we think the brake is qualified. It meets the qualifications as shown by laboratory tests, not flight tests if you please.

COMPTROLLER GENERAL'S REPORT

Comptroller General's Report to the Honorable William Proxmire United States Senate

Review of the Qualification Testing of Brakes for the A–7D Department of the Air Force Department of the Navy Defense Supply Agency B–167023

D I G E S T

WHY THE REVIEW WAS MADE

At the request of Senator William Proxmire, the General Accounting Office (GAO) has reviewed certain aspects of the qualification tests of brakes for the A–7D aircraft. These tests were performed by the B.F. Goodrich Company under a subcontract with LTV Aerospace Corporation, Vought Aeronautics Division (LTV/VAD), the prime contractor with the Navy for the A–7D aircraft to be used by the Air Force.

FINDINGS AND CONCLUSIONS

The results of GAO's review indicated that:

- In some instances Goodrich's test procedures for the four-rotor brake did not appear to comply with specification requirements or normal industry practice,
- Goodrich's qualification report on the results of testing the four-rotor brake contained some discrepancies that might be considered significant,
- Opinions differed as to the danger to the pilot and the potential damage to an aircraft due to brake failure. No significant aircraft damage due to the use of the four-rotor brake had been reported,
- Goodrich offered to, and did, replace the four-rotor brake with a new five-rotor brake without any apparent increase in cost to the prime contractor or the Government. We were advised that the change did not cause any delays in the delivery or testing of the aircraft,
- The prime contractor's procedures and those of the Defense Contract Administration Services District (DCASD) were inadequate to protect the Government's interests in the qualification tests of the four-rotor brake, and
- The Department of the Air Force protected the Government's interest by withholding approval of the qualification report.

6.5

The Bay Area Rapid Transit (BART) Incident

ROBERT M. ANDERSON, JR.
JAMES OTTEN
DAN E. SCHENDEL

On the morning of March 2, 1972, Holger Hjortsvang was busily at work in his office. Hjortsvang, at the age of 61, was employed as an engineer for the Bay Area Rapid Transit System (commonly known as BART), a modern rail transit system in the San Francisco area. Suddenly the phone rang. It was the secretary of Hjortsvang's superior, Mr. Ray, calling. The message was that Mr. Ray wanted Hjortsvang to come to his office for a meeting of some kind.

When he arrived, Hjortsvang found himself in a meeting not only with Mr. Ray, but also with Mr. Kramer, another of Hjortsvang's superiors. These two men informed Hjortsvang that the General Manager of BART had made the decision that Hjortsvang must either resign or be dismissed. According to Hjortsvang, however, neither man would give any specific reasons for the decision. At first Hjortsvang refused to resign, but he changed his mind after considering the benefits that he would receive. A security officer accompanied Hjortsvang back to his office so that he could collect his belongings and then leave the premises.

Later that morning, around 11:00 A.M., Hjortsvang received a phone call from Max Blankenzee. Blankenzee, at the age of 30, was also employed as an engineer at BART, and worked in the same department with Hjortsvang. Blankenzee was calling to say when he would be returning to the office after a meeting with a BART contractor. Hjortsvang relayed the bad news that he had just been dismissed, and told Blankenzee that Blankenzee's superiors wanted to see him when he returned.

Soon after Blankenzee returned to his office, he was summoned to a meeting with Mr. Ray and Mr. Kramer, his superiors. He too was told that he must either resign or be dismissed. According to Blankenzee, when he asked why he had to resign or be fired, Mr. Ray said that it had to do with "your affiliation with Burfine, I guess." Edmund Burfine was a private consultant who had written a report on alleged safety problems in BART's Automatic Train Control System, and had presented the report to BART's Board of Directors. Management was upset by this because it had not commissioned the report, and Burfine would not say exactly who had commissioned it. At any rate, Blankenzee refused to resign, and was fired. Blankenzee was then accompanied back to his office by the head of security, and was instructed to gather his belongings and vacate the premises.

The next morning Robert Bruder, 49 years of age, who was also an engineer with BART, but who was not in the same department with Hjortsvang and Blankenzee, was called in for a meeting with his superiors. When Bruder was asked if he had any involvement with Burfine, he answered that he had not. His superiors responded by saying that they had good reason to believe that he had in fact been involved with Burfine, and they proceeded to ask for his resignation. When Bruder declined to offer his resignation, he was fired.

BART's dismissal of these three engineers—Hjortsvang, Blankenzee, and Bruder—marked the culmination of one chain of events within BART, and marked the beginning of another chain of events which ultimately involved hundreds of interested people and organizations. The first chain of events centers around the three engineers' acts of "organizational disobedience." (An act of organizational disobedience, as we define it, is the deliberate violation of organizational policies, procedures, or accepted organizational standards of behavior; it is an action that an employee takes with the intention of overturning some organizational policy which the employee thinks is harmful to the public welfare or to the organization's welfare.) The second chain of events is comprised of reactions to and consequences of the dismissal of the three engineers for their acts of organizational disobedience. In the following pages we shall be tracing out the various links of these two chains of events.

THE EARLY HISTORY OF BART

The most appropriate place to begin is with the early history of BART. The formative idea for the BART system goes back to 1947. In that year a joint Army-Navy Review Board recommended construction of a tunnel under San Francisco Bay for the use of a high-speed electric train system. Actually the idea for such a system is a very natural one when you consider the geography of the Bay Area. San Francisco, shaped like a blunt-ended thumb and surrounded on three sides by water, is the center of tourism and commerce in the nine-county area surrounding the Bay.

In the years following World War II, intense political pressures were mounting for the creation of a transportation system that could reach out from San Francisco to serve the spreading cities and suburban communities in the Bay Area. Responding to these political pressures, the California State Legislature created the San Francisco Bay Area Rapid Transit Commission to undertake a detailed study of the transportation needs of the Bay Area. After six years of deliberation, this 26-member Commission recommended the formation of a nine-county rapid transit district which would coordinate its plans with the Bay Area's total plan for economic and social development. In accordance with the Commission's recommendation, the California State Legislature created BART in 1957. However, the plan for a nine-county BART system was not to survive. By 1962 only three counties remained in the BART plan—San Francisco, Alameda, and Contra Costa.

In late 1962 the voters of these three counties approved a design for BART and a proposal for bonding support for the system's construction. The difficult task of recruiting then began. BART's management decided not to hire its own technical and support staff to design and build the system. Instead, it formed a consortium of three outstanding engineering firms to design and build the system. The consortium was called Parsons, Brinckerhoff, Tudor and Bechtel (PBTB), and was comprised of Parsons, Brinckerhoff, Hall and MacDonald of New York; Tudor Engineering of San Francisco; and the Bechtel Corporation also of San Francisco. B. R. Stokes, who was General Manager of BART from 1963 to 1974, said of BART's recruiting plan: "We did not want to build a huge staff for the construction effort, that would later have to be disbanded... The decision was made to rely very heavily on... consultants in the various disciplines and to only build a staff based on largely permanent needs as we transitioned into an operating status." As a result of this recruiting decision, the BART staff was rather small, and perceived itself as a permanent group with a long-term commitment to both the construction and operation of the BART system.

THE ROLE OF THE THREE ENGINEERS IN BART

Construction of the BART system finally began in 1966. It was in September of that same year that Holger Hjortsvang came to work for BART. He was hired as a Train Control and Communications Engineer. In the beginning Hjortsvang received drafts of the specifications for the Automatic Train Control System from PBTB, which he would read and comment on regarding clarity, additions, and corrections. Later, after Westinghouse Electric Corporation had been awarded the contract for the Automatic Train Control System, Hjortsvang began to oversee Westinghouse's activities.

In 1969 Hjortsvang's superiors approved his request to accept an invitation from Westinghouse for BART engineers to go to Pittsburgh and work with the Westinghouse software design group in developing the Automatic Train Control System. Hjortsvang worked for ten months with the Westinghouse designers in Pittsburgh before resuming his regular job activities at BART.

According to Hjortsvang, it was during his stay at the Westinghouse facility that he first became concerned about the Automatic Train Control System and about general organizational problems at BART in trying to monitor the activities of PBTB and Westinghouse. At this time Hjortsvang prepared five or six reports detailing his concerns, and sent these to his superiors. These reports were acknowledged, but Hjortsvang indicated that there was "no response in reality. Nothing changed because of my reports." He added that Mr. Wargin, one of his superiors, "seemed to agree with what I had to say, but said, 'That's nothing. There's nothing I can do about it. BART's policy is that this is PBTB's business and not our business. Let's not rock the boat.'"

In November, 1969, a few months after Hjortsvang left for Pittsburgh, Robert Bruder joined the BART organization. His job involved coordinating train control and communication contracts. Whereas Hjortsvang worked in the Operations Division at BART, Bruder worked in the Construction and Engineering Division. In his position, Bruder had to maintain contact with the Operations Division at BART and with outside contractors, especially PBTB and Westinghouse. Bruder felt that he did not have a very good working relationship with his supervisor at BART, Mr. Fendel. On many occasions Bruder urged Mr. Fendel to do something about the lack of test scheduling by Westinghouse and PBTB, but, according to Bruder, Mr. Fendel did not respond because he felt that such matters were not the responsibility of his department. Bruder said that Mr. Fendel's reaction to his concerns could be summed up in this statement: "Hey, thank God it's not in our group, it's downstairs or in operations."

In May, 1971, Max Blankenzee came to work for the BART organization. His position was that of a senior programmer, and he worked in the Operations Division with Hjortsvang. Concerning his job responsibilities, Blankenzee recalled that he was told that he would have "to do the hardware and the software, more or less work along with Westinghouse, stay abreast of them, get involved with the development in such a way by trying to find out what Westinghouse is doing so that later on we could maintain the central computer complex." His analyses of some of BART's technical problems were shared by his colleague Hjortsvang, with whom he worked on a daily basis.

All three engineers—Hjortsvang, Bruder and Blankenzee—claimed that within BART there was a freedom to determine one's own work activities and an absence of specific job descriptions and assignments. Looking back, Blankenzee speculated: "I think if BART would have had a little bit better controlled environment, they probably not only would have seen these things we've talked about, but they would have stopped us, would have stopped us long before we ever got started on it (the organizational disobedience)."

RISING CONCERNS

Through 1970 and 1971 the character and complexion of the BART effort slowly changed. The massive construction phase of BART development was slowly coming to an end, and connections were being made between the various structural components of the system. Now BART was beginning the great effort to use the subway, surface and aerial rails which had been laid in order to test the prototype car, and to install and check out the Automatic Train Control System. As summer turned to fall in 1971, BART experienced a full quota of technical and managerial problems. Of course, the BART system was very innovative in its technical design as well as in its managerial and contract-monitoring methods. Consequently, management was not really surprised when problems arose with the Automatic Train Control System.

Through 1970, the concerns of Hjortsvang and Bruder persisted about the Automatic Train Control System, the lack of adequate schedules for testing, and the failure of BART management to monitor the work of PBTB and Westinghouse. Sometime early in 1971, after Hjortsvang had not received an adequate response from his superiors to the concerns he had been expressing, he decided to go "outside normal channels" to present his concerns to a Director of BART. A third party arranged for Hjortsvang to meet with Mr. Blake, one of BART's Directors. However, the meeting never took place because Hjortsvang changed his mind. Later Hjortsvang said that the reason he had changed his mind was that "I did not want to . . . start anything, anything that could be dangerous for me personally."

Several months later Hjortsvang tried another approach. He set up a luncheon meeting with Robert Bruder and Jay Burns, two other BART engineers. Bruder later recalled the meeting in this way: "I think at the time even he (Hjortsvang) wanted to go to some member of the Board. He wanted us to support him and go with him. To ring the bell or whatever you want to call it." But Bruder and Burns both refused to support Hjortsvang. When Hjortsvang was rebuffed by these two men, he again lost interest in pursuing his concerns. "I sort of gave up," he said. "I had to do what I'm supposed to do here and not concern myself with BART's management."

Meanwhile Blankenzee's concerns about the Automatic Train Control System began to intensify. In the fall of 1971 he sent numerous memoranda to management expressing his concerns. He was afraid that he and others at BART would inherit a system from PBTB and Westinghouse which they could not understand and could not maintain. Blankenzee's assessment of the BART situation re-awakened the concerns of Hjortsvang and Bruder.

In November, 1971, Hjortsvang distributed an unsigned memorandum to as many levels of the BART organization as he could reach. In this memorandum he proposed the creation of a Systems Engineering Department. In the memorandum he stated: "Most of the personnel can be extracted from existing departments, but a number of new specialists should be hired." Later management was to interpret this remark as indicating that Hjortsvang was seeking power for himself in the form of a department which he would head.

THE BURFINE REPORT

In November, 1971, the three engineers began trying to arrange a meeting with a BART Director to express their concerns. Contact was made with Mr. Blake, one of BART's Directors. There was talk of a possible meeting between the three engineers, Mr. Blake, and Mr. Bianco, another BART Director. According to the three engineers, Mr. Blake thought that an outside consultant should be obtained before any such meeting took place. Presumably an outside consultant would be able to strengthen the case the three engineers wanted to make.

Blankenzee contacted an independent consultant, Edmund Burfine, to confirm the engineers' views. After spending one day discussing the BART situation with the three engineers, Burfine produced a seven page report entitled "Review of BART Operations." It was dated January 12, 1972. Finally, after all the effort to get an independent consultant to back up the engineers during their proposed meeting with Mr. Blake and Mr. Bianco, that meeting never materialized.

What happened instead was that another BART Director, Daniel Helix, became interested in meeting with the three engineers. In early January, 1972, Hjortsvang and Blankenzee met with Mr. Helix in a local union headquarters in Oakland. Hjortsvang said of this meeting: "My first comment was that this meeting was confidential and I pointed out to Mr. Helix that we had some concerns about

BART's management, and we would like him to know about it and . . . perhaps use his influence as a Director to get these problems solved in the interests of BART and the taxpayers . . . But I pointed out very strongly that I had a family and I was sixty years old and I was not going to expose myself to any open fight against BART. And he assured me that he would respect that and that nothing would happen to me or Blankenzee. During the meeting Hjortsvang gave Mr. Helix some copies of memoranda he had written about the need for a Systems Engineering Division at BART. Two days after the meeting, the *Contra Costa Times* published a story about BART's internal problems in which Hjortsvang memoranda were reproduced in their entirety.

BART management immediately started trying to discover who had been releasing information to Mr. Helix and the *Contra Costa Times*. Hjortsvang, Blankenzee, and Bruder were each asked whether they had any responsibility for this release of information, and each firmly denied responsibility.

On February 22, 1972, Burfine, at the behest of Mr. Helix, presented his report to the Engineering Committee, a subcommittee of the BART Board of Directors. Blankenzee summarized what happened at this meeting by saying: "Burfine was slaughtered." Three days later, on February 25, 1972, Burfine made his presentation to the BART Board of Directors. The Board voted 10 to 2 to reject the Burfine report. B. R. Stokes, General Manager of BART, said of the Burfine report: "My first reaction to the report was that it was a ridiculously unprofessional, incomplete . . . almost silly effort in terms of anything smacking of professional engineering." Interestingly enough, Burfine asked the BART Board to pay for his effort in preparing the report. The Board declined to do so.

THE FIRING OF THE THREE ENGINEERS

Soon after the Board meeting at which the Burfine report was presented and rejected, management learned that Hjortsvang, Blankenzee, and Bruder were involved with Burfine. As described earlier, the three engineers were dismissed on March 2 and 3, 1972.

General Manager Stokes said of the dismissal: "I specifically authorized it, and directed it after satisfying myself that we had no other recourse, after repeated exhortations for people to come in and talk, after the material continued to flow into the press, and after continued frustration at getting . . . the havoc that was being caused among the staff."

Apparently there were at least four major considerations in management's decision to dismiss the three engineers. First, the three engineers' acts of organizational disobedience were, in Mr. Ray's words, "causing a lot of dissension . . . Work wasn't getting done. It was causing a lot of disruption." Second, management was upset by the fact that the three engineers had lied about their involvement with Burfine and Mr. Helix. Mr. Tillman, one of Bruder's superiors, remarked: "I'm not used to having people tell me one thing while they're doing something else." Third, management found that the three engineers were guilty of repeated insubordination in their activities. And, fourth, management believed that the three engineers were not acting out of concern for the public interest or for BART's best interests when they engaged in organizational disobedience against BART, but rather management thought that the three engineers were acting out of self-interest. General Manager Stokes recalled the last few days before the dismissal in this way: "At this point, some of the true motives (of the three engineers) were beginning to come out . . . it was not so much that anything was wrong with what was going on at the moment, but (that) it could be done better with a . . . Systems Engineering Group, (and the) three of them would be very very highly . . . involved at a very high level (in this group)."

INVOLVEMENT BY SOCIETIES

One society which became involved with the three engineers was the California Society of Professional Engineers (CSPE). Bill Jones, president of CSPE, tried to arrange a meeting with BART management to discuss the dismissal of the three engineers. But BART management would not agree to such a meeting. Mr. Hammond, Assistant General Manager of BART, commented: "I don't believe professional organizations ought to act like unions; they ought to act like professional organizations . . . They saw three engineers who were fired and they felt, like a union, 'they can't fire our members.'"

The three engineers filed suit against BART, and Bill Jones instructed the CSPE attorney to give legal assistance to the engineers. Although the engineers asked for a total of $875,000 in damages, they eventually made an out-of-court settlement with BART under which each engineer was awarded $25,000.

Especially active in its support of the three engineers was the Diablo Chapter of the CSPE. The Diablo Chapter was spurred on in its support by two of its members, Roy Anderson and Gil Verdugo. These two men were instrumental in bringing about an investigation of BART by the California State Legislature. This investigation eventually led to great changes in BART, including the ouster of Stokes as General Manager.

Another society which became involved with the three engineers was the Institute of Electrical and Electronics Engineers (IEEE). After much internal struggle, the IEEE filed a friend of the court brief in the lawsuit brought by the three engineers against BART. Rather than support the particular position of the three engineers, the brief proposed a definition of a proper standard of ethical behavior for an employed engineer, thus indirectly supporting the three engineers.

THE AFTERMATH

The reaction of the engineers to being dismissed was first surprise and then anger. They felt that they had been led to believe by Mr. Helix that there would be no reprisal for the actions they had taken. Furthermore, they viewed their actions as morally right, and in light of this their dismissal seemed to them unjust. As time moved on, the sudden loss of income and the inability to find new positions became the central experiences of their lives.

For a period of about fourteen months after the dismissal, Hjortsvang had a total income of $5,000, and when he finally did get employment his starting salary was

$2,000 below what he had earned at BART. It took Blankenzee fifteen months after the dismissal to find a job with which he was fully satisfied. In this period, he experienced several months of unemployment or partial employment, and a cross-country move to Rochester, New York, which threatened the stability of his family. The impact of being fired by BART had its greatest effect on Bruder. He worked for a circus for a week, receiving about $20 that week for "putting up the middle of the ring." He held a sales job in which he lost more money than he made. After holding a low-paying consulting job for three months, he found a satisfactory job in November, 1972, with Singer Business Machines. All three engineers felt that the BART experience had a negative influence on their chances for finding employment.

Bruder later recalled with some bitterness the promises that Mr. Helix, the BART Director, had made to the three engineers when they were undertaking their acts of organizational disobedience: "(Helix) said, 'Gentlemen, don't worry... It's still up to you. You will not be in trouble... You won't starve." Obviously Mr. Helix was only partly right.

6.6

Engineering Ethics:
The Amicus Curiae Brief
of the Institute of Electrical
and Electronics Engineers
in the BART Case

Editor's Note: [*] The following is a copy of the brief filed by the attorneys for IEEE (Frank Cummings and Jill Cummings for Gall, Lane, and Powell; and Robert G. Werner) in the case of Hjortsvang vs. San Francisco Bay Area Rapid Transit District and ten Does. The brief was filed on January 9, 1975 in Superior Court of California for the County of Alameda. Three editorial changes have been made: The title page has been omitted; the page footnotes have been numbered and placed at the end of the transcript; the date and signatures at the end of the brief (see above) have been omitted.

The case involved 3 engineers (Holger Hjortsvang, Max Blankenzee, Robert Bruder) who were dismissed after having called attention to poor engineering practices in the BART development project (See the CSIT Newsletter beginning with the September 1973 issue, and Spectrum for 10/74). The case has since been settled out of court. It is unusual that such a brief is filed at the trial level. However, the court (upon application to file by IEEE Attorneys) decided that there were important matters of principle involved, and ordered the brief to be submitted.

The brief itself puts forward an important concept: that the engineer acting professionally to defend the public interest is protected by an implicit contractual clause from arbitrary reprisals by his employer. This could constitute a significant shield for the defense of the ethical practitioner.

The Institute, by intervening in this case, has recognized that it has a role to play in promoting high standards of professional conduct among engineers. It is important that this responsibility be institutionalized through the establishment of procedures that can be invoked at the outset of such cases (see CSIT Newsletter 12/73). Proposals along these lines are now being considered by the Ethics and

[*] Editor, IEEE *CSIT Newsletter*

Employment Practices Committee of the IEEE's US Activities Board. For the IEEE Board of Directors resolution regarding IEEE interventions as amicus curiae (adopted December 5–6, 1974), see page 6 of this issue [December 1975].

STATEMENT

This brief is filed as *amicus curiae* because, on the basis of the pleadings, it is clear that rulings in this case will involve important questions concerning the proper ethics of an engineer in the employ of a public employer.

The Institute of Electrical and Electronics Engineers ("IEEE") is the largest engineering society in the nation and has a direct concern with the establishment, maintenance, and recognition (including governmental and judicial recognition) of ethics within the engineering field.

This brief is submitted with two limited aims: first, to inform this Court of the existence and terms of established standards and codes of ethics for engineers, in the employment context generally and particularly in the context of public employment;° and, second, to seek the Court's recognition that such standards and codes are relevant and material to this case for the reasons discussed below. †

SUMMARY OF ARGUMENT

This Court is expected to rule, as the trial proceeds, on questions of law, and this *amicus curiae* brief is addressed solely to those rulings.

Within that framework, we urge this Court to rule:

1. *As to Admissibility of Evidence.* That evidence of professional ethics of engineers, as outlined herein and as further developed by the parties, is relevant, material, and admissible;

2. *As to Any Motions for Judgment.* That, in consideration of any motion to dismiss or for judgment by this Court, the Court should rule that an engineer is obligated to protect the public safety, that every contract of employment of an engineer contains within it an *implied term* to the effect that such engineer will protect the public safety, and that a discharge of an engineer solely or in substantial part because he acted to protect the public safety is a breach of such implied term; and

3. *As to Jury Instructions.* In any charge to the jury herein, this Court should instruct the jury that if it finds, based upon the evidence, that an engineer has been discharged solely or in substantial part because of his bona fide efforts to conform to recognized ethics of his profession involving his duty to protect the public safety, then such discharge was in breach of an implied term of his contract of employment.

°IEEE, moreover, is familiar with and can supply expert evidence concerning the ethical codes of engineers.

†IEEE takes no position on the merits and the claims, as IEEE has no direct evidence to offer as to what the claimants did, what defendants did, or why.

We base this position upon the cases, statutes and ethical codes discussed below.

POINT I

Professional Ethics are Material and Relevant

California judicially recognizes that an employee may not be arbitrarily discharged where the discharge would be inconsistent with the public good, even if his employment contract is terminable at will. In *Petermann v. International Brotherhood of Teamsters*, 174 Cal. App. 2d (1959), it was held that an employer may not discharge an employee because the employee refuses to commit perjury. The public has too great a stake in the integrity of the judicial process to permit such a discharge.[1]

In *Petermann*, the District Court of Appeal for the Second District noted that the contract of employment did not provide for any fixed period of duration and that such a relationship is generally terminable at will, "for any reason whatsoever." But it also noted that such a right of discharge "may be limited by statute" or "by considerations of public policy." The Court then said at page 188:

> By 'public policy' *is intended that principle of law which holds that no citizen can lawfully do that which has a tendency to be injurious to the public or against the public order* ... (emphasis by the Court).

The Court then noted that, because the State has a declared policy against perjury, 'the civil law, too, must deny the employer his generally unlimited right to discharge an employee whose employment is for an unspecified duration, when the reason for the dismissal is the employee's refusal to commit perjury." The Court said that "the law must encourage and not discourage truthful testimony. The public policy of this state requires that every impediment, however remote to the above objective, must be struck down when encountered." *Id.* at 188, 189. The lower court having dismissed, the Court of Appeal reversed.

When questions of public safety are at stake, an engineer's code of ethics stands in the same position as the laws against perjury. If a code of ethics properly requires the protection of the public, a discharge because an employee insisted on following that code would be inconsistent with the public good. Thus compliance with such a code must be deemed an *implied term* of the employment contract.[2]

California statutes clearly recognize an engineer's obligation to protect the public. California Government Code, Section 835 waives the State's sovereign immunity and makes a public entity liable for conditions dangerous to the public. Section 840.2(b) of the same Code makes a public *employee* liable if he fails to take adequate measures to protect the public from such conditions. That section obviously encompasses any and all engineers engaged in public employment.

The same recognition is reflected in California statutes governing licensing[3] of professional engineers, including electrical and mechanical engineers. California Business and Professional Code Section 6730 states that the purpose of that Code is "to safeguard life, health, property and public welfare." And Section 6775 provides that a licensed engineer may be disciplined—indeed his registration may be revoked—for *"negligence," "incompetency in his practice,"* or if he "has not a *good character."*

What is *"negligent,"* under ordinary common law principles, is determined by the scope of the negligent person's duties, and those duties are in part determined by what is generally recognized to be ethical. *"Incompetency in his practice"* involves failure to adhere to generally accepted standards of conduct and must be taken to include ethical standards, if those standards are widely publicized and generally recognized. And, most important, the notion of *"good character,"* particularly in a professional sense, certainly involves adherence to generally accepted ethical standards, and particularly standards of professional ethics.

California law, then, mandates adherence to ethical and moral standards. Engineers have adopted (see Point II below) proper ethical codes to complement statutory codes. We urge this Court on the *Petermann* principle to recognize (1) that an engineer has an overriding duty to protect the public, and (2) that California law, including statutes and case law, supports the drafting of ethical codes, makes the terms of generally accepted professional ethics relevant and material in a case such as this, and effects a legally enforceable incorporation of such codes into engineering contracts of public employment, insofar as such codes are widely acknowledged to be necessary for the protection of the public.

POINT II

Engineering Professional Codes Require Protection of the Public

1. A Common Thread: The Duty to Protect the Public. The various professional engineering societies have, for many years, adopted and published Codes of professional ethics. Such codes contain at least one common thread—that the engineer owes an overriding duty to protect the public safety.

For example, the *Canons of Ethics for Engineers* was prepared and adopted by the Engineers' Council for Professional Development ("ECPD") in 1946.[4] These Canons were then adopted by the Board of Directors of the National Society of Professional Engineers ("NSPE") in October 1946, and were published in NSPE's Journal, *"The American Engineer,"* in its November 1947 issue.

Section 4 of these Canons provided:

> He [the engineer] will have due regard for the safety of life and health of public employees who may be affected by the work for which he is responsible.

This code has an even longer history, having been discussed initially in the May 1935 issue of *"The American Engineer,"* although the code was formally adopted in 1946, in a form differing from the present code.[5]

NSPE's own code of ethics (distinct from ECPD's) was adopted in 1964, and published in the September 1964 issue of *"The American Engineer."*[6] This code provided, in Section 2:

> Section 2—The Engineer will have proper regard for the safety, health, and welfare of the public in the performance of his professional duties. If his engineering

judgement is overruled by nontechnical authority, he will clearly point out the consequences. He will notify the proper authority of any observed conditions which endanger public safety and health.

    a. He will regard his duty to the public welfare as paramount.

    b. He shall seek opportunities to be of constructive service in civic affairs and work for the advancement of the safety, health and well-being of his community.

    c. He will not complete, sign, or seal plans and/or specifications that are not of a design safe to the public health and welfare and in conformity with accepted engineering standards. If the client or employer insists on such unprofessional conduct, he shall notify the proper authorities and withdraw from further service on the project.

We emphasize in this regard the code's injunction to the engineer that he must "notify the proper authority" of anything he observes which may "endanger public safety." We think it fair to say that the ultimate proper authority in the case of public employment is the public itself.

ECPD, meanwhile, adopted revised Canons in September 1963, which stated, in the very opening paragraph:

> 1.1—The Engineer will have proper regard for the safety, health and welfare of the public in the performance of his professional duties.

These Canons were adopted by a variety of professional engineering societies. The American Society of Mechanical Engineers, whose membership now totals close to 70,000, ratified these canons in 1963, and they were published in ASME's magazine, *"Mechanical Engineering."*

The same principles are carried forward to the current day. For example, a set of "Guidelines to Professional Employment of Engineers and Scientists" published by the IEEE Board of Directors in its national monthly magazine, *Spectrum*, in April, 1973,[7] contains the following paragraph:

> The professional employee should have due regard for the safety, life, and health of the public and fellow employees in all work for which he/she is responsible. Where the technical adequacy of a process or product is involved, he/she should protect the public and his/her employer by withholding approval of plans that do not meet accepted professional standards and by presenting clearly the consequences to be expected if his/her professional judgement is not followed.

2. General Acceptance and Publication of the Common Thread. Because the cited codes have been widely circulated and generally endorsed, it seems eminently reasonable to conclude that every engineer is aware of his obligation to the public. The guidelines published by IEEE, for example, have also been endorsed by over twenty societies.[8]

Even before the engineer's obligation to serve the public was fully codified in writing, moreover, there was an historical recognition of that obligation, discussed in professional journals.[9]

CONCLUSION

Based upon the foregoing, we submit and we urge this Court to acknowledge that an engineer has an overriding obligation to protect the public.
Specifically, we urge this Court:

1. To rule that evidence of professional ethics is relevant, material and admissible in this case; and

2. To rule, as to any motions for judgement or any jury instructions, that an engineer is obligated to protect the public safety, that an engineer's contract of employment includes as a matter of law, an implied term that such engineer will protect the public safety, and that a discharge of an engineer solely or in substantial part because he acted to protect the public safety constitutes a breach of such implied term.

Notes

1. See also *Slochower v. Board of Higher Education of the City of New York*, 350 U.S. 551 (1956).

2. This court may, but need not, decide the extent to which the principles of this case would be applicable in the case of a private employer. The complaint in this case alleges that a public employer discharged public employees because those employees informed the public of a danger to the public safety. In a very real sense, the public at large was the "employer" of the plaintiffs herein; whatever may be the limits of the duties of public disclosure by the engineer in private employment, there is clearly a higher duty in the case of public employment.

3. Not all members of IEEE or other professional engineering societies are (nor are they all required to be) licensed to practice engineering in their home states. The ethical standards covering both licensed engineers and other engineers are the same, and this is particularly true where both types of engineers are working together on the same project, as was the case, we understand, in the BART situation.

4. ECPD is an organization founded by a group of professional engineering societies, whose participants and affiliates now include the American Institute of Aeronautics and Astronautics, the American Institute of Chemical Engineers, the American Institute of Industrial Engineers, the American Institute of Mining, Metallurgical and Petroleum Engineers, the American Nuclear Society, the American Society of Agricultural Engineers, the American Society of Civil Engineers, the American Society for Engineering Education, the American Society of Mechanical Engineers, the Institute of Electrical and Electronics Engineers, National Council of Engineering Examiners, the Society of Automotive Engineers, National Institute of Ceramic Engineers, and the National Society of Professional Engineers.

5. The ethical proposal originally published by NSPE in the May 1935 issue of *"The American Engineer"* included the following: "The engineer shall at all times and under all conditions seek to promote the public welfare by safeguarding life, health and property."

6. NSPE, when it published its code in 1964, had membership of 62,038 engineers, and its journal was circulated, in addition, to over 1,000 libraries and institutions. Its membership today is approximately 70,000 engineers.

7. A much earlier code, adopted and published by the American Institute of Electrical Engineers (IEEE's predecessor) in 1912 provided: "An engineer should consider it his duty to make every effort to remedy dangerous defects in apparatus or structures or dangerous conditions of operation, and should bring these to the attention of his client or employer." The "employer," in a case such as this, is first the public entity and ultimately the California general public which is the entity's own employer. IEEE supplemented the 1912 code in 1974 by a new code which includes the following: "Engineers shall, in fulfilling their responsibilities to the community: (1) protect the safety, health and welfare of the public and speak out against abuses in these areas affecting the public interest..."

8. The endorsing societies include: American Association of Cost Engineers, American Institute of Aeronautics and Astronautics, American Institute of Chemists, American Institute of Industrial Engineers, American Institute of Professional Geologists, American Nuclear Society, American Society of Agricultural Engineers, American Society of Engineering Education, American Society of Civil Engineers. American Society of Mechanical Engineers, American Society of Quality Control, Data Processing Management Association, Engineering Societies of New England, Inc., Engineers' Council for Professional Development, Engineers' Joint Council, Institute of Electrical and Electronics Engineers, Instrument Society of America, Institute of Traffic Engineers, National Association of Corrosion Engineers, National Institute of Ceramic Engineers, National Society of Professional Engineers, Society for Technical Communications, Society for Experimental Stress Analysis, Society of Fire Protection Engineers, Society of Women Engineers, Technical Association of the Pulp & Paper Industry.

9. The code of ethics of the NSPE, for example, was discussed initially in the May, 1935, issue of the *American Engineer* although that code was first formally adopted in 1946 (in a form differing little from the present code).

6.7

The Case of the DC-10 and Discussion

FAY SAWYIER

A ghastly accident which took many hundreds of lives recently was preventable. Not only once, but several times prior to the crash of the Turkish Airlines DC-10 near Paris on March 3, 1974, there had been specific signals of imminent danger. The warnings were ignored by some, not fully understood by others, never explained or presented to others, and perhaps falsified and/or erased by yet others. Of the persons so involved, most were highly trained engineers, and a few were government officials or corporate officers. None of them could plead ignorance, since each was explicitly charged with guaranteeing the integrity and safety of the aircraft. The evidence of likely danger was pressing, plain, and had been accumulating for several years; it may, therefore, properly be expected that these men ought to have known any relevant information. Our concern here is how did it come about that the individuals singly and collectively responsible for this aircraft should have failed in so discouraging a variety of ways. In what follows, the chronology of events (A) should support the claim that this accident could have been prevented at any one of a number of times. Since it was not prevented, in the discussion section (B) we consider a number of factors which contributed to the failures or inadequacies of appropriate response. In particular, we shall take brief notes of (1) the economic world of the aircraft industry, (2) the problems of international competition without internationally equivalent standards for training and maintenance, (3) the character and personality of some of the central figures in the story, (4) the problem of adequate quality control, (5) the status and function of the F.A.A. during this period, and (6) the intense secrecy and the exclusiveness of this highly technical and competitive industry. Even though support of commercial aviation both in tax dollars and in customers comes from ordinary citizens, neither their own common sense nor that of their representatives was ever allowed exercise in this, their common interest.

Reprinted by permission of the author from notes prepared for the Center for the Study of Ethics in the Professions, Illinois Institute of Technology, December 1976, pp. 5.1–5.23.

A

1933.   Commercial aviation got underway with the first of the Douglas Commercial ("DC") series. The compromising of the regulatory agency also started, since it (precursor to the F.A.A.) had two potentially inconsistent assignments. One was to push and promote the American aviation industry as such; another was to provide impartial and detailed inspections, certifications, and "airworthiness directives" (if needed). Even the inspection crews were contaminated at the start by the routine presence on them of members of the very companies whose product was to have been impartially investigated. This was the so-called "Designated Engineering Representative" operation.

Until 1953.   Douglas held unquestioned lead, but in the early fifties made the gross miscalculation that the public was not ready to accept jet travel.

1958.   The Boeing company put its 707 (jet) into commercial service and soon overtook Douglas in sales.

By 1966.   The Douglas company was in desperate financial straits. The corporate officers frequently misrepresented the company's economic position and this culminated in a major Wall Street scandal concerning lies associated with a debenture issue.[1] American Airlines invited proposals for a new plane, wide-bodied like the 747 but capable of using almost any runway in a major city.

1967.   As arranged by the firm of Lazard Freres, the McDonnell aircraft company took over Douglas. When Lockheed announced in September that it would take orders from airlines for a wide-bodied "air bus" (the 1011 later called the Tristar), the new management at McDonnell-Douglas was forced to make a similar proposal; orders were lagging on their DC-8 and -9 models and Boeing had captured the long-distance market for wide-bodies with the 747. Now if Lockheed collected the short-haul market too, McDonnell-Douglas perceived economic ruin. Airlines naturally liked wide-bodies and airbuses since they significantly increased the possible passenger load.

1968.   Three engineers from RLD (the Dutch equivalent of the FAA) issued serious and repeated warnings about the integrity of the passenger compartment floors of the Jumbo jets.[2]
   American Airlines insisted to Douglas on a change in design for the actuators in closing the cargo doors from hydraulic to electrical. Electrical actuators were cheaper (weighed less) but were not so safe.[3]
   The Convair division of General Dynamics won the sub-contract from McDonnell-Douglas to build the fuselage of the DC-10, including the cargo doors.

1969.   The spate of briberies of other countries to buy this or that aircraft was increasing. In connection with one of these (a Lockheed bribe) McDonnell-Douglas lost its sale to All-Nippon Air and began a desperate search for new customers. This resulted eventually in the sale to Turkey of the DC-10.
   Douglas asked Convair to draft a Failure Mode and Effects Analysis (FMEA) for

the lower cargo door system. Prior to certification, the FAA must be given an FMEA for those systems critical to safety. Convair found nine possible failure-sequences which could lead to a Class IV hazard.[4] But "neither the FMEA draft nor anything seriously resembling it was shown to the FAA by Douglas, who, as lead manufacturers, made themselves entirely responsible for the certification of the aircraft. (Indeed, under the terms of the subcontract, General Dynamics was forbidden from contacting the FAA about the DC-10) . . . FMEAs submitted by Douglas to the FAA leading up to certification of the DC-10 do not mention the possibility of Class IV hazards arising from the malfunction of lower cargo doors."[5]

1970. A prototype DC-10 was unveiled.

The RLD stepped up its warnings, and at the ICAO meeting in Montreal made explicit mention of the cargo-door danger. A model of a DC-10 in fact "blew" its rear cargo door during testing and with a consequent collapse of the passenger floor.[6] Convair and Douglas argued, but failed to resolve, which company should bear the expense of redesigning.

1971. Although reports of problems (both on paper and in physical testing) with the rear cargo door were increasing, and RLD was now complaining directly to Douglas about it, changes were not made and all discussions were kept "in house"; none was sent to the FAA. The aircraft was certified.

1972. Many DC-10s had been built but unsold or undelivered. These sat at Long Beach, California and among them numbers 29 (involved in the Paris crash) and 47 (found to have been unmodified even months after the Paris crash).

June 12. The rear cargo door of an American Airlines DC-10 did "blow" over Windsor, Ontario, with resultant collapse of part of the passenger floor. Catastrophe was averted only by the extraordinary poise and skill of the pilot.[7]

June 19. Since the FAA was now of necessity involved, the regional branch, headed by Basnight, worked out the details of an Airworthiness Directive. But this was never issued. Instead, an agreement was arranged between McGowan (of Douglas) and Shaffer (head of the FAA)[8] later called the "Gentlemen's Agreement." This expression denotes any arrangement (usually verbal) not having the force of law and resting only on mutual trust. But the expression has come to have the connotation of a snobbish pact between friends to subvert or evade legal requirements, as was the case, for example, in "Gentlemen's Agreements" to evade the outlawing of restrictive (discriminatory) covenants in the rental and sale of property.

June 20. Basnight wrote a memorandum detailing his role, opinions, and dismay at this agreement . . .

June 27. Applegate, an engineer at Convair, wrote out and filed away his own shock at the protracted disregard both of safety and even of routine procedures, culminating in the "Agreement."[9]

All suggested modifications thus became only "in house" requests. Meanwhile at Long Beach even these suggestions were not carried out on numbers 29 and 47, although these two aircraft were stamped by their inspectors (by "Q" for qualified and "A" for approved work *done*). In short, these two aircraft were embodied lies.[10]

1973. #29 arrived in Turkey in the last week of 1972. During the following months THY (Turkish Air Lines) was presumably trained and drilled in all aspects of maintenance and control, but records show that this was not the case.[11]

1974. The system for closure of the door on #29 was sublet to SAMOR and the individual doing the closure was an ex-Algerian named Mahmoudi, who knew several languages but not English. All instructions on the craft were printed in English. After the crash Sanford Douglas attempted to put the entire blame on this very Mahmoudi. Final inspection was to have been done either by the Flight Engineer (who didn't) or by D. Zeytin (THY's head mechanic at Orly) who was on holiday.

So, that evening the door was not properly closed. It blew after nine minutes of flight; the cabin floor collapsed taking out all or most of the controls to ailerons and rudder. The plane crashed into a forest at nearly 500 mph and there were no survivors.

B-1

Forbes magazine stated that "... the DC-10 was a gamble that McDonnell-Douglas could not afford to take and could not afford to lose."[12] Comparable remarks had been made, and with comparable justification, about Boeing (which put 95% of its corporate capital into the development and construction of the 747) and about Lockheed.

> ... Lockheed and Douglas did more than simply enter a contest which at best only one of them could win, and which, according to the economic rules obtaining at the time, neither of them could afford to lose. They did so in each case with accounting systems which began from the proposition that each company had already attained considerable success in a race which, realistically, it was imminently possible that both would lose. There was, to put it mildly, a considerable premium on being the first firm actually to get an airbus into service so that it might start earning money and attracting further orders.[13]

The crucial feature of this desperate economic gamble against tremendous odds was that it was perceived only as such by the top management and everyone else involved in the work. That is, since we are primarily interested in the various factors which contributed to the array of oversights, errors and falsifications respecting the DC-10, our concern with economics is to try to evaluate how the great pressure of this economic gamble might warp the individual judgment of the men involved.[14] Consideration should be given, too, to plausible solutions for the ever-present danger of warped judgment in so signal a public business as the aerospace industry. Shall loyalty to a company be downgraded? Shall unlimited competition in this area be prevented? Shall provision be made for regular and more disinterested checks by non-company persons?

B-2

Nowadays in this industry, competition is international; a premium is put on sales not only to domestic airlines but to foreign companies as well. The probability always exists that some foreign airlines will have flight, engineering and maintenance standards which we perceive to be inadequate in one or more respects. For a variety of reasons, this was the case with Turkish Airline (THY) to whom Douglas sold—on whom they most urgently pressed the sale of—a group of DC-10s. Efforts by Douglas to train could and did meet with objections from THY, a chief one being the added expense. This increased cost, supplemented by the threat of offending national pride by over-solicitious training, resulted in training which failed to meet the minimum standards. Even the problems of language are not inconsiderable for, as we have noted, most instructions continue to be written in English and the measurements specified in pounds and inches (for example "psi").

What solution might one propose? A policy of refusing to sell to an "underdeveloped" nation would meet with some obvious objections. What are the moral obligations of a manufacturer of a product to insure that the use of his product be understood? How credible is the rejoinder that if we do not sell to them, someone else even less responsible will? And what of the persons doing the actual training while noting that it is going poorly; what options have they to make their doubts effectively known and acted on? Would different personnel, i.e., those trained in some traditions of the culture in which they are now assigned training jobs, make a difference? Or is the sole significant factor still the frenzied competition?

Here follows a cast of characters, most of whom will be identified only with respect to their jobs; in a few cases a slightly more extended identification will be offered. With the exception of a baggage handler for THY and three inspectors at Douglas' Long Beach plant, all of the characters in the narrative were from similar backgrounds. That is, they were middle-class, white, Anglo-Saxon men. These very classifications have long tended to characterize those who are dominant in the engineering profession generally and especially so in the parts of it associated with heavy industry. It is also true that most were Republicans and very few had been educated in or even exposed to enterprises other than business and technology. The issues of racism (anti-Semitism as well as anti-Black-ism) and of sexism are clear and manifest. What may not be so clear is one of the probable consequences of having an entire industry controlled by persons whose views might be expected to converge so habitually, and with so little self-questioning. An important factor in good judgment as well as in safety was omitted, namely, "input" from persons of genuinely diverse backgrounds and beliefs.

In Alphabetical Order

Applegate	Director of Product Engineering at Convair
Basnight	Head of the Western Region of the FAA

Brizendine	After 1973, president of Douglas
Curtis	Vice-president of Convair
Donald Douglas	Founder and longtime president of Douglas
Evans	the "A" stamp inspector at Long Beach plant
Haughton	Chairman of Lockheed Corporation
Hurt	Program Manager at Convair for DC-10 support system
Lewis	From McDonnell, later head of Douglas Division, now head of General Dynamics
McCabe	Manager of flight training at Douglas, went to Turkey
McDonnell, James	Founder of McDonnell Aircraft
McDonnell, Sanford	Nephew and successor of James
McCormick	American Airlines pilot at "Windsor" accident
Mahmoudi	Baggage handler for THY at Paris
Miller	Former head of NTSB's Bureau of Aviation Safety who objected repeatedly to the political interference from Nixon appointees and resigned
Newton	A Black, "Q" inspector at Long Beach
Noriega	Also a "Q" inspector at Long Beach
Shaffer	Nixon appointed head of FAA who negotiated with McGowen the "Gentlemen's Agreement."

Both Donald Douglas and James McDonnell have been described as "patriarchal idividualists."[15] Douglas went to the Naval Academy and to M.I.T.; he founded his own company in 1929 and in early years had extensive connections with the military. He continued to dominate the Company (and his own son) until well into old age.

> The Douglas Aircraft Company in the sixties was a jovial . . . ship. All major executives were expected to gather at midday for martinis in the Heritage Room . . . This midday court in the Heritage Room became the main-decision-making forum . . . if forum is the right word in a situation where one man's writ dominates almost completely.[16]

McDonnell also attended M.I.T. and came from a fairly well-to-do family. He made his initial fortune and the success of his aircraft company from government contracts. He is said to have been highly interested in attempting a scientific proof of the soul and of immortality. The intensity of his conviction about his own virtue and his sense of personal responsibility may have obscured his perception of a need for outside checks as well as of a growing tendency among others in his company to diffuse and finally to refuse such individual responsibility.

Jackson McGowen had been with Douglas since his graduation in 1939 from Indiana Technical College. His chief talents and principal occupation were salesmanship. It was he who instigated the Gentlemen's Agreement by telephoning Shaffer. And Shaffer himself was " . . . identified totally with the industry"[17] which, as head of the F.A.A. it was his job to inspect and regulate. He had worked with aircraft manufacturers in developing new bombers while he was in the Air Force. Later he became an executive with TRW, Inc., a large engineering sub-contractor with the aerospace industry. He held strongly throughout his period as administrator to the merits of voluntary compliance, even extending this to a cheerful faith that aircraft manufacturers were best left to their own devices. He perceived his job as one of contending against persons who raised questions either about air safety or about the effects of airliners on the environment.

We must take note of the fact that after he learned that even after Windsor he was not to proceed with the Air Worthiness Directive against Douglas, Basnight wrote and filed away a memorandum covering his concerns (about the cargo door and the floor) and spelling out his outrage at the infamous Agreement. At approximately the same time, Applegate sent a memorandum to his immediate chief, Hurt, going over once again the clear dangers in the design of the parts of the airplane which Convair had sub-contracted to build. The dates of these memoranda, respectively, were June 20 and June 27, 1972; "Windsor" had occurred June 12, 1972. It is appropriate to our present purposes to speculate some about these memoranda. Why were they written at all? To keep the record clear? To file and thus preserve some evidence of what had been going on as well as of the serious doubts in the minds of these two individuals about the airworthiness of the DC-10? To make a record of the personal objections of these men? (i.e., to try to keep their personal records clean in the event of future developments)? But it is also appropriate to consider why nothing more than filing a memorandum was done. In Basnight's case, his authority had been undercut. In Applegate's, he had written the memorandum specifically to call the attention of his company (and of his immediate superior) to his concerns. What would have been the consequences for either or both had they gone to Congress or to the Press for example? Hurt's reply to Applegate's memorandum was to counter it with mention of the tremendous cost that would probably be born by Convair (instead of by Douglas, where, in all justice, it belonged) if it were they who "blew the whistle." Was Applegate convinced or simply over-ruled? In cases in which a highly competent engineer clearly perceives a serious hazard and as clearly calls the attention of his superiors to it and is over-ruled on economic/legal grounds rather than on grounds of engineering design, what can he do? What ought he do? It is important to note that not only was the public not informed of any of this, but that in many aspects of concern with design-safety problems even the F.A.A. was never alerted.

Obviously, one problem is to work towards a system such that non-company concerns will somehow weigh more heavily with serious and competent persons. Just to declare what Basnight or Applegate ought to have done is insufficient; our assignment is, in a sense, to work out ways likely to facilitate doing more. And if one's response is "get a man with more moral courage," then one is obligated to try to specify how to do this. Or if one takes another line and remarks that extremes of moral courage ought not to be an essential ingredient in the design of so major and hazardous a commercial product, then what *other* methods can supplant courage?

B-4

One method, it would seem, is to insure that one has extremely tight provisions for quality control. In principle Douglas had this, but in fact the control at Long Beach over whether or not even the "in house" modifications had actually been made, was lax. "Q" stamps and "A" stamps certifying that work had been properly done were affixed to aircraft on which no work at all had been done.[18] In the cases of the three inspectors at Long Beach it is still unclear whether they had been robbed (of their personal seals) or bribed or confused—or all three, at the same time. What is clear is that the stamps of approval ("Q" and "A") had been affixed to aircraft which had not been worked on in these particulars. One unquestionable factor was the tremendous pressure that had built up by then to get those planes sold and delivered. A litany of reminders of the urgency of this were daily experiences at Long Beach. It may be that reasonable counter-measures against this particular kind of failure should begin by devising a way to insulate at least the inspection and repair divisions from precisely that kind of competitive-economic pressure. Again, as in the cases of Basnight and especially of Applegate, what would have happened to any of these three inspectors had they delayed the production line? And were they being asked to approve "work-properly-done" faster than one might reasonably expect possible? Moreover, how well and how repeatedly were they made aware of responsibilities which transcended company interests?

B-5

An increasingly popular solution to many problems of public responsibility and safety is the governmental regulatory agency, a surrogate of sorts for our properly self-interested concerns for our safety, as Congress is surrogate for our expression of political claims and opinions. Examples that come readily to mind are the F.D.A. (regulating things we eat or otherwise take into our bodies), the E.P.A. and of course the F.A.A. In each instance a central charge of the Agency is to oversee and regulate the products of commercial enterprises. To some degree, the rational thinking behind the creation of these agencies is similar to that of the Founding Fathers, that power should be balanced by (potentially) opposed power for the best interests of the people. It has been objected[19] that one unfortunate consequence of regulatory agencies is that individual human beings are encouraged, by the very existence of these agencies, to behave like functionaries, to abjure personal responsibility on the grounds that "Jack will do it" or that "the E.P.A. can worry about that so I needn't."

In addition, we must consider that if the regulatory agency is to fulfill its charge, it must in a real and continuing sense *be* "opposed". The F.A.A. is an agency compromised in that crucial aspect from its inception, inasmuch as it has always had the dual assignment of promoting the American aviation industry *and* of regulating it. Industries are, understandably, hostile to the concept of being regulated at all, more especially by antagonistic outsiders.[20] But where the well-being and the lives of the public are at stake (indeed the very public on whose dollars these same industries rely), then perhaps we ought to recognize but over-rule the objections of industry.

B-6

Finally we come to the question of privacy and exclusiveness and thus to the virtually paranoid perceptions of a private "club" engaged in what it sees as a life or death struggle to survive.[21] I begin by explaining the use of the adjective "paranoid" in this context. We have a paranoid position whenever we are locked into a circle of judgments and perceptions each of which only reinforces the others and all of which are dominated by and contribute to the belief of being surrounded by enemies. Other persons and alien beliefs are all seen as "they" or "them" and as necessarily hostile. An individual whose perceptions and behavior are under this sort of dynamic principle is considered crazy, indeed dangerously so. Why? Because nothing outside of his own desperate commitment to his fearful, fighting goal can really be experienced by others.[21] A symptom of this is, of course, systematically warped or distorted perception and judgment.

In some ways the aircraft industry has taken on paranoid characteristics. Exclusiveness and privacy have contributed to this and an attempt at rationalization of such behavior has been sought in the rubric that the laity cannot possibly understand (and, a fortiori, judge) matters so highly technical. Consequently the whole issue of professional or peer review, insofar as it means that no one outside the (engineering) profession is competent to judge, has excluded not only those who might have pirated trade secrets etc. but also the objective input of the consuming public. This exclusiveness has been aggravated by the similarity of background and training of the relevant persons, as indicated in B-3. And it has been alarmingly enhanced by the cultivated stance of loyalty to an embattled super-person, the company. The odds are, therefore, that no one will assert, and if they did no one else would hear, the common sense objection that this is crazy, where "this" refers to any of the tightly-bound spiral of decisions affecting policy. Accordingly, we must take another look at the basis of any rejection of information when buttressed by "they wouldn't understand." In the first place the reason may be plainly false, for perhaps laypersons would indeed understand or could be helped to. In the second place it is "they" too whose lives (as well as dollars) are at stake. Thirdly, this fosters the dangerous climate of secrecy which, almost unaided, breeds a kind of paranoia. Merely to consider something as a secret is to suggest that there are others, potential or actual enemies, who are engaged in trying to harm us. Finally, the very restrictiveness of highly placed corporate persons and engineering members of major industries greatly diminishes the availability of contrary and informed alternative positions. If part of the moral obligations of persons is to keep informed especially on matters in which they are major agents, then this obligation is not met wherever secrecy and exclusiveness are the prevailing structure.

This last point makes connection, too, with the relevance to engineering education of thorough-going exposure to non-engineering views on the encounter between human intelligence and human lives. A purely technical education therefore may prove a training ground for the very sort of paranoia of which we have been speaking.

Footnotes

1. " . . . [A] marathon lawsuit launched . . . debentures" *H. H. Levy v. Douglas Aircraft Company Inc., et al.*, U.S. District Court, SD New York, September 30, 1975 (Civ. 3382).

2. Even a layperson can easily grasp the basic structural problem as can be seen from the following very crude design-caricature, on which students and Technically-Trained faculty are urged to improve.

Curved surfaces have been known (since the Arch) to distribute stress better than flat ones. (Consider flat-feet). Yet, passenger cabin floors of planes are flat surfaces.

Crucial control systems ought not to traverse a vulnerable surface. Those of the 747 are along the "ceilings"; in the DC-10 they run along the flat floor.

Since a doubling of width, say, increases the volume eightfold, the quantity of pressurized air in a Jumbo Jet is much greater than in the smaller ones. Yet devices to cope with the associated stresses remained basically unchanged.

If there is a hole in the pressurized hull (by a "blown" door for example) one section will depressurize and the other (here the passenger shell) will now present a huge pressure differential against the flat floor. And the result is—at best!—that instantly one is faced with a hull like this, in effect:

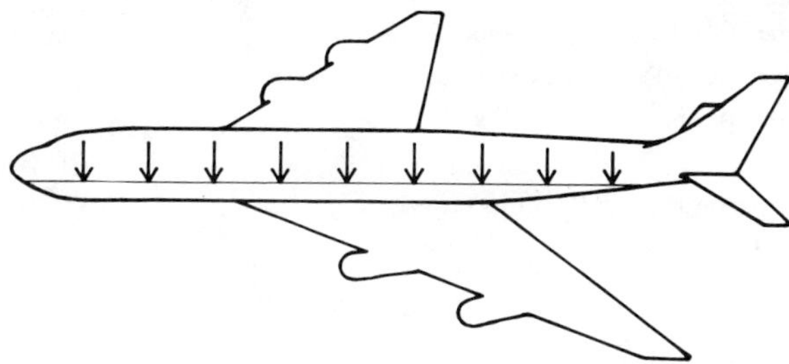

which at high altitudes and speeds is extremely dangerous!

3. Once more the ordinary person's common sense is roughly adequate to understand that since hydraulically misclosed doors leak or "bleed" continually, the chances of a catastrophic door blowout and subsequent decompression are lessened. Design sketches are available for interested students.

4. That is, a hazard likely to result in destruction of the Craft and loss of lives.

5. The starting point was seen as a failure of the locking-pin system, due to the jamming of the locking tube or of one or more of the locking pins. In that case, said the FMEA: "Door will close and latch, but will not safety lock." There should of course be a warning against this, and if it works properly: "Indicator light [in the cockpit] will indicate door is unlocked and/or open." But one of the ground rules of the FMEA was that circuit failures in the indicator system might well go undetected, in which case: "Indicator light will indicate normal position." If that happened, malfunction of the electric latch actuators could produce a situation in which "door will open in flight—resulting in sudden depressurization and possibly: structural failure of floor; also damage to empennage (tail flying surfaces) by expelled cargo and/or detached door. *Class IV* hazard in flight."

The difference between this and what actually happened over Windsor is: first, the FMEA envisaged failure in a spring-loaded, rather than a hard-driven, locking-pins system; second, the failure mentioned was one of inadvertent electrical reversal of the latches rather than a failure of the latches to go "over-center" in the first place. But it was a powerful demonstration that the door design was potentially dangerous without a totally reliable fail-safe locking system.

But neither this FMEA draft nor anything seriously resembling it was shown to the FAA by Douglas, who, as lead manufacturers, made themselves entirely responsible for certification of the airplane. (Indeed, under the terms of the subcontract, General Dynamics *was forbidden* from contacting the FAA about the DC-10.) Our evidence is drawn from documents produced by Douglas and testimony given in the complex of compensation lawsuits which resulted from the Paris crash (*Hope v. McDonnell Douglas et al.*: Civ. No. 17631, Federal District Court, Los Angeles, California). Evidence given by J. B. Hurt, Convair's DC-10 support program manager during the litigation, was that Douglas never replied to the Convair FMEA.

FMEA's submitted by Douglas to the FAA, leading up to certification of the DC-10, do *not* *mention* the possibility of Class IV hazards arising from malfunction of lower cargo doors.

The Documentary Warning of the dangers of depressurization were followed in 1970 by a physical manifestation at Long Beach. But even this it seems could not dent the self-assurance of the Douglas design team.

By May 1970, the first DC-10 (Ship 1) had been assembled at Long Beach and was going through ground tests to prepare for the maiden flight scheduled for August. On May 29, outside Building 54 the air-conditioning system was being tested, which involved building up a pressure-differential inside the hull of four to five pounds per square inch. Suddenly the forward lower cargo door blew open. Inside, a large section of the cabin floor collapsed into the hold.

6. *The Last Nine Minutes*, Mara Johnston, William Morrow & Co., Inc. Page 117, New York, 1976.
7. McKnight had trained himself to fly the Ship (turn and up/down) using only engines. Since he was himself concerned about decompression—caused loss of control cables.
8. "The Basnight Memorandum," Appendix D in *Destination Disaster*, Eddy, Potter and Page, Quadrangle/N.Y. Times Book Co. 1976.
9. "The Applegate Memorandum," pp. 183–185, *Destination Disaster* (op. cit.).
10. *The Last Nine Minutes*, p. 207.
     Destination Disaster, Ch. 13 and see especially p. 235.
11. *Destination Disaster*, pp. 208, 210, 213–4.
12. *Destination Disaster*, p. 66.
13. Ibid., page 82.
14. No evidence of any women involved has emerged.
15. *Destination Disaster*, page 33.
16. Ibid., page 44.
17. Ibid., page 43.
18. Ibid., all of chapter thirteen.
19. Thomas Martin, President of I.I.T.
20. This applies even when survival is couched in excitedly patriotic terms.
21. See section B-6.
22. Paranoid individuals have a particularly hard time in therapy to perceive the therapist's interest and concern as friendly or even benign.

THE APPLEGATE MEMORANDUM

27 June 1972

Subject: DC-10 Future Accident Liability

The potential for long-term Convair liability on the DC-10 has caused me increasing concern for several reasons.

1. The fundamental safety of the cargo door latching system has been progressively degraded since the program began in 1968.
2. The airplane demonstrated an inherent susceptibility to catastrophic failure when exposed to explosive decompression of the cargo compartment in 1970 ground tests.
3. Douglas has taken an increasingly "hard-line" with regards to the relative division of design responsibility between Douglas and Convair during change cost negotiations.
4. The growing "consumerism" environment indicates increasing Convair exposure to accident liability claims in the years ahead.

Let me expand my thoughts in more detail. At the beginning of the DC-10 program it was Douglas' declared intention to design the DC-10 cargo doors and door latch systems much like the DC-8s and -9s. Documentation in April 1968 said that they would be hydraulically operated. In October and November of 1968 they changed to electrical actuation which is fundamentally less positive.

At that time we discussed internally the wisdom of this change and recognized the degradation of safety. However, we also recognized that it was Douglas's prerogative to make such conceptual system design decisions whereas it was our responsibility as a sub-contractor to carry out the detail design within the framework of their decision. It never occurred to us at that point that Douglas would attempt to shift the responsibility for these kinds of conceptual system decisions to Convair as they appear to be now doing in our change negotiations, since we did not then nor at any later date have any voice in such decisions. The lines of authority and responsibility between Douglas and Convair engineering were clearly defined and understood by both of us at that time.

In July 1970 DC-10 Number Two[*] was being pressure-tested in the "hangar" by Douglas, on the second shift, without electrical power in the airplane. This meant that the electrically powered cargo door actuators and latch position warning switches were inoperative. The "green" second shift test crew manually cranked the latching system closed but failed to fully engage the latches on the forward door. They also failed to note that the external latch "lock" position indicator showed that the latches were not fully engaged. Subsequently, when the increasing cabin pressure reached about 3 psi (pounds per square inch) the forward door blew open. The resulting explosive decompression failed the cabin floor downward rendering tail controls, plumbing, wiring, etc. which passed through the floor, inoperative. This inherent failure mode is catastrophic, since it results in the loss of control of the horizontal and vertical tail and the aft center engine. We informally studied and discussed with Douglas alternative corrective actions including blow out panels in the cabin floor which would provide a predictable cabin floor failure mode which would accommodate the "explosive" loss of cargo compartment pressure without loss of tail surface and aft center engine control. It seemed to us then prudent that such a change was indicated since "Murphy's Law"[†] being what it is, cargo doors will come open sometime during the twenty years of use ahead for the DC-10.

Douglas concurrently studied alternative corrective actions, in house, and made a unilateral decision to incorporate vent doors in the cargo doors. This "bandaid fix" not only failed to correct the inherent DC-10 catastrophic failure mode of cabin floor collapse, but the detail design of the vent door change further degraded the safety of the original door latch system by replacing the direct, short-coupled and stiff latch "lock" indicator system with a complex and relatively flexible linkage. (This change was accomplished entirely by Douglas with the exception of the assistance of one Convair engineer who was sent to Long Beach at their request to help their vent door system design team.)

---

[*] We have been unable to establish whether Applegate's reference to an accident involving Ship Two, in July 1970, is a mistake on his part or whether there were *two* blowout incidents. Certainly Ship One was damaged on May 29, 1970, in circumstances very similar to those described by Applegate.

[†] Murphy's Law: "If it can happen, it will." Also known sometimes as the totalitarian or Hegelian law of physics, from Hegel's view that all that is rational is real: that is, if you can think of it, it must exist. This is a case where implausible philosophy makes for good engineering.

This progressive degradation of the fundamental safety of the cargo door latch system since 1968 has exposed us to increasing liability claims. On June 12, 1972 in Detroit, the cargo door latch electrical actuator system in DC-10 number 5 failed to fully engage the latches of the left rear cargo door and the complex and relatively flexible latch "lock" system failed to make it impossible to close the vent door. When the door blew open before the DC-10 reached 12,000 feet altitude the cabin floor collapsed disabling most of the control to the tail surfaces and aft center engine. It is only chance that the airplane was not lost. Douglas has again studied alternative corrective actions and appears to be applying more "band-aids." So far they have directed us to install small one-inch diameter, transparent inspection windows through which you can view latch "lock-pin" position, they are revising the rigging instructions to increase "lock-pin" engagement and they plan to reinforce and stiffen the flexible linkage.

It might well be asked why not make the cargo door latch system really "fool-proof" and leave the cabin floor alone. Assuming it is possible to make the latch "fool-proof" this doesn't solve the fundamental deficiency in the airplane. A cargo compartment can experience explosive decompression from a number of causes such as: sabotage, mid-air collision, explosion of combustibles in the compartment and perhaps others, any one of which may result in damage which would not be fatal to the DC-10 were it not for the tendency of the cabin floor to collapse. The responsibility for primary damage from these kinds of causes would clearly not be our responsibility, however, we might very well be held responsible for the secondary damage, that is the floor collapse which could cause the loss of the aircraft. It might be asked why we did not originally detail design the cabin floor to withstand the loads of cargo compartment explosive decompression or design blow out panels in the cabin floors to fail in a safe and predictable way.

I can only say that our contract with Douglas provided that Douglas would furnish all design criteria and loads (which in fact they did) and that we would design to satisfy these design criteria and loads (which in fact we did). There is nothing in our experience history which would have led us to expect that the DC-10 cabin floor would be inherently susceptible to catastrophic failure when exposed to explosive decompression of the cargo compartment, and I must presume that there is nothing in Douglas's experience history which would have led them to expect that the airplane would have this inherent characteristic or they would have provided for this in their loads and criteria which they furnished to us.

My only criticism of Douglas in this regard is that once this inherent weakness was demonstrated by the July 1970 test failure, they did not take immediate steps to correct it. It seems to me inevitable that, in the twenty years ahead of us, DC-10 cargo doors will come open and I would expect this to usually result in the loss of the airplane [Emphasis added]. This fundamental failure mode has been discussed in the past and is being discussed again in the bowels of both the Douglas and Convair organizations. It appears however that Douglas is waiting and hoping for government direction or regulations in the hope of passing costs on to us or their customers.

If you can judge from Douglas's position during ongoing contract negotiations they may feel that any liability incurred in the meantime for loss of life, property and equipment may be legally passed on to us.

It is recommended that overtures be made at the highest management level to persuade Douglas to immediately make a decision to incorporate changes in the DC-10 which will correct the fundamental cabin floor catastrophic failure mode. Correction will take a good bit of time, hopefully there is time before the National Transportation Safety Board (NTSB) or the FAA ground the airplane which would have disastrous effects upon sales and production both near and long term. This corrective action becomes more expensive than the cost of damages resulting from the loss of one plane load of people.

F. D. Applegate

Director of Product Engineering

6.8

The Browns Ferry Case

VIVIEN WEIL

On February 2, 1976, three engineers in General Electric Co.'s nuclear energy division resigned and made statements to the press and on TV declaring their concern for the effects on the public of technical flaws in the nuclear power program. The three engineers, Dale G. Bridenbaugh 44 years old, Richard B. Hubbard 38 years old, and Gregory C. Minor 38 years old, had each joined GE at the age of 22. They were managers in the areas of performance evaluation and improvement, quality assurance, and advanced control and instrumentation respectively [see "Three Leave GE . . . ," 4.7].

On January 13, 1976, Robert D. Pollard, a nuclear safety engineer and project manager for the Nuclear Regulatory Commission, acting without knowledge of the decisions of the three GE engineers, had given notice of his resignation to be effective February 15. He had expressed his concerns about nuclear power plant safety in a CBS interview recorded on January 13, but not aired until February 8.

What led to these concerns and the four startling resignations which involved substantial personal sacrifice? The engineers cited a number of specific unresolved safety problems in commercial nuclear power plants. Prominent among them were hazards revealed by the Browns Ferry Plant fire of March 22, 1975. The fire, which started in the electrical control cables from the use of a candle to detect air leaks, burned uncontrolled for 7½ hours. The two operating GE nuclear reactors were at full power when the fire began. One of them went dangerously out of control for several hours and was not stabilized until a few hours after the fire was put out. The reactor's sophisticated emergency safety devices failed totally. The unit was in the end controlled by some available equipment which was not part of the elaborate safety apparatus, and which emerged from the fire undamaged as a matter of random chance.

The accident was a case of common-mode failure, a type of accident assumed to

---

Reprinted by permission of the author from notes prepared for the Center for the Study of Ethics in the Professions, Illinois Institute of Technology, January 1977, pp. 1–11.

402

be highly unlikely, in fact, not "credible." Harry J. Green, Superintendant of Browns Ferry, said after the fire, "We had lost redundant components that we didn't think you could lose." The record shows, however, that there was extensive official fore-knowledge of safety deficiencies in Browns Ferry and that the very combination of problems responsible for the accident had been identified by Federal safety authorities but left uncorrected.

The responsibility for designing and maintaining nuclear power plants and for assessing and guaranteeing the safety of their operation rests to an important degree with engineers, individually and collectively, in the industry and in the regulatory agency. Failures by engineers at many different levels to anticipate consequences, to establish safety criteria, to meet applicable criteria, and to respond to recognized situations of non-compliance led to the Browns Ferry fire.

We are left with the question of what made possible all these failures. Our concern is to discover where and how engineers fell short in discharging professional and moral responsibilities. The actions of the four engineers who resigned out of moral and professional concerns raise important questions. Were these men morally required to take a course of action such as they pursued in resigning and "going public"? Was it a professional obligation? Or did their actions exceed what was morally and/or professionally required of them? If so, how should we regard their actions—heroic, morally creditable, emulable, fool hardy, or unnecessary?

All these questions have particular urgency if Dale Bridenbaugh was correct when he said in his letter of resignation, "In the past we have been able to learn from our technological mistakes. With nuclear power we cannot afford that luxury."

The narrative which follows consists of (A) a chronology of events and (B) a brief discussion of certain general matters including (1) the economic setting of the nuclear power industry, (2) the problems posed by increases in knowledge, especially of hazards and safety requirements, (3) the existence of a network of scientists and engineers heavily invested (emotionally and otherwise) in the nuclear power industry, (4) the problem of adequate quality control, (5) the status and function of the NRC and its ancestor the AEC, and (6) the problem of access by the public to information about nuclear power and its industrial development. The operations of the industry go on largely hidden from the public and to some extent from the NRC. The industry's cover is that divulging requested information would cost a company competitive business advantages. Ordinary citizens have been the consumers of nuclear energy (35–50% of energy use in Illinois is nuclear), and they have paid in tax dollars for the development of the nuclear industry without knowledge of the risks and costs.

A: CHRONOLOGY

1954. The Atomic Energy Commission begins to regulate the commercial nuclear power industry. The Commission has the dual roles of promoting and regulating commercial nuclear power plants. This situation is to lead to conflicts over maintaining development schedules and resolving known safety problems.

1958. Commercial nuclear power gets underway with the installation and start-up of the first large-scale commercial nuclear power plant, Commonwealth Edison's Dresden 1 near Chicago. Dale G. Bridenbaugh is the field engineer for that project.

In the 1960's. Section III of the hallowed ASME codes, originally developed to protect the public from boiler explosions, is further developed for application to nuclear power plant components. However, these codes do not apply to some safety-related equipment. Present Nuclear Regulatory Commission (one of the two agencies into which AEC was split in 1975) requirements for equipment not covered by the ASME codes are less stringent than those for ASME boiler code items.

1963. AEC's Division of Operational Safety warns that the combustibility of polyurethane foam constitutes a fire hazard. Nevertheless this is the material later used in parts of Browns Ferry's electrical system.

1965. During construction of the Peach Bottom plant a serious electrical cable fire erupts. The fire is the first of a series over several years which involves major damage to important cable installations. These fires make plain the capacity of electrical cable fires to cause failure of important safety systems.

1966. Construction begins on the Browns Ferry Nuclear Power Plant near Decatur, Alabama. It is intended as a model for future U.S. power production and is to supply electricity for about two million people. The plant is to be ten times the size of any plant already in operation. Indeed, it is to become one of the world's largest electrical generating facilities.

1967. Browns Ferry goes through a major Federal safety review and is granted a Federal construction permit.

1969. AEC adopts an *industry committee's* vague design standard for electrical cables. The need for physical separation of cables is admitted, but there is a failure to specify how to achieve it. On July 3, F. U. Bower, an AEC inspector monitoring Browns Ferry, sends the AEC a memo in which he notes, among other items, the need for specific criteria for cable separation. He points out the incongruity of requiring the spending of immense sums on specific safety systems in case of accident without providing equivalent criteria for the electrical cable installation.

1970. In January, after a five day inspection of Browns Ferry, five AEC inspectors report deficiency in quality control over cable separation, and other deficiencies as well. The AEC adopts an addition to its regulations to minimize the danger of fires. However, there are no specific provisions for achieving cable separation, that is, as to how much, which cables, the design of cable spreading rooms, etc.

1971. Fire erupts at Indian Point 2, before the plant is in operation. In the AEC Review, the conclusion is that there is an urgent need "to re-evaluate

previously approved cable separation criteria for this facility and other facilities."
In October, three AEC inspectors, including F. U. Bower, warn about safety
problems at Browns Ferry in their evaluation report.

1972.   In January, the new head of Region II, Norman Mosely, sends a memo to
AEC headquarters supporting Bower's report, and he puts as his first regulatory
question, "What enforceable requirements exist for separation of redundant
component instrumentation and wiring?" When Browns Ferry is under review by
AEC's Committee on Reactor Safeguards, the Assistant Manager of Power for the
TVA urges deferring safety improvements that could interfere with the schedule
for starting up. In December AEC safety reviewers criticize electrical cable
separation at Browns Ferry, but they defer needed improvements to unit 3. They
allow serious compromises with the safety of units 1 and 2.

1973.   In June, Browns Ferry is issued a license by the AEC for commercial
operation. In November, Manning Muntzing, AEC's Director of Regulation, speaks
personally with Browns Ferry officials about serious deficiencies in their Quality
Assurance program. The AEC regulatory position is that the company operating a
nuclear power plant should be self-regulating. The detailed implementation is also
left up to the companies. Quality Assurance programs are the companies' devices
for implementing safety guidelines and for checking up on implementation.
However, Browns Ferry is extended a grace period of several years to upgrade its
Quality Assurance program. Browns Ferry is allowed to operate during that
interval without Quality Assurance programs considered essential to nuclear
safety. Requiring Browns Ferry to meet new separation criteria for its electrical
system would involve extensive rewiring and construction of redundant systems.
Such efforts would entail substantial expenditures and delays in going into
operation.

1974.   In March, Charles E. "Doc" Murphy supervises pre-operational testing at
Browns Ferry. On August 1, Browns Ferry goes into full operation after Murphy
warns AEC headquarters about the electrical cable installation of the plant. The
warning is ignored. The AEC thus overlooks warnings since 1969 about dangers of
electrical cable fires arising from poor control of combustible materials, inadequate fire prevention programs, and poor separation of redundant circuitry.

1975.   On March 22, in the course of plant modification at Browns Ferry a
candle which is being used to detect air leaks ignites polyurethane foam. The foam
is employed to plug leaks where electrical cables pass through the wall between
the cable spreading room and the reactor building. The fire which erupts causes
extensive damage to electrical power and control systems. This damage interferes
with normal and standby cooling systems. The capability for monitoring the plant's
status is also impeded. It is a matter of random chance that unit 1 is brought under
control. A potentially catastrophic radiation release is avoided "by sheer luck."
Units 1 and 2 are put out of service for many months. Coincidentally, over the
course of the year Dale Bridenbaugh has discussions with colleagues and his boss
in which he talks about his concerns about safety in the nuclear power plant
program.

1976.   In February, Bridenbaugh, Hubbard, and Minor resign from their nuclear plant management positions at GE and Pollard resigns from his project management post at the Nuclear Regulatory Commission. They give as their reasons their concerns about known hazards of a serious nature which are left uncorrected.

B: DISCUSSION

(1)

The nuclear power industry is involved with a very complex product which is extremely expensive; indeed, its development has required billions of dollars of government expenditure as well as private resources. Solving the technical and safety problems which are gradually revealed requires additional huge expenditures. The very work of uncovering the problems is exceedingly costly in money and technical skills. Plant shutdown is also an extremely expensive proposition, involving hundreds of thousands of dollars per day. The momentum of heavy investment needed to initiate a nuclear power project suffices to carry it forward in a headlong way with continuing large expenditures to salvage the original investment. Furthermore, power plants, which take about seven years to build, must be large to be economical. Though larger plants with their greater fuel loads and higher power present more dangers, the huge plants and giant plant constellations, such as, at San Jose, California, employ large numbers of workers and bring in substantial tax revenues to the state and community. These advantages help to explain the reluctance of legislators and NRC officials to put any brakes on nuclear power development.

In addition, the risk of oil boycotts has tended to support nuclear power development. The pressures of international competition have played a similar role. There is intense effort to avoid large escalations of cost, including strategies for dealing with safety problems that avoid the costs of correcting them. Nevertheless, the costs are so high and the safety problems so persistent that there is increasingly, evidence of retrenchment and retreat from nuclear power by large private utilities and big reactor producers.

The economic considerations come to distort the thinking of those engineers and others—who make safety decisions in the industry and in the regulatory agency. The rush to retrieve investment and avoid more expense is so great that Bridenbaugh testified, "I am convinced that economic considerations cause us to have a cloudy view of the decisions that are made." Determinations about start-up or continuing operation are often made in a highly pressured atmosphere in which threatened economic losses from delay or shutdown loom large.

Our concern is to understand how the perceived economic pressures affect the thinking and decision-making of those involved and to see what measures might insure decision-making based on technical and safety considerations. Are there devices for making engineers' decisions more independent of company loyalty and agency coziness? Can we find feasible ways of introducing disinterested judgement and the opinions of a better informed public into the decision-making process?

(2)

An important feature of development in the nuclear industry lies in the possibility of succeeding generations of reactors taking advantage of advances in technology, increased knowledge of hazards, and development of safeguards. Construction and operation are and have been carried forward on the basis of theoretical projections rather than empirical testing and large-scale mock-ups. It therefore happens that when plants go into operation, and/or mock-ups are made, and testing is finally carried out, flaws and hazards are revealed in operating reactors. The problem arises of bringing already operating units into conformity with present standards. Since that task involves huge expenditures in testing and highly expensive back-fitting of older or already operating units, there is great resistance to efforts to produce general conformity with current standards. In general, the prospect of huge expenses making nuclear power less and less economical has slowed responses to known deficiencies and safety problems. This even applies to incorporating new data into units under construction, as at Browns Ferry. During the last four years of construction of that plant there were repeated stabs at showing the need for adequate electrical cable separation. In the end, TVA succeeded in deferring the inclusion of this safeguard and, thus, units 1 and 2 were vulnerable to the common-mode failure actually suffered in the fire of 1975. This "penny-wise, pound-foolish" attitude, which F. U. Bower commented on in the Browns Ferry case, may not be uncommon.

What remedies can we find for this situation? How can we insure that decisions about whether to back-fit are made on technical and safety grounds, minimizing the biases caused by economic and political pressures? Whose responsibility is it to see that empirical testing and production of mock-ups precedes construction? By now there is a considerable body of data indicating that serious problems show up in construction and operation which were not anticipated in the theoretical studies. Since the risks to health, safety, and property of the public are so great, decisions about whether to "grandfather" (exempt older units from current standards) should at least be made systematically and according to clear criteria.

What channels could be developed to enable engineers with access to problem situations to bring their information and concern to bear on the decision-making process? What about ordinary citizens living in ignorance of risk they might not choose? Bridenbaugh, Hubbard, Minor, and Pollard resigned partly out of the frustration of being unable to impress the problems they were familiar with upon the consciousness of others in the industry, the regulatory agency, and the public. In January 1975 Pollard attempted to learn—from his superiors and from the NRC Counsel—if any such channels were available to him. This effort was not successful. He later said, "I would still be working at NRC if I thought that the public in general was aware of all the problems." Bridenbaugh's testimony was similar: "I have one suggestion ... that would be if a way could be developed whereby people in the industry who do have specific concerns could express those without having to quit to do it, that would be a very valuable thing to do."

(3)

So far in our story we have identified the economic investment in nuclear power. There exists an economic community, so to speak of those with a hefty financial stake in nuclear development. There is another involved community (which has some over-lap with the first group); this second group is known as the "nuclear fraternity." In the fraternity are very dedicated people, many of whom have spent most of their careers in the development and operation of nuclear power. Included are physicists such as Nobel-laureate Hans Bethe, a well known proponent of nuclear power, and academics such as Dr. Norman Rasmussen of MIT who directed the recent and highly controversial reliability study of nuclear plants. Not so well known but very committed to the development of nuclear power are a few thousand physicists and engineers in positions in industry, government (especially the regulatory agency), universities, and technological institutes.

The fraternity originated in the military, in the atomic and hydrogen bomb projects. It was shrouded in military secrecy for a long time and has bred close bonds of support and mutual protectiveness. An "old-boy network" has grown in which men move smoothly back and forth between agencies such as AEC, now NRC, and the large corporations which dominate the private nuclear domain, such as GE, Westinghouse, and Bechtel. For example, Robert Hollingsworth, the former general manager of the AEC, became a top official at Bechtel. Likewise, W. Kenneth Davis, a vice president of Bechtel, was formerly head of AEC's Reactor Development Division.

Members of the fraternity command respect through personal prestige, connection with high-status institutions, like MIT and Cal Tech and through associations such as the American Physical Society and the American Nuclear Society. They may be assumed to be sincere in their support of commercial nuclear power development, and not necessarily motivated by personal or financial gain. Their investment derives, often enough, from career commitment and the fascination with the prospects and problems of harnessing atomic power.

However, we should bear in mind that their careers and reputations are bound up with the nuclear power program. There is a tendency in the fraternity, as in other professions, to "rally round" to the extent of covering up (whether wittingly or not) flaws, errors, and problems. Those on the outside are kept in the dark. Members are inclined to underrate the powers of comprehension and the critical judgment of those outside. They prefer to believe that ordinary citizens, when apprised of problems, will react hysterically.

As a result of all these factors, momentous decisions which affect all our lives are made within a relatively closed circle. Given this type of situation, it is not surprising that the efforts of knowledgeable and crusading outsiders are sometimes needed to correct the insulated judgment of those within the circle. (e.g. Ralph Nader vis-à-vis the auto industry). Without adequate outside checks or channels for dissenting judgments within the industry, the public is at the mercy of those with vested interests within the industry.

It seems that the consequences from cover-up, delayed imposition of standards, etc., are so potentially catastrophic that engineers must seriously think about their own personal responsibility for such harm. Charles E. "Doc" Murphy, the Federal official supervising pre-operational testing at Browns Ferry, authored the memo to AEC shortly before the plant went into operation warning of the electrical cable installation. Murphy had been discussing the problem with AEC officials since 1970. He has said that he did not expect a fast response but wanted to prod the AEC to develop adequate safety standards governing electrical cable installation. He received no response at all. Exactly a year after he wrote the memo, he was the first NRC official informed of the accident at Browns Ferry. His response was, "Oh my God!"

Our concern is with methods for encouraging individual engineers to view their professional responsibilities more independently, as autonomous agents. Would portable insurance and pension benefits protect engineers so that unusual moral courage would not be required to speak up or pay attention to warnings? Are there feasible schemes for the profession, through its national associations, to back up the engineer who reveals or responds to problems and thus to encourage responsible behavior? Can we find ways for staff employees to function more adequately as responsible, autonomous professionals?

(4)

There is a standard two-fold solution to the technical and safety problems of the nuclear industry and other industries—government regulation and Quality Assurance programs. This section will be addressed to Quality Assurance and a later section will explore government regulation.

NRC sets general regulations and standards for the company to follow but leaves detailed implementation up to the company. Checks on implementation are also carried out by the company; NRC inspectors check only about 1 or 2% of safety-related activities at a particular plant. Each plant is supposed to establish its own management system to assure conformity with applicable safety requirements. This is the "operating quality assurance" program, and it is supposed to yield the unprecedented meticulous care required for safe nuclear plant operation.

Yet plant management at Browns Ferry was so unreceptive to Quality Assurance that even after the fire in July of 1975, Norman Mosely, head of NRC's Region II (covering Alabama), said, "NRC, quite candidly, is trying to ram quality control down TVA's throat." Recall that the AEC had issued Browns Ferry a licence to operate allowing it to defer its Quality Assurance program in order to keep on schedule.

In theory, TVA safety reviews should have detected the fire hazards associated with the construction work going on. However, in violation of NRC requirements, TVA had no written procedures governing the work, no review of the work was carried out by the plant safety review committee, and there was no safety evaluation of the leak testing. In addition, no independent quality audits were carried out while work proceeded to determine if there was conformity with applicable requirements. As a result, management permitted an extensive un-supervised work program with unmonitored safety implications to go on in the electrical cable spreading room beneath the control room. In that room were the controls for the two operating units. This work project made use of an open flame and highly combustible polyurethane foam. There were numerous small fires before March 22, including two on March 20, one so large that dry chemicals were required to extinguish it. These fires were not properly reported, and no safety review of their significance was conducted. All these failures to write procedures, supervise, review, monitor, and report were failures of Quality Assurance.

How is it that professional engineers on the plant staff failed to insist upon a proper Quality Assurance Program? Were the economic imperatives to keep on schedule such that it didn't occur to them? Were potential dissenters worried about being and appearing to be team players? Can we find devices for reminding professional engineers of their responsibilities which may go beyond company interest? Can conscientious engineers produce a climate in which the over-loyal company engineer may feel pressured to reflect upon his/her actions?

(5)

As we have seen, there are serious obstacles to genuine independence on the part of the regulatory agency. Two factors primarily vitiate the independence of the agency. One is the promotional role which the regulatory agency has had from the outset. The splitting of the AEC early in 1975 into ERDA and the NRC ostensibly separated promotional from regulatory functions. However, this division has not succeeded in insulating the regulatory function adequately from the pressures of cost and schedule, according to the testimony of former NRC manager Pollard.

The other factor is the interchange of personnel between the industry and the regulatory agency. Agency officials who anticipate more lucrative jobs in the industry may not be prepared to make the technically based, independent safety decisions required by law when these decisions are unwelcome (i.e. are costly, cause delay) to the industry. This problem is a general one across many government regulatory agencies. One solution might be to establish high enough professional standards for regulatory agency personnel so that the prestige of connection with the agency offsets the lures from industry.

Obviously, industry resists regulation, and the factors noted support that resistance. Nevertheless, there are examples of effective regulation.

The ASME codes were initially developed in 1911 to protect the public from boiler explosions in public facilities, such as office buildings. These codes command universal respect and are more strict than comparable NRC codes. For example, there is a disciplined program of third-party inspection required by the ASME codes absent from the NRC Regulations for non-code safety-related items.

Another example of effective regulation comes from Underwriters Laboratory, Inc. (UL) founded in 1894. Many household electrical appliances receive the third-party review required for listing by UL. NRC, by contrast, does not require independent third-party evaluation and product proof testing of the Class I safety-related electrical equipment which controls and protects a nuclear power plant. Electrical appliances such as a toaster or hair dryer receive more stringent safety checks than the electrical equipment which controls a nuclear power plant.

It would seem that a more educated public and professionals more conscious of their responsibility not to cause harm would agree in insisting on more stringent codes and more careful oversight of the industry to insure conformity. If the result would be serious economic problems for the industry, these problems must be publically confronted.

(6)

This brings us to our final concern: exclusiveness, restrictiveness, and secrecy in the industry vs. the need for an informed public making responsible choices about life and death matters. With some exceptions, the NRC does not require plant owners to report field failures. There is an informal arrangement for such reporting, but this set-up permits excessive filtering and omission of data. The whole question of "trade secrets" in an industry with such potential for catastrophic accidents deserves investigation.

Here's an illustration of the problem. In the autumn of 1974, GE undertook a Nuclear Reactor Study of Boiling Water Reactors. The director of the study, Dr. Charles E. Reed, a vice-president of GE and former MIT faculty member, admitted before the Congressional Joint Committee on Atomic Energy that the study dealt with over-all design considerations, plant components, test facilities, and management and organization. However, he said in the report, "although in the course of the Study Group's review nuclear safety aspects were considered, this study was not a safety review." On the grounds that the report was a sensitive document "from a competitive standpoint," GE did not make the report available to the NRC. It merely conducted an in-house review which concluded that there were no reportable deficiencies not previously reported to NRC. Only after Bridenbaugh, Hubbard, and Minor, who had participated in the study, revealed its existence and its safety significance before the Joint Congressional Committee, was the report made available to the NRC. The study was not to be made public at all, however, and it was to be available to the Congressional Committee only via an NRC report after a review of the study by that agency. Reed repeatedly defended this secretiveness, saying that there was no new safety-related information in the report.

We have already encountered the relevant assumption operative in the industry: the public cannot comprehend the issues in nuclear plant safety, and if ordinary people were informed about the risks and costs, they would hysterically reject nuclear power altogether. The advantages in the alternative of public debate and informed public support of perhaps a modified schedule in nuclear power development are thus lost. Instead, the nuclear power community proceeds feeling embattled and estranged from the public and constantly on guard against the leakage of any negative data. They are deprived of the common sense, diversity of outlook, and cool judgement which might come from public discussion.

If nuclear power is an "unfinished engineering dream," a most promising way to a happy completion is to enlarge the perspectives and sharpen the moral and professional consciousness of engineers in training and to raise the level of literacy of ordinary citizens about these momentous projects.

6.9

Whistleblowing &
Professional Responsibilities

SISSELA BOK

"Whistleblowing" is a new word in the glossary of labels generated by our increased awareness of the ethical conflicts encountered at work. Whistleblowers sound an alarm from within the very organizations in which they work, aiming to spotlight neglect or abuses that threaten the public interest.

The stakes in whistleblowing are high. Take the nurse who alleges that physicians enrich themselves in her hospital through unnecessary surgery; the engineers who disclose safety defects in the braking systems of a fleet of new rapid-transit vehicles; the Defense Department official who alerts Congress to military graft and overspending. All know that they pose a threat to those whom they denounce and that their own careers may be risked.

Moral conflicts on several levels confront anyone who is wondering whether to speak out about abuses or risks or serious neglect. In the first place, he/she must try to decide whether, other things being equal, speaking out is in fact in the public interest. This choice is often made more complicated by factual uncertainties: Who is responsible for the abuse or the neglect? How great is this threat? And how likely is it that speaking out will precipitate changes for the better?

In the second place, a would-be whistleblower must weigh the responsibility to serve the public interest against the responsibility owed to colleagues and the employer institution. This conflict between responsibilities is reflected in contradictory messages within many professions. The professional ethic requires collegial loyalty, while the codes of ethics often stress responsibility to the public over and above duties to colleagues and clients.

PROFESSIONAL LOYALTY VS. RESPONSIBILITY

The question for professions, then, is how to resolve, insofar as it is possible, the conflict between professional loyalty and professional responsibility toward the outside world. The same dichotomy arises to some extent in all groups, but professional groups often have special cohesion and claim special dignity and privileges. The strain between the ideals of public service and collegiality in the professions can therefore be especially strong. They add to the pressure on would-be whistleblowers.

The plight of whistleblowers has come to be documented by the press and described in a number of books. Evidence of the hardships imposed on those who chose to act in the public interest has combined with a heightened awareness of professional malfeasance and corruption to produce a shift toward greater public support of whistleblowers. Public service law firms and consumer groups have taken up their cause; institutional reforms and legislation have been proposed to combat illegitimate reprisals. Some would encourage ever greater numbers of employees to ferret out and publicize improprieties in the agencies and organizations where they work.

Given the indispensable services performed by so many whistleblowers—as during the Watergate period and after—strong public support is often merited. But the new climate of acceptance makes it easy to overlook the dangers of whistleblowing: instances of error or malice; work and reputations unjustly lost for those falsely accused; privacy invaded and trust undermined. There comes a level of internal prying and mutual suspicion at which no institution can function. It is a fact, as well, that the disappointed, the incompetent, the malicious, and the paranoid all too often leap to accusations in public. Worst of all, ideological persecution throughout the world traditionally relies on insiders willing to inform on their colleagues or even on their family members, often through staged public denunciations or press campaigns.

NO IMMUNITY

No society can count itself immune from such dangers, but neither can it risk silencing those with a legitimate reason to blow the whistle. How then can we distinguish among different instances of whistleblowing? A society that fails to protect the right to speak out, even on the part of those whose warnings turn out to be spurious, obviously opens the door to political repression. But from the moral point of view there are important differences among the aims, messages, and methods of dissenters from within.

The public debate over whistleblowing is already under way. In the press, articles, and books, these problems have been described, and a number of remedies proposed. Institutional and legislative proposals are being made. Still lacking is work of a fact-finding, comparative, and analytical nature on which practical decisions might be based; as well as the opportunity for individuals to give careful thought in advance to how they might respond, in their own working lives, to the conflicts where whistleblowing is one of the alternatives. To what extent can these needs be met through teaching?

TEACHING ETHICS

Case studies of the dilemmas individuals face with respect to whistleblowing are interesting from the point of view of teaching applied ethics. They are concrete, striking, and reminiscent of experiences all have had since childhood with conflicts of loyalty and the burden of deciding whether or not to "tell on" a friend. A careful study of the moral arguments for and against whistleblowing in individual cases, and of the possible alternatives, will allow a better understanding of what is at issue in such situations and how they arise; it can allow a more considered choice of what stance to take if similar situations were to arise at work, as well as an inquiry into what might be done at an earlier stage to achieve the desired result without the costs of whistleblowing.

Such teaching, moreover, may contribute to a broader analysis of practices of dissent and of conflicting responsibilities. It can allow both a problem-solving approach to immediate conflicts and a critique of long-standing practices with their underlying assumptions.

These broader purposes are best served by going beyond the dilemmas presented for individual agents, and looking, as well, at whistleblowing from the perspective of organizations and of professions. Such an inquiry could be undertaken either in a particular institutional or professional perspective—say in a business school or a school of engineering—or in a still more general perspective that cuts across all forms of work.

What questions might such a broader inquiry raise? And how might they be approached in teaching and research? What institutional arrangements might best cope with the uncovering of wrongdoing and neglect? And how might professions best cope with the need to reconcile loyalty to the profession and responsibility toward the public interest?

What changes, outside and inside business organizations, government agencies, and other places of work, might serve to protect the right of dissenters, cut down on endless breaches of loyalty and on false accusations, while assuring public access to needed information?

ROLE OF ETHICS CODES

Requirements to disclose deserve careful study in courses on professional ethics, as do the corresponding parts of the various codes of ethics. What role do such codes play with respect to the conflicts engendered by dissent? How have they changed in this regard? Ought they to be further changed, made more specific, or provide for exceptional cases? The weighing of responsibilities—to colleagues, to the profession itself, to clients, to innocent bystanders, and to the public—presents unusually difficult problems for most professions; these problems are still far from being resolved.

Teaching can also explore the different conceptions of such responsibilities in the professions. Some place the highest value on service; others on profit; still others on accountability to the public. How do these differences affect the view of the legitimacy of whistleblowing? Do some positions limit this legitimacy?

The questions of obedience and dissent are receiving increased attention in classes on ethics in the military academies. It is clearly not enough to insist on obedience at all times in bureaucracies and the military. Obedience cannot be the right response to requests to cover up for the use of deficient machinery, or for a planned usurpation of power, or for the bombing of a neutral country. More careful distinctions will have to be drawn, and, once again, the codes and documents setting forth professional standards may have to be revised in the light of these distinctions.

COURSES ON PROFESSIONAL ETHICS

Comparing different fields of work regarding whistleblowing forms part of a larger set of issues that ought not to be ignored in courses on professional ethics: Must professionals sometimes act in such a way as to breach their personal ethical standards? Might professional and personal ethics actually conflict?

These questions merit serious debate not only in courses on professional ethics but also in pre-professional ones. Students should have an opportunity to think in advance of what will be expected of them and how it might conflict with what they expect of themselves. Such consideration will in turn without a doubt affect the professional practices themselves—if in no other way than to draw distinctions in what is now all too often an impossible tangle of diverging views. These distinctions will be clarified, too, if they are looked at from a perspective that includes a number of kinds of work; there is danger, otherwise, that the sense of mission and high purpose within any one calling might serve to undercut criticism and to reinforce the status quo.

Perspective will be gained, too, if the class discussions of whistleblowing are placed in the context of the larger political and moral issues—dissent and free speech, fairness in accusation and refuge in anonymity, institutional changes and professional responsibilities.

Such issues may be raised at all levels of training, be they general, pre-professional, or professional. But at every level, it is important that instructors who take up questions of whistleblowing not do so in ignorance of the contexts in which such conflicts spring up. It would be helpful, too, especially in courses preparing for a particular professional experience, if practitioners could join the students at times to convey a sense of the practical difficulties that arise, as well as of the problems that seem excruciating on paper but that can, in fact, easily be averted.

Whistleblowing is a new word for an ancient practice. It is becoming more prominent now, as we learn to spot it in many circumstances and as organizations grow in size and number. It needs careful study and offers the opportunity for a close look at individual choices as well as at possible institutional changes and professional standards. The teaching of ethics can benefit from and contribute to such inquiries and can help seek ways to protect dissent and encourage criticism while cutting down on erroneous or harmful resorts to the panic button.

Selected Additional Readings

Anderson, R. M., Perrucci, R., Schendel, D. E., and Trachtman, L. E., *Divided Loyalties*, Purdue Research Foundation, West Lafayette, Indiana, 1980.

Ashkinazy, A., "Are Engineers Responsible for the Uses and Effects of Technology," *Professional Engineer*, August 1972, pp. 46–47.

Bok, S., "Whistleblowing and Professional Responsibility," *New York University Education Quarterly*, Vol. 11, No. 4, 1980, pp. 2–10.

Carey, W. N., "Competitive Bidding for Professional Services Not in the Public Interest," *Civil Engineering*, May 1954, pp. 50–52.

Chalk, R., and von Hippel, F., "Due Process for Dissenting 'Whistle-Blowers,'" *Technology Review*, June/July 1979, pp. 49–55.

Heilbroner, R. L., et al., *In the Name of Profit*, Doubleday & Company, Inc., New York, 1972.

Peters, C., and Branch, T., *Blowing the Whistle: Dissent in the Public Interest*, Praeger Publishers, New York, 1972.

Raven-Hansen, P., "Dos and Don'ts for Whistleblowers: Planning for Trouble," *Technology Review*, May 1980, pp. 34–44.

Unger, S., "Ensuring the Right of Professional Dissent: A Review of a Proposed New NRC Policy," *Technology and Society*, Committee on Social Implications of Technology, IEEE, March 1980, pp. 7–8.

Westin, A. F., *Whistle-Blowing: Loyalty and Dissent in the Corporation*, McGraw-Hill Book Company, New York, 1981.

Codes of Ethics
and Enforcement

The careful textbook measures
(Let all who build beware!)
The load, the shock, the pressure
Materials can bear
So when the buckled girder
Lets down the grinding span
The blames of loss, or murder,
Is laid upon the man.
Not on the Stuff—the Man!

Rudyard Kipling
Hymn of Breaking Strain

A man should be upright; not be kept upright.

Marcus Aurelius
Meditations II

There are many ways of looking at professional codes of ethics. Some of these express more humor and cynicism than understanding. There is the view that a profession has a code of ethics for the same reason phone company employees wear identification badges—it's how one recognizes the real thing. Another view sees codes as the equivalent of a Mercedes and a house in the country—they show that one has arrived. A third view, generally not publicly held by members of professions, is that codes are like sheepskins—primarily of use to wolves.

These three views are *humorous* precisely because they are *distortions*—because we recognize them as either relatively harmless (in the case of the first and second view) or pernicious (in the third) distortions of a more fundamental function and use. Codes of ethics exist because there is a real need for a statement of the intentions and objectives of the group advocating the code. Equally important is the need for the less knowledgeable individual to feel the security of trust and confidence in an organized profession committed to the concepts expressed by a code of ethics.

Adherence to a code of ethics is traditionally taken to be a fundamental aspect of a profession. It is traditionally used as a distinguishing characteristic for all groups that claim professional status. A professional code of ethics extends the moral rules, moral ideals, and obligations that apply to all rational beings to include those special concepts that are accepted by professionals because they are professionals. A common view is that a code of ethics is rather like a contract between the profession and society. In return for certain privileges and an elite status, a profession agrees to place the public interest first. In such an understanding, the code itself becomes a clause in the contract. This was the position put forward by IEEE in the BART case.

The first article in this chapter is an excerpt from a widely known classic statement of the reason for a code of ethics in engineering. Daniel Mead in "Why a Code of Conduct?" argues that an explicit statement of principles of conduct and their application is a necessary part of the attainment and maintenance of character. The professional codes of ethics, he says, should be taken as guidelines for each person's formulation of his own personal code.

The texts of four current codes of ethics (Engineers' Council for Professional Development, National Society of Professional Engineers, Institute of Electrical and Electronics Engineers, and the American Public Works Association) are included in this chapter. Most of the larger engineering societies have adopted the ECPD code though some have made minor changes in the wording of the "Suggested Guidelines" (ASCE, for example). These four examples represent the codes of ethics in use by a broad spectrum of engineering and engineering related societies.

Understanding codes of ethics as, in part, a clause in a contract and, in part, the outward sign of a certain kind of character gives us a basis for the examination and evaluation of particular codes. We can ask whether the provisions are consistent with each other, for an inconsistent contract cannot be adhered to. We can ask whether the code indicates principles by which conflicting provisions can be harmonized. We can ask what kind of personal character is implied by the provisions of the code. We can turn this around and ask whether a given code adequately expresses and delineates the desired character. We can take this further still and ask on what basis, for what reasons, is the character embodied in a given code desirable—pragmatic, ethical, legal, etc.

Another issue that the engineering codes of ethics raise is their diversity and number. On the one hand, different kinds of engineers do different sorts of things, so one might expect that there would be different codes with different provisions depending on the nature of the work. However, insofar as the codes are expressions of a certain kind of personal character, one might expect that all professional codes would be essentially the same. In "Proposed: A Single Code of Ethics for All Engineers" Oldenquist and Slowter present arguments for the adoption of a single code. They have analyzed several existing codes to define a group of core concepts. They conclude with a proposal that a convention be called to formulate a single code embodying these concepts and consistent with ethical theory.

In "When in Rome Do as the Romans," Colin Lauchlan points out that the expectation of behavior for professionals is more than simply adherence to the terms of a contract. A code of ethics defines a standard of behavior that is above the mere terms of an agreement or contract and transcends national boundaries, codes of law, and the varied customs and religious beliefs of our world. Ethics is the concern for standards of interpersonal behavior and must be applicable to all people, in all places, and at all times. Professional codes of ethics expand upon these concerns of all rational human beings, extending them to those people who profess to be professionals. A code of ethics is intended not only to define certain modes of behavior in contractual terms but also to be a code of conduct that expresses a certain kind of character—a character that engenders trust and confidence.

The last two articles dealing with codes in this chapter address the issue of personal character as the underlying meaning of codes of ethics. McMinn suggests in "Ethics Spun from Fairy Tales" that ethics and codes of ethics are an extension of the human propensity, as seen also in fairy tales, to articulate guides and models for behavior. A number of short cases are included in McMinn's article to illustrate some of the engineering ethics dilemmas and how they would be resolved through reliance upon codes of ethics. Romano uses a whimsical tale in "The Fable of the Consulting Engineer" to illustrate the many functions a code of ethics can and will perform throughout the life of an engineer as his role in engineering practice changes.

The enforcement of codes of ethics requires self-regulation by the profession with the full cooperation of its members. The alternative is control by others—most likely by government in our society. Individuals are reluctant to "blow the whistle" on fellow practitioners that violate accepted standards of conduct and to penalize those found to have violated these standards. This reluctance is aggravated by the internal conflict that exists in most codes of ethics through the inclusion of indefinite provisions, the attempt to protect the interests of both the individual and society simultaneously, and, unfortunately, ignorance of the codes by many practitioners. The net result has been an appreciable hue and cry by the public that many professions are simply not doing the job they should with enforcement. Numerous examples of this concern involving the recognized professions are publicized each year. Public concern has been expressed by a reduction of self-regulation through the inclusion of lay members on licensing and regulatory boards, legal questions of the propriety of certain code provisions, and a general loss of confidence in the professions.

The process of professional society enforcement of a code of ethics starts with the individual members who accepted freely an obligation to follow the code when they entered the professional organization. Part of that obligation is to act in accord with the code as an individual *and* to call any violation to the attention of the society. Experience has indicated the great majority of code enforcement cases do arise in this fashion, though in recent years more and more have resulted from public sources. Each professional society has its own procedure for code enforcement. All are established to protect the rights of the individual by following a due process of law procedure for investigation, indictment, hearing, and judgment.

It is not always recognized that professional society enforcement actions do not carry the weight of the law. Penalties for code violations are limited to admonishment or reprimand, suspension of membership for a fixed period of time, or expulsion from the society. To the nonprofessional these penalties may appear inconsequential, but they are serious indeed to those individuals who desire the respect and confidence of their peers. Indeed, in some cases, the penalties for ethical code violations may seriously impair the development of a professional career. This is particularly true for engineers working directly or indirectly for public agencies. Thomas Jefferson put it this way, "When a man assumes a public trust, he should consider himself public property." (1) Does the public wish damaged or questionable property?

There is one aspect of professional conduct enforcement that does have legal significance. Most state licensing groups have either rules of conduct or grounds for disciplinary action that have been developed under the legal act that established the licensing procedure. These rules are very similar, in most cases, to professional society codes of ethics and are legally enforceable. The reporting, investigation, hearing, and penalty procedures of licensing boards are similar to those of the professional societies, but the penalties are more significant. The potential effect of loss of license to practice for a specific period or permanent revocation of a license can severely limit the opportunity for employment or ability to practice as a professional.

Enforcement is attracting more and more attention from the various engineering societies. All have increased their involvement in this important and delicate function. ASCE, for example, reports that it considers approximately thirty cases per year, and ASME has recently reorganized its procedures for handling code violation questions. IEEE has assumed a leadership role in positive action to support individuals that have acted in accord with code provisions. Several societies have established an advisory service for members with code of ethics related questions, whether they are concerned with a personal ethical dilemma or recommended procedures to follow when violations are noted. All organizations following this approach report it is being used with apparently very satisfactory results.

In the final analysis, a code of ethics is not a set of laws enforceable in the courts. Rather they are a guide to the individual in the daily practice of a profession and this guide is based on a voluntary "adherence to principles which transcend the practitioners' own immediate interest." (2) The enforcement of the code is an attempt, though perhaps an imperfect one, to recognize the need or even the necessity for the individual to act responsibly beyond the expedient action self-interest may dictate. (See Phoebe Cary's poem "They Didn't Think"—the closing item in this chapter.)

The selections included in this chapter are intended to define the argument and procedures for enforcement of professional codes. McMinn describes the procedure used by one society (ASCE) when presented with an apparent violation of its code of ethics. Pletta presents the argument for self-regulation and indicates the NSPE approach to enforcement. He suggests several possible actions, including an expanded role for education in ethics, to reduce enforcement problems in both detail and extent. Bagley outlines the revised ASME procedures for dealing with unethical engineers in his article, "Ethics, Unethical Engineers, and ASME."

There is a reverse side to the issue of enforcement that has been largely neglected. If the societies as a matter of course investigate alleged violations and take action to punish confirmed violations of their codes of ethics, should they not investigate situations where it is alleged that a member has suffered because of actions upholding the codes of ethics and take actions to support and defend members to whom this has occurred? The selections in this chapter "Professional Responsibility and the Dispatch of Police Cars—A Case Study" and "MCC Report in the Matter of Virginia Edgerton" illustrate the way one professional society is extending their ethics activities to protect their members as well as to censure the transgressors. This case study describes how IEEE responded to the request for help from a member who was fired for acting according to the specific charge to protect the health, safety, and welfare of the public contained in the IEEE code of ethics.

The enforcement question is not an easy one when the problem is considered more than superficially. Legal, moral, and societal questions become entangled in the analysis. It is quite likely that current practices have major drawbacks. Enforcement activities, however, do help create an awareness of ethical principles among the members of professional societies.

References

1. Thomas Jefferson, Remarks to Baron von Humboldt as quoted in *Bartlett's Familiar Quotations*, Fourteenth Edition, Little, Brown and Company, Boston, 1968, p. 472.
2. R. Walton, "Some Neglected Aspects of Information Processing Ethics," *Business & Professional Ethics*, Winter, 1974, p. 4.

7.1

Why a Code of Conduct?

DANIEL W. MEAD, ASCE

The question is sometimes asked as to why a code of conduct or ethics is desirable or necessary for engineers or other professional men. Is it not sufficient to say that "Every professional man should at all times think and act in accord with the highest principles of personal and professional honor"? or is not the Golden Rule "Do unto others as you would have others do unto you" sufficient for such a code? Either would be sufficient if the individual were always capable of applying either rule under the manifold conditions under which such rules should be applied.

This is the weakness of all brief rules of conduct. In themselves they contain no information as to how they should be applied in detail to professional conduct which, in the writer's judgment, is a serious mistake if they are offered to students or young engineers to guide them in their professional work. The writer believes that in the teaching of the application of the best principles of conduct the rule must be as specific as possible so as to apply as nearly as practicable to specific conditions.

Laws are but rules of conduct established by legislative action and enforced by the courts. These laws are supposed to represent the views of the majority of the people in the community in which they are established as to what individual action should or should not be allowed. There are thousands of these laws on the federal and state statute books; and there is scarcely one of them that under certain specific conditions may not be unfair and unjust to some individuals. Such laws frequently become archaic and are repealed or, if negligently left on the statute books, their absurdity becomes apparent and embarrassing, and public opinion demands and secures their repeal. The fact that the world is changing does not destroy the necessity for law, but it does necessitate its readjustment from time to time as conditions develop.

It is a fact with which every one is familiar that an individual may strictly observe the laws of the land and yet be an undesirable citizen and a poor neighbor. The idea that each individual can and should establish for himself rules of conduct for such relations as are not covered by law and without reference to the experience or opinions of others seems equally as absurd as would a similar attempt to establish principles of law. Laws must be established by the majority action of a legislative body, and rules of professional conduct must be based on the concurrent opinions of the members of a profession.

It is true of established rules of conduct, and also of rules of law, that few ethical or legal principles are universally applicable, and that in certain cases each individual must depend upon his common sense and conscience as to what his conduct should be under the limitations of the conditions under which his conduct must be exercised.

Common sense and conscience are the results of early training, of the personal influence of those with whom one comes in contact, of education, of experience, and of such reflection as the individual may give to these factors. On this account common sense and conscience are limited in their application to the breadth of the experience on which they are based, and can be applied successfully only within such limits. When that experience is extended, common sense and conscience are sometimes of benefit by analogy but are very often apt to be mistaken because of the limitation of experience.

A young man leaving college is plunged frequently into a life so different from his previous experience that he needs the advice of the men who have already had the experience that is about to become his. Rules and principles are then of value in reflecting the conclusions of those who have gone before. Just as precedent in the design and construction of engineering structures is of value when similar structures are being designed and constructed, so principles of conduct established by the experience of one's predecessors are of value in the consideration of one's own line of action. The writer will agree that precedents are not to be blindly followed; but to ignore the experience of others, either in engineering design or in conduct, is a dangerous and often a serious mistake. Regarding this point, Raymond Moley observes:[*]

> What are principles that men live by? What, for that matter, is the meaning of principle itself? Reduce the question of principle to the case of an individual and one of the problems he faces. A man does not govern his life by chance. He learns, as the years pass, and profits by what he learns. He learns that there are some things that he cannot eat without distress. He learns that there are some games he cannot play. He learns that there are ways of doing his work better. He learns how to conduct his relations with other people. Out of the accumulation of individual experience he creates rules for himself. As time goes on, those rules become principles of living. He finds that by observing and respecting them he saves himself untold trouble and discomfort. He doesn't have to argue out thousands of individual decisions with himself. He depends upon his principles. Ultimately he lives not only with but by them.

It is unlikely that a man of advanced age and long experience would, in his ordinary relations in life, find it necessary to study a code of conduct in order to determine what his own conduct should be; neither would he ponder very deeply concerning the various possible outcomes of his action. His answer would be given at once, based on his established principles, and would probably be correct. If,

---

[*]"Indispensable Principles," by Raymond Moley. An address delivered on Constitution Day—September 17, 1940—to the Union League Club of Chicago.

however, the conduct concerned new conditions entirely beyond and different from his previous experience, then comes the necessity for due consideration; and in such cases the opinion of those who have had similar experiences and have reached definite conclusions cannot safely be ignored.

A person does not seek legal advice concerning his ordinary conduct in everyday life, but when he enters into new legal relations with which he is not familiar it is the part of wisdom to seek advice.

The Golden Rule is, and should be, in general, the basis of almost all ethical conduct. The Golden Rule is as sound today as it was in the days of Christ. It may puzzle the philosopher who desires to live by the Golden Rule as to what he should do if he captures a thief who is attempting to plunder his home. However, the practical man who also desires to live by this rule would recognize his obligations to society and would at once hand the malefactor over to the police.

So far as the writer's information goes, a code of ethics was first established for physicians about 400 B.C. by Hippocrates, a famous Greek physician often called "The Father of Medicine"; and the writer understands that Hippocrates' "oath of service" is still used in some medical schools. This would seem to indicate that a "changing world" has not seriously affected the code of Hippocrates in more than twenty-three hundred years. In a similar manner, although certain purely technical requirements of a code of conduct may change with changing conditions, the fundamental principles of good conduct and good ethics are unchangeable and eternal.

Legal ethics is of many years standing. Today thousands of codes have been adopted by professional societies, technical and business organizations and associations, and numerous books have been published discussing the basis of business and professional relations contained, or those which should be contained, in these codes. Is the assumption warranted that all this effort is wasteful and unnecessary and that each individual in all these diverse interests is qualified to evolve from his own reflections all of the principles of conduct which he should exercise under each of the thousands of circumstances in which he may happen to be or into which he may possibly enter from an entirely different environment?

REQUIREMENTS OF THE PROFESSION

The profession of engineering calls for men with honor, integrity, technical ability, businesss capacity, and pleasing personalities. The engineer is entrusted with investigating and reporting on the advisability and feasibility of public, semi-public, and commercial works. On the basis of his reports, important works are inaugurated in which large expenditures are involved, and securities sold and purchased. Often the limited means of widows, orphans, and other dependent persons are invested on the basis of his knowledge and good faith, and the results of financial failure of such projects are very serious. The engineer is also entrusted with designing and constructing important public and commercial undertakings in which his honesty of purpose and sincerity of statement must be above suspicion. He acts as a professional adviser, and his advice must be honest and disinterested; he must exercise judicial functions between his clients and contractors, manufacturers, and material supply companies, and his decisions must be fair and impartial.

He has moral responsibilities to the public and to his superiors, to his associates and subordinates, which should be exercised with the highest ideals. His work carries with it grave responsibilities to the public which demand his conscientious consideration; and he cannot properly perform his numerous functions unless his ability, conduct, and motives are such as to command the highest respect and confidence.

PRINCIPLES APPLICABLE TO ALL PROFESSIONAL POSITIONS

It may appear to the younger members of the profession that such requirements apply only to the heads of the profession, but it must be remembered that the profession is not static. The business and engineering world is full of examples of men who have started in the humblest positions and who by means of the proper development of their knowledge, intelligence, character, personality, and ability have become the heads of industry and of the professions in the United States. Opportunities are open to every man who has the personality, initiative, ability, and determination to strive for better things; and the hope of advancement should spur every young man to fit himself for the higher positions which soon will become available for new men, as the retirement of those of advanced age, or as changed conditions, make necessary any advancement in the personnel of each industry. The younger man should recognize that his future is involved and he should prepare himself for those higher responsibilities.

7.2

ECPD
Code of Ethics of Engineers*

THE FUNDAMENTAL PRINCIPLES

Engineers uphold and advance the integrity, honor and dignity of the engineering profession by:

  I. Using their knowledge and skill for the enhancement of human welfare;

  II. Being honest and impartial, and serving with fidelity the public, their employers and clients;

  III. Striving to increase the competence and prestige of the engineering profession; and

  IV. Supporting the professional and technical societies of their disciplines.

THE FUNDAMENTAL CANONS

1. Engineers shall hold paramount the safety, health and welfare of the public in the performance of their professional duties.

2. Engineers shall perform services only in the areas of their competence.

3. Engineers shall issue public statements only in an objective and truthful manner.

4. Engineers shall act in professional matters for each employer or client as faithful agents or trustees, and shall avoid conflicts of interest.

5. Engineers shall build their professional reputation on the merit of their services and shall not compete unfairly with others.

6. Engineers shall act in such a manner as to uphold and enhance the honor, integrity and dignity of the profession.

---

*Approved by the Board of Directors, October 5, 1977.

7. Engineers shall continue their professional development throughout their careers and shall provide opportunities for the professional development of those engineers under their supervision.

SUGGESTED GUIDELINES FOR USE WITH
THE FUNDAMENTAL CANONS OF ETHICS

1. Engineers shall hold paramount the safety, health and welfare of the public in the performance of their professional duties.

   a. Engineers shall recognize that the lives, safety, health and welfare of the general public are dependent upon engineering judgments, decisions and practices incorporated into structures, machines, products, processes and devices.

   b. Engineers shall not approve nor seal plans and/or specifications that are not of a design safe to the public health and welfare and in conformity with accepted engineering standards.

   c. Should the Engineers' professional judgment be overruled under circumstances where the safety, health, and welfare of the public are endangered, the Engineers shall inform their clients or employers of the possible consequences and notify other proper authority of the situation, as may be appropriate.

      (c.1) Engineers shall do whatever possible to provide published standards, test codes and quality control procedures that will enable the public to understand the degree of safety or life expectancy associated with the use of the design, products and systems for which they are responsible.

      (c.2) Engineers will conduct reviews of the safety and reliability of the design, products or systems for which they are responsible before giving their approval to the plans for the design.

      (c.3) Should Engineers observe conditions which they believe will endanger public safety or health, they shall inform the proper authority of the situation.

   d. Should Engineers have knowledge or reason to believe that another person or firm may be in violation of any of the provisions of these Guidelines, they shall present such information to the proper authority in writing and shall cooperate with the proper authority in furnishing such further information or assistance as may be required.

      (d.1) They shall advise proper authority if an adequate review of the safety and reliability of the products or systems has not been made or when the design imposes hazards to the public through its use.

      (d.2) They shall withhold approval of products or systems when changes or modifications are made which would affect adversely its performance insofar as safety and reliability are concerned.

    e. Engineers should seek opportunities to be of constructive service in civic affairs and work for the advancement of the safety, health and well-being of their communities.

    f. Engineers should be committed to improving the environment to enhance the quality of life.

2. Engineers shall perform services only in areas of their competence.

    a. Engineers shall undertake to perform engineering assignments only when qualified by education or experience in the specific technical field of engineering involved.

    b. Engineers may accept an assignment requiring education or experience outside of their own fields of competence, but only to the extent that their services are restricted to those phases of the project in which they are qualified. All other phases of such project shall be performed by qualified associates, consultants, or employees.

    c. Engineers shall not affix their signatures and/or seals to any engineering plan or document dealing with subject matter in which they lack competence by virtue of education or experience, nor to any such plan or document not prepared under their direct supervisory control.

3. Engineers shall issue public statements only in an objective and truthful manner.

    a. Engineers shall endeavor to extend public knowledge, and to prevent misunderstandings of the achievements of engineering.

    b. Engineers shall be completely objective and truthful in all professional reports, statements, or testimony. They shall include all relevant and pertinent information in such reports, statements, or testimony.

    c. Engineers, when serving as expert or technical witnesses before any court, commission, or other tribunal, shall express an engineering opinion only when it is founded upon adequate knowledge of the facts in issue, upon a background of technical competence in the subject matter, and upon honest conviction of the accuracy and propriety of their testimony.

    d. Engineers shall issue no statements, criticisms, nor arguments on engineering matters which are inspired or paid for by an interested party, or parties, unless they have prefaced their comments by explicitly identifying themselves, by disclosing the identities of the party or parties on whose behalf they are speaking, and by revealing the existence of any pecuniary interest they may have in the instant matters.

    e. Engineers shall be dignified and modest in explaining their work and merit, and will avoid any act tending to promote their own interests at the expense of the integrity, honor and dignity of the profession.

4. Engineers shall act in professional matters for each employer or client as faithful agents or trustees, and shall avoid conflicts of interest.

    a. Engineers shall avoid all known conflicts of interest with their employers or clients and shall promptly inform their employers or clients of any business association, interests, or circumstances which could influence their judgment or the quality of their services.

b. Engineers shall not knowingly undertake any assignments which would knowingly create a potential conflict of interest between themselves and their clients or their employers.

c. Engineers shall not accept compensation, financial or otherwise, from more than one party for services on the same project, nor for services pertaining to the same project, unless the circumstances are fully disclosed to, and agreed to, by all interested parties.

d. Engineers shall not solicit nor accept financial or other valuable considerations, including free engineering designs, from material or equipment suppliers for specifying their products.

e. Engineers shall not solicit nor accept gratuities, directly or indirectly, from contractors, their agents, or other parties dealing with their clients or employers in connection with work for which they are responsible.

f. When in public service as members, advisors, or employees of a governmental body or department, Engineers shall not participate in considerations or actions with respect to services provided by them or their organization in private or product engineering practice.

g. Engineers shall not solicit nor accept an engineering contract from a governmental body on which a principal, officer or employee of their organization serves as a member.

h. When, as a result of their studies, Engineers believe a project will not be successful, they shall so advise their employer or client.

i. Engineers shall treat information coming to them in the course of their assignments as confidential, and shall not use such information as a means of making personal profit if such action is adverse to the interests of their clients, their employers, or the public.

    (i.1)   They will not disclose confidential information concerning the business affairs or technical processes of any present or former employer or client or bidder under evaluation, without his consent.

    (i.2)   They shall not reveal confidential information nor findings of any commission or board of which they are members.

    (i.3)   When they use designs supplied to them by clients, these designs shall not be duplicated by the Engineers for others without express permission.

    (i.4)   While in the employ of others, Engineers will not enter promotional efforts or negotiations for work or make arrangements for other employment as principals or to practice in connection with specific projects for which they have gained particular and specialized knowledge without the consent of all interested parties.

j. The Engineer shall act with fairness and justice to all parties when administering a construction (or other) contract.

k. Before undertaking work for others in which Engineers may make improvements, plans, designs, inventions, or other records which may justify copyrights or patents, they shall enter into a positive agreement regarding ownership.

l. Engineers shall admit and accept their own errors when proven wrong and refrain from distorting or altering the facts to justify their decisions.

m. Engineers shall not accept professional employment outside of their regular work or interest without the knowledge of their employers.

n. Engineers shall not attempt to attract an employee from another employer by false or misleading representations.

o. Engineers shall not review the work of other Engineers except with the knowledge of such Engineers, or unless the assignments or contractual agreements for the work have been terminated.

   (o.1) Engineers in governmental, industrial or educational employment are entitled to review and evaluate the work of other engineers when so required by their duties.

   (o.2) Engineers in sales or industrial employment are entitled to make engineering comparisons of their products with products of other suppliers.

   (o.3) Engineers in sales employment shall not offer nor give engineering consultation or designs or advice other than specifically applying to equipment, materials or systems being sold or offered for sale by them.

5. Engineers shall build their professional reputation on the merit of their services and shall not compete unfairly with others.

a. Engineers shall not pay nor offer to pay, either directly or indrectly, any commission, political contribution, or a gift, or other consideration in order to secure work, exclusive of securing salaried positions through employment agencies.

b. Engineers should negotiate contracts for professional services fairly and only on the basis of demonstrated competence and qualifications for the type of professional service required.

c. Engineers should negotiate a method and rate of compensation commensurate with the agreed upon scope of services. A meeting of the minds of the parties to the contract is essential to mutual confidence. The public interest requires that the cost of engineering services be fair and reasonable, but not the controlling consideration in selection of individuals or firms to provide these services.

   (c.1) These principles shall be applied by Engineers in obtaining the services of other professionals.

d. Engineers shall not attempt to supplant other Engineers in a particular employment after becoming aware that definite steps have been taken toward the others' employment or after they have been employed.

   (d.1) They shall not solicit employment from clients who already have Engineers under contract for the same work.

(d.2)   They shall not accept employment from clients who already have Engineers for the same work not yet completed or not yet paid for unless the performance or payment requirements in the contract are being litigated or the contracted Engineers' services have been terminated in writing by either party.

(d.3)   In case of termination of litigation, the prospective Engineers before accepting the assignment shall advise the Engineers being terminated or involved in litigation.

e.   Engineers shall not request, propose nor accept professional commissions on a contingent basis under circumstances under which their professional judgments may be compromised, or when a contingency provision is used as a device for promoting or securing a professional commission.

f.   Engineers shall not falsify nor permit misrepresentation of their, or their associates' academic or professional qualifications. They shall not misrepresent nor exaggerate their degree of responsibility in or for the subject matter of prior assignments. Brochures or other presentations incident to the solicitation of employment shall not misrepresent pertinent facts concerning employers, employees, associates, joint ventures, or their past accomplishments with the intent and purpose of enhancing their qualifications and work.

g.   Engineers may advertise professional services only as a means of identification and limited to the following:

(g.1)   Professional cards and listings in recognized and dignified publications, provided they are consistent in size and are in a section of the publication regularly devoted to such professional cards and listings. The information displayed must be restricted to firm name, address, telephone number, appropriate symbol, names of principal participants and the fields of practice in which the firm is qualified.

(g.2)   Signs on equipment, offices and at the site of projects for which they render services, limited to firm name, address, telephone number and type of services, as appropriate.

(g.3)   Brochures, business cards, letterheads and other factual representations of experience, facilities, personnel and capacity to render service, providing the same are not misleading relative to the extent of participation in the projects cited and are not indiscriminately distributed.

(g.4)   Listings in the classified section of telephone directories, limited to name, address, telephone number and specialties in which the firm is qualified without resorting to special or bold type.

h.   Engineers may use display advertising in recognized dignified business and professional publications, providing it is factual, and relates only to engineering, is free from ostentation, contains no laudatory expressions or implication, is not misleading with respect to the Engineers' extent of participation in the services or projects described.

    i. Engineers may prepare articles for the lay or technical press which are factual, dignified and free from ostentations or laudatory implications. Such articles shall not imply other than their direct participation in the work described unless credit is given to others for their share of the work.

    j. Engineers may extend permission for their names to be used in commercial advertisements, such as may be published by manufacturers, contractors, material suppliers, etc., only by means of a modest dignified notation acknowledging their participation and the scope thereof in the project or product described. Such permission shall not include public endorsement of proprietary products.

    k. Engineers may advertise for recruitment of personnel in appropriate publications or by special distribution. The information presented must be displayed in a dignified manner, restricted to firm name, address, telephone number, appropriate symbol, names of principal participants, the fields of practice in which the firm is qualified and factual descriptions of positions available, qualifications required and benefits available.

    l. Engineers shall not enter competitions for designs for the purpose of obtaining commissions for specific projects, unless provision is made for reasonable compensation for all designs submitted.

    m. Engineers shall not maliciously or falsely, directly or indirectly, injure the professional reputation, prospects, practice or employment of another engineer, nor shall they indiscriminately criticize another's work.

    n. Engineers shall not undertake nor agree to perform any engineering service on a free basis, except professional services which are advisory in nature for civic, charitable, religious or non-profit organizations. When serving as members of such organizations, engineers are entitled to utilize their personal engineering knowledge in the service of these organizations.

    o. Engineers shall not use equipment, supplies, laboratory nor office facilities of their employers to carry on outside private practice without consent.

    p. In case of tax-free or tax-aided facilities, engineers should not use student services at less than rates of other employees of comparable competence, including fringe benefits.

6. Engineers shall act in such a manner as to uphold and enhance the honor, integrity and dignity of the profession.

    a. Engineers shall not knowingly associate with nor permit the use of their names nor firm names in business ventures by any person or firm which they know, or have reason to believe, are engaging in business or professional practices of a fraudulent or dishonest nature.

    b. Engineers shall not use association with non-engineers, corporations, nor partnerships as 'cloaks' for unethical acts.

7. Engineers shall continue their professional development throughout their

careers, and shall provide opportunities for the professional development of those engineers under their supervision.

a. Engineers shall encourage their engineering employees to further their education.

b. Engineers should encourage their engineering employees to become registered at the earliest possible date.

c. Engineers should encourage engineering employees to attend and present papers at professional and technical society meetings.

d. Engineers should support the professional and technical societies of their disciplines.

e. Engineers shall give proper credit for engineering work to those to whom credit is due, and recognize the proprietary interests of others. Whenever possible, they shall name the person or persons who may be responsible for designs, inventions, writings or other accomplishments.

f. Engineers shall endeavor to extend the public knowledge of engineering, and shall not participate in the dissemination of untrue, unfair or exaggerated statements regarding engineering.

g. Engineers shall uphold the principle of appropriate and adequate compensation for those engaged in engineering work.

h. Engineers should assign professional engineers duties of a nature which will utilize their full training and experience insofar as possible, and delegate lesser functions to subprofessionals or to technicians.

i. Engineers shall provide prospective engineering employees with complete information on working conditions and their proposed status of employment, and after employment shall keep them informed of any changes.

7.3

National Society of Professional Engineers
Code of Ethics for Engineers*

PREAMBLE

Engineering is an important and learned profession. The members of the profession recognize that their work has a direct and vital impact on the quality of life for all people. Accordingly, the services provided by engineers require honesty, impartiality, fairness and equity, and must be dedicated to the protection of the public health, safety and welfare. In the practice of their profession, engineers must perform under a standard of professional behavior which requires adherence to the highest principles of ethical conduct on behalf of the public, clients, employers and the profession.

I. FUNDAMENTAL CANONS

Engineers, in the fulfillment of their professional duties, shall:

1. Hold paramount the safety, health and welfare of the public in the performance of their professional duties.
2. Perform services only in areas of their competence.
3. Issue public statements only in an objective and truthful manner.
4. Act in professional matters for each employer or client as faithful agents or trustees.
5. Avoid improper solicitation of professional employment.

---

*NSPE Publication No. 1102. As revised, January 1981.

II. RULES OF PRACTICE

1. Engineers shall hold paramount the satefy, health and welfare of the public in the performance of their professional duties.

    a. Engineers shall at all times recognize that their primary obligation is to protect the safety, health, property and welfare of the public. If their professional judgment is overruled under circumstances where the safety, health, property or welfare of the public are endangered, they shall notify their employer or client and such other authority as may be appropriate.

    b. Engineers shall approve only those engineering documents which are safe for public health, property and welfare in conformity with accepted standards.

    c. Engineers shall not reveal facts, data or information obtained in a professional capacity without the prior consent of the client or employer except as authorized or required by law of this Code.

    d. Engineers shall not permit the use of their name or firm name nor associate in business ventures with any person or firm which they have reason to believe is engaging in fraudulent or dishonest business or professional practices.

    e. Engineers having knowledge of any alleged violation of this Code shall cooperate with the proper authorities in furnishing such information or assistance as may be required.

2. Engineers shall perform services only in the areas of their competence.

    a. Engineers shall undertake assignments only when qualified by education or experience in the specific technical fields involved.

    b. Engineers shall not affix their signatures to any plans or documents dealing with subject matter in which they lack competence, nor to any plan or document not prepared under their direction and control.

    c. Engineers may accept an assignment outside of their fields of competence to the extent that their services are restricted to those phases of the project in which they are qualified, and to the extent that they are satisfied that all other phases of such project will be performed by registered or otherwise qualified associates, consultants, or employees, in which case they may then sign the documents for the total project.

3. Engineers shall issue public statements only in an objective and truthful manner.

    a. Engineers shall be objective and truthful in professional reports, statements or testimony. They shall include all relevant and pertinent information in such reports, statements or testimony.

    b. Engineers may express publicly a professional opinion on technical subjects only when that opinion is founded upon adequate knowledge of the facts and competence in the subject manner.

c. Engineers shall issue no statements, criticisms or arguments on technical matters which are inspired or paid for by interested parties, unless they have prefaced their comments by explicitly identifying the interested parties on whose behalf they are speaking, and by revealing the existence of any interest the engineers may have in the matters.

4. Engineers shall act in professional matters for each employer or client as faithful agents or trustees.

a. Engineers shall disclose all known or potential conflicts of interest to their employers or clients by promptly informing them of any business association, interest, or other circumstances which could influence or appear to influence their judgment or the quality of their services.

b. Engineers shall not accept compensation, financial or otherwise, from more than one party for services on the same project, or for services pertaining to the same project, unless the circumstances are fully disclosed to, and agreed to, by all interested parties.

c. Engineers shall not solicit or accept financial or other valuable consideration, directly or indirectly, from contractors, their agents, or other parties in connection with work for employers or clients for which they are responsible.

d. Engineers in public service as members, advisors or employees of a governmental body or department shall not participate in decisions with respect to professional services solicited or provided by them or their organizations in private or public engineering practice.

e. Engineers shall not solicit or accept a professional contract from a governmental body on which a principal or officer of their organization serves as a member.

5. Engineers shall avoid improper solicitation of professional employment.

a. Engineers shall not falsify or permit misrepresentation of their, or their associates', academic or professional qualifications. They shall not misrepresent or exaggerate their degree of responsibility in or for the subject matter of prior assignments. Brochures or other presentations incident to the solicitation of employment shall not misrepresent pertinent facts concerning employers, employees, associates, joint ventures or past accomplishments with the intent and purpose of enhancing their qualifications and their work.

b. Engineers shall not offer, give, solicit or receive, either directly or indirectly, any political contribution in an amount intended to influence the award of a contract by public authority, or which may be reasonably construed by the public of having the effect or intent to influence the award of a contract. They shall not offer any gift, or other valuable consideration in order to secure work. They shall not pay a commission, percentage or brokerage fee in order to secure work except to a bona fide employee or bona fide established commercial or marketing agencies retained by them.

III. PROFESSIONAL OBLIGATIONS

1. Engineers shall be guided in all their professional relations by the highest standards of integrity.

   a. Engineers shall admit and accept their own errors when proven wrong and refrain from distorting or altering the facts in an attempt to justify their decisions.

   b. Engineers shall advise their clients or employers when they believe a project will not be successful.

   c. Engineers shall not accept outside employment to the detriment of their regular work or interest. Before accepting any outside employment they will notify their employers.

   d. Engineers shall not attempt to attract an engineer from another employer by false or misleading pretenses.

   e. Engineers shall not actively participate in strikes, picket lines, or other collective coercive action.

   f. Engineers shall avoid any act tending to promote their own interest at the expense of the dignity and integrity of the profession.

2. Engineers shall at all times strive to serve the public interest.

   a. Engineers shall seek opportunities to be of constructive service in civic affairs and work for the advancement of the safety, health and well-being of their community.

   b. Engineers shall not complete, sign, or seal plans and/or specifications that are not of a design safe to the public health and welfare and in conformity with accepted engineering standards. If the client or employer insists on such unprofessional conduct, they shall notify the proper authorities and withdraw from further service on the project.

   c. Engineers shall endeavor to extend public knowledge and appreciation of engineering and its achievements and to protect the engineering profession from misrepresentation and misunderstanding.

3. Engineers shall avoid all conduct or practice which is likely to discredit the profession or deceive the public.

   a. Engineers shall avoid the use of statements containing a material misrepresentation of fact or omitting a material fact necessary to keep statements from being misleading; statements intended or likely to create an unjustified expectation; statements containing prediction of future success; statements containing an opinion as to the quality of the Engineers' services; or statements intended or likely to attract clients by the use of showmanship, puffery, or self-laudation, including the use of slogans, jingles, or sensational language or format.

   b. Consistent with the foregoing, Engineers may advertise for recruitment of personnel.

   c. Consistent with the foregoing, Engineers may prepare articles for the lay or technical press, but such articles shall not imply credit to the author for work performed by others.

4. Engineers shall not disclose confidential information concerning the business affairs or technical processes of any present or former client or employer without his consent.

   a. Engineers in the employ of others shall not without the consent of all interested parties enter promotional efforts or negotiations for work or make arrangements for other employment as a principal or to practice in connection with a specific project for which the Engineer has gained particular and specialized knowledge.

   b. Engineers shall not, without the consent of all interested parties, participate in or represent an adversary interest in connection with a specific project or proceeding in which the Engineer has gained particular specialized knowledge on behalf of a former client or employer.

5. Engineers shall not be influenced in their professional duties by conflicting interests.

   a. Engineers shall not accept financial or other considerations, including free engineering designs, from material or equipment suppliers for specifying their product.

   b. Engineers shall not accept commissions or allowances, directly or indirectly, from contractors or other parties dealing with clients or employers of the Engineer in connection with work for which the Engineer is responsible.

6. Engineers shall uphold the principle of appropriate and adequate compensation for those engaged in engineering work.

   a. Engineers shall not accept remuneration from either an employee or employment agency for giving employment.

   b. Engineers, when employing other engineers, shall offer a salary according to professional qualifications and the recognized standards in the particular geographical areas.

   c. Engineers in sales employment shall not offer, or give engineering consultation, or designs, or advice other than specifically applying to the equipment being sold.

7. Engineers shall not compete unfairly with other engineers by attempting to obtain employment or advancement or professional engagements by taking advantage of a salaried position, by criticizing other engineers, or by other improper or questionable methods.

   a. Engineers shall not request, purpose, or accept a professional commission on a contingent basis under circumstances in which their professional judgment may be compromised.

   b. Engineers in salaried positions shall accept part-time engineering work only at salaries not less than that recognized as standard in the area.

   c. Engineers shall not use equipment, supplies, laboratory, or office facilities of an employer to carry on outside private practice without consent.

8. Engineers shall not attempt to injure, maliciously or falsely, directly or indirectly, the professional reputation, prospects, practice or employment of other engineers, nor indiscriminately criticize other engineers' work. Engineers who believe other engineers are guilty of unethical or illegal practice shall present such information to the proper authority for action.

   a. Engineers in private practice shall not review the work of another engineer for the same client, except with the knowledge of such engineer, or unless the connection of such engineer with the work has been terminated.

   b. Engineers in governmental, industrial or educational employ are entitled to review and evaluate the work of other engineers when so required by their employment duties.

   c. Engineers in sales or industrial employ are entitled to make engineering comparisons of represented products with products of other suppliers.

9. Engineers shall accept personal responsibility for all professional activities.

   a. Engineers shall conform with state registration laws in the practice of engineering.

   b. Engineers shall not use association with a nonengineer, a corporation, or partnership, as a "cloak" for unethical acts, but must accept personal responsibility for all professional acts.

10. Engineers shall give credit for engineering work to those to whom credit is due, and will recognize the property interests of others.

   a. Engineers shall, whenever possible, name the person or persons who may be individually responsible for designs, inventions, writings, or other accomplishments.

   b. Engineers using designs supplied by a client recognize that the designs remain the property of the client and may not be duplicated by the Engineer for others without express permission.

    c. Engineers, before undertaking work for others in connection with which the Engineer may make improvements, plans, designs, inventions, or other records which may justify copyrights or patents, should enter into a positive agreement regarding ownership.

    d. Engineers' designs, data, records, and notes referring exclusively to an employer's work are the employer's property.

11. Engineers shall cooperate in extending the effectiveness of the profession by interchanging information and experience with other engineers and students, and will endeavor to provide opportunity for the professional development and advancement of engineers under their supervision.

    a. Engineers shall encourage engineering employees' efforts to improve their education.

    b. Engineers shall encourage engineering employees to attend and present papers at professional and technical society meetings.

    c. Engineers shall urge engineering employees to become registered at the earliest possible date.

    d. Engineers shall assign a professional engineer duties of a nature to utilize full training and experience, insofar as possible, and delegate lesser functions to subprofessionals or to technicians.

    e. Engineers shall provide a prospective engineering employee with complete information on working conditions and proposed status of employment, and after employment will keep employees informed of any changes.

By order of the United States District Court for the District of Columbia, former Section 11(c) of the NSPE Code of Ethics prohibiting competitive building, and all policy statements, opinions, rulings or other guidelines interpreting its scope, have been rescinded as unlawfully interfering with the legal right of engineers, protected under the antitrust laws, to provide price information to prospective clients; accordingly, nothing contained in the NSPE Code of Ethics, policy statements, opinions, rulings or other guidelines prohibits the submission of price quotations or competitive bids for engineering services at any time or in any amount.

Statement by NSPE Executive Committee

In order to correct misunderstandings which have been indicated in some instances since the issuance of the Supreme Court decision and the entry of the Final Judgment, it is noted that in its decision of April 25, 1978, the Supreme Court of the United States declared: "The Sherman Act does not require competitive bidding."

It is further noted that as made clear in the Supreme Court decision:

1. Engineers and firms may individually refuse to bid for engineering services.

2. Clients are not required to seek bids for engineering services.

3. Federal, state, and local laws governing procedures to procure engineering services are not affected, and remain in full force and effect.

4. State societies and local chapters are free to actively and aggressively seek legislation for professional selection and negotiation procedures by public agencies.

5. State registration board rules of professional conduct, including rules prohibiting competitive bidding for engineering services, are not affected and remain in full force and effect. State registration boards with authority to adopt rules of professional conduct may adopt rules governing procedures to obtain engineering services.

6. As noted by the Supreme Court, "nothing in the judgment prevents NSPE and its members from attempting to influence governmental action...."

Note: In regard to the question of application of the Code to corporations vis-a-vis real persons, business form or type should not negate nor influence conformance of individuals to the Code. The Code deals with professional services, which services must be performed by real persons. Real persons in turn establish and implement policies within business structures. The Code is clearly written to apply to the Engineer and it is incumbent on a member of NSPE to endeavor to live up to its provisions. This applies to all pertinent sections of the Code.

7.4

IEEE
Code of Ethics for Engineers

PREAMBLE

Engineers affect the quality of life for all people in our complex technological society. In the pursuit of their profession, therefore, it is vital that engineers conduct their work in an ethical manner so that they merit the confidence of colleagues, employers, clients and the public. This IEEE Code of Ethics is a standard of professional conduct for engineers.

ARTICLE I

Engineers shall maintain high standards of diligence, creativity and productivity, and shall:

1. Accept responsibility for their actions;
2. Be honest and realistic in stating claims or estimates from available data;
3. Undertake engineering tasks and accept responsibility only if qualified by training or experience, or after full disclosure to their employers or clients of pertient qualifications;
4. Maintain their professional skills at the level of the state of the art, and recognize the importance of current events in their work;
5. Advance the integrity and prestige of the engineering profession by practicing in a dignified manner and for adequate compensation.

ARTICLE II

Engineers shall, in their work:

1. Treat fairly all colleagues and co-workers, regardless of race, religion, sex, age or national origin;

2. Report, publish and disseminate freely information to others, subject to legal and proprietary restraints;
3. Encourage colleagues and co-workers to act in accord with this Code and support them when they do so;
4. Seek, accept and offer honest criticism of work, and properly credit the contributions of others;
5. Support and participate in the activities of their professional societies;
6. Assist colleagues and co-workers in their professional development.

ARTICLE III

Engineers shall, in their relations with employers and clients:

1. Act as faithful agents or trustees for their employers or clients in professional and business matters, provided such actions conform with other parts of this Code;
2. Keep information on the business affairs or technical processes of an employer or client in confidence while employed, and later, until such information is properly released, provided such actions conform with other parts of this Code;
3. Inform their employers, clients, professional societies or public agencies or private agencies of which they are members or to which they may make presentations, of any circumstance that could lead to a conflict of interest;
4. Neither give nor accept, directly or indirectly, any gift, payment or service of more than nominal value to or from those having business relationships with their employers or clients;
5. Assist and advise their employers or clients in anticipating the possible consequences, direct and indirect, immediate or remote, of the projects, work or plans of which they have knowledge.

ARTICLE IV

Engineers shall, in fulfilling their responsibilities to the community:

1. Protect the safety, health and welfare of the public and speak out against abuses in these areas affecting the public interest;
2. Contribute professional advice, as appropriate, to civic, charitable or other non-profit organizations;
3. Seek to extend public knowledge and appreciation of the engineering profession and its achievements.

7.5

American Public Works Association Code of Ethics

Recognizing their responsibilities to the people, desiring to inspire public confidence and respect for government, and believing that honesty, integrity, loyalty, justice, and courtesy form the basis of ethical conduct, members of the American Public Works Association:

- Uphold the Constitution, laws and regulations of their country and all other applicable units of government.
- Put public interest above individual, group, or special interest and consider their occupation an opportunity to serve society.
- Recognize that government service is a public trust that imposes responsibility to conserve public resources, funds, and materials.
- Recognize that political (policy) decisions are the responsibility of the people's elected representatives but that identification and communication of technical and administrative alternatives and recommendations as a basis for decision making are the responsibility of public works officials, professional engineers, or other administrators.
- Never offer, give, nor accept any gifts, favors, or service that might tend to influence them in the discharge of their duties.
- Never use their position to secure advantage or favor for themselves, their family, or friends.
- Never disclose confidential information gained by reason of their position, nor use such information for personal gain.
- Never make recommendations, while employed by a public agency, on any matter that involves a business in which they have a direct or indirect financial interest.
- Never engage in supplemental employment, business or professional activity which impairs the efficiency of their services; or while employed by a public agency become involved in work which could come before their agency for review or inspection.

- Recognize that it is not in the public interest for officials of public agencies to select and retain professional engineering services on the basis of price alone and that consideration must also be given to experience, technical expertise, availability, and other qualifications.

- Do not attempt either falsely or maliciously to injure the reputation, business, or employment status of any individual.

Proposed: A Single Code of Ethics for All Engineers

ANDREW G. OLDENQUIST
EDWARD E. SLOWTER, P.E.

The engineering profession needs a unified code of ethics, and it needs one now. At a time when scandals have shocked the nation and the motives of professions have come under suspicion, the engineering profession needs to speak with one voice, clearly and authoritatively, regarding the nature of its dedication to integrity, competence, and the public interest. The fact that there exist today nearly as many codes of ethics as there are engineering societies promotes neither credibility nor clarity on these matters. What is noteworthy is that at the present time the opportunity for achieving a unified ethical code is as great as the need.

INCEPTION OF PROJECT

In the autumn of 1977 the authors applied for participation in the National Project on Philosophy and Engineering Ethics, which was to begin with a workshop at Rensselaer Polytechnic Institute the following summer. The project was funded by the National Endowment for the Humanities and administered by the Center for the Study of the Human Dimensions of Science and Technology at RPI. Applications had to be made by two-person engineer/philosopher teams, each team submitting as part of the application a project proposal concerned with some aspect of engineering ethics. The idea was to combine the skills and expertise of engineers and those of professional philosophers who specialize in ethics, so as to enhance the likelihood of developing conceptually sound, justifiable ethics projects that also were responsive to the realities of the engineering profession. Eventually 18 team proposals were accepted from the large number that were submitted.

Our application was one of those accepted; our project idea was to attempt to develop a set of core ethical concepts that could form the basis for a unified code of ethics for engineers. The Summer Institute at RPI consisted of a two-week program for engineers on philosophical ethics, a concurrent two-week institute for philosophers on the engineering profession, and a one-week joint program on project development. There also were talks by outside consultants about specific cases, about other engineering ethics programs, and on related matters. We received some limited funding through the National Project on Philosophy and Engineering Ethics to pursue our team project and will be attending the follow-up conferences this June [1979] and in the summer of 1980.

PHILOSOPHICAL BASIS FOR AN ETHICAL CODE

There are obvious reasons why a single, specific code of ethics is needed for engineers. The public's health, welfare, and safety depend on engineers' knowledge, competence, and integrity. Closely related to this, engineers do not merely possess special knowledge and capabilities; like judges and physicians they have exclusive stewardship over their special knowledge. Professional integrity is especially vital for any professional group that is self-regulating on expert matters and to whom society has given stewardship over important knowledge and activities. A code of ethics serves to remind individuals how important integrity is in a self-regulating profession. It lays out the specific matters deemed most important in the collective wisdom of the profession, and, in solemnly promulgating and enforcing the code, notice is served that its elements are to be taken seriously.

There are at least two premises that underlie the basic ethical content of the existing engineering codes, and we believe they are crucial to formulating the core concepts of a unified code for the profession. First, engineers owe respect to the same fundamental ethical principles as do all people. The differences in what ethics requires of engineers come not from different principles but from how these principles apply in the context of their special expertise and position of public trust. Second, at the level of these basic principles, matters of interpersonal morality are not just matters of local custom or individual feelings and attitudes, but are capable of being acknowledged as rational and justifiable by the great majority of reasonable persons. If one did not believe this, it would be difficult to attribute any authority to an ethical code or justify imposing it on engineers, except, perhaps, as an expression of the desires and power of the persons who wrote it.

A philosophical basis for ethical judgments that are rational and objective, as well as distinct from judgments based on custom, convention, and self-interest, is usually taken to include the following.

A. Ethical judgments are universal.

B. Rational ethical judgments require factual accuracy; ethical judgments based on ignorance of the facts are defective.

C. Ethics is made for humankind, not the other way around, and therefore ethics is primarily concerned with what harms or benefits people.

D. Judgments made from an ethical point of view are all-things-considered judgments; they are not based on just one of several competing perspectives or on the good of just one individual or one special group.

E. Ethical considerations concern the good of everyone alike and therefore they override self-interest as well as group business interests.

These marks of the ethical rule out cynical, self-interested, and relativistic concepts of what is acceptable. If they did not, a code of ethics for engineers (or for anyone) could not claim any moral authority and it could say no more than, "Do whatever it is that you think is right." If a code did no more than express the will of an engineering society, it could say only, "Do these things or we will expel you from the group."

MORAL AUTHORITY

Moral authority is not the same as power; in choosing the language in which they have formulated their codes the engineering societies have assumed that right and wrong are not just matters of enforced custom or local group egoism. Ethical considerations are social, not egoistic. They therefore can require moderate sacrifices of self-interest, which we might think of as a price we are obligated to pay for living in a civilized society; for a civilized society is one in which people make mutual concessions. The rationality of this flows in large part from the universality of our ethical judgments.

Ethical considerations place restrictions on the self-interested policies of groups as well as individuals. It is true that what protects the good of engineers as a group often also will protect or serve the public good. Nevertheless, if a code is to be perceived as an ethical one it must keep these considerations distinct and must appeal only to ethical principles. The public assumes that any group will make policies that serve its own interests except when it feels constrained morally or legally; it is on the lookout for elements of a code of "ethics" that really are simply self-serving. Hence the public credibility of a code of ethics depends on an acute sensitivity to the difference between actions that are unethical and actions that are uncustomary, disloyal, or financially disadvantageous to the engineering profession.

CORE OF ETHICAL CONCEPTS

For these reasons a core of basic ethical concepts will comprise a statement that is briefer than most of the present engineering codes. The concepts concern matters of honesty, fairness, regard for the public interest, and several aspects of professional competence. Their relative importance is mentioned only in the case of the public health, safety, and welfare, which are called "paramount," on the ground that where lives are at stake our public obligations override competing ones. In this respect the proposal follows most of the existing codes.

There are real disadvantages to each engineering society having its own code of ethics. The different codes are likely to be perceived as "relative" to their societies and binding only on the societies that wrote them, with the consequence that the moral authority of all of them is diminished. It seems to be a psychological fact that ethical diversity tempts people to view each different code as merely the expression of the customs of that group, changeable and serving the interests of the group. If the public views these codes as saying that one sort of thing is right for the electrical and electronics engineers, and another sort of thing is right for the civil engineers, and still another for the members of NSPE, it will be more likely to conclude that none of these things is "really right."

And what is the engineer who belongs to more than one society to think, when behavior that is not mentioned by one is prohibited by another? Engineers of course will make up their own minds, but in this situation, a likely result will be that they take the codes less seriously. Therefore if it is true, as we think it is, that basic ethical principles apply to everyone, and that they already are present, in different modes of expression, in the existing engineering codes, a unified set of core concepts and eventually a unified code is both possible and desirable.

A single ethical code for engineers will have greater authority and credibility; it will provide a more readily acceptable basis for developing detailed guidelines for enforcement; and what it presents to the public will be more readily perceived as ethics rather than custom. Business, government, and the general public will much more easily come to know, and know what to expect from, an ethical code for engineers if they do not have to wonder whether a particular guideline applies only to civil engineers or only to electrical and electronics engineers.

TABLE 1. Core Concepts in Engineering Ethics

I. *The public interest*
   - **A.** Paramount responsibility to the public health, safety, and welfare, including that of future generations.
   - **B.** Call attention to threats to the public health, safety, and welfare, and act to eliminate them.
   - **C.** Work through professional societies to encourage and support engineers who follow these concepts.
   - **D.** Apply knowledge, skill and imagination to enhance human welfare and the quality of life for all.
   - **E.** Work only with those who follow these concepts.

II. *Qualities of truth, honesty, and fairness*
   - **A.** Be honest and impartial.
   - **B.** Advise employer, client or public of all consequences of work.
   - **C.** Maintain confidences; act as faithful agent or trustee.
   - **D.** Avoid conflicts of interest.
   - **E.** Give fair and equitable treatment to all others.
   - **F.** Base decisions and actions on merit, competence, and knowledge, and without bias because of race, religion, sex, age, or national origin.
   - **G.** Neither pay nor accept bribes, gifts, or gratuities.
   - **H.** Be objective and truthful in discussions, reports, and actions.

III. *Professional performance*
   - **A.** Competence for work undertaken.
   - **B.** Strive to improve competence, and assist others in so doing.
   - **C.** Extend public and professional knowledge of technical projects and their results.
   - **D.** Accept responsibility for actions and give appropriate credit to others.

CODE CAN'T BE "TRIBAL"

If it is to have moral authority and win respect, an ethical code cannot be perceived as "tribal," that is to say, as holding only for some particular group of engineers and expressing their special interests. But even a unified ethical code for all engineers must appeal to universal moral principles and not to principles that just aim at the good of engineers. There is no doubt that most of the basic principles in the existing codes are universal in this sense. However, mixed with these ethical principles are a number of rules and customs concerning business practices and political convictions which, at least in the eyes of much of the public, seem more designed to protect the interests of engineers than to serve the general good. They render codes that contain them "tribal" and this diminishes their moral authority.

For example, prohibitions against competitive bidding, advertising, and collective bargaining are not universal ethical principles, and it is controversial whether or not they serve the general good in addition to protecting the financial interests and the prestige of engineers. No stand is hereby implied on whether or not prohibitions of advertising or collective bargaining are good for the engineering profession or whether or not they are good for American society: it is only that they are not basic ethical principles; they appear to many people to be self-serving, and they are matters of political controversy.

Consequently a core of basic ethical concepts should include only rules whose violation clearly implies unethical behavior. This contrasts with other kinds of violations that offend the majority of engineers because they go against traditional business practices, predominant political orientations, engineers' economic interests, or conceptions of dignity and prestige. Some of these latter matters may well belong in a second statement that each engineering society, in its own way, may advocate in guidelines that are supplementary to (and consistent with) a unified ethical code. Thus one might envisage a supplementary document, e.g., "Ideals and Guidelines for Professional Engineers," that NSPE would promote to protect the long-range, rational interests of professional engineers and that perhaps also contained guidelines on issues that were particularly pertinent to its membership.

HISTORY OF PAST CODES

The fact that there is no single code of ethics for the engineering profession has been a source of concern, as well as an unattainable goal, for many engineers and their professional societies for at least the past 60 years. In the early 1920's ASME produced a Code of Ethics for Engineers, and the American Association of Engineers developed Principles of Conduct. Unfortunately, neither was widely accepted. By 1947, ECPD developed Canons of Ethics which eventually had acceptance by a significant number, but not a preponderance, of engineering societies; these canons were updated in 1963, and again some acceptance was obtained; more recently, a major effort by ECPD produced a Code of Ethics for Engineers in 1974 which contained three levels of specificity: Fundamental Principles; Fundamental Canons; and Suggested Guidelines.

Although a number of societies participated in the preparation of this three-level code and it seemed to offer opportunity for universal acceptance of at least the Fundamental Principles, it has not secured the support of a majority of the professional societies, and the goal of a universal code continues to elude the profession.

PAST FAILURES

In examining the reasons for past failures to secure broad acceptance of an overall code, there undoubtedly have been organizational and personal conflicts which interfered with the attainment of this goal, but perhaps the most obvious stumbling block has been the "NIH" (Not Invented Here) concept. One organization is usually reluctant to endorse or accept a concept produced by another organization; this has led to a number of codes only slightly different in wording. The only obvious way to overcome this difficulty appears to be to use an ad hoc group, with representatives from all major societies participating in the development of a universal code that is not identified with any specific group. This mechanism proved highly successful in developing "The Guidelines to Professional Employment for Engineers and Scientists" and has worked to a lesser extent in developing endorsement and support of various legislative positions.

Perhaps another reason for the failure to achieve a unified code was the lack of public pressure on the profession. For at least the first half of the 20th century, the products and actions of the engineer were so well received and perceived by the public that, at least in the eyes of the engineering profession, there was no great need to demonstrate its ethical standards in a convincing way. Developments of the past 30 years, including consumerism, environmentalism, public revulsion at kickbacks and fraud, massive layoffs of engineers, whistle blowing, etc., have all too clearly demonstrated that the engineering profession must get its ethical house in order if it is to have the public support it needs. Our professional societies have sensed this need, and a number of societies are in the process of considering revision of their codes. The present situation is one that should encourage cooperative action and emphasize the importance of service to the public by the engineering profession.

DEVELOPMENT OF CORE CONCEPTS

As the codes of the various engineering societies are examined, a few characteristics stand out:

1. A number of codes are identical or very similar to either the 1963 or 1974 ECPD code or are similar in many ways to one another.
2. Most codes are primarily concerned with ethical principles, but practically all contain elements relating to custom, professional courtesy, or business practice.
3. The differences among the various codes seem mainly to be in the more detailed statements of common or very similar basic principles.

These characteristics lead to the view that all of the codes contain a relatively few ethical concepts which should be identifiable and ultimately acceptable to the engineering profession. On this basis, it appears that the essential or core concept can be stated in three broad areas:

1. The public interest
2. Qualities of truth, honesty, and fairness
3. Professional performance

When a fairly large number of current codes were examined in this light, and together with other suggestions from many sources, there appeared to be fewer than 20 concepts which would cover all of the ethical principles enumerated in these codes. These concepts are listed in Table I.

As an example of the method by which the various codes were analyzed, Table II shows the extent to which the current [1978] NSPE code represents, in its various sections, the core concepts listed in Table I.

Several points are obvious from this comparison:

1. The preamble and the various sections cover all of the core concepts except that relating to professional society support of those who act in accord with them. The concept of working through professional societies to promote these ethical concepts seems so important to the operation of an ethical code that it must be stated explicitly.
2. Some of the core concepts have multiple references in the NSPE code because of their importance or complexity. These are:
   a. Public health, safety, and welfare
   b. Avoidance of conflicts of interest
   c. Fair and equitable treatment of others
   d. Refusing to pay or accept bribes, gifts, or gratuities
   e. Objectivity and truthfulness in discussions
   f. Improving competence and assisting others to do so
   g. Extending public and professional knowledge of technical projects
   h. Accepting responsibility and giving credit to others

3. Two sections of the NSPE code, (1(f), relating to participation in collective action, and 3(c), relating to advertising for recruitment of personnel) do not really deal with a question of ethics and should not be included in an ethical code.
4. Some of the NSPE sections may embody more than one of the core concepts, but reference to the one considered most pertinent was usually viewed as adequate for this analysis.

As part of this study, each of the major codes was analyzed in this fashion and the results were used in developing the core concepts which have been presented. The list of concepts does not contain non-ethical material, and it is, we believe, adequately inclusive. But it is brief enough to provide a common basis from which a single, acceptable code of ethics could result.

TABLE II. Comparison of the 1978 NSPE Code of Ethics with the Core Concepts

Core Concepts

Section of NSPE Code	I. The Public Interest					II. Qualities of Truth, Honesty, & Fairness								III. Professional Performance				Other Considerations
	A	B	C	D	E	A	B	C	D	E	F	G	H	A	B	C	D	
Preamble Sections	2 2a 2b 2c	2 2c		x	13	x 13a	x 1c	1 7	1d 1g 4a 8 8a 8b 8c 10 11e 14b 14c 14d	7a 9 9a 9b 9c 11 11a 11d 12 12a 12b 12c 15e	5 5a	10a 10b 11b 11c 11f	1a 1e 3 3a 3b 5a	6	x 15 15a 15b 15c 15d	3d 4 15 15b	1b 3d 13b 14 14a	1f 3c

ETHICAL CODE CONVOCATION

Now is the time to assemble an ad hoc convocation of representatives of all the engineering bodies concerned with ethics. Starting with the ethical core concepts, these representatives should be able to put these concepts into specific words for a single code of ethics for the engineering profession which could be received and acted upon by the sponsoring societies by the fall of 1979. We now have the necessary interest on the part of the public and the profession, we have groups already working toward this goal, and we have had the successful experience of getting large numbers of our societies to agree on common goals. Why not end 60 years of frustration and develop a single Code of Ethics for Engineers?

The number of basic ethical principles involved in all of the codes of engineering ethics that have ever been written can be consolidated into a brief statement of a single code which can satisfy the ethical needs of all engineers; if desired, this code can be supplemented by details of the operation of the code in any particular discipline, but the ethical principles can cover the whole profession. With a cooperative spirit and input from all of the engineering bodies concerned with ethics, a truly unified code of ethics can be created by building on the work that has been done by our fellow engineers in the past. We must merge our individual parochial interests into a truly universal profession-wide ethical code by our own actions, or we will find a code of some sort forced upon us by others.

When in Rome Do as the Romans: Should the Professional and Business Ethics of American Engineers be a Function of the Country in Which They Practice?

A. COLIN LAUCHLAN

In a speech in Paris in 1952, Dr. Albert Schweitzer defined ethics as "the name we give to our concern for good behavior. We feel an obligation to consider not only our own personal well being, but also that of others and and of human society as a whole." Defined as such, ethics rises above laws, customs or religious practices of any one country. When one stoops to the approach of trying to get away with what is legally permissible—or of engaging in illegal activities which a foreign government might turn a blind eye to—then one begins to compromise one's character principles. A man cannot take red-hot coals to his chest and expect not to be burnt!

Ethical principles, like the laws of physics, are above the laws of national governments. The law of gravity exists even though we might be unaware of it or choose to ignore it, and it brings a penalty upon us if we defy it. In the same way, just because the legal systems of certain countries do not adhere to certain ethical principles does not mean that those principles are no longer in effect. Failure to adhere to them results in certain penalties which must be paid. The recent payoff scandals involving several U.S. corporations have brought the question of business ethics to the public's attention. In seeking to justify these unethical activities, the allegation has been made that in some situations, or in certain countries, it is common practice to bribe officials and that "slush funds" are a necessity if a company wishes to win a contract. Everyone does it and it is only the unlucky ones who get caught. Such an argument is deceiving for, although customs and practices differ in various countries, no country extols the virtues of bribery and corruption, or denounces honesty. This attitude of "doing as the Romans" results in a situation where the company that engages in the most unethical or immoral conduct sets the business standards at the lowest demominator. It is also an attitude that denies the reality of penalties incurred when natural laws are violated.

MANIFOLD PENALTIES

The example of those corporations that were involved in foreign bribes and payoffs make it clear that in the long run unethical practices do not pay. The penalties have been manifold. The careers of the corporation officials involved were suddenly interrupted by their resignations. Bad publicity causes several problems for the company involved. A loss of investor confidence impairs its ability to raise capital in the public market, and every business deal that that company engages in becomes suspect. Public officials in any country would be reluctant to be associated with any contract or sale involving a company convicted of dishonesty. In view of this it can be seen that a reputation for honesty, far from being a liability, is an important competitive asset, and that no company could maintain a significant position internationally without such a reputation. The penalties extend to other areas as well. Revelations of overseas payoffs compound international tensions and weaken American foreign policy. They further the disillusionment and suspicion that the American public has of big business. This could result in a call for more stringent federal regulations of corporations, which would be both costly and annoying to business.

The penalties paid by anyone who engages in any unethical practice must also be examined. Ethics is a matter of what a person is, not where he is. When a person engages in less than desirable practices he suffers a deterioration of his moral fiber and character. The ability to distinguish between right and wrong becomes blurred and practices that once seemed shady, but were engaged in because everyone else did, now become acceptable and commonplace. Attention is diverted from improving one's product or service, which is what every business ultimately depends on, to devising schemes to undercut competitors. Because unethical activities in another country have become acceptable to the individual, they also become acceptable here in the United States—as long as one is not caught. Thus, it should not have come as a shock to find that congressmen and senators here in the U.S. have been the subjects of "influence buying." One wonders whether, perhaps, the American companies which have engaged in such practices did not learn them from their subsidiaries in Rome, which were just doing what the Romans did. As the moral fiber of the nation deteriorates, as reflected by the character of its people, so the ability of that nation to function as a free society is reduced. Coercive laws are instituted to regulate every facet of life giving rise to a government which becomes larger, more expensive, and more inefficient. A free society depends on individual honesty. Honesty is the basis of all ethical behavior.

MAINTAIN TRUST AND CONFIDENCE

Today, engineers must seek to display the highest standards of conduct and professionalism. The engineering profession has enjoyed considerable status in the eye of the public, but this esteem can only continue while the public has trust and confidence in the profession. Reports of unethical practices, whether at home or abroad, can only lead to cynicism and mistrust from the American public. Society is increasingly charging professionals with the responsibility of weighing the impact of their actions. Engineers can no longer hide behind the anonymity afforded by working for a large corporation, but must consider their ethical duty to put the safety of the public whom they serve above the interest of the corporation or other institution that either employs them or contracts for their services.

Failure to do so will result in increasing public oversight of engineers' activities and a lessening of independence. For this reason, the courageous examples of many engineers who have risked their career rather than engage in unethical activities deserve our praise, while those who have stooped to practices which discredit the profession should be roundly condemned.

To suggest that it is permissible for American engineers or American businesses to engage in unethical practices in order to do business in foreign countries, while submitting to ethical conduct at home, is both totally inconsistent and ridiculous. Ethics does not depend on circumstances, it is a reflection of what we are. Thus, our concern should be not whether our actions are merely legal but what is best and right. The future of America as a free society depends on it.

Ethics Spun from Fairy Tales

JACK McMINN, F. ASCE

INTRODUCTION

In 1956 I walked into the office of the late Bill Ellison, structural engineer partner with Art Sedgwick in what was then the firm of Ellison and Sedgwick in San Francisco. I was a young engineer trying to start a consulting practice in soils engineering, making cold calls on potential structural engineer clients.

Ellison was a serious but kindly, white-haired gentleman who worked in the typical no-nonsense environment of a structural engineering office of 20 years ago. He was clearly no novice to his profession, but he gave the impression of a man who, after I left his office, would roll up his sleeves, climb back on his drafting stool, and go to work.

My visit with Ellison was only one of a number of such contacts I made trying to develop some business. But I particularly remember him for the very candid response he gave to my expressed concern about whether these business development contacts were ethical. Ellison looked me straight in the eye and said: "Ethics are rules old men make to keep young men from getting any business."

Little did Bill Ellison realize how prophetic that remark was. Fifteen years later the Department of Justice filed a restraint of trade action, and obtained a consent decree under which ASCE was forced to remove from its Code of Ethics an article which prevented the submission of priced proposals on engineering projects.

You may still have professional concerns and personal opinions about the ethics of submitting priced proposals, but let me assure you it is no longer a valid basis for determining whether someone is qualified for membership in ASCE. And for DoJ, this was only a modest beginning.

EVERYBODY'S TALKING ABOUT IT

Old ethical values and standards are crumbling in so far it can be shown that they constitute restraint of trade.

In 1975 the Department of Justice again came to ASCE and said we want access to all ASCE files that contain material pertaining to Articles 3, 5, and 7 of ASCE's Code of Ethics. No explanation was given, but it soon became clear that concern over restraint of trade was again the immediate cause for DoJ's action.

In January 1976 the Attorney General in the State of Arizona notified the Arizona Section of the Attorney General's contention that Articles 3, 5, 6, and 7 of ASCE's Code of Ethics constitute a restraint of trade in Arizona.

When a Code of Ethics has only nine articles, removal of four of them has a substantial impact. In fact, for ASCE, it was pretty much back to the drafting board with only one drafting instrument—the *GOLDEN RULE*.

With the letter from Arizona's Attorney General came a neat package. It contained:

1. A draft summons
2. A draft complaint
3. A draft stipulation
4. A draft judgment in the form of a consent decree
5. A threat that failure to negotiate a settlement with the Attorney General would expose ASCE to several things:
   a. A prolonged, expensive, and most likely losing court battle.
   b. The possibility that a court might order a less acceptable settlement than a negotiated compromise.
   c. The naming of individuals as defendents who would be subject to civil penalties.
   d. And the possibility that successful litigants in restraint of trade actions might obtain treble damages from named dependents.
6. Finally, the packet contained a summary statement to the effect that ASCE had 4 to 6 weeks to obtain counsel and to reach a mutually acceptable solution with the State Attorney General. Justice is swift in Arizona. Let's look at articles 3, 5, 6, and 7, the challenged articles.

· Article 3 says that an ASCE member should not take any positive steps to try to supplant another engineer once definite steps have been taken to hire that other engineer.
· Article 5 says that an ASCE member should not review the work of another engineer for the same client except with the knowledge of the first engineer unless the first engineer's contract has been terminated.
· Article 6 says "no self-laudatory advertising."
· Article 7 says "no moonlighting."

ASCE's constraints on advertising in practice have been treated as a ban on anything but professional cards in trade magazines. ASCE's code in fact permits display advertising of a *factual* nature in *dignified* publications. DoJ feels it has a particularly strong case against constraints on advertising. That is why you have seen much in the news media about DoJ actions against lawyers, doctors, CPA's and any other professional group that restricts advertising by its members. Informed opinion says it is only a matter of time, possibly quite a short time, before this ethical constraint is removed from ethical codes of the *so-called* learned professions.

POST-WATERGATE CREDIBILITY

Ironically, and to some extent paradoxically, all of this DoJ activity to weaken ethical constraints comes at a time when many professions and businesses are seeking to improve ethical standards, seeking to strengthen credibility in a post-Watergate society.

- Engineers, civil engineers in particular, are reeling from revelations of many reported instances of bribes and payoffs, related to engineering contracts, reaching as high as the Office of the Vice President of the United States.
- Bankers are reeling from revelations that prestigious banks and bankers in New York City were less than candid, or less than careful, with regard to full disclosure of financial information in securities purchases and reofferings for New York City and related public agencies.
- The association executive must give thoughts to ethics if he is to be more than an automaton, says a recent issue of *Association Management* magazine.
- Doctors are wrestling with the ethics of maintaining life-support systems and drug therapy for patients who are beyond hope of recovery, with the ethics of altering behavioral patterns through drugs and surgery, and with the ethics of creating behavioral patterns through genetics. *Bioethics* in medicine it is called.
- All professions, most noticeably the medical profession at the moment, are struggling with the financial and ethical consequences of professional liability judgments that threaten the integrity, the stability, the very existence of private professional practice.
- Engineering professionals in private practice meanwhile are struggling with the financial and ethical consequences of both regulation and competition from federal and state bureaucracies.

ETHICS SPUN FROM FAIRY TALES

There seems little we can do for the moment about the problem of professional liability other than to paper our trails with risk-reducing reports, plans, specifications, contracts and disclosure documents.

We can, however, develop a *new* code of ethics that is consistent with the best interest of both the engineering profession and the public it serves. To do so we must go back to basics.

Last December the *New Yorker* magazine published a delightful article by Bruno Bettelheim, a psychoanalyst. (Now available as a book entitled *The Uses of Enchantment.*) The article was particularly refreshing in that it exploded the myth that fairy tales are somehow damaging for young children. It made a strong case for the use of fairy tales, for the use of enchantment in helping young children to form systems of values, indeed to help them find meaning in life.

Bettelheim's definition of finding meaning in life is beautifully simple.

> To find meaning in life a child must become able to transcend the confines of a self-centered existence and believe that he will make a significant contribution—if not right now, then at some future time.

I particularly encourage any of you who are in the process of raising young children to read the article. But my purpose here is not to send all of you back to Grimm's volume one. My purpose is to stress that ethical and moral values are formed at a very early age.

In Bettelheim's words,

> The child must be helped to bring order into the turmoil of his feelings. He needs—and the point hardly requires emphasis at this moment in our history—a moral education that subtly, by implication only, conveys to him the advantages of moral behavior, not through abstract ethical concepts but through that which seems tangibly right and therefore has meaning for him. The child can find meaning through fairy tales.

WHAT IS A CODE OR SYSTEM OF ETHICS?

For a clue to answering the question, "*What is a code of ethics?*", look at the two unwritten articles in all codes of ethics:

- Love thy neighbor as thyself (Leviticus XIX, 18; and Matthew XIX, 19).
- Do unto others as you would have them do unto you (Golden Rule).

Sigmund Freud, a skeptic, reflected on loving thy neighbor in his paper on "Civilization and its Discontents." To Freud loving thy neighbor made no sense at all in a romantic or personal sense, but if you and I think of loving thy neighbor as a matter of business ethics, ethical love rather than in terms of romance or personal attachment, it makes good sense.

The attorneys, in *their* code of ethics, take an approach slightly different from that of the engineers. They first state a brief ethical principle, and call it a *Canon*. They follow that with a more detailed explanation of the basic principle, and call the explanation *Ethical Considerations*. Finally, for those attorneys who are really obtuse, or who simply don't want to understand, the code spells out in detail some examples of nonconforming behavior and calls these *Disciplinary Rules*.

The suggestion that some attorneys (and engineers) simply may not want to understand, brings to mind a column by Charles McCabe in the San Francisco Chronicle. In it, McCabe philosophized on a brief passage written by Charles Kingsley, a 19th-Century novelist and Cambridge professor. "*The two brief paragraphs,*" McCabe contended, *contain almost the wisdom of a lifetime.*"

The ideas expressed are relevant to this discussion of ethics. Kingsley wrote:

The human race may for all practical purposes be divided into three divisions. First, the honest men who mean to do right, and do it; second, the knaves who mean to do wrong, and do it; and third, the fools who mean to do whichever of the two is pleasanter at the moment.

And these last may be divided into black fools and white fools. The black fools are they who would rather do wrong than right, but dare not unless it is the fashion; while the white fools are they who would rather do right than wrong, but dare not do it unless it is the fashion.

For another clue to answering the question, What is a code of ethics?, look at what is left of ASCE's Code of Ethics if we acquiesce to Arizona's Attorney General and remove articles (3, 5, 6 and 7) held in restraint of trade.

- Article 1 says act only as a faithful agent or trustee for your client or employer.
- Article 2 says accept remuneration for services only from your client or employer.
- Article 4 says do not attempt to injure, falsely or maliciously, the professional reputation, business, or employment position of another engineer.
- Article 8 says do not exert undue influence or offer, solicit or accept compensation for the purpose of affecting negotiations for an engineering engagement.
- And Article 9 says do not act in any manner derogatory to the honor, integrity or dignity of the engineering profession.

On the one hand the challenged articles (3, 5, 6 and 7) speak basically to business related matters—and imply downright unfriendly behavior, possibly even scandalous behavior, but not malicious behavior.

On the other hand the remaining articles speak directly to malicious, felonious, in short even criminal behavior—behavior that clearly is not in the best interest of either the profession or the general public.

REWRITING ASCE'S CODE

It was with something like this in mind when, in the Fall of 1973, an ASCE Subcommittee on the Code of Ethics recommended that the code be rewritten to reflect these principles:

- Stress the responsibility of the professional engineer to the public and to advancement of human welfare.
- Restrict the code to matters of moral and ethical concerns, as constrasted with matters related to how to operate a business.

The subcommittee's report noted that even then the Department of Justice was starting to move against professional societies in restraint of trade matters. It also noted that the consensus opinion of subcommittee members was that now (September 1973) was the time to start rewriting ASCE's code due to the lead time required for consideration and adoption of such an important document. Justice may not always move swiftly, but it has moved swifter than ASCE in this situation.

At about the same time (September 1973) the Engineers' Council for Professional Development (ECPD) had a committee developing a Code of Ethics to which, it was hoped, all professional societies might subscribe.

At that time ASCE's code subcommittee had some differences with the ECPD code committee, but these were resolved and now the ECPD/ASCE Code of Ethics is in force.

ASCE's Board of Direction has adopted ECPD's statement of Fundamental Principles, ECPD's Fundamental Canons (Code of Ethics), and new "Guidelines to Professional Pratice under the Fundamental Canons."

The guidelines are lengthy, containing more than half again as many words as the current guidelines. The proposed code appears free of restraint of trade problems, and the guidelines provide that engineers may advertise in a professional way. References to supplanting one another, to reviewing the work of others, and to taking advantage of a salaried position have been deleted from the canons and guidelines.

Will we now have a simon-pure code, certified for all times by the Department of Justice. Or are we looking through 1976 glasses with a particular mind set, which will be altered in another ten years by changing judicial, legislative, perhaps even professional standards? Whatever ensues, for the present we have a new useful code of ethics.

THE COURSE OF A PROFESSIONAL CONFLICT ACTION

When I was a student at Cornell, in the early 1940's, the civil engineering school administered examinations under the honor system. Students taking tests were monitored only by their conscience and their peers—there was no instructor or professor in the room.

My first exposure to professional conduct actions was as a member of the three-person Honor Committee that acted as judge and jury in alleged violations of the honor system. Taking exams in an honor-system setting was a pretty heavy load for many people. Being asked, as a member of the Honor Committee, to judge your peers, was at times an almost unbearable responsibility.

On one occasion a member of my fraternity was brought before the Honor Committee on which I sat. Conflict of interest? You bet! Mixed emotions? Definitely! But we were less sensitive to conflicts of interest 30 years ago, and there was no easy way out. The evidence was not disputable, and the integrity of the honor system was maintained. But passing sentence in that case was one of the more uncomfortable things I did as an undergraduate.

More recently, in one of the professional conduct actions brought to ASCE's Board of Direction during my three-year term as a director, the accused was an acquaintance of the president of my firm, who is, in fact, my employer. Prior to the date scheduled for a board hearing of the case, the accused called my associate and asked him to intercede with me on his behalf. The action incensed me and destroyed my ability to view the case with detachment. Consequently, when the case was heard, I abstained from voting on findings or disciplinary action. This is a possible course of action for any board member, and generally it will not influence the outcome, since the judge and jury consists of a 28-person board.

I cite these examples to illustrate the drama, emotion, and human interest in the procedure I am about to describe—a procedure that might otherwise sound very sterile.

The procedure is very carefully drawn from ASCE's constitution and bylaws, very carfully prescribed in writing with assistance from legal counsel, and very carefully administered by Society staff, officers, and the Committee on Professional Conduct. This is essential to protect the rights of the accused, and to protect the Society and individual members from possible libel actions by members accused of violating the Code of Ethics. Confidentiality is stressed at all stages of the proceedings.

At this point I must stress three things:

- First, the National Board of Direction is the only entity that can make a finding that ASCE's Code of Ethics has been violated, and prescribe disciplinary action.
- Second, The Committee on Professional Conduct, appointed by the board and consisting of former board members, is the only entity that can direct a formal investigation, and recommend that the board conduct a hearing or that the charge be dropped without a hearing.
- Third, the only way other ASCE members can or should get involved, aside from bringing possible code violations to the Executive Secretary's attention, is in response to a request for assistance from the Committee on Professional Conduct.

Initiation of Proceedings. A disciplinary proceeding by the Board of Direction against a member originates only upon recommendation by the Committee on Professional Conduct (CPC), or upon the written request of ten or more ASCE members. Before the case is presented to the board, however, often long before, a complaint alleging misconduct or information indicating possible misconduct comes to the attention of CPC. The case may come to CPC's attention in various ways, through news stories, in letters from members, in petitions from groups of members, or by word of mouth.

CPC Review and Recommendations. CPC reviews the complaint or other information and decides whether the preliminary charges or evidence justify further investigation. Because of potential and sometimes unwarranted embarrassment or harassment for the accused, because of the time involvement for staff and officers and members, and because of costs, particularly legal costs associated with each investigation, the decision to proceed with an investigation is never made lightly.

If CPC decides that the preliminary charges or evidence warrant further investigation, CPC has several alternative ways to accomplish its investigation, but always stressing confidentiality.

- A member or subcommittee of CPC may conduct the investigation.

- CPC may ask an ASCE staff member to conduct the investigation.
- CPC may ask an appointee or committee of an appropriate ASCE section to assist with the investigation.
- CPC may designate legal counsel to assist with the investigation.

Except in the special case where 10 or more Society members petition for a board hearing of a case, CPC completes its investigation and makes one of two possible recommendations:

1. That facts do not support the charges or are inconclusive, and that the case should be dropped, or possibly tabled if it appears that more evidence may be available at some future date.
2. That the facts do seem to support the charges, and that the board should schedule a professional conduct hearing.

At this stage of the proceedings, the board notifies the member of charges against him and invites him to present his defense at the scheduled hearing.

Also at this stage the member has the right to resign from membership. This would be a resignation with prejudice towards readmission, notice of which may be published in *Civil Engineering* magazine.

Board Hearing.   Unless the member resigns, the board proceeds with a formal hearing. The accused and his counsel may be present throughout the hearing which includes:

- An introductory statement by the presiding officer describing the proceedings, and cautioning board members to limit their deliberations to charges contained in the formal statement of charges.
- An opening statement by CPC of the charges and a recommendation for some specific board action.
- An opening statement by the accused or his counsel, if present.
- Presentation of evidence by CPC.
- Presentation of defense, if accused is present.
- Questions by the board addressed to CPC or the accused.
- Closing statements by CPC and the accused.
- Conclusions of the hearing followed by closed deliberation and action by the board.

Disciplinary Action.   If the board finds that the evidence sustains the charges, it can expel the member, suspend him for a stated period, or give him a written admonition. The board also decides whether to publish notice of its action in *Civil Engineering* magazine.

So there you have a carefully structured, legalistic process that is often touched with drama, strange turns of events, poignancy, and at times suspense.

CASE EXAMPLES

Board hearings during the last three years have dealt with bribes and payoffs to influence the selection of engineers, sometimes in high places; with attempts to supplant another engineer; with illegal political contributions; and with lesser matters.

To avoid unnecessary exposure my discussion of case examples will center on some hypothetical situations and on a survey of reactions by some ASCE members to these situations.

First, though, the matter of political contributions will serve to illustrate the potential to violate federal law, and hence ASCE's Code of Ethics. And the violation may be due to nothing more than naivete.

By now most of you know that the U.S. Code (Title 18, Section 610) prohibits any contribution or expenditure *by a corporation* in support of a political party committee or candidate in a federal election. The penalty for violation of Section 610 is imprisonment for not more than one year and a fine of not more than $1,000 for officers and directors who consent to the contribution. The corporation is liable for up to $25,000 and officers and directors who commit a willful violation are liable for a fine of up to $50,000. An ASCE member responsible for such a corporate practice would be in violation of Article 9 (conduct unbecoming a professional engineer) and in certain cases Article 8 (undue influence to solicit work).

But did you know that the U.S. Code (Title 18, Section 611) more specifically prohibits *an individual or corporation* from making a contribution to a political party, committee, or candidate in a federal election while the *individual or corporation* has or is negotiating a personal services contract with the United States or any department or agency thereof? The penalty for violation of Section 611 is a fine of not more than $25,000 and imprisonment for not more than five years. Again an ASCE member convicted of such an act would be subject to a professional conduct action. More than one engineer has unwittingly come a cropper on this federal law.

What follows are some case examples dreamed up by the Professional Activities Committee of the Wisconsin Section, ASCE, followed in each instance by some responses from Section members, and by my own opinion.

Case Study No. 1.   George Ord, a city employee, is the resident engineer for a large sewer contract. With extensive field engineering experience, George is able to suggest techniques and procedures that save both time and money, although the work is done strictly according to the plans and specs.

At Christmas time, George receives a case of good Scotch from the contractor with his greeting card attached.

May George accept the gift?

At the Wisconsin Section meeting 79 percent of the 130 people present believed George should not accept the gift. Yet significantly, 19 percent believed the gift could properly be accepted.

—A consultant said *"No, this is a substantial gift, not a token of frienship or goodwill."*

—One government employee believed it OK saying *"anything under $100 is OK."*

—Another consultant said, *"No gifts should be accepted."*

—Still another consultant said *"Yes, if not more than $25."*

—An industry participant said *"no,"* but commented, *"it has been historic practice in the construction industry to present such gifts as a bottle of liquor at Christmas."*

No too, in my view, said Article 2 of ASCE's earlier Code of Ethics: *"It shall be considered unprofessional to accept remuneration for services rendered other than from his client or his employer."* If remuneration means anything of value, then the case of Scotch is remuneration.

No, also says the new Procurement Regulations of the Environmental Protection Agency, if this happens to be an EPA-financed sewer project. A section of the regulations titled "Code or Standards of Conduct" says, *"The grantee's officers, employees or agents shall neither solicit nor accept gratuities, favors, or anything of monetary value from contractors or potential contractors."* Here we have a federal agency, perhaps reacting to Watergate, prescribing ethics.

It's tough, but the only safe course is California's Governor Brown's zero-gratuity policy.

Case Study No. 2.   As plant engineer for Lotsa Chrome, Inc., Jim Smith knows that the manufacturing process results in periodic discharges of cadmium and chrome in Deadfish Creek in concentrations which may cause serious long-term health effects for downstream water users.

Because Lotsa Chrome, Inc., is marginally profitable, management has made a policy decision to close the plant when and if waste water controls are required.

When questioned by the Department of Natural Resources (DNR), Jim's boss understates the levels of cadmium and chrome.

Must Jim provide DNR the correct information?

At the Wisconsin Section meeting 46 percent voted *NO* on the question, but many of these believed Jim had an obligation to take affirmative action other than directly with DNR. Fifty-two percent voted *YES*, but many of these indicated their *YES* vote meant a correction of the data by other than direct reporting to DNR.

—One participant commented after voting NO, *"Jim has an obligation to put as much pressure as possible on his employer. If not supplied to DNR, he should change employers."*

—Another also voted NO, but said *"Jim should notify DNR as to the falseness of the report only."*

—Still a third observed, voting YES, *"otherwise where is your conscience?"*

—A No voter stated, *"The information is privileged."*

—A government employee voted YES and stated, *"If Jim was a witness to a hit-and-run accident involving a Lotsa Chrome truck, must he report this to the authorities? Yes, of course."*

Here we have much more of a moral-economic dilemma for Jim Smith, our case-study engineer. ASCE's earlier code spoke only indirectly, though really quite clearly to the situation when it said in Article 9: *"It shall be considered unprofessional to act in any manner derogatory to the honor, integrity, or dignity of the engineering profession."* If Smith cannot convince his boss of the error of his way, under ASCE's code Smith's only choice seems to be to resign his position, and to then go about notifying the proper authorities about the facts as he knows them.

Under the new ECPD/ASCE Code Jim's choices are about the same, but they are spelled out in more detail.

Canon 6 says: *"Engineers shall associate only with reputable persons or organizations."*

Canon 1 says: *"Engineers shall hold paramount the safety, health and welfare of the public in the performance of their professional duties."*

And the new Guidelines to Practice Sections 1c(c.3) say: *"Should the engineers' professional judgment be overruled under circumstances where the safety, health and welfare of the public are endangered, the engineers shall inform their employers of the possible consequences and notify other proper authority of the situation, as may be appropriate."*

"Should engineers observe conditions which they believe will endanger public safety or health, they shall inform the proper authority of the situation."

Jim Smith's choice is clear, though perhaps a tough one.

Case Study No. 3.   Wisconsin Case Study No. 3 again involves a gratuity, and the comments for Case Study No. 1 apply.

Case Study No. 4.   City Engineer Paul Hopeful is convicted of filing a fraudulent Federal Income Tax Return.

Is there a Code of Ethics violation?

The surprise at the Wisconsin Section meeting was that 7 percent believed there was no code violation.

—One participant stated, *"I don't believe that the code should influence our entire lives, if it has nothing to do with our professional actions or decisions."*

—A consultant concluded, *"He violated the express letter of the Code, but I find it difficult to sit in judgment of his ethics in his engineering profession."*

At the Wisconsin Section meeting, however, 90 percent believed that this was a code violation. Three percent abstained from voting.

Here again we have to go to the general article in ASCE's earlier Code of Ethics, Article 9, which said: "It shall be considered unprofessional to act in any manner derogatory to the honor, integrity or dignity of the engineering profession."

The new ECPD/ASCE Code is not very explicit on this question which has both legal and moral aspects. The new code does say in Fundamental Principle No. 2: *"Engineers uphold and advance the integrity, honor and dignity of the engineering profession by being honest and impartial, and serving the public, their employers and clients."*

In any case, Paul Hopeful should abandon all hope. He clearly has violated any professional code of ethics.

Case Study No. 5.   Wisconsin Case Study No. 5 again involves gratuities, apparently gross gratuities, and the comments for Case Study No. 1 apply.

Case Study No. 6.   Jack Jones is chief engineer for Excellent Company, a manufacturer of recreation vehicles.

Jack, who is very photogenic, is asked to appear in TV ads endorsing the company's snowmobiles.

The ASCE Code of Ethics provides that a member shall not endorse products or processes in commercial advertisements.

Is there an ethical distinction since Jack is responsible for the engineering quality of the snowmobile?

By a two to one vote, the participants at the Wisconsin Section meeting believed that it was acceptable practice for Jack to appear in the TV commercials.

—One government employee observed that *"if it's in the Code, it should be removed."*

—A consultant believed *"there is a definite need for a change in the code."*

—But one consultant stated, *"It is wrong to do what he did."*

—Still another participant believed that *"it comes down to the public viewpoint rather than a question of ethics and, if the general public thinks there is a violation, he brings dishonor to his profession."*

—An educator stated, *"There is no ethical wrongdoing."*

The earlier ASCE Code was quite specific in Article 9 (4), where it said: *"(The professional engineer) shall not endorse products or processes in commercial advertisements."*

Jack Jones perhaps can be a model on TV, if his employer doesn't object, but he cannot endorse an engineered product.

Case Study No. 7. Tom Wellborn is retained by the Town of Easy Acres to design a swimming pool.

Steve Petite, a salesman for Clear Water, Inc., provides Tom a layout for the circulation and filtration equipment which includes equipment produced by Clear Water Inc.

Tom reproduces the layout on his plans.

Is this practice acceptable?

At the Wisconsin Section meeting 74 percent of those voting believed the practice is acceptable. Yet most qualified their YES vote by conditions:

—*"If 'or equal' is stated."*

—*"providing his research confirms the plans are OK."*

—*"if he reviewed others and feels this is best."*

—One attendee voted NO observing *"I do not agree to using my P.E. signature on any design not mine."*

—A consultant who also voted NO stated, *"The design is not his, similar to plagiarism."*

The earlier ASCE Code was not very specific on this point, but did say in Article 2(1): *"(The professional engineer) shall not accept compensation from more than one interested party for the same service, or for services pertaining to the same work under circumstances where there may be a conflict of interest without the consent of all interested parties."*

The new ECPD/ASCE Code is more specific in two instances. Article 2c says: *"Engineers shall not affix their signatures and/or seals to any engineering plan or document dealing with subject matter in which they lack competence by virtue of education or experience, nor to any such plan or document not prepared under their direct supervision and control."*

Article 4d says: *"Engineers shall not solicit nor accept financial or other valuable considerations, including free engineering designs, from material or equipment suppliers for specifying their products."*

Tom Welborn jumped off the deep end, and the code of ethics will not come to his rescue.

Case Study No. 8. Doug, Tom, and Bill graduated from State U in 1940 and were fraternity brothers and close friends. Doug and Tom eventually became partners in a consulting firm while Bill, who was class president, continued a political career and is now governor. Doug and Tom personally contributed $500 and $2,000, respectively, to Bill's 1973 campaign. During 1974, their firm received three design contracts from state agencies involving $750,000 in fees over a two-year period.

These facts were recently exposed by the press.

Has the Code of Ethics been violated?

At the Wisconsin Section meeting 64 percent believed under the facts stated there was no code violation. A high of 8 percent abstained from voting.

—One consultant who believed the code was violated stated, *"They should consider the possible dishonor if they are awarded a contract."*

—A student voted NO but asked, *"What was the intent."*

—Another consultant believed it OK, *"if the contracts were awarded properly."*

—A participant who voted YES believed it was a form of bribery.

Article 8 (1) of the earlier ASCE Code said: *"(The professional engineer) shall not make political contributions for the purpose of influencing the selection of engineers on future engagements."*

Article 5a of the new ECPD/ASCE Code says: *"Engineers shall not give, solicit, or receive either directly or indirectly, any commission, political contribution, or gift or other consideration in order to secure work, exclusive of securing salaried positions through employment agencies."*

I believe, however, that in the case of Doug, Tom, and Bill, more evidence is needed. Was the size of the contribution unusually large or illegal? Did other engineers make contributions of a similar nature to other candidates, or to the same candidate? Was there any relationship between the size of the contribution and the size of the fee? Was there any valid criticism of the quality of the engineering work? Were Doug and Tom getting substantially more state work than other consultants?

ASCE encourages engineers to participate in the political process, and to support the candidates of their choice. In a situation like Doug, Tom and Bill find themselves, however, some special discretion is warranted.

It is worth noting that none of the case examples selected by the Wisconsin Section relate to articles 3, 5, 6 and 7 in the earlier ASCE Code. These are the articles which are challenged as being in restraint of trade.

PROBLEMS OF ENFORCEMENT

Problems with respect to code enforcement fall into several categories.

With respect to the code itself there is the continuing problem of omissions, ambiguities, and changing conditions and philosophies in the setting in which engineering is practiced.

· The threat of libel actions by those disciplined tends to deter code enforcement in some instances.

· Lack of peer support, and the need for better education of practicing engineers with respect to the nature of the code of ethics, influence the effectiveness of the code.

Code Omissions, Ambiguities, and Changing Conditions.   For the immediate future the problem of omissions, it is hoped, will be taken care of by the new ECPD/ASCE Code. I have said enough about that elsewhere.

Some ambiguities are apparent in the case examples. When is a political contribution a bribe or payoff. Is a free lunch a gratuity? What about a case of Scotch or a discounted set of tires? But the most interesting ambiguity to me is a footnote to the earlier ASCE code which it had been proposed to carry over to the ASCE version of the ECPD code. It is a clear condonement of situation ethics. That footnote says:

> On foreign engineering work, for which only U.S. engineering firms are to be considered, a member shall order his practice in accordance with the ASCE Code of Ethics. On other engineering works in a foreign country he may adapt his conduct according to the professional standards and customs of that country, but shall adhere as closely as practicable to the principles of this code.

Large international engineering corporations presumably are adapting their conduct to the professional standards and customs of the Arab nations when they comply with the boycott of Israel. The U.S. Department of Justice successfully sued one such corporation for being in restraint of trade, while the U.S. Department of State reportedly felt that such a suit would adversely affect U.S. foreign policy. Meantime ASCE has adopted a policy statement which says ASCE endorses the principle of complete freedom of movement in international engineering and construction by individuals and organizations of all nations, and another position statement which says ASCE opposes anti-boycott legislation. And the U.S. Department of Justice is suing ASCE with regard to certain articles of the Code of Ethics which allegedly restrain trade. Maybe the way to unscramble all this is for ASCE to remove from the code all ethical constraints that relate to business practices including the footnote about foreign engineering work. Then enforce the ECPD/ASCE Code of Ethics, and let the chips fall where they will.

Some have even suggested that ASCE go a step farther, and have a code of ethics for guidance only, with no provision for enforcement or disciplinary procedures.

The Threat of Libel.   The threat of of libel comes to a professional society from two directions in code enforcement matters. First, and most basic, the procedure for investigating, hearing, and acting on disciplinary matters must be structured carefully to maintain confidentiality and to protect the rights of every accused. A finding of guilt in a code violation must be supported with indisputable evidence.

Interestingly the other libel threat comes straight out of Roberts *"Rules of Order"* where it says:

> If (after trial) a member is expelled, the Society has the right to disclose the fact that he is no longer a member—circulating it only to the extent required for the protection of the Society or, possibly, of other organizations. Neither the Society nor any of its members has the right to make public the charge of which an expelled member has been found guilty, or to reveal any other details connected with the case. To make any of the facts public may constitute libel. A trial by the Society cannot legally establish the guilt of the accused, as understood in a court of law; it can only establish his guilt as affecting the Society's judgment of his fitness for membership.

To do other than what is stated in Roberts *"Rules of Order"* could evidence malicious intent, and so establish a case for libel. So the Society must tread the fine line that separates libel from consideration of public interest and the need to inform Society members about what constitutes a breach of the code of ethics.

Informing Members about the Code. ASCE must keep members informed about its Code of Ethics both to avoid inadvertent breaches, and to help members understand the purpose of, and need for, the code. The Society needs to keep members informed of professional conduct actions to establish credibility for the code and to act as a limited deterrent. Only through an informed membership can breaches of the code be brought to the attention of Society headquarters so that disciplinary actions can be initiated.

Publications in February 1976 *"Civil Engineering"* magazine of the proposed ECPD/ASCE Code of Ethics was an important effort to inform Society members about how a code evolves and of what it consists. Interesting comments were elicited.

It is very difficult for members to develop an informed opinion on the practicability of the code unless they have served on the Board of Direction or the Committee on Professional Conduct and have seen what is involved in applying the code and in maintaining it in the face of challenges from defendants and from the Department of Justice.

Case examples probably come the closest to demonstrating how the code works. But if they approach reality too closely, they are loaded with the danger of exposing the Society or its members to libel actions.

Another important effort to inform Society members about engineering ethics, codes of ethics, and responsibility of the engineering profession to the public occurred in March, 1977, possibly at Ohio State University. ASCE sponsored, with other organizations, a two-day specialty conference on education for ethics and professionalism.*

ETHICS AND PROFESSIONAL LIABILITY

Too little concern about professional liability clearly can lead a firm or individual into a professional liability exposure. Paradoxically, too much concern about professional liability can lead a firm or individual into a professional liability

---

*Author is apparently referring to the ASCE Specialty Conference on Ethics, Professionalism and Maintaining Competence, held at Ohio State University, March 10–11, 1977.

exposure. And somewhere in the mix is an ethical as well as an economic dilemma.

Overly-Cautious Design. The most direct example of this ethical and economic dilemma is the situation where an engineer is forced, by professional liability concerns, to develop an overly-cautious, and so overly-costly design. The ethical dilemma is clear. For his own peace of mind, and his own economic security, the engineer has developed a design that may have unreasonable cost implications for the client. Is that in the public interest?

Perhaps nuclear power plant design is a case in point, though the question there seems to be are they safe at any cost.

Fees Based on Construction Cost. This again presents an ethical as well as an economic dilemma. Can an engineer design objectively, and arrive at the most cost-effective solution, when his fee increases directly with the construction cost? Many professionals agree that proper incentive is lacking, and many are finding other bases for establishing fees. Many public agencies simply are not permitting consulting agreements based on a percent of the construction cost, and instead requires contracts based on cost reimbursement, fixed price, or per diem arrangements.

Fees based on a percent of construction cost due have the advantage of ease of administration. The method is at times useful in special situation, particularly where a small firm does not have the complex accounting system necessitated by other fee arrangements.

Fees Based on a Finding of Feasibility. This may present an economic dilemma for some, but the ethical implications are clear. *"Guidelines to Practice for the proposed ECPD/ASCE Code of Ethics"* summarize the matter in Article 5e: *"Engineers shall not request, propose or accept professional commissions on a contingent basis under circumstances under which their professional judgments may be compromised, (or when a contingency provision is used as a device for promoting or securing a professional commission)."*

Restraint of Trade. Restraint of trade represents a special kind of professional liability threat. It stems, not from error or omission, but from commission of some business practice which can be shown to restrain trade. It represents a financial liability only when someone can shown that they experienced financial damage due to the particular restraining practice. Since the penalty is the award of triple damages, it can be a significant financial liability.

Generally the liability action takes the form of a class-action suit against a group of firms. The construction industry has seen some examples of this in the cases where public agencies have filed group actions against electrical equipment manufacturers and construction materials suppliers.

The threat at the moment is that some group action could be filed against ASCE and its members because of certain articles in the code which restrain trade in our profession.

Engineering News-Record of January 6, 1977, reports that the State of New Jersey has settled out of court a $61.7-million antitrust suit brought against 15 individuals and 12 companies accused of conspiring to rig highway bids from 1957 to 1970.

Law Violations. Any infraction of civil or criminal codes by professional engineers generally carry with them both financial liability and ethical conduct penalties. From an ethical standpoint law violations fell under Article 9 of ASCE's earlier code which relates to conduct unbecoming a professional engineer, and now fall under similar provisions in the new code.

Ethical conduct disciplinary actions have ensued from such things as income tax evasion, bribes and payoffs, and illegal political contributions.

SUMMARY

New ECPD/ASCE Code. If you are an ASCE member I urge you to read the proposed new *Code of Ethics* including the "*Guidelines to Practice.*" It is something everyone should do at least once. Your continued membership in ASCE is in fact contingent on your complying with the code.

The "*Fundamental Principles*" and the "*Fundamental Canons*" are brief. The proposed "*Guidelines to Practice*" are quite long, half again as long as the earlier guidelines. The hope is that they are more comprehensive, easier to understand, and easier to interpret and enforce.

DoJ Threat. Some people feel very threatened by the Department of Justice intrusion into ASCE's Code of Ethics. But what really is the impact of DoJ's activities? DoJ doesn't say you must advertise in self-laudatory fashion, or that you must submit priced proposals on all projects, or that you must review the work of other engineers without their knowledge, or that you must try to supplant other engineers on projects where they are established. DoJ doesn't even recommend moonlighting.

DoJ merely says that you must not conspire with others not to do those things, because that restrains trade, and restraint of trade is illegal under antitrust laws.

In any field there will always be those who cut corners and who follow devious business practices. And there are clients who will hire them once. But by and large good people will continue to follow wholesome business practices and provide quality services. My firm is in a field, municipal financing consulting, where there are no prescribed business ethics. In fact there is no association to prescribe them. And yet most people in our field meet ASCE's ethical standards.

To paraphrase Kingsley: There are honest men who mean to do right, and do it; there are dishonest men who mean to do wrong, and do it; and there are those who will do whatever is pleasanter at the moment.

I am convinced that removing a particular article from the "*Code of Ethics*" will make no substantial difference in the way engineers behave.

The code and guidelines must, however, formulate certain standards of behavior that stress public interest, set the moral tone for the profession, enhance relations with other professionals, increase the pleasures derived from the practice of engineering, and encourage excellence.

Thoughts on Engineering Ethics. Participating in professional conduct hearings, as a member of ASCE's Board of Direction, had its inspiring moments. Here are some ideas expressed in various ways at those hearings.

First, here is an exerpt from a paper titled "*Blueprint for Scandal: The Engineering Business in New Jersey*":

> Thus, the major political scandals involving New Jersey's most prominent political machines all provide central roles for professional engineers. It could just be a remarkable coincidence. On the other hand, it would seem difficult to squeeze graft out of construction contracts where able, honest engineers ride herd on price and performance. The licensed professional engineer often occupies a position with great potential for uplifting the moral tone of government affairs. It is also a position which can be misused for personal gain and public harm.

On another occasion the question arose in a board hearing as to what happens when an ASCE member is convicted of a violation of the Code of Ethics and is expelled or suspended from membership in the Society? One of the directors pointed out that ASCE can go no further than to make a determination as to whether the member has violated the Code of Ethics. Beyond that it is up to engineers and others among whom the suspended or expelled member works to decide the significance of the action. Is such an engineer fit to lead others, to be associated with other engineers in practice, to participate with other engineers in joint ventures? Only interaction with other professionals, and with the public at large, can decide such questions.

Throughout the professional conduct hearings in which I participated I was tremendously impressed by the high value placed by most accused members on their membership in ASCE. Evidence of this ranged from batteries of lawyers and character witnesses amassed by some, to poignant pleas by others. It was at times most impressive, at times intimidating, and at other times very moving. But it was always a reminder of the value each of us should place on our membership in ASCE or in any similar organization.

And finally a parting thought on ethics from William Shakespeare, better than whom no one has ever said it: "*This above all: to thine own self be true, and it must follow, as the night the day, thou canst not then be false to any man.*"

7.9

The Fable of the Consulting Engineer

JAMES A. ROMANO, P.E.

As all good fables should, this one starts with "Once upon a time."

Once upon a time, Childe Engineer, right at the beginning of his traverse and travail across the vast, uncharted wasteland that lies between the campus and maturity, came upon a copy of the NSPE Engineers' Creed. The cursive type, dimly printed on glazed paper, was hard to read but Childe Engineer finally deciphered it; he learned that the Creed called upon him:

- To give the utmost of performance.
- To participate in none but honest enterprise.
- To live and work according to the laws of man and the highest standards of professional conduct.
- To place service before profit, the honor and standing of the profession before personal advantage, and the public welfare above all other considerations.

Childe Engineer was impressed; overwhelmed is a better word. It was as if trumpets had sounded, calling on him to don shining armor, pick up a spear, mount a white steed, and go forth to slay all sorts of dragons. Visions rose before him, visions in which logic and goodness—but especially logic—triumphantly prevailed 100 percent of the time over ignorance, fuzzy thinking, and venality.

In pursuit of those visions, Childe Engineer went forth and offered his services to a number of engineering organizations but, alas, few offered him place. And of the few that offered him place, none seemed to appreciate that, under his burden of callow youth, under his untested store of knowledge, he had the noble purity of a Sir Galahad and, if he were only called upon, he could even find the lost Holy Grail.

After several years of erratic wandering in the wilderness, our Childe Engineer chanced upon a fair damsel who turned out to be the girl he dated in high school— or was it college? One thing led to another and Childe Engineeer proposed a

476

permanent arrangement, *with* benefit of the clergy. Now the fair damsel—like most of the fair damsels who marry engineers—kept hidden beneath her lovely exterior a very practical nature; the lady could add two and two together several times and get a fairly consistent answer. So, she sighed prettily and said, "OK, Childe Engineer, but first you have to cut out this wandering from job to job, find yourself, and come up with the scratch to pay the rent and buy the groceries at the A & P."

Taken aback by the common sense that came so trippingly off the tongue of his fair damsel, Childe Engineer sulked in his tent for a while and then, on a day when no dragons presented themselves for slaying, he betook himself to the Wizard, from whom he sought advice. Now the Wizard was capable of changing his form, sometimes appearing as the Dean of a College of Engineering, sometimes as an uncle, sometimes as a priest or minister, sometimes as a barber. On this particular day, he appeared to Childe Engineer in the guise of Your Friendly Bartender.

"Childe," he said, after hearing our hero's tale of woe, "Childe, if you really are hung up on that girl, you'd better get a job with a consulting engineering firm. They will give you a chance to stick to that Engineers' Creed you're always mumbling about, and pay you a decent salary while you're doing it. And who knows? If you make some money for the consultant, he might even let you in on part of the action!"

Then did Childe Engineer, with lagging steps (for he was loath to swallow his pride and to abandon his complete faith in his store of knowledge, which he confused with wisdom) hie himself to the offices of Williams & Works where he was graciously received by the head guru himself, Scarred Veteran Engineer Everett Thompson.

After some preliminary sparring, in which Childe Engineer seemed to be confessing his brilliance rather than suing for a position, and in which Scarred Veteran Thompson mostly smiled and listened patiently, Childe Engineer finally got to the point. He asked Veteran Engineer Thompson what a consulting engineering organization does and whether there was a place in it for his spear. In reply, Scarred Veteran Thompson said to Childe Engineer:

"Putting it as simply as possible, consulting engineers are in the business of solving other people's problems. Sometimes we are called upon because a client has no one on his staff who can solve his problem. Or he may not even have a staff. In any event, it is up to the consulting engineer to isolate the problem, break it down into its parts, and then come up with a solution or solutions. Another client may have engineers on his staff but, because they are temporarily swamped with other problems, he will call on a consulting engineer to solve his problem. Other clients favor a philosophy of not doing engineering on an in-house basis, maintaining a small group of staff engineers to work with outside consultants. But regardless of the capability of a client, his condition, or his philosophy, the consulting engineer is in the business of solving other people's engineering problems."

Childe Engineer was ecstatic! Imagine being part of such a noble adventure: solving other people's problems! He was sure he had found the place for his talents, his ideals, his longing to do the right thing, a place where he could be fulfilled and still come up with the scratch for the rent and the groceries. He could have his cake and eat, too.

Borne along on this euphoric cloud, he scarcely heard, and certainly didn't heed, the parting words of the Scarred Veteran. It did not get through to him that Veteran Engineer Thompson said, just before sending him to personnel for in-processing: "My Son, there are problems as well as pleasures. There are realities other than the verities of logic and technical knowledge. There are living, breathing, irrational people to be dealt with. There is an ever-present, ever-disruptive force—politics—that must be acknowledged, contended with, and accommodated. There is also perversity: clients are not always enchanted by our personalities or our solutions. People are not always grateful for the help we give them in solving their problems."

But Childe Engineer didn't hear the admonitory words of Guru Thompson; he could see his way clear, all the way to his own nirvana.

And so it came to pass that, in the next 10 years, Childe Engineer wed his fair damsel, begat a number of miniature damsels and engineers, invested in dental braces and dancing lessons for his offspring, bought a home he couldn't afford, cut the grass faithfully, and let his golf game go to pot. He was diligent in his work at Williams & Works, and was ultimately promoted to Project Engineer by Veteran Thompson.

In the next decade, Project Engineer acquired his own office and his own secretary. He was happy, or contented, about 84.2 percent of the time. In those other hours he uneasily and dimly began to perceive that his shining armor was getting a bit dented, a trifle corroded, a little rusty. He also had to reassess his perception of dragons, for he found that dragons don't always sit still or charge straight at you so that they can impale themselves neatly on the spear of logic. He also found that his charger, like Project himself, tended to breathe a bit heavy and might even turn into a plow horse.

Even more disturbing to Project Engineer, however, were the rumblings and mutterings of thunder he heard from over the horizon: from the drafting room, the accounting department, the administrative staff, his peers, and from other professions. He tried to ignore such rumblings, finding comfort in the esteem in which he was held by his now matronly but still fair damsel, his progeny, and his secretary.

His secretary, a woman of strong opinions, fiercely and loyally explained away any of his alleged weaknesses and mistakes. She characterized him, in his role of consulting engineer, as one who had voluntarily assumed the role of employer, assumed the task of counseling those who needed his expertise, and still maintained his professional status as an engineer. For her, he moved in a sort of nimbus as a member of a profession dispensing service based on judgment; one who backed that service with education, ability, and experience; one who competed for those professional assignments while staying within the letter and the spirit of a professional code of ethics and within his own personal code.

While Project Engineer appreciated, and basked in, the plaudits of his secretary, he was aware that others, in and out of his place of employ, were more ambivalent, or disparaging, or even hostile. He had heard that, on occasion, the Chief Draftsperson had characterized him as "not knowing as much as he thinks he does" and as "always wanting to be right." The Editor of the in-house newsletter said of Project, "He is uninformed, administratively inept, and dreadfully ignorant of the basic concept of human motiviations." The Personnel Manager, in an unguarded and unbuttoned moment, scathingly assessed Project as "one who wraps himself in a blanket of self-pity and, feeling hurt and persecuted, finds contentment in a cocoon of irresponsible withdrawal from society."

A psychologist was so unkind as to say that Project, like other engineers, "froze at the emotional age of seven-and-one-half and grew up frustrated, unloved, and rejected, always needing recognition and acceptance." In another encounter, a sociologist friend (?) told Project, above the din of a cocktail party, "It is virtually impossible to name one engineer who is nationally known for his contributions in areas of social concerns."

On those days when one or more such opinions reached Project's ears—ears almost inundated by the flood of paper across his desk—he would skulk home and barricade himself behind the newspaper with a goblet of J&B on the rocks to comfort him. He would not venture outdoors for fear that small boys would throw rocks at him and large dogs nip at his legs.

Near the end of his second decade with Williams & Works, Project was rescued from his winter of discontent, from his slide toward the abyss of self-pity, from his slough of despond. He began to get outside himself and engineering; he began to mingle, physically and mentally, with the rest of the world. He was pushed mightily in this new direction by the scrivenings of Philosopher-Engineer Samuel C. Florman, who suggested that maybe, just maybe, the engineer *doesn't* have all the answers, that maybe, just maybe, the weight of the world doesn't rest entirely on the engineer's shoulders. Philosopher Florman really stung Project when he observed that engineers must have confidence that other members of society are doing their jobs.

Project pondered on such things for many weeks. Finally he said out loud to himself one morning as he was shaving: "Maybe the engineer can't be God." Then he smiled and he relaxed and the tenseness went out of him. It was the beginning of wisdom.

As he rounded out his 20 years of service with Williams & Works, many of the movers and shakers in the organization noted the new aplomb in Project, his growing serenity, and his flashes of wisdom. In recognition, he was invited to become a Partner in the firm. It was what he had wanted, even coveted, for a long time but he nonetheless played out the obligatory charade, humbly pointing out his deficiencies but not to the point where he could not accept the invitation. Veteran Thompson and the other Partner Engineers knew what they were doing, of course. They needed all the help they could get to keep clients satisfied while collecting money due from them in sufficient volume to pay the bills. They also needed help to ward off litigation, to pay for insurance, to say nothing of complying with laws, coping with auditors and—oh, yes—settling momentous internal disputes about the sizes of offices, the shapes of wastebaskets, the number of telephones, the assignment of the parking spaces at the front of the building, who gets the corner office, etc., etc., etc.

Partner Engineer learned about those earth-shaking problems but they became as gnats are to the vultures when he was ushered and initiated into the arcane aspects of being a full-fledged, duly certified, genuine manager of a consulting engineering organization. He felt the tremendous pressure to change the procedures for obtaining assignments. He learned that the traditional ways were being challenged, sometimes with good reason and sometimes not. He was beset with propositions to combine design and construction and others to combine design and construction with financing, all advocated in the hope that the alchemy of synergism would shorten construction time, reduce costs, and foster the application of new ideas, materials, and techniques.

Partner Engineer also learned to live with lawyers at his side. No longer would a firm handclasp or a simple letter suffice to bind a contract between the client-owner and the engineer. He learned that only lengthy, formal documents studded with "whereases," "now therefores," and "parties of this and that other part" would suffice, but only if they were drawn up, reviewed, commented upon by batteries of attorneys.

Despite the advice proffered by one of Shakespeare's characters, "First thing, we kill all the lawyers," Partner Engineer learned to endure and even to enrich lawyers, for litigation had become a way of life for engineering organizations. In the halcyon days when Partner was Childe Engineer, engineers were seldom sued, but it had become quite the vogue to sue and be sued. Partner Engineer's computer told him that, in any one year, he could expect to be involved in one active lawsuit for each million dollars of services his organization performed. He was greatly saddened to realize that acrimony and bile had seeped into the engineer-client relationship and were threatening to destroy it. He deplored the adversary relationship that had developed with many of the people and clients his firm was pledged to help. Partner Engineer despaired of these persistent trends, for he was convinced his clients would be served best in a personal, professional relationship. He shrank from the idea that a client could turn into yet another dragon with whom he would have to joust.

His despair was not alleviated by his induction into the mysteries of professional liability insurance. When he was Project Engineer, he looked upon professional liability insurance as a strictly private affair, as protection his organization bought for itself on its own initiative. He was astonished to learn it was not uncommon for a prospective client to insist or require that his organization carry professional liability insurance, almost as if it were planned from the start to sue the engineer for something at some time in their relationship. The dimensions and the cost of the insurance staggered him. He was astounded to learn that his organization had found it necessary to gradually increase the professional liability insurance it carried until the face amount was as much as 20 percent of the total billing value of the work performed each year. More to his dismay, Partner Engineer found that, even after a sizable deductible was absorbed by his firm, the annual premium was about one percent of the total billing value of the services performed in the year.

Partner Engineer also learned to share a great portion of his time with government auditors. In fact, he spent almost as much time with auditors as he did with lawyers. Bemused by the thought processes of these functionaries, a paraphrasing of Alfred, Lord Tennyson's "Charge of the Light Brigade" kept intruding into his consciousness, epitomizing the unequal battles between engineers and auditors:

> Auditors to the right of them,
> Auditors to the left of them,
> Auditors in front of them,
> Blue penciled and excised;

> Assailed with threats of living hell,
> Boldly they rode—I know not how well,
> Into the jaws of debt,
> Into the mouth of insolvency,
> Rode the profession.

In his natural zeal to be fair, Partner Engineer defended auditors for a while; were they not professionals simply trying to do their job? His defense was shattered, however, when Scarred Veteran Partner dryly asked, "Did you ever hear of an auditor from the government who told you to add some charges to those you claim? Did you ever hear of an auditor who didn't feel he simply *had* to find something to *disallow* in your charges?"

Sensing that he had captured Partner Engineer's mind as well as his ear, Scarred Veteran threw in the gritty observation that "unless and until engineers resist unreasonable regulations and audits, they don't deserve to call themselves professionals; they will be no more than technicians carrying out the dictates of others."

As Partner Engineer toiled in the inner chambers of his organization, he grew in wisdom but he was increasingly saddened by the persistent intrusion of crassness into his idealistic vision. He was dismayed that his overtures to do good were often misunderstood and frequently rebuffed; he was disheartened when personal pique subverted logic; he was hurt when good was repaid with evil; he was confused to observe that some people could forgive him for being *wrong* but could not bring themselves to forgive him when he was *right*.

Partner was suffering from mid-life crisis. He toyed with the notion of choosing another way of making a living; he wondered about the need for a midcourse correction. He wondered where and how his spear had ceased to be exclusively directed at tangible technical problems and had become principally tilted at elusive dragons such as auditors, depressions, people problems, recessions, earnings, litigation, politics, insurance, etc.

His disenchantment grew as he came to realize he had excluded himself from that intellectual and philosophical discourse in which the values and goals of our society are shaped. He had devoted himself to technical pursuits only to find himself unable to respond when his professional product was misunderstood, misused, and misdirected by forces beyond his control. If. If. If only he had gained a liberal education along with his technical education.

Partner Engineer's dream of Camelot was fast receding, along with his hairline. His brow was perpetually furrowed, he grumbled much, his food did not satisfy him.

Then one day, when things seemed at their darkest, Partner Engineer had an experience that led him to understand, accept, and to even enjoy the perpetual balancing act in which he was cast, an act which compelled him to keep one foot in Camelot and the other in the hurly-burly arena of the real world, an act which required part of his intellect to contemplate Utopia while simultaneously another part coped with the mundane; an act which subjected him to warring, conflicting concepts but required him nonetheless to remain serene and judicious.

On that fateful day, weary from jousting with auditors and lawyers, Partner Engineer delivered of himself a mighty oath: "By the great Bent of Tau Beta Pi," he swore, "I've been trapped, bamboozled, gulled, and plain deceived by the Engineers' Creed. I've tried to do all it had me pledge—*give, participate, serve, put service before myself*—and what has it gotten me? Headaches, heartaches, a 25-handicap at golf, a small bank balance, a blunt and bent spear, rusty and dented armor, paranoia that sees dragons everywhere, a library of moldy *Consulting Engineer* magazines, and a museum piece—my slide rule!

"Cursed is the day I found the Engineers' Creed, and cursed am I who found that scrap of glazed paper and strove to live up to the impossible Engineers' Creed inscribed thereon!"

No sooner had Partner Engineer uttered that terrible blasphemy than he was engulfed by a cloud of smoke, smoke reeking corrosively of equal parts of brimstone and noxious fumes from Milt Lunch's cigars. The smoke cleared a bit and, through streaming eyes, Partner Engineer perceived the old Wizard in the cloud, staring and smiling. This time, the Wiz chose to appear in a composite, kaleidoscopic combination, sometimes resembling Mark Twain but blending with, or giving way to, Teddy Roosevelt, or Herbert Hoover, or Sam Florman, or any and all of the presidents of the National Society of Professional Engineers rolled into one.

"Nay, my Son," said the Wizard, "Curse not, but rather bless the day you took the Engineers' Creed to be your personal testament. Despite obstacles physical and mental, real or perceived, despite difficulties logical or political, selfless or selfish, you have persevered. You have tried, you have given your best. You have tried to provide food and shelter in abundance. You have striven to restore the purity of our air and water, to heal the blight of our cities, to harness our rivers, to develop power from a myriad of energy sources. You have made herculean efforts to control floods and to minimize danger from other natural occurrences.

"That you fell short at times, that you did not live up to the high expectations you set for yourself is not as important as having done the right thing, at the right time, for the right reasons.

"Though it troubles you that your education has not encompassed the arts and philosophy, do not despair. Your very awareness of the lack of such formal training has made you alert to the need to add taste and sensitivity to your arsenal of technical brilliance. Though your fellow man does not always appreciate your technical abilities and few consider you to be an intellectual, you have earned the accolade of all; you are a useful person.

"You do these things, you are this kind of person, because you have tried to live up to the Engineers' Creed. You will continue to do as you have done and to be as you have been because you must. You are indeed the prisoner of the Engineers' Creed but you are also its guest.

"And so my son, though you see yourself as flawed at times, though you sometimes have fallen short of your own high expectations, though you are scarred by the slings and arrows of life, you are not venal; you have raised yourself up, you are not beaten. You are not just a Consulting Engineer; you are an Engineer; you are a Man!"

With that, the Wizard faded from sight. The brimstone odor disappeared, and only Milt Lunch's acrid cigar smoke lingered. As for Partner Engineer, he lived happily forever after—at least it seemed forever to his Junior Partners.

7.10

Ethical Standards for the Engineering Profession: Where is the Clout?

DAN H. PLETTA, P.E.

If ethical codes are to be complied with voluntarily, or enforced for those few who choose to violate them, all members of the profession, and of society as well, must be aware of the existence of these standards of conduct and of the consequences violations may ensure. Hopefully, most members will choose to obey. Policing any code—or law—becomes impossible when a large minority chooses to "disobey."

Precious little effort has been expended so far to acquaint engineering students with the need for or existence of ethical codes, even when the students' education extends through the doctorate. Quite probably, the pride in belonging to a public spirited profession would also generate a group willing to abide by its ethical codes. Two national efforts among engineers, the "Ritual of the Calling of an Engineer" founded in Canada in 1926 and the "Order of the Engineer" founded in the United States in 1970, do endeavor to "instill a consciousness of belonging to one another, to themselves as individuals, and to those whom they serve." These organizations require only a dignified public ceremony at which the engineers pledge to abide by their obligation and to wear an iron ring on the small finger of the working hand. No action is taken against those who later may violate this obligation.

But if the public is to be protected by legal constraints or by professional devotion, the members of society and of that profession must be willing at times to "blow the whistle" on those few offenders who choose not to conform but who did manage to survive its rigorous admission standards. Initiating such charges requires very mature judgments.

Every engineer must avoid all risks that may cause heavy loss of life. But there is some risk in every design. Safety must be balanced against economic feasibility. Engineers of the future will at times need to adopt an adversary role to protect the public health and safety and its purse and resources, if this civilization is to survive. Admittedly, technology creates ecological and social problems. Only technology is able to solve these problems. Engineers can help effectively only if they become activists and leaders in society.

But the status quo always changes with time. The current favored status of professions may get worse. Medicine's shift from private practice to clinics to socialized care is an example. Our professional curricula are geared to the past and present. Their time horizon is too biased. Future engineering curricula will have to forsake the baccalaureate as the first designated degree not only because of the expansion of knowledge but also because of the renewed emphasis of education for practice—as distinct from research—and for professional societal service.

Some professional societies, like those of economists, have no ethical codes. They believe that civil law should be used to punish those who violate legal constraints and that the sole purpose of corporations, in our free enterprise system, should be restricted to that of maximizing profits within the legal constraints imposed by society.

The point here is, who blows the whistle? Who is to determine when a fellow professional, corporate superior, or public servant pursues programs that are not in the public interest or are even unlawful? These actions could involve bribery or extortion connected with public construction or procurement, or accepting substantial financial favors as a purchasing agent, or producing unsafe consumer products, or plagerism, or making slanderous statements, or attempting to supplant other firms or individuals already operating under contract. If professionals fail to report violations, they fail to accept that obligation expected of them by the public.

Frequently, these violations of laws or ethical codes are discovered by our free press. Sometimes charges are brought by one or more members against their professional peers. Perhaps only a minor fraction of all violations are uncovered. Nevertheless, all charges must be investigated by those professional societies whose members are involved. Such investigations can be time-consuming and lengthy. They should be thorough. They may involve delays caused by court cases on which evidence is based. They will involve considerable expense.

In NSPE special local committees of the state societies conduct the investigations and take final action. When complaints are lodged directly with the National Society, the case is first referred to the appropriate state society for comment. The national board then reviews the evidence and the state society's recommendation. It then makes its own recommendation to the state society but may elect to publish the results of its action in *Professional Engineer* magazine. The final action by the state society may or may not coincide with that of the national board. Complaints involving only members at large, who do not belong to any state society, are handled entirely at the national level. Since 1947, there has been only one hearing before the national board.

It is difficult to determine how many NSPE cases are handled on only a state basis. Frequently, they are handled informally and the members may be allowed to resign before formal charges are filed. It has been estimated that 150 cases per year are considered. Of the 52 member societies, 20 have never processed any disciplinary case, and 15 are very aggressive. For instance, in 1974 the Florida Engineering Society investigated 45 cases and the Iowa SPE about 30.

The discussion so far indicated that engineering societies could only expel, suspend, or reprimand violators of their codes. They cannot prevent a member—or non-member—from practicing. They could censure corporations or governmental bureaus for conduct not in the public interest or they could refuse membership to employees of such organizations. They never have. Perhaps they never will. But it is possible that the future may see some of the societies joining in class action suits restricted to unsafe products or public projects.

Hopefully, such legal entanglements can be avoided by prudent technological leadership properly advising corporations and governmental bureaus of the engineer's responsibility to the public. Whether or not the engineer is excluded from licensure because of industrial and governmental exemptions may not apply in such suits. At any rate, it may be far cheaper to adopt policies that involve corporate social responsibility than pay legal fees involved in the class action quagmire.

But what of those engineers who must be registered to practice? Their licenses may be revoked by the registration board in all states for gross misconduct—but this is not defined in the laws. Some states have amended their laws so that these boards may adopt rules for suspension, but not all of these have done so. Even in Maryland, the state board is still pursuing the bribery charges that caused Vice President Agnew's resignation and has not revoked any licenses. The National Council of Engineering Examiners is unable to estimate the number of cases investigated or of licenses revoked by the 52 state boards because of legal prohibitions on the release of such information.

In substance then, it might be concluded that engineers have had their licenses revoked far less frequently than they have been disciplined with expulsion, suspension, or reprimands by their professional societies. A cynic might conclude that either few engineers are bad, or that few have whistles or want to (or know how to) blow them.

A more realistic conclusion indicates that codes of ethics must exist and be enforced by engineering societies and state registration boards if the profession is to fulfill its obligation to the public. An unwillingness of the profession to discipline those of its members whose practice is not in the public interest can only result ultimately in its being policed by non-engineers. This could happen. Legislation is now being considered in some states that would require a majority of the members on registration boards to be non-professional lay persons.

Hopefully, engineers will respond to their public responsibilities and keep their own house in order. Industry did not. Some industries neglected the health and safety of their employees long enough for the Federal government to create the Occupational Safety and Health Administration (OSHA). And all industries must now contend with just one more bureaucracy.

Enforcement will be far easier if engineers are first educated so as to emphasize the need for ethical behavior and of the public purpose of their profession. They are not now. No suggestion is made here to crowd another course into the four-year engineering curriculum. If we consider demands ten years into the future, a six-year curriculum may be needed to educate truly technically literate engineers who are aware of the profession's increasing obligations to society and of its dire need for active leadership.

Engineers have the logic for such a role. They need now to be motivated for it. Perhaps rigorous admission qualifications for engineering practice, selecting only those willing and able to fulfill that role will accomplish that purpose. Engineering leaders will definitely be needed to stop wasting the public's material resources, to design for permanence rather than obsolescence, and look after its health and safety as well.

7.11

Ethics,
Unethical Engineers,
and ASME

FENTON BAGLEY, P.E.

Jim Smith, an engineering consultant, bribes state officials in order to get a new government contract.

Mary Jones declares herself a registered engineer in the state of Kentucky and an expert in systems management. Neither is the case.

Contrary to state law, Bill Williams orders the dumping of his plant's untreated wastewater into the river. His subordinates complain to him and he fires them. The practice is continued, though corporate records are altered to suggest compliance with the law.

Alice Witherspoon leaves the R&D Division of XYZ, Inc,. and joins ABC, Inc. She brings along secret designs and data for the manufacture of cheap photovoltaic cells and gives them to ABC, Inc., for a "modest financial incentive."

All are ASME members. All have violated ASME's Code of Ethics, and, in some cases, have also run afoul of the law. What can ASME do to rid itself of these wrongdoers and alert the profession to their deeds? A procedure has been devised within ASME to deal with these cases. The procedure is bulwarked by the ECPD Canon of Ethics which, in turn, are clarified by criteria for the enforcement of the Canons.

WHAT IS THE CODE OF ETHICS?

Since the early 1920s, ASME has supported a uniform code of ethics for all engineers. The Society participated with four others in drafting such a code, but was the only one to adopt it in 1922. In 1947, ASME adopted the ECPD Canons of Ethics as part of its constitution. This document was revised several times by ECPD, most recently in 1974. In its last revision, fundamental principles and canons are stated as general principles. These are accompanied by a set of specific guidelines to help interpret and implement the principles and to offer a more complete statement of what constitutes ethical or unethical conduct. In March

1975 ASME ratified the principles and canons. However, the Society spent another two years reviewing the supplementary guidelines, which we call criteria for enforcement of the canons. The society adopted these in 1976.

Most discussions during the two-year review of the criteria centered on questions of advertising and competitive bidding. Several specific statements against both were deleted on the advice of ASME counsel. The rationale was that these might be·construed by the U.S. Department of Justice as a conspiracy to limit the individual's freedom to practice. The current thrust of anti-trust law is that no combination of individuals (such as ASME) can regulate an individual's right to practice as he or she sees fit. And, in fact, competitive bidding clauses in NSPE's ethics code are being challenged by the Justice Department on these very grounds.

Also eliminated from the ASME criteria for code .enforcement is reference to restrictions on moonlighting. There was a feeling that these are conditions of employment between the employer and employee and so should not be a part of our Code.

The age-old argument against this code, and, in fact, against any other ethics code you might mention, is that you can't legislate morality. A code is merely a piece of paper. People must develop their own personal ethical standards. A set of principles won't stop those intent on wrongdoing. All this, of course, is true. Yet a written code is a necessity. It attests to the standards held by the profession in general. It offers guidelines for action, and support for those engineers who, in the face of the unethical practices by others, would follow their conscience and fulfill their social responsibilities.

PUTTING TEETH IN THE CODE

ASME's procedures for dealing with unethical engineers can "put teeth" in our Code of Ethics. Under the auspices of the Professional Affairs and Ethics Committee (PAEC), these procedures have been and are being implemented.

PAEC consists of 25 ASME members (ideally, at least two from each Region) nominated by the Regional vice presidents for a 5-yr. term. Of these, 15 should be from the consulting field and the rest from industry, education, and government. There is no remuneration from ASME for time or expenses. Obviously, it does take dedication to cover expenses out of your own pocket. Still, those of us practicing engineering in private industry should not shirk our responsibility to provide our insight on such an important committee.

The mechanics of PAEC activities on ethics cases is very straightforward. First, an individual or group within or outside of ASME files a written complaint stating all known facts with Rogers Finch, ASME's Executive Director and Secretary. Finch will then determine if the alleged violator is currently a member of the Society. If so, and if the violation is of a trivial nature, he will correspond with the parties involved and dispose of the matter to the satisfaction of both. All such cases will be kept on file in case the individual is involved in similar activities in the future. If there is some question about the seriousness of the matter, Finch will send the PAEC chairman a copy of the complaint so that he and/or PAEC may render an opinion. Then the matter is either dropped and the complainant so advised or the alleged violator is advised of the charges preferred and a member of ASME staff is designated to investigate the situation. A full report is prepared by the investigator and sent to the chairman of PAEC. If the charges are substantiated the chairman of PAEC then appoints a hearing board. The hearing board consists of three to five members of PAEC from the geographical area where the alleged unethical conduct occurred. The expenses of these individuals will be covered by ASME for the duration of the proceedings.

All parties will be represented at the hearing. The accused, who may elect to be represented by counsel, can provide a defense and can question witnesses. The hearing board will reach a decision by secret ballot during executive session. If the accused is found guilty, he or she will be dropped from ASME membership, suspended for a given length of time, or censured, depending on the seriousness of the offense. An appeal may be made within 30 days of the notice of the decision. The Executive Committee of Council then acts upon the decision of the hearing board and/or the appeal as they see fit. *Mechanical Engineering* will carry a notice of the disposition of the case.

ASME cannot keep the unethical individual from practicing engineering. Our actions do not have the force of law. However, we can and will warn the public of the problem. If the engineer is registered, and when the case involves criminal wrongdoing, we can bring the matter before the relevant State Board of Registration and provide them with the facts.

IF IT'S IN THE COURTS

What should ASME do if a case involving unethical action by an engineer is under litigation? Recently, ASME suspended action on an ethics case that was in the courts. However, this precedent has not been turned into policy. Rather, following the advice of legal counsel, PAEC will consider situations on a case-by-case basis and act as seems fitting in each particular case. The reasons to delay action in such a case might be several: ASME would be trying to reach a judgment on the same set of facts as are before the courts. However, ASME has far less power than the courts to discover the facts and it has fewer resources to devote to investigation. Then too, the existence of court action is likely to make ASME's effort to find the facts more difficult. Ongoing litigation limits the number of involved people who are willing to discuss the case, and the amount of readily available information on the case. In view of these limitations, it could be very embarrassing to ASME to reach a verdict different from that of the courts.

The reasons to act in an ethics case, even though it is in the courts, are also several: Deference to the courts may be seen as an abdication of our responsibility to police our profession. Engineering societies, including ASME, do not have an admirable history in the prosecution of ethics cases. This must change. But if we let litigation be another barrier to action, we may never fulfill our responsibilities in this area. As a professional society carrying out its responsibility to the public, ASME can and should pursue ethics cases independent from any pending or ongoing civil action. Besides, civil actions take years, so if ASME waits for them to be resolved the Society may be dealing with a "dead issue." At the same time, if

ASME could discover these cases early and act promptly, the question of conflict with litigation could be a moot point.

WHAT CAN YOU DO?

How does all this affect you the individual, and presumably ethical engineer? What is your responsibility? If you know of a case of unethical action by a fellow engineer, it is your responsibility to write ASME and report the matter. Engineers have been reluctant to report on or testify against their counterparts. But if we are to rid our profession of malpractice, we must rid ourselves of this reluctance. To further support ASME efforts in this area, you might contact ASME or your Regional vice president and offer to serve on PAEC. New members are constantly needed and are difficult to recruit.

PAEC welcomes, in fact solicits, your ideas and comments on the code of ethics and the prosecution of ethics cases. It wants your answers to questions like: Does the current code of ethics cover all issues an engineer might face? If not, how should it be revised to address these issues? How can we learn about ethics problems in a timely manner? Should we, like ASCE, hire a full-time investigator to ferret out and/or investigate these matters? Should a formal engineering education include more emphasis on ethics?

7.12

Professional Responsibility and the Dispatching of Police Cars— A Case Study [1]

1. INTRODUCTION

On May 25, 1977, IEEE member, Virginia Edgerton, a senior information scientist employed by the City of New York, telephoned the chairman of CSIT's Working Group on Ethics and Employment Practices, having been referred to the committee by IEEE Headquarters. She said that she had encountered a situation that might lead to the degradation of a data processing system (called SPRINT) used to dispatch police cars in response to emergency calls, and that her immediate superior, who disagreed with this assessment, refused to have the problem studied. Ms. Edgerton sought advice from the committee.

The problem involved predicting the effects on the SPRINT system if additional real-time tasks were to be executed on the same machine. The working group chairman referred Ms. Edgerton to Dr. Howard Eskin, Manager of Systems Programming, at the Columbia University Computer Center, in order to obtain a preliminary assessment of the technical aspects of the situation.

He reported after meeting with her that the matter was complex, that Ms. Edgerton was raising a legitimate issue and that definitive answers could not easily be found. Shortly afterward, Ms. Edgerton submitted a memorandum to her superior, Project Director, Sarwar A. Kashmeri, outlining the danger as she saw it. He rejected it. Two weeks later, expressing concern for the public safety, Ms. Edgerton circulated a revised version of this memorandum to the members of the Criminal Justice Coordinating Council, the organization employing her.

Following this action, she was summarily discharged by Mr. Kashmeri on grounds of insubordination. Ms. Edgerton then asked our committee to formally

---

[1] A draft of this report was sent to the individual who requested assistance (V. Edgerton), to the technical supervisor directly involved (S. A. Kashmeri) and to the principal officials in overall charge of the project (R. J. McGuire, R. M. Morgenthau and H. Sturz). It was then revised in the light of their responses.

investigate the situation, signed a waiver letter in accordance with our procedures, and sent us various documents bearing on the case. We then wrote to the Project Director Kashmeri, asking for his version of the matter.

No response was received and so, after a suitable interval, we wrote to District Attorney Robert M. Morgenthau, CIRCLE Project Chairman, asking him to look into the situation and to respond to the prima facie case presented by Ms. Edgerton. This letter was answered by Mr. Kashmeri (acting, he stated, at Mr. Morgenthau's request). A letter to Mr. Gerald Hecht, Director of the Probation Department, for which Ms. Edgerton did some work during her tenure with the CJCC elicited the response that she had been "extremely diligent" in carrying out these responsibilities, evidencing "loyalty and enthusiasm."

A letter to Captain[2] R. J. Noonan, the police official in charge of the division that operates SPRINT has not been answered.

2. BACKGROUND

SPRINT is a presently operational police and emergency on-line dispatching system that accepts as inputs (from a police terminal) New York City street addresses and responds (typically within seconds) with street coordinates and the location of the nearest patrol car. Police dispatchers, upon receipt of emergency calls for assistance, enter the addresses given and use the output to direct the nearest patrol cars to the scene. The system has been successfully operating for several years to reduce response times and may thereby be presumed to have helped save lives in critical situations. It is operated by the NYC Police Department on a pair of IBM 370-158 computers, one of which is used for back-up and test purposes.

PROMIS is another on-line system, intended for use by prosecutors to keep track of various data pertinent to cases scheduled for trial. Under the aegis of the NYC Criminal Justice Steering Committee (CJSC), a project called the Criminal Justice Information Systems (CJIS) or CIRCLE project (these terms appear to be synonymous) was established to install the PROMIS system for use by the offices of the various District Attorneys of NYC. The project Chairman is Robert M. Morgenthau (Manhattan District Attorney), the Project Director (and technical manager) is Sarwar A. Kashmeri. Note that the SPRINT system and host computer are under the jurisdiction of the police department—not the CIRCLE project, while PROMIS is under the jurisdiction of the CIRCLE staff.

Ms. Edgerton has had 13 years of experience in the data processing field, a good deal of this time in responsible positions involving the installation and use of on-line systems. She was hired early in 1977 by the CIRCLE project as a consultant with the title of Senior Information Scientist (See Appendix A for job description).

3. THE TECHNICAL ISSUE

As indicated in the introduction, the technical issue is whether the host computer currently being used by the SPRINT system could also handle the PROMIS system

---

[2]Now Deputy Inspector.

without appreciably increasing the response time to SPRINT outputs. More precisely, the question is whether a serious analysis should be undertaken to estimate the likelihood of such an overload condition occurring. Ms. Edgerton does not assert that it *would* necessarily occur, but rather that it is her professional judgment that the possibility cannot be excluded a priori on the basis of existing data. She contends that, in view of the likely consequences to human life of an increase in response time to emergency police calls, a reasonable study is essential.

Dr. Eskin, on the basis of the limited information supplied to him, informs us that the matter is quite complex, requiring an assessment of the hardware and software involved, measurements of the load imposed by SPRINT, and a prediction of the amount of activity (at its peak) likely to be generated by users of PROMIS. No formal techniques exist for solving such problems, he stated. One can only study all of the data carefully, watch such installations (or similar ones) operate and then make estimates based on experience and intuition. He agrees that an off-the-cuff estimate in such a case could easily be wrong.

The point might be raised that the overload question could be dealt with by assuming that the load can be handled satisfactorily and then observing what actually occurs as various features of PROMIS are put into operation, and as terminals are added to extend the community of users. When an overload condition is reached, load could be removed and other arrangements made to deal with the situation. The problem is that overloading of such interactive systems can occur with very little warning, and the effects can be quite intermittent when the overload is marginal. In a situation, where commitments have been made, substantial sums spent on hardware and software, and personal prestige invested, one can easily imagine a prolonged period of degraded SPRINT operation during which those responsible might procrastinate about admitting that they had erred. If only inconvenience or marginal monetary losses were involved, this would not necessarily be too serious. But where individual human lives are thereby endangered, such an approach is highly questionable. It therefore would seem prudent to give the matter careful consideration before making a decision.

4. THE SALIENT EVENTS

Ms. Edgerton's duties included reviewing and evaluating computer programs and projects and doing feasibility studies for the CIRCLE project. At an early stage, after learning that the plan was to run PROMIS on the SPRINT computer, she recognized the possibility of overload and attempted to obtain the data necessary to study the problem. These attempts were unsuccessful, and the Project Director, according to her account, tried to dissuade her from pursuing the matter. She claims that no other efforts had been made or were planned to do such a study.

Mr. Kashmeri, in his response to our letter to Mr. Morgenthau (Appendix B), asserts that the issues she raised "were at the time (and still are) under continuing discussion with the computer staff of the New York City Police Department and members of the CIRCLE Committee." He does not however mention any specific individuals or documents. Our understanding is that no members of the CIRCLE Committee are computer experts, and that other than Mr. Kashmeri and Ms. Edgerton, no computer experts were employed by the Committee. Furthermore,

Ms. Edgerton informs us that she knows of no police department employees who were senior systems analysts expert in on-line systems.

Ms. Edgerton maintains that Mr. Kashmeri refused to allow her to circulate memoranda on the overload question to others involved in the project (that is, the CIRCLE Committee). She was summarily dismissed when she did circulate such a memorandum.

There seems to be no dispute as to immediate reason for her termination. In his letters both to her and to us, Mr. Kashmeri states (the quote is from the latter—see Appendix B):

> The termination . . . was effected because (1) her distribution of the memorandum to the members of the CIRCLE Committee was in direct violation of policy established by me, and (2) against expressly given orders that all communications sent to the members must be approved by the Project Director.

5. THE ISSUE OF PROFESSIONAL RESPONSIBILITY

A key aspect of this case is the conflict between the concept of hierarchical responsibility—the chain of command—and the idea that a professional has certain direct responsibilities that cannot be delegated. In the present instance, the particular responsibility involved from the engineer's point of view is embodied in Article IV, item 1 of the IEEE Code of Ethics. "Engineers shall . . . Protect the safety, health and welfare of the public and speak out against abuses in these areas affecting the public interest." (Virtually all codes of ethics for engineers and scientists contain similar provisions.)

A distinguishing characteristic of a profession is that practitioners, in important aspects of their work, cannot allow their professional judgements to be peremptorily overridden by organizational superiors. The head of surgery in a hospital, for example, cannot order a staff surgeon to use a particular technique in a certain operation when the surgeon considers it to be inappropriate. A lawyer who believes that it is in the best interest of his client to advise a plea of guilty cannot properly give contrary advice when ordered to do so by the head of his law firm.

The obligation entails more than simple refusal to participate directly in acts one judges to be improper. Professionals must view the consequences of their acts broadly. Where it is clear that simple abstention, which usually implies a resignation, is unlikely to avert the likelihood of serious harm being done, then a more active stance is required. This might entail appeals within the organization to higher ranking technical managers or to top officials in the organization. Professional societies, governmental regulatory agencies, volunteer public interest groups, legislators, the press or other outside groups might be alerted. The kind of action appropriate to a given situation depends on such factors as the seriousness of the matter, the degree of certainty felt by the professional, and the structure of the organizations involved. If the only legitimate recourse were the principled resignation, then, apart from the fact that this may often be ineffectual, the high personal cost to the individual will result in the quiet acceptance of many very serious errors.

On the other side is the concept of a manager making a decision for his organization and taking full responsibility for the consequences. This may be considered an appropriate mode of operation in the great majority of cases. However, when a professional within the organization who has responsibilities in an area affected by that decision feels strongly that an error with potentially serious consequences is being made, then means should exist for having that decision reviewed by other qualified people. Any organization using technology in a manner that may affect the public safety or welfare is operating in an irresponsible manner if it does not provide for such reviews. The situations contemplated here are clearly *not* analogous to those faced by an army on a battlefield or a ship at sea, where, on the basis of extreme time constraints, rational cases can be made for demanding immediate, unquestioning obedience to orders from superiors; though even there exceptional situations sometimes arise. Where an individual repeatedly challenges managerial decisions and competent technical reviews subsequently find these challenges to be without merit, then that person's professional competence may be legitimately questioned, as it should be in any instance of serious technical error. For this reason, and because of the natural reluctance of most people to "make waves," competent managers need not fear that their positions would be made untenable by the overzealous application of the aforementioned principle by subordinate professionals.

6.　THE RESPONSE

As stated in our opening footnote, a draft of this report was sent to the principals involved. The only response from a NYC official came from Deputy Mayor for Criminal Justice Herbert Sturz (Appendix C). Note that he assumed office in 1978, many months after the events related here.

Mr. Struz states his understanding that the matter raised by Ms. Edgerton has been, and continues to be, under consideration. If indeed this important and difficult technical question has been seriously studied, there would surely exist at least one technical memorandum discussing the issues. No reference to such a document appears in Mr. Sturz's letter. Nor does he name any individual who made such a study. Similar omissions also characterize the earlier letter (Appendix B) from Mr. Kashmeri.

While Mr. Struz expresses general agreement with our view of the responsibility of professionals in matters of the kind involved here, he does not comment on the treatment meted out to Ms. Edgerton when she exercised that responsibility.

7.　CONCLUSIONS

On the basis of the information and principles discussed above, the investigating committee has arrived at the following conclusions:

1.　Ms. Edgerton encountered in the course of her professional work a situation that might reasonably be considered as entailing a risk to the public safety.

2.   She made a professional judgement that a detailed study of this problem was necessary.

3.   No appropriate formal study was under way.

4.   Ms. Edgerton conveyed her judgement to her superior in a written report, and this was rejected out of hand, with a warning not to take the matter further.

5.   She then circulated a report on the subject to the next level of management in her organization.

6.   Her action was in full accord with the letter and spirit of the IEEE Code of Ethics.

7.   She was peremptorily discharged on the basis of the acts cited above.

8.   This discharge constituted seriously improper treatment of a professional.

9.   Ms. Edgerton's action (at considerable personal sacrifice) on behalf of the public safety was in the highest tradition of professionalism in engineering.

Investigating Committee (IEEE-CSIT Working Group on Ethics and Employment Practices).
Stephen H. Unger (Chairman)
R. Jeffrey Bogumil
Joseph S. Kaufman

APPENDIX A

Duties of Position (Virginia Edgerton)

Will serve as senior information scientist for the criminal justice steering committee grant program. Will assist the director by reviewing computer projects, evaluating computer programs and doing feasibility and cost/benefit studies. Will work with criminal justice agency personnel to create short and long range systems plans and do the detailed work on individual installation plans. Will be responsible for keeping the citywide information system plan current and liaise with DCJS to coordinate local and state planning.

APPENDIX B

August 12, 1977

Mr. Stephen Unger
229 Cambridge Avenue
Inglewood, N.J. 07631

Dear Mr. Unger,

Your follow-up letter of August 4, 1977 addressed to Mr. Robert M. Morgenthau has been received and I have been asked to direct the following reply to your office.

The termination of Ms. Virginia Edgerton from the CIRCLE Project was effected because 1) her distribution of the memorandum to the members of the CIRCLE Committee was in direct violation of policy established by me and 2) against expressly given orders that all communications sent to the members must be approved by the Project Director. Also, as her termination letter of June 24, 1977, of which I believe you have a copy states, the issues raised in her memorandum were at that time (and still are) under continuing discussion with the computer staff of the New York City Police Department and members of the CIRCLE Committee.

I would like to assure you that the policy makers in the City of New York are as concerned with public safety and the treatment of technical professionals as is your Committee. It is however imperative that an employee who is in a highly professional capacity, and has the exposure that accompanies a position dealing with top level policy makers, follow expressly given orders and adhere to established policy.

We hope that the above will provide satisfactory answers to the questions that you have raised.

Sincerely yours,

Sarwar A. Kashmeri
Director, CIRCLE Project

SAK/rs

cc: Hon. Nicholas Scoppetta
Deputy Mayor for Criminal Justice

Hon. Robert M. Morgenthau
Chairman, CIRCLE Committee

Hon. Michael J. Codd
Vice-Chairman, CIRCLE Committee

Hon. Cesar A. Perales
Director, CJCC

APPENDIX C

February 16, 1978

Professor Stephen H. Unger
Chairman, Working Group on
Ethics and Employment Practices
229 Cambridge Avenue
Englewood, New Jersey 07631

Dear Professor Unger:

Thank you very much for sending to me your draft report in the matter of Virginia Edgerton. I appreciate the opportunity that you have given me to comment upon the report. As it happens, I am unable to make any contribution to the report in terms of factual material because, at the time the events described in the report occurred, I did not occupy my present office as Deputy Mayor for Criminal Justice and, accordingly, I have no first hand knowledge as to any of the facts concerning the dismissal of Ms. Edgerton.

For what it may be worth, I agree with your general conclusion that a professional person has a special responsibility to seek to be heard by principal policy makers on issues directly affecting the public safety. As your report makes clear, the pivotal issue in this regard is whether, notwithstanding the alleged frustration of Ms. Edgerton's efforts to communicate directly with members of the CIRCLE Committee, the SPRINT degradation issue was receiving attention by New York City policy makers. My understanding, which admittedly is secondhand, is that this issue was considered by the City in its planning for the development of a criminal justice information system. It is also my understanding that the issue raised by Ms. Edgerton is continuing to receive attention by the City and, of course, now that I am in office as Deputy Mayor for Criminal Justice, I will be concerned with this issue also.

Certainly, we do not intend to develop a criminal justice information system at the expense of any existing computer application which may affect the public safety.

Again, I thank you for the opportunity to examine your report and I commend you for your professional and thorough approach to the matter.

Sincerely yours,

Herbert Sturz

7.13

MCC Report
In the Matter of Virginia Edgerton
(IEEE-7366040)[1]

THE COMPLAINT

Ms.Edgerton was engaged as a consultant by the Criminal Justice Coordinating Council of the City of New York in December 1976, on a per diem basis, which employment was approved by the Office of the Mayor in January 1977. Her duties as Senior Information Scientist included the review, evaluation, feasibility analysis and development of computer programs and associated plans, including liaison activities with the police department and the Criminal Justice Steering Committee, each of which groups were utilizing or planned to utilize computer facilities available to the city for programs in operation or contemplated. SPRINT, an on-line police emergency dispatch system was in operation. PROMIS, a second on-line system, was in development for use by district attorneys throughout the New York City to aid in the prosecution of current litigation. The latter program was the responsibility of the Criminal Justice Steering Committee, the project chairman of which was Robert M. Morgentheau and whose project director was Sarwar A. Kashmeri, Ms. Edgerton's immediate supervisor.

Ms. Edgerton, during the course of her work, determined that concurrent use of the computer facilities available to the SPRINT and PROMIS programs raised, in her judgment, important questions pertaining to the possible degradation of the performance of the police dispatch program. These concerns were expressed to Mr. Kashmeri in a June 3, 1977 memo from Ms. Edgerton.

---

[1] This matter was initially brought to the attention of the Committee on Social Implications of Technology, Working Group on Ethics and Employment Practices (IEEE-CSIT), in 1977. Ms. Edgerton, by letter dated 6/25/77 requested the assistance of CSIT (Exh. 10). The Working Group, composed of Joseph S. Kaufman, R. Jeffrey Bogumil and Stephen H. Unger, Chairman, investigated the matter extensively, and submitted its report to the Executive Committee of IEEE. The Executive Committee on May 21, 1978, referred the matter to the Member Conduct Committee established pursuant to Bylaw 112 (as amended) which was adopted by the Board of Directors in February 1978. This is the first member request for support submitted to the MCC.

By memo of June 17, 1977 to the Criminal Justice Steering Committee transmitting a copy of the memo to Kashmeri, Ms. Edgerton advised its members of her continuing concern for the public safety in the light of her evaluation of the possible consequence of overloading the computer facilities when the PROMIS program was fully developed and in operation.

Mr. Kashmeri, by letter to Ms. Edgerton dated June 24, 1977, terminated her employment, effective June 21, 1977. The stated reasons were that distribution of her memo of June 17 to the Steering Committee violated his policy that all such memos must be approved by him and the matters raised therein were then under discussion by the police department and members of the CIRCLE (Criminal Justice Information Systems) project. Shortly thereafter, Ms. Edgerton requested the project chairman for a hearing on the matter of her discharge. There is no indication that such a review was afforded Ms. Edgerton.

THE IEEE CODE OF ETHICS

The foregoing circumstances suggest that the relevant portions of the code are:

1. Article I, §3: engineers shall "undertake engineering tasks and accept responsibility only if qualified by training or experience, or after full disclosure to their employers or clients of pertinent qualifications";
2. Article I, §2: engineers shall be "honest and realistic in stating claims or estimates from available data";
3. Article III, §5: engineers shall: "assist and advise their employers or clients in anticipating the possible consequences, direct and indirect, immediate or remote, of the projects, work or plans of which they have knowledge"; and
4. Article IV, §1: engineers shall "protect the safety, health and welfare of the public and speak out against abuses in these areas affecting the public interest."

DISCUSSION

Ms. Edgerton's education and employment history, prior to her retainer by New York City in 1977, show her to have had more than a decade of relevant experience in computer programming and application, including seven years management responsibility with various employers. The City, itself, had been her employer for over a year in 1972–1973. That the City anticipated benefits from her work was made clear by Mr. Kashmeri's letter to her dated December 23, 1976, which declared "the project and the city... will benefit from the background and expertise that you will bring to this senior technical position." It is the opinion of the Member Conduct Committee (hereafter "MCC") that Ms. Edgerton was professionally qualified by training and experience to undertake the tasks assigned to her and accept the responsibilities of a senior information scientist, thus satisfying the provisions of Article I, §3 of the Code of Ethics.

The Code also requires IEEE members to be "honest and realistic in stating claims or estimates from available data" (Article I, §2), coupled with the attendant responsibility to her employer to "assist and advise . . . in anticipating the possible consequences, direct or indirect, immediate or remote, of the projects, work or plans of which [she had] knowledge." (Article III, §5). From the information available to the MCC it is reasonable to conclude that Ms. Edgerton in her capacity as senior information scientist has adhered to these two obligations in the discharge of her professional responsibilities. She was selected to join the staff of the CJIS Project, of which Mr. Kashmeri was the Director, and Mr. Morgentheau the Chairman.[2] By memo of June 3, 1977 Ms. Edgerton, on the basis of information available from her activities and consultation with others not employed by the City, advised the project director of the possible consequences which she anticipated from the proposed joint use of available computer facilities for both the SPRINT and PROMIS programs. In so doing Ms. Edgerton clearly was following the mandates of Articles I (Section 2) and III (Section 5) of the Code of Ethics.

We do not address the issue whether Ms. Edgerton's judgment was formed from more reliable or different information than that available to others working within the police department, for the CIRCLE (CJIS) project or other groups with which Ms. Edgerton had contact. Nor is it relevant to the Code of Ethics whether judgment of other responsible or cognizant persons differed. It was Ms. Edgerton's responsibility under the Code to advise her employer, the project director, of her judgments with respect to possible consequences, remote, immediate, direct or indirect. This she did.

The remaining question relates to Ms. Edgerton's conduct following the submission of her memo of June 3, 1977 to the project director. As discussed below, it is the opinion of the MCC that Ms. Edgerton adhered to the Code in her attempts to bring her concern to the attention of the Criminal Justice Steering Committee.

Two elements are undisputed. Ms. Edgerton transmitted to the Steering Committee a memo substantially in the form submitted to the project director and shortly thereafter her employment was terminated by written notice from the project director. Was the transmittal to the Steering Committee consistent with the Code of Ethics? As an employed professional and IEEE member, Ms. Edgerton had the obligation in fulfilling her responsibilities to the New York community, to "protect the safety, health and welfare . . . in these areas affecting the public interest." (Article IV, §1).

This provision does not, of course, deal with procedures or management policies operative within the administrative organization of the City of New York. Rather, it pertains to IEEE member conduct. Further, "public interest" is undefined. The material reviewed by the MCC discloses,[3] however, that the computer facilities and associated programs in operation and under development during the period Ms. Edgerton was employed by the City, were used by the police department in

---

[2]In the relevant documentation, CJIS apparently is the Criminal Justice Information System, which is also referred to as the CIRCLE Project. For the purpose of this report, this project was chaired by Morgantheau and directed by Kashmeri.

[3]We accept as authentic all the documents furnished to the MCC for its consideration.

dispatching personnel in response to emergencies and was intended to aid, among others, the district attorneys in their work as litigators in the courts. The communications show that in one manner or another the scope of Ms. Edgerton's responsibilities were related to the activities of the Deputy Mayor for Criminal Justice, Mr. Morgantheau, the Criminal Justice Coordination Council and its Steering Committee, officers of the police department and state and local planning related to these groups. In this context, then, we conclude the "public interest" means the interest of the citizens to be served by effective law enforcement on the streets and in the courts. Though neither a police officer, nor a district attorney Ms. Edgerton's responsibilities and professional judgments did (or could) affect that public interest (immediately, remotely, directly, or indirectly). Indeed, the MCC is of the opinion that Ms. Edgerton's responsibilities placed her at the interface between the potentially divergent interests or competing needs for computer access of the police and the court. This circumstance enhances the importance and relevance of Article IV, Section 1 of the Code of Ethics.

It is not to be inferred from this discussion that there are "abuses" existent in the programs within the City with which Ms. Edgerton had contact. Rather, Ms. Edgerton has, under Article IV, Section 1, a responsibility to the community to protect the safety and welfare. By her distribution of her memo to the Steering Committee she endeavored to have her views of an important, potentially adverse consequence affecting the public safety considered by the committee within the organizational framework of the City that had responsibility for the computer application contemplated.

In our view this action by Ms. Edgerton was reasonable in light of her apparent inability to resolve the matter with the project director. By first submitting her memo to the project director, the director was afforded an opportunity to consider her analysis of the potential problem. The Steering Committee was the cognizant and responsible group closest in relationship to the activities of both Ms. Edgerton and the project director. The memo initially submitted to the project director indicates that copies were directed to the official responsible for the relevant computer operations within the police department, an attorney assisting in the coordination and the committee responsible for programs under development for use by the district attorneys. At no time, apparently, was the subject of Ms. Edgerton's memo discussed among the project director, the author, and others to whom it was directed. There is no indication that Ms. Edgerton was requested to pursue the matter further, and little to suggest that the project director or other undertook to do so.[4]

The police commissioner, the director and the Chairman of the CIRCLE (PROMIS) project and the deputy mayor for criminal justice were asked to comment on the report prepared by the Working Group (IEEE-CSIT) prior to its submission to the IEEE Executive Committee. Only the deputy mayor replied. He

---

[4]The Chairman of the Working Group on Ethics and Employment Practices by letter to the police officer to whom her initial memo was sent sought information relevant to the assertion by the project director that the issues raised by Ms. Edgerton were under continuing discussion with the police computer staff and members of the CIRCLE (PROMIS) committee. It is our understanding no reply (written or verbal) was made. Similar inquiries were transmitted to the project director and the Chairman of CIRCLE. Only the director replied, but he did not provide further information as to the nature, extent or outcome of discussions to which his letter discharging Ms. Edgerton referred.

responded that he was not the deputy mayor during Ms. Edgerton's employment and the period relevant to the inquiry, and declined to make any contributions to the report.

CONCLUSION

The MCC concludes that Ms. Edgerton has adhered to the IEEE Code of Ethics. It is our opinion (1) that her professional training and experience qualified her to discern the potential for degradation of the police–emergency dispatch system, (2) that she undertook reasonably to inform the project director of her concern, and (3) that her communication of this same concern to the Criminal Justice Steering Committee represented a good faith attempt to protect the community interests served by the computer applications about which she was informed. We believe the attempts were appropriately directed to those persons which were in part or whole responsible for the ultimate compatibility of the systems involved. Ms. Edgerton's adherence to the Code has jeopardized her livelihood. Moreover, it is our opinion that the action by those responsible for her employment termination compromised the discharge by her of her professional responsibilities.

COMMENT

The stated reason for Ms. Edgerton's termination was her distribution of the memorandum to the members of the CIRCLE committee "in violation of policy established by" the project director, and against express "orders that all communications sent to the members must be approved by the Project Director." The fact of termination and the reasons stated therefore are not in dispute. The Code of Ethics becomes relevant in this matter as the basis upon which to ascertain the reasonableness of the IEEE member's conduct, not that of the project director.[5] Because we have determined that Ms. Edgerton's actions are consistent with the Code, however, there is presented a conflict of "policies," whose principles guiding professional activities in an employment relationship which we endorse, and those "policies" which guide the administration of, communications by, and supervision of employed persons.

This is not a circumstance in which the IEEE member, dissatisfied with the consideration or treatment afforded by supervisory personnel, took the issue outside of the confines of the employer's organization either in search of relief of the member's personal grievance or to remedy a potential detriment to the public interest through publication in the media or otherwise. Under the circumstances. we are of the opinion that Ms. Edgerton's action was demonstrably a more professional approach in her relation to the employer. With benefit of hindsight, it is possible, of course, to suggest Ms. Edgerton might have persisted in her efforts to resolve the matters satisfactorily with the project director, thus eliminating the need to solicit participation by the Steering Committee, which effort led to her

---

[5]The Code of Ethics would be relevant to the project director (if an IEEE member) should a complaint be submitted alleging violation of the Code.

discharge. In a similarly facile manner, it can be suggested that the project director could have undertaken to deal constructively, and more responsively, to the substance of Ms. Edgerton's professional judgments. Neither approach, however, addresses what we perceive to be the focal point in this matter. Was it reasonable to resolve what apparently was a matter of divergent judgment by the discharge of Ms. Edgerton? We conclude that it was not. We have found Ms. Edgerton to have acted in a manner consistent with the Code of Ethics. We have found no indication that the discharge was influenced by any circumstance other than the distribution of the memorandum. The prohibition against such distribution to the Steering Committee stemmed from a policy prescribed by the project director. No matters have been brought to our attention that explain the need for, the purpose or efficacy of such a policy. Neither has it been shown that such a policy was existent elsewhere among the relevant committees, directors or staff serving to implement the subject computer applications.

Finally, we believe the circumstances of the situation described herein indicate the present need of employers to develop a means whereby professional employees can raise and be afforded review of their judgments, responsibly formulated, so as to avoid their summary discharge for violation of "policy," when the result of such policy serves to prevent the dissemination and reasonable consideration of professional opinions related to the successful functioning of systems or equipment involving safety and welfare considerations, directly or indirectly, affecting the public interest of a community of citizens to be served by such systems or equipment.

Member Conduct Committee
J. F. Fairman, Jr.,
Chairman
R. F. Cotellessa
R. W. Sears

June 1978

Extract from:
"They Didn't Think"

PHOEBE CARY

Once a trap was baited
With a piece of cheese;
It tickled so a little mouse
It almost made him sneeze;
An old rat said, "There's danger,
Be careful where you go!"
"Nonsense!" said the other,
"I don't think you know."
So he walked in boldly—
Nobody in sight;
First he took a nibble,
Then he took a bite;
Close the trap together
Snapped as quick as wink,
Catching mousey fast there,
'Cause he didn't think.

The Poetical Works of Alice and Phoebe Cary: Household Edition, Alice and Phoebe Cary, Houghton Mifflin, Boston, 1882, p. 382.

Selected Additional Readings

Anderson, R. W., "Professionalism, Ethics and the Public Welfare," *Proceedings*, Conference on Engineering Ethics, American Society of Civil Engineers, May 1975, pp. 8–16.

Austin, R. W., "Code of Conduct for Executives," *Ethics for Executives Series*, pp. 19–27. Reprinted from *Harvard Business Review*, September–October 1961.

Boulden, L., "The Perils of Integrity," *Automation*, March 1975, pp. 42–47.

Bowie, N. E., "Business Codes of Ethics: Window Dressing or Legitimate Alternative to Government Regulation?" *Ethical Theory and Business*, by T. L. Beachamp and N. E. Bowie, Prentice-Hall, Inc., Englewood Cliffs, N.J., 1979.

Caplan, Arthur, "Cracking Codes," Hastings Center Report, August 1978, p. 18.

Clayton, L. A., "Professional Ethics in Private Practice," *Issues in Engineering*, April 1981, pp. 105–110.

Cebik, L. B., "On the Nature of Ethical Codes: An Outline of a Cookbook," paper presented at the National Conference on Engineering Ethics, Renssalaer Polytechnic Institute, June 20–22, 1980.

DeGeorge, R. T., "Ethical Responsibilities of Engineers in Large Corporations," *Business and Professional Ethics Journal*, Fall 1981, with "Commentary" by H. T. Mankin, pp. 1–17.

Draft Report, USAB Ethical Conduct Activities Task Force, "Proposed Procedures for Handling Alleged Infractions of the IEEE Code of Ethics by Members," *Technology and Society*, Committee on Social Implications of Technology, IEEE, September 1977, pp. 10–12.

Drucker, P.F., "Ethical Chic," *Forbes*, September 14, 1981, pp. 160–173.

Flores, A., "Engineering Ethics," *Business & Professional Ethics*, September 1977, pp. 1–3.

Goldman, S. L., and Cutcliffe, S. H., "Responsibility and the Technological Process," *Technology in Society*, Vol. 1, Pergamon Press, New York, 1979, pp. 275–286.

Hodges, M., "Professional Ethics: Two Models for Understanding," *Proceedings*, Frontiers in Education Conference, 1981, Institute of Electrical and Electronics Engineers, pp. 134–138.

Kipnis, K., "Engineers Who Kill: Professional Ethics and the Paramountcy of Public Safety," *Business and Professional Ethics Journal*, Fall 1981, with "Commentary" by J. F. Fairman, pp. 77–91.

Konald, D. E., "Codes of Medical Ethics, Part I—History," *Encyclopedia of Bioethics*, Vol. 1, edited by W. Reich, Free Press, New York, 1978.

Lockhart, T. W., "Professional Societies and the Enforcement of Professional Codes," *Business & Professional Ethics*, Spring 1980, pp. 1–3.

Luegenbiehl, H. C., "Engineering Profession and Professionals,"*Proceedings*, Frontiers in Education Conference, 1981, Institute of Electrical and Electronics Engineers, pp. 190–195.

Luegenbiehl, H. C., "Ethics and Education for Professionalism," *Mechanical Engineering*, November 1981, p. 99.

Matthews, K. W., "Why We Need a Code of Ethics," *The Chartered Mechanical Engineer*, November 1968, pp. 454–456.

McCabe, J., "Some Ethical, Moral and Legal Factors of Professional Conduct," *Ethics, Professionalism and Maintaining Competence*, ASCE Specialty Conference, March 1977, pp. 38–50.

Murphy, A., "The Socio-Technical View," *Mechanical Engineering*, October 1975, p. 81.

Newton, L. N., "Lawgiving for Professional Life: Reflections on the Place of the Professional Code," *Business and Professional Ethics Journal*, Fall 1981, with "Commentary" by D. Wilson, pp. 41–57.

Pavlovic, K. R., "A Common Interest," *Proceedings*, Frontiers in Education Conference, 1981, Institute of Electrical and Electronics Engineers, pp. 196–201.

Task Committee on Professional Civic Involvement, *Action Program for Elimination of Unethical Practices in the Engagement of Professional Services*, American Society of Civil Engineers, November 1974.

Unger, S., "Engineering Societies and the Responsible Engineer," *Annals of the New York Academy of Sciences*, Vol. 196, Article 10, 1973, pp. 433–437.

Unger, S., "Engineering Ethics and the IEEE: An Agenda," *Technology and Society*, Committee on Social Implications of Technology, IEEE, June 1980, pp. 16–17.

Professional Registration and Maintenance of Competence

The Legislature finds that if incompetent engineers performed engineering services, physical and economic injury to the citizens of the state would result and, therefore, deems it necessary in the interest of public health and safety to regulate the practice of engineering in this state.

Section 1. Chapter 81-471
Florida Statutes

Thus, to meet his obligations to perform efficiently and safely, the engineer has need to keep informed of technical development....
UNESCO Helsinki Conference, 1972
Continuing Education of the Engineer

The registration or licensing of engineers is a legal requirement in all the states and territories of the United States. The particular conditions of practice requiring registration are specified by the various laws establishing licensing requirements, but, in all cases, the laws clearly state that the purpose of registration is to protect the public health and safety.

The issue of professional registration arises from the obvious danger to the public presented by the incompetent practice of certain vocations. The issue is sharpened when the vocation involved also lays claim to professional status. The earliest known laws to ensure the competence of a professional were those promulgated by Roger of Normandy (A.D. 1140) providing for examination and certification by their peers for those who wished to practice medicine (1).

The concept of engineering registration as it is known today originated in Wyoming in 1907 when Governor Clarence Johnson prevailed on the legislature to pass a law requiring those persons practicing as engineers or surveyors to be licensed for such practice. Governor Johnson's action was intended to relieve the danger to the public which resulted from incompetent or unscruplous individuals defrauding the general public in the design of dams, canals, and reservoirs that were inadequate and often unsafe. The legal basis for this action was the recognised common-law concept of the police power of the state which permits a state to impose restraining actions on its citizens in the interest of the public safety, health, and welfare. From this beginning there was a steady increase in the number of states requiring registration until the Montana Law was passed in 1947. Hanna (2) reports that today over 400,000 engineers are registered. This number is far less than the number of persons considering themselves engineers.

There is no unanimity regarding registration. The registered engineer sees his license as a mark of legal recognition of his competence and professional stature. The unlicensed engineer considers registration unessential for certain kinds of practice of engineering within a corporation or government entity. Further, he may feel his degree and work record are adequate and more relevant demonstrations of his competence. Registration is often considered as an unwarranted intrusion into private affairs that can deprive a citizen of his right to earn a livelihood. The courts tend to view registration as a proof of qualification of expert knowledge and the basis for recognition of the right to offer professional services to the public under the title of engineer. The public, on the other hand, is rarely aware of the registration law or the reasons for it.

A license is granted under most state laws only after a specified period of recognized engineering education (usually a four-year degree program in engineering), a period of experience under the direction of a registered engineer (typically, an additional four years), and rigorous examination that is most commonly of two days duration. In addition, the candidate must be of good moral character, present letters of recommendation, and may be subject to oral examination. Unfortunately, each state provides for its individual licensing requirements though many have chosen to use a common written examination provided by the National Council of Engineering Examiners (NCEE). The latter organization consists of representatives of each of the various state licensing boards as established by the individual state's laws.

A license in one state entitles the holder to practice engineering in that state only. Through the use of the principle of comity (friendliness or consideration for others), a licensed engineer may be granted a license in another state upon application if the requirements fulfilled for the initial license are equal to those enforced in the second state. Not all states follow this procedure because of particular local registration requirements.

Licensing laws frequently include a number of exemptions to the registration requirement. In most instances, the exemptions pertain to engineers in industrial practice where their duties are limited to the design and manufacture of products offered for sale by the company (the so-called industrial exemptions). In other instances, public officials and others may be included in the exemption.

In the article selected to open this chapter, Milton S. Fine, exeuctive director of the National Council of Engineering Examiners, advocates registration for all engineers whether required by law or not. In Fine's opinion, registration demonstrates responsibility to the public and to employers and thus is a bond to the accepted meaning of the learned profession.

The second selection is an extract from the State of Florida Registration Law of 1981. A close reading of this law, which is typical of state registration laws, reveals that it provides a definition of engineering for legal purposes, a clear statement of why registration is required, the qualifications and procedures for attaining an engineering license, the organization of a licensing board, the licensing exemptions permitted for the practice of engineering, and other questions of significant interest to any engineer concerned with registration. The law serves to illustrate those points raised in this introduction and to provide a background against which to consider the remaining articles in the chapter.

In a provocative article, "Legislators Play Dealers' Choice with P. E. Registration Laws," Milton Lunch, general counsel for NSPE, discusses several controversial questions. These include the industrial exemption clauses in registration laws which grant exemption from licensing requirements for engineers that do not offer their services directly to the public; eductional requirements or whether an individual should be licensed on the basis of examination without prior formal engineering education; the use of lay people on boards of engineering examiners rather than the more common practice of forming these boards of professional peers only.

Articles by M. J. Kolhoff and G. J. Kettler present the arguments for and aginst the industry exemption to professional registration. Both Kolhoff and Kettler are licensed engineers, both are employed by industrial organizations, yet each reaches a different conclusion on the question of the industry exemption. In "Getting Your P.E. Why? How?" J. D. Constance discusses these important questions from the point of view of an engineering practitioner advocating professional registration. Registration is both a legal and personal question. Some engineers must be registered to practice; others want to be registered for a sense of professional accomplishment and recognition; and still others have no desire to test their qualifications against registration requirements. It is at present, however, the case that registration is a valuable asset for advancement and professional recognition in consulting, government practice, and in many industrial assignments.

One of the most hotly debated issues in professional registration, as well as in engineering education, is the need for formal requirements for the maintenance of competence to practice engineering. Ingersoll has presented a comprehensive commentary on this subject in his paper, "Qualifications for Continued Practice." Continuing education has long been accepted as a part of the professional's obligation. If professionals do not maintain current knowledge of the state of the art, they become less and less valuable to society, their profession, and to themselves. But how should competence be maintained? This is the crux of the discussion. Some favor formal education through classes, short courses, workshops, etc., while others believe that daily practice and association with professional and technical organizations will achieve the same end.

Several states, notably Iowa in 1977, have modified their registration laws to require continuing education to specified minimum levels in order to renew state licenses for engineering practice. Limited experience with the Iowa law appears to indicate that the profession in that state has accepted the requirement with a recognition of the benefits it ensures to the practitioner as well as to the public.

Another approach to solve the maintenance of competence dilemma is illustrated by the penultimate article in this chapter. The Florida Society for Professional Engineers has established a voluntary program for continued professional development that parallels the more formal requirements of the Iowa law. This article describes the procedures and successes of the voluntary program.

Many engineering organizations have established a policy on continuing education or on maintaining competence. The position paper issued by the National Council for Engineering Examiners is presented in summary to illustrate the general nature of the stand of professional societies on this question.

References

1. T. E. Stivers, "Professional Licensing: The Ancient Rite of Protecting the Public," *Consulting Engineer*, September 1975, pp. 41–44.
2. W. J. Hanna, "Examinations for Registration," *Engineering Issues*, October 1978, pp. 257–261.

8.1

Registration Viewed as Bond to Practice of Learned Profession

MORTON FINE, P. E.

As of 1978, statistics show that there are 320,000 registered engineers in the United States. Of these, maybe 20 percent (or 64,000) have legal need for registration. These are in: (1) private practice; (2) state and town practice: (3) other positions requiring registration as a prerequisite. Therefore, about 256,000 engineers in this country have sought registration voluntarily though not legally required to do so.

Their reasons: (1) future legal need; (2) most likely to pass the exam early in their career; (3) ego symbol; (4) hedge against removal of industrial and other exemptions; (5) peer identification especially for those coming into practice of engineering through non-traditional routes. In other words, as originally set up by state statutes in accordance with the definition of the practice of engineering, many practicing engineers have chosen to not hide behind the exemptions and have indeed become licensed.

In accordance with the ECPD definition of the engineering profession, they consider that their practice does indeed come under the statute and contributes to the safeguarding of life, health, and property, and promotes the public welfare. It is, therefore, in the public interest, they believe, to be licensed. These licensees submit to the process of licensure voluntarily.

The practice of engineering in the state statutes and model law encompasses all those whose work fits the ECPD definition of the engineering profession: the exemption, if it exists, is purely a legal exclusion. It is not an expression that the practice of engineering in the exempted areas does not affect the safeguarding of life, health, and property, nor that such practice is not involved with the public interest. The exemption is a political fact of life, recognition that the statute could not have passed in the first place if the exemption were not included.

Regarding the industrial exemption and the continual controversy about its removal, I think we have been duped by a myth. In a recent detailed review of exemptions in state statutes, I have determined that not more than half (and more likely less) have a so-called industrial or manufacturing exemption or its equivalent. And yet, I have not seen any overt attempts by any state boards to delve deeply into the prctice of engineering in industry on a witch hunt for conformance to state laws (with or without the exemption) except in obvious blatant circumstances which required action.

Industry has been left alone to determine at what levels of responsibility registration becomes desirable, and the engineers themselves, as employees, have determined at what point in their careers they want to identify with the rest of the profession. And perhaps this is the way it should be where, without bureaucratic intervention, the desirability of conformance to an idealistic statute (no exemption) overcomes the legal screen in the statute.

I previously made reference to one of the reasons for registration in industry on a voluntary basis as being "peer recognition, especially for those coming in through non-traditional routes." It is perhaps especially true in industry that engineering is not so homogeneous a profession as law and medicine for which the education and training are highly structured.

Although most engineers come into practice through the traditional route of graduation from an engineering curriculum plus engineering experience, there are many who come from the areas of math, science, chemistry, physics, and computer science, to name a few, but end up actually practicing engineering. For them and the profession, the ability to identify with the profession they adopted is most appropriate.

The engineering statutes provide an opportunity for those engineering practitioners from diverse backgrounds to meet the criteria at a minimum uniform entry level of competence.

Certainly there is another reason for suggesting that all those who practice engineering by its basic definition, but not in a legally required environment, seek registration. We who are engineers definitely consider ourselves as members of a profession. I believe also that the general public has the perception that engineering is a profession and that it is relatively high on the list of professions and occupations in terms of public trust. The definition of practice includes "for the benefit of mankind" and the statutes speak of the "public welfare." Although there are some employers who do not support the principle of engineering registration and use the legal screen of exemption, they cannot and indeed, do not, prevent their engineers who know they are practicing engineering, from voluntarily seeking registration.

On the contrary, responsible employers encourage registration. The attainment of registration automatically encompasses a legal responsibility to be guided by rules of professional conduct as well as other precepts of moral behavior under the penalty of disciplinary action. For the employed engineer and especially the registration-exempt one, this adds other dimensions to responsibilities—one to the employer, and another to the public. Responsibility is one of the hallmarks of the true professional.

In conclusion, I would merely restate my original conviction that all engineers who are in fact practicing engineering in accordance with the broad definition, should become registered, whether the motivation is mandatory or voluntary. In so doing, the engineer attests to the conviction that what is done in practice not only serves a client or an employer but also implies a dedication to public trust. Registration is a bond to a learned profession.

8.2

Extracts from
Florida Registration Law, 1981

STATE OF FLORIDA

471.001 Purpose. The Legislature finds that, if incompetent engineers performed engineering services, physical and economic injury to the citizens of the state would result and, therefore, deems it necessary in the interest of public health and safety to regulate the practice of engineering in this state.

471.003 Qualifications for practice, exemptions.—

[1] No person other than a duly registered engineer shall practice engineering or use the name or title of "registered engineer" in this state.

[2] The following persons are not required to register under the provisions of ss. 471.001–471.039 as a registered engineer.

   [a] Any person practicing engineering for the improvement of, or otherwise affecting property legally owned by him, unless such practice involves a public utility or the public health, safety, or welfare or the safety or health of employees. This paragraph shall not be construed as authorizing the practice of engineering through an agent or employee who is not duly registered under the provisions of ss. 471.001–471.039.

   [b] A person acting as a public officer employed by any state, county, municipal, or other governmental unit of this state when working on any project the total estimated cost of which is $10,000 or less.

   [c] Regular full-time employees of a corporation not engaged in the practice of engineering as such, whose practice of engineering for such corporation is limited to the design or fabrication of manufactured products and servicing of such products.

   [d] Regular full-time employees of a public utility or other entity subject to regulation by the Florida Public Service Commission, Federal Energy Regulatory Commission, or Federal Communications Commission.

[e] Employees of a firm, corporation, or partnership who are the subordinates of a person in responsible charge, registered under ss. 471.001–471.039.

[f] Any certified full-time faculty member teaching the principles and methods of engineering design in any college or university located in the state, as of July 1, 1979, and any such faculty member initially employed after July 1, 1979, for a period of 2 years from the date of employment.

[g] Any person as contractor in the execution of work designed by a professional engineer or in the supervision of the construction of work as a foreman or superintendent.

[h] A registered land surveyor who takes, or contracts for, professional engineering services incidental to his practice of land surveying and who delegates such engineering services to a registered professional engineer qualified within his firm or contracts for such professional engineering services to be performed by others who are registered professional engineers under the provisions of ss. 471.001–471.039.

[i] Any electrical, plumbing, mechanical or air-conditioning contractor whose practice is the design and fabrication of electrical, plumbing, air-conditioning and mechanical systems which he installs by virtue of having qualified under Chapter 489 (Contracting) or any special act or ordinance, when working on any construction project requiring electric service of less than 600 amps in residential and less than 800 amps three-phase in commercial or industrial, or requiring a plumbing system of less than 125 fixture units, or requiring air conditioning and refrigeration equipment to serve an occupant content of less than 100 persons, or has a value of $10,000 or less.

471.005 Definitions.—As used in ss. 471.001–471.039:

[1] "Board" means the Board of Professional Engineers.

[2] "Department" means the Department of Professional Regulation.

[3] "Engineer" includes the terms "professional engineer" and "registered engineer" and means a person who is registered to engage in the practice of engineering under ss. 471.001–471.039.

[4]

[a] "Engineering" includes the term "professional engineering" and means any service or creative work, the adequate performance of which requires engineering education, training, and experience, in the application of special knowledge of the mathematical, physical, and engineering sciences to such services or creative work as consultation, investigation, evaluation, planning, and design of engineering works and systems, planning the use of land and water, teaching of the principles and methods of engineering design, engineering surveys, and the inspection of construction for the purpose of determining in general if the work is proceeding in compliance with drawings and

specifications; any of which embraces such services or work, either public or private, in connection with any utilities, structures, buildings, machines, equipment, processes, work systems, projects, and industrial or consumer products or equipment of a mechanical, electrical, hydraulic, pneumatic, or thermal nature, insofar as they involve safeguarding life, health, or property, and including such other professional services as may be necessary to the planning, progress, and completion of any engineering services.

[b]   A person shall be construed to practice or offer to practice engineering within the meaning and intent of ss. 471.001–471.039, who practices any branch of engineering; who, by verbal claim, sign, advertisement, letterhead, card, or in any other way, represents himself to be an engineer or, through the use of some other title, implies that he is an engineer or that he is registered under ss. 471.001–471.039; or who holds himself out as able to perform, or does perform, any engineering service or work or any other service designated by the practitioner which is recognized as engineering.

[5]   The term "engineer intern" means a person who has graduated from, or is in the final year of, an engineering curriculum approved by the board and has passed the fundamentals of engineering examination as provided by rules adopted by the board.

[6]   "License" means the registration of engineers or certification of businesses to practice engineering in this state.

[7]   "Certificate of authority" means a license to practice engineering issued by the department to a corporation or partnership.

471.007   Board of Professional Engineers.—

[1]   There is created in the Department of Professional Regulation a Board of Professional Engineers. The board shall consist of nine members, seven of whom shall be registered engineers and two of whom shall be lay persons who are not and have never been engineers or members of any closely related profession or occupation. Of the members who are registered engineers, three shall be civil engineers, one shall be either an electrical or electronic engineer, one shall be a mechanical engineer, one shall be an engineering educator, and one shall be from any discipline of engineering other than civil engineering.

[2]   Initially, the Governor shall appoint four members for a term of 4 years, three members for a term of 3 years, and two members for a term of 2 years. Thereafter, members shall be appointed for 4-year terms.

471.013   Examinations; prerequisites.—

[1]

[a]   A person shall be entitled to take an examination for the purpose of determining whether he is qualified to practice in this state as an engineer if the person is of good moral character and:

1.   Is a graduate from an approved engineering curriculum of 4 years

or more in a school, college, or university which has been approved by the board and has a record of 4 years of active engineering experience of a character indicating competence to be in responsible charge of engineering; or

2. Is a graduate of an approved engineering technology curriculum of 4 years or more in a school, college, or university within the state university system, having been enrolled or having graduated prior to July 1, 1979, and has a record of 4 years of active engineering experience of a character indicating competence to be in responsible charge of engineering; or

3. Has, in lieu of such education and experience requirements, 10 years or more of active engineering work of a character indicating that the applicant is competent to be placed in responsible charge of engineering; provided, however, that this subparagraph does not apply unless such person notifies the department before July 1, 1984, that he is engaged in such work on July 1, 1981.

   The board shall adopt rules providing for the review and approval of schools or colleges and the courses of study in engineering in such schools and colleges. The rules shall be based on the educational requirements for engineering as defined in s. 471.005. The board may adopt rules providing for the acceptance of the approval and accreditation of schools and courses of study by a nationally accepted accreditation organization.

[b] A person shall be entitled to take an examination for the purpose of determining whether he is qualified to practice in this state as an engineer intern if he is in the final year of, or is a graduate of an approved engineering curriculum in a school, college or university approved by the board.

[2]

[a] The board may refuse to certify an applicant for failure to satisfy the requirement of good moral character only if:

1. There is a substantial connection between the lack of good moral character of the applicant and the professional responsibilities of a registered engineer; and

2. The finding by the board of lack of good moral character is supported by clear and convincing evidence.

[b] When an applicant is found to be unqualified for a license because of a lack of good moral character, the board shall furnish the applicant a statement containing the findings of the board, a complete record of the evidence upon which the determination was based, and a notice of the rights of the applicant to a rehearing and appeal.

471.015 Licensure.—

[1] The department shall license any applicant who the board certifies is qualified to practice engineering and who has passed the licensing examination.

[2] The board shall certify for licensure any applicant who satisfies the requirements of s. 471.013. The board may refuse to certify any applicant who has violated any of the provisions of s. 471.031.

[3] The board shall certify as qualified for a license by endorsement an applicant who:

  [a] Qualifies to take the examination as set forth in s. 471.013, has passed a national, regional, state, or territorial licensing examination which is substantially equivalent to the examination required by s. 471.013, and has satisfied the experience requirements set forth in s. 471.013; or

  [b] Holds a valid license to practice engineering issued by another state or territory of the United States, if the criteria for issuance of such license were substantially identical to the licensure criteria which existed in this state at the time the license was issued.

[4] The department shall not issue a license by endorsement to any applicant who is under investigation in another state for any act which would constitute a violation of ss. 471.001–471.039 or of Chapter 455 until such time as the investigation is complete and disciplinary proceedings have been terminated.

471.031 Prohibitions; penalties.—

[1] No person shall knowingly:

  [a] Practice engineering unless the person is registered pursuant to ss. 471.001–471.039;

  [b] Use the name or title "registered engineer" when the person is not registered pursuant to ss. 471.001–471.039;

  [c] Present as his own the registration of another;

  [d] Give false or forged evidence to the board or a member thereof for the purpose of obtaining a registration;

  [e] Use or attempt to use a registration which has been suspended, revoked, or placed on inactive status;

  [f] Employ unlicensed persons to practice engineering; or

  [g] Conceal information relative to violations of ss. 471.001–471.039.

[2] Any person who violates any provision of this section is guilty of a misdemeanor of the first degree, punishable as provided in s. 775.082, s. 775.083, or s. 775.084.

471.033 Disciplinary proceedings.—

[1] The following acts constitute grounds for which the disciplinary actions in subsection [3] may be taken:

  [a] Violation of any provision of s. 471.031 or 455.227(1);

  [b] Attempting to procure a license to practice engineering by bribery or fraudulent misrepresentations;

  [c] Having a license to practice engineering revoked, suspended, or otherwise acted against including denial of licensure, by the licensing authority of another state, territory, or country;

[d] Being convicted or found guilty, regardless of adjudication, of a crime in any jurisdiction which directly relates to the practice of engineering or the ability to practice engineering;

[e] Making or filing a report or record which the licensee knows to be false, willfully failing to file a report or record required by state or federal law, willfully impeding or obstructing such filing, or inducing another person to impede or obstruct such filing. Such reports or records shall include only those which are signed in the capacity of a licensed engineer;

[f] Advertising goods or services in a manner which is fraudulent, false, deceptive, or misleading in form or content;

[g] Fraud or deceit, negligence, incompetence, or misconduct, in the practice of engineering.

[h] Violation of chapter 455;

[i] Practicing on a revoked, suspended, or inactive license; or

[j] Affixing or permitting to be affixed his seal or his name to any plans, designs, drawings, or specifications which were not prepared by him or under his responsible supervision, direction or control.

[2] The board shall specify, by rule, what acts or omissions constitute a violation of subsection [1].

[3] When the board finds any person guilty of any of the grounds set forth in subsection [1], it may enter an order imposing one or more of the following penalties:

[a] Denial of an application for licensure.

[b] Revocation or suspension of a license.

[c] Imposition of an administrative fine not to exceed $1,000 for each count or separate offense.

[d] Issuance of a reprimand.

[e] Placement of the licensee on probation for a period of time and subject to such conditions as the board may specify.

[f] Restriction of the authorized scope of practice by the licensee.

[4] The department shall reissue the license of a disciplined engineer or business upon certification by the board that the disciplined person has complied with all of the terms and conditions set forth in the final order.

8.3

Legislators Play Dealers' Choice with P.E. Registration Laws

MILTON F. LUNCH

Supreme Court Justice Louis Brandeis once observed that the United States could be compared to the operation of 48 laboratories in which each state may experiment with new or different concepts of government and social policy. Through this mode of operation, he observed, faulty ideas can be tried and discarded, and new ideas found workable can be adopted, refined, and adapted by the other states.

Thus it has been for the past 68 years since the first engineering registration law was adopted, and thus it remains today. The only difference is that the tempo of experiment and change has accelerated as society has grown more complex, with the engineering profession often leading the way through its own diversity and expanding scope of activities.

Changes in the laws governing the offer of, and practice of, engineering have developed over the years under the adversary system by which a democratic nation seeks the optimum solution to its never-ending problems. This system often is confused with the idea that "argument" is bad and that it is better to achieve change through "consensus." But the adversary system, properly understood, has little to do with "argument"; it is only a means for differing views, often influenced by self-interest, to challenge the basis of views held by others, through which a pragmatic resolution emerges.

In applying this historic and traditional process to the development of state engineering registration laws, we can note the gradual transition, for example, from strongly divided views on the issue of corporate practice. Only about 15 years ago a substantial segment of the profession fought the idea that engineering could properly be conducted through the corporate form, fearing that control by an artificial "person" (a corporation) could have an adverse effect on the quality and ethical standards of the profession. But gradually a pattern emerged that permitted engineering firms to enjoy the benefits of tax laws and other legal advantages of the corporate form without the loss of ethical or professional standards. This was accomplished by the development of carefully drafted language and controls, utilizing the NCEE "model law" as the vehicle to provide guidance to the state societies and state registration boards.

Today, only a handful of states do not permit corporate practice of engineering, and even those bars are illusory for practical purposes in light of the enactment of state professional corporation laws in all states. Interestingly, the state professional corporation laws were spearheaded by the other professions (medicine and law), which have been cited by the opponents of engineering corporate practice as the sworn defenders of only the proprietorship and partnership form of practice.

Other significant issues now confront the engineering profession. Among the foremost is the question of professional registration, in all its aspects and applications.

MANDATORY DEGREE REQUIREMENT

Initially, the engineering laws followed the hallowed Horatio Alger principle that the poor and uneducated could aspire to and achieve the heights in business, politics, or the professions through diligence and hard work. That great American tradition has yielded, however reluctantly, to the sterner requirements of broader and more sophisticated and complicated requirements for the practice of all progressions, engineering being no exception. Consequently, all of the professions have moved slowly but surely into formal educational requirements. The "good old days" when a person could be registered as an engineer solely on the basis of experience as evaluated by a state board have long since given way to a mandatory examination requirement in the absence of an engineering degree, then into a mandatory examination requirement even with a degree (some few states still permit registration by only oral examination for those with degrees).

Now the trend is towards a mandatory degree requirement as a basis to take the examination. Only about a dozen state laws have been amended to include the degree requirement, and in each of those instances, a grace period of several years was included before the requirement became fully effective. It would appear that the pattern has been set for more of the state laws to include a degree requirement as the long-established practice clause is phased out.

EIT OR INTERN

Since the engineer-in-training concept was introduced into the registration laws some 30 years ago, the recognition has grown that the designation EIT was a poor description of an engineering graduate who is employed and paid as an engineer and expected to perform as an engineer, both technically and professionally. The "in-training" reference has been felt by many to be demeaning and not truly descriptive of the status. It has been suggested that such persons be labeled as "engineers" while those fully registered be distinguished by the designation of "professional engineer." But this idea has not been accepted in any state law.

A somewhat more promising approach has been to drop the EIT in favor of "intern engineer," borrowing from the medical profession. Thus far, only New York has done this and reports a large upswing in applications for "intern" certification as the first step on the registration road. A few other states are considering the same change. None of the states have yet been willing to opt for the approach recommended by NSPE several years ago–that the EIT concept be dropped altogether as an anachronism of the past "apprenticeship" way to achieve professional status; that in its place the profession accept a bachelor's degree in engineering and passage of the standard written examination as the basis for entry into the professional ranks as a *professional* engineer, but without authority to offer services to the public or to certify engineering documents until the registrant had completed four years of engineering experience acceptable to the state board. In the absence of acceptance of the NSPE proposal, the lack of enthusiasm for the "intern" title, and the continued insistence on maintaining a title distinction between the "engineer" and the "professional" engineer, this item remains high on the list of unfinished business.

PROFESSIONAL SCHOOLS OF ENGINEERING

Related to the broader aspects of the title question is the developing debate over professional schools of engineering. While this may be considered an issue primarily for the educators, its future development will have a distinct bearing on the registration process. After numerous articles, panels, and forums on the subject, the idea of moving towards professional schools (à la medicine and law) remains uncertain in the absence of any consensus on definition. In general, however, the basic idea is to require a pre-engineering higher educational requirement, probably starting with a minimum of two years; the separation of the engineering school from the undergraduate program of the university; and engineering faculty jurisdiction over the educational program and degree requirements, as well as the appointment, promotion, tenure, and activities of the faculty.

If the professional school movement achieves acceptance, the expected result would be the emergence of engineers with broader exposure to the liberal arts, who are more mature and more highly trained in both the fundamentals of engineering and some higher degree of knowledge in selected areas of specialization. That eventual result, if achieved, could have substantial impact in eliminating the apprenticeship concept, even if some "intern" period were retained. In the medical profession, as an example, the medical school graduate, while an "intern," is to the public, the nurses, and other doctors, a doctor of medicine.

INDUSTRY EXEMPTION

An issue that has aroused intense interest and broad support among registered engineers is the elimination of the so-called "industry exemption." For practical and political reasons, most, but not all, of the state laws have included an exemption for engineering related to the design and manufacture of products.

Now, a movement is underway to eliminate these exemptions on the ground that faulty design of certain products by unqualified personnel is as much a matter warranting legally-binding qualification as the design of a bridge or water system. Not surprisingly, industry is fighting to retain its special exemption status in the state laws. So far only Montana has been able to achieve a breakthrough. The Montana legislature recently adopted revisions to its registration law, using the precise language found in the NCEE model law definition of the "practice of engineering," to extend jurisdiction to include " ... industrial or consumer products or equipment of a mechanical, electrical, hydraulic, pneumatic, or thermal nature, insofar as they involve safeguarding life, health, or property... "

It might be necessary, particularly in the heavily industrialized states, to make special temporary provisions in the law to achieve the coverage of products of the type described. These might include an exemption for current employees of a company designated as engineers, and performing engineering work, or a new "grandfather" provision for non-registered employees who may be qualified for registration, but who feel too long removed from academe to undertake the examination.

The removal of the industry exemption, as it may occur state-by-state, will pose new responsibilities and burdens on the state boards. Numerous questions of interpretation, not all of which can be answered at this time, will arise. Which products will fall within the definition? What state law will apply to the various components designed or manufactured in different states and assembled totally or in part in still other states? How will the registration requirement apply to foreign imports? The answers to such questions will require time and experience. Meanwhile, however, the societies in many states will be pressing for a change in the law to eliminate the present loophole in light of growing demands for control over the safety aspects of products.

LAY MEMBERS

The registration boards themselves have not been immune to attack and criticism, particularly following the recent scandals involving engineers' and architects' relations with public officials in Maryland and elsewhere. Official spokesmen for the Department of Justice and the Federal Trade Commission have launched direct attacks on the basic concept of state licensing of the professions and occupations. They have charged that licensing laws and procedures often are used as a means to hold down the numbers who may legally provide the particular services, as a device to stifle competition. There is not a shred of evidence to support such a charge in the case of engineering licensure, but the state boards are vulnerable to the claim that they have moved too slowly and too cautiously to suspend or revoke the licenses of wrongdoers.

This probably accounts for the trend towards placing lay members on the licensing boards. There has been initial resistance to this practice from the engineering profession, particularly with regard to the concept of giving the lay member a vote on the technical qualifications of applicants. However, opposition has softened considerably as experience has indicated that the lay members will not be authorized to invade technical areas beyond their competence. There is, in fact, substantial support for the idea that lay members on the boards can be a good thing for the profession in terms of showing that there is nothing secret or sinister in the way the boards handle their business. Also, the lay member can be a valuable ally in helping the public understand the true purpose and public benefit of the registration law and to serve as a "watchdog" to prod the board or even "blow the whistle" if the boards do not act with reasonable diligence and promptness against those who should not be allowed to continue in the profession.

RULES OF PROFESSIONAL CONDUCT

If, as now seems beyond doubt, the state boards will be expected to take on the added burden of policing the profession for both legal and ethical transgressions of a serious nature, more state boards must develop guidelines and rules of both substance and procedure. The substantive guidelines have been well developed by NCEE in the form of suggested rules of professional conduct which are based, in turn, on those parts of the code of ethics that directly relate to protecting the public health, safety, and welfare. Some 16 state boards have now adopted the NCEE-type of rules, some with variations or deletions. In some instances, the registration law does not clearly give the boards the authority to adopt such binding rules, even though all the state laws give the boards disciplinary authority for "misconduct" by licensees. To remove any doubt of the authority of the board, however, it is to be expected that the state laws will be amended to prescribe specifically the right and duty of the board to adopt rules of professional conduct, and to spell out clearly that the violation of such rules may be grounds for suspension or revocation of the license of the individual involved.

Whatever doubt may have existed as to the legal right of the state boards to adopt disciplinary rules for ethical misconduct has been dispelled by the recent decision of the U.S. Supreme Court in the celebrated Goldfarb Case. In that decision, holding that a mandatory fee schedule for legal services issued by a *private* bar association was a violation of the antitrust law, the Court went on to state clearly:

"We recognize that the states have a compelling interest in the practice of professions within their boundaries, and that as part of their power to protect the public health, safety, and other valid interests they have broad power to establish standards for licensing practitioners and regulating the practice of professions."

And, significantly, the Court added that the states under their sovereign power may decide that "forms of competition usual in the business world may be demoralizing to the ethical standards of a profession," citing an earlier Supreme Court decision in 1952 upholding a state prohibition of advertising of professional services.

Some of the state boards that have adopted binding rules of professional conduct based on the NCEE model have held back from including the ban on competitive bidding for engineering services for fear that such a ban might run afoul of the antitrust law. But that concern no longer can be a reason for the boards to defer action on the competitive bidding issue in view of the Supreme Court's affirmation of the "Parker Doctrine" in the Goldfarb Case. Under the 1943 Parker case, the Court held that the Sherman antitrust law does not apply to the states or to state agencies, and in now reaffirming that position the Supreme Court said that state action of an anticompetitive nature (e.g., the ban on competitive bidding) is not a violation of the Sherman Act if it is clearly established that the state agency action was mandated by state law. To meet that test, therefore, it will be highly desirable for the state registration laws to give undoubted authority to the state boards to adopt rules of professional conduct, and for those board rules to spell out clearly the activities that are forbidden to licensees in order to protect the public interest.

In almost every state, there are discussions of changes that the state society or its members might like to see enacted. The Texas Society of Professional Engineers recently put through an amendment to give the state board the right to increase substantially the annual renewal fees in order that the board might have the funds it needs to implement a vigorous enforcement program. In other states efforts are being made to eliminate or restrict exemptions for those holding public office in an engineering capacity. In still others work is underway to better define the meaning of "responsible charge" of engineering work. Each state is, in effect, a laboratory, and from its experimentations and experiences the profession is able to analyze, debate and act upon the constant wave of revisions of the state laws, which are intended to better serve the public health, safety, and welfare.

8.4

For the Industry Exemption...

M. J. KOLHOFF, P.E.

Those of us who aspire to the development and unity of engineering as a learned profession generally view our goals as similar to those of our medical and legal associates. We feel a special kinship for those in the medical profession, for both medicine and engineering are the application of science and technology for the service of mankind.

In contrast to our medical brethren we engineers stand divided. We have been unable to form a strong and comprehensive unity organization. We cannot even agree upon a code of ethics. We have NSPE's Code of Ethics, ECPD's Fundamental Canons, NCEE's Rules for Professional Conduct, IEEE's new Code of Ethics and variations of these in other societies. Now we are diverted by an in-house quarrel over whether or not industrial firms and the 80–85 percent of the nation's engineers employed therein should be brought under state-by-state occupational regulation. This is a far cry from emulating the unity and intellectual and ethical ideals of medicine.

THE FOCUS OF DEBATE

Just what is this debate that seems to demand resolution among engineers before we can get on with the pursuit of learned profession aspirations? We hear the emotionally popular phrase "Eliminate the industry exemption," but the focus is upon extending state occupational regulation beyond that of service to the public where both the medical and engineering professions are now regulated. What is the central issue of the debate? It is the proposition that engineering activities within industry must be regulated by the state in order to adequately assure the safety of the products purchased by its citizens. The specific target is the internal engineering that is ancillary to the design, manufacture, sale, service, and repair of products of the state's industries.

Unlike the existing safeguards of product safety legislation, or standards, or the legal precedents of the manufacturer's responsibility for the quality and safety of

526

his products, the proposed legislation does not govern the safety of products. It regulates engineering performed by manufacturers within the state. It is not clear how this will be linked to the safety of products purchased by the state's citizens.

NATURE OF STATE REGULATION VIA LICENSING

To fully understand what is proposed it is essential to consider just what state regulation of a professional activity is. Among the freedoms traditionally cherished by all Americans are those of seeking education in fields of one's own choice and of employing the skills to produce a livelihood. These and other freedoms are inviolate except as their exercise threatens to injure the life, health or property of other citizens. In these circumstances the police power of the state is overriding, and states may restrict or withdraw such freedoms for the general good of the whole citizenry.

It is under this principle that states impose regulations and licensing for certain learned professions and skilled trades. The object is to protect citizens from possible fraud and other forms of economic or physical harm that might stem from services by unscrupulous or unqualified practitioners. In particular, this protection has been applied where the public is unable to adequately evaluate the qualifications of the practitioners that they may employ.

NATURE OF INDUSTRY EXEMPTION

Let us examine the nature of the so-called industry exemption. In fact, it is an essential boundary that protects the rights of the 80–85 percent of all U.S. engineers who are not engaged in public service or in providing engineering services to the public on a professional-to-client basis.

The way that the promoters of state regulation have chosen to phrase the legislation has been to start with an all-encompassing description of engineering rather than with the service rendered, as it is done in medicine. Literally, if there were no delimiters, the engineering legislation would encompass any kind of activity that is a creative utilization of knowledge of engineering, science, and mathematics. In contrast, the regulated scope of the physician's practice of medicine is that of diagnosing and treating human illnesses, injuries or other physical conditions—strictly a physician-to-patient or to-public relationship.

MAJOR CONCERNS FOR ENGINEERS IN INDUSTRY

In assessing the nature and implications of the drive for state regulation of engineering within the manufacturing industry I have probed for the answers to these key questions.

1. Are we engineers in industry, and the manufacturing corporations in which we function, posing a serious threat of injury to life, health, and property of

the state's citizens—a threat for which there is inadequate legal protection?
I do not find that there is convincing evidence of this.

2. Would the proposed legislation appreciably reduce any residual product-
related threat to life, health, or property of the citizens of the state wherein
the engineering is regulated? Again, I do not find convincing evidence of
this.

3. For what other purposes does anyone advocate the imposition of such state
regulation? In searching for the answer I find some self-serving purposes
like those of enhancing the public recognition of engineering, limiting entry
to the profession, and other occupational restraints that resemble a legally-
imposed closed-shop. However, these are not constitutionally valid purposes for state regulation.

I do not find the answers to any of these questions to be satisfactory. The only
professionally palatable concept raised by the assessment is that the unification
and strengthening of engineering as a recognized learned profession is a valid
goal. That is why I have chosen to step aside from the questions of motivation,
whatever they may be, and concentrate on another significant, but generally
unasked question: What would be the burdens imposed on engineers, corporations, and the public if the state seriously attempted to control the safety of
products purchased by its citizens through regulation of engineering in the state's
manufacturing corporations?

BURDENS ON ENGINEERS IN INDUSTRY

State regulation of engineering in manufacturing corporations, even if it only
requires licensing of engineers in responsible charge, is equivalent to licensing for
every engineer who aspires to accept engineering responsibility in his corporation.
One modest burden is that of obtaining and maintaining a license, with the
associated cost and loss of time and effort expended to qualify. The vast majority of
U.S. engineers practice in industry, and only a small fraction are voluntarily
licensed in the states in which they practice. The proposed regulation would
extend this burden to the balance of this majority.

License fees are eyed hungrily as a new tax base. One state recently proposed to
raise the fee to $150–$200. This was pushed back to $35, but who knows for how
long. Engineers will become especially vulnerable if licensing is universal, so there
is no longer a question of discriminatory taxation of the small percentage of the
profession that must be licensed. It is entirely plausible that fees could bound back
to the $200-year level, raising the bill to $150,000,000–$180,000,000 for all old
and new licensees.

A legal review of state occupational licensing statutes shows, at best, minimal
safeguards for the rights of licensed individuals. Generally, a license is regarded as
a privilege to which no individual has a right. The doctrine of separation of powers
makes licensing exclusively a legislative function, separate from the judiciary
where the traditions of equity and fair play are assured. These elements combine
to permit a licensing authority to revoke an occupational license with mere notice

to the licensed individual plus the opportunity for a minimal due-process hearing at which even hearsay evidence may be admissible.

In deference to the many honorable engineers who serve on state boards, we should note that licensing power is delegated generally to responsible individuals appointed by state governors, is exercised with care, and is seldom knowingly abused. Nonetheless, those subject to mandatory licensing should recognize that their licenses can be withdrawn without a hearing in a court of law and that time and expense can be required to reverse such an action through a court of law.

CORPORATION AS AN ENFORCER

Compliance with the proposed legislation may place the corporate entity in the unsavory role of enforcer if it is to avoid being an accessory to illegal practice or malpractice. The corporation must assure that every engineer to whom responsibility may be assigned obtains and renews licenses in the proper states. Each significant product fault (like those leading to product recall or involved in a liability claim) may be investigated to identify the engineers at fault, so records must be kept for state board of examiners' review and possible disciplinary action.

This creates a new focus on pinpointing an individual responsible for each judgment and decision. To limit their individual responsibilities, licensed engineers may wish to separately seal the documentation of design, manufacture, quality audit, customer application and use of the product. The seal carries implications of sworn attestation to personal knowledge of the engineering details.

Regardless of the financial indemnification that may be provided by the manufacturing corporation to protect the engineers, each responsible engineer in this chain must be gravely concerned about being found at fault and thus risk losing his or her license to practice. All in all, this injects an unsavory relationship and separation of interests between engineers and between engineers and their corporation.

PUBLIC'S LOSS OF ENGINEERING INNOVATION AND PRODUCTIVITY

Turning to another area of concern, a disturbing implication of the blame-pinpointing characteristic of state regulation is the potential destruction of the system of teamwork and integration that has given birth to the enormous engineering advances of this century. Within the structure of the manufacturing corporations, scientists in fields like metallurgy, nuclear and solid state physics, and physical and organic chemistry have functioned in teams with engineers of all disciplines to bring about the exquisite integration of sophisticated product designs.

Through the mirror of past experience we can also foresee the bickering between the designing engineers, manufacturing engineers, quality assurance engineers, application engineers and service engineers as each tries to narrow his scope of legal accountability. Worse still, such preoccupation of each team member diverts professional concern away from the focus on overall product safety and other attributes of quality. Instead of having many professional guardians of total product safety we may have none—again the public loses.

WORRISOME UNCERTAIN PUBLIC BURDENS

There will be another uncertain but extremely worrisome array of consequences growing out of this approach to adding more state protection of citizen safety. Since there are no precedents, the state's implementation measures have to be deduced from the safety-intent and engineering-target of the proposed legislation.

To make this deduction we note that those products having substantial engineering content that are manufactured in one state are predominantly sold in other states or exported. Some products are not sold at all within the state where manufactured. With our system of free competition and interstate commerce, all but an insignificant portion of the engineered products purchased by citizens of one state are manufactured in other states or imported. What will the legal sanctions be to make regulation of engineering in the state an effective added guardian of product safety for the state's citizens?

When I examine the engineering performed in manufacturing corporations I do not find that the life, health or property of the state's citizens is seriously threatened or without adequate legal protection. When I examine the potential implementation of the proposed legislation I do not find promise of noticeable improvement in citizen safety—worse than that, it may reduce overall safety. When I examine the burdens such legislation would impose upon engineers in industry I find the burdens to be substantial and totally unreasonable—and liability shared with the corporation, personally, is a totally new concept and public exposure for them. When I examine the burdens that regulation of engineering would place upon manufacturing corporations I find them to be substantial and immediately reflected in unpleasant implications for the engineers of the corporations and other citizens of the state.

These observations do not alter my conviction that engineers in industry should become registered, voluntarily, to establish their credentials for positions of public trust (like serving as engineering advisers for government agencies or committees at local, state and national levels) and to open the options of choosing to enter private practice or join an engineering firm that offers services to the public.

However, these observations do establish my conviction that a state occupational regulatory system, set up to assure that those who offer engineering services to the public do possess certain minimum qualifications, is a totally inappropriate mechanism for improving the safety assurance of products purchased by the state's citizens. A serious regulatory attempt to make such a mechanism work would be costly as well as counter-productive.

8.5

Against the Industry Exemption...

G. J. KETTLER, P.E.

The stated purpose of the registration of engineers is to protect the health, safety, and welfare of the public by providing a recognizable method for determining the qualifications of persons practicing engineering. Thus an individual or public entity requiring the services of an engineer for the design of a structure, product, or system could be assured of a practitioner with the necessary knowledge and ethical qualifications. The industry exemption is one of a series of exclusions that permits a manufacturing enterprise to exempt its engineers and practices from the scope of the registration law.

It would seem a contradiction to have a law to protect the public from incompetent design while excluding the majority of the engineers, i.e., those working in industry. The reason for the industry exemption is a simple, hard political fact. The exemption was a political condition for enactment of the registration laws in the first place. Industry had enough political influence and lobbying effort to demand and obtain exemption for its engineers and engineering activity as a condition necessary to enactment of the laws.

Although the ultimate goal is that all engineers eventually meet their obligation for professional registration, the discussion here involves only those "in responsible charge" for the design, production, and maintenance of structures, products, and processes that affect the health, safety, and welfare.

The lesson that the federal government has made obvious in the last few years by the enactment of some 30 major laws on health, safety, and consumer protection, is that the old saying "Let the buyer beware" has passed into history. The contention that the public needs the protection when selecting structure designers but not product designers seems a little contradictory. In our increasingly technological society, the ordinary consumer can no more judge whether a complex product is properly designed than he can determine the structural integrity of a building.

The reason for the removal of the industry exemption is to assure qualified and responsible design, production, and maintenance of products, systems, and work places to all our citizens. While many fine companies have excellent records in design and safety, there are many others who need strong encouragement or even compulsion to provide safe products.

Why then do some companies oppose the elimination of the industry exemption? I believe that is a matter of education to the benefits of registration and cooperative actions by all sides to solve or minimize the problems. Some of the questions or statements that industry has raised are as follows.

1. "One test for registration won't assure technical qualifications." It must be remembered that the engineering registration tests verify only a basic knowledge of engineering and scientific principles (which, by the way, isn't a bad start on design qualifications). Professional registration is an attitude as well as, or even more important than, a simple technical qualification. By becoming subject to legal registration, we "go the extra mile" and accept the responsibility for our work.

2. "Leave us alone" or "we don't need another law." This is the same argument that was applied to the first child labor laws, and work-hour laws in the 19th century up to the recently enacted Occupational Safety and Health Act. This is no more valid a reason now than it was then.

3. "We are already doing the job." In 1970 the President's Commission on Product Safety chronicled a set of shocking statistics: some 20 million Americans are injured each year as a result of incidents connected with consumer products. Incidents connected with industrial products account for an additional seven million injuries each year. This hardly sounds like doing the job of providing safe products.

4. "We (industry) are responsible and liable." This is true, but the remedies for failure and injury provided by the legal system do little to reduce suffering and bring victims back to life. It seems that the results are being attended to rather than eliminating the causes—the old horse and the barn door story. Proper design in the beginning could greatly reduce the injury problem and most likely save the large insurance and legal fees. Nearly 300,000 product liability cases were filed in 1969, and the number is constantly increasing.

5. "Let us choose and evaluate our own employees." There is no restriction imposed by the registration law requiring that a certain individual be hired. Qualifications for those making final design decisions are the only requirement and that should also be the policy of any responsible company. Over 350,000 engineers are now registered; over half of those eligible. There should be no shortage of talent available. The registration boards evaluate the registration candidate much more thoroughly than most companies do when hiring.

6. "Engineering is too varied a field for common registration." The registration process only approaches the basic knowledge and understanding in the field and the professional attitude. Special requirements and job assignments would still be the employer's right and responsibility.

7. "There is no need to have all the engineers in the design process registered." This is true. Only those in responsible charge making the final

decisions are the subject of this discussion. Some of the company-labeled design engineers may actually be technicians not qualified for registration, but titles are a completely different subject.

8. "How would the 'person in responsible charge' and other terminology be interpreted? In most cases the person in charge is established by existing and commonly used company procedures. Team or product leaders and drawing sign-off procedures are now commonly used and would change little. Some special cases may be ruled on by the registration boards or the courts. The conjecture that engineers will try to hide behind organizations, bicker among themselves, or try to pin blame on someone else if registered is not consistent with the professional approach and is a poor example of a company's view of its engineers.

9. "Standards and codes are sufficient to protect the public." Those familiar with OSHA and other government-developed codes would point out several faults with that system.

10. "How would a multistate industry be expected to operate under the many varied state laws?" The present operations of many of the larger consulting companies cross multiple state lines successfully. Common sense application of the laws and reciprocity provisions should solve most of the problems including the transfer of engineers.

11. "What about presently employed engineers who are not qualified for registration or are not willing to pass the registration exam?" First, if not qualified, what are they doing making designs that may endanger the consumer or employees? Those who are qualified could be registered by several methods including testing, special "grandfather" provisions, allowing present employees to retain positions (which doesn't really approach the competence problem), and many other methods. This will be a difficult problem that will require state-by-state and individual evaluations.

12. "This will expose the engineer to personal liability and arbitrary rulings by registration boards." The lawyers in a liability case will be after the settlement payment not the individual engineer. On the second point, the record of the state registration boards proves this allegation to be highly unlikely.

13. "Why isn't the engineering profession more like the medical profession?" This statement, I believe, is somewhat overworked. The very unity pointed out for the doctors gives them special control over the education process and entry into it—something companies have been opposing for engineering. Since the product and system designs are eventually sold as products, they do have a direct public impact. One decision of one product design engineer could affect more people than the doctor in his lifetime of practice—something to keep in mind when choosing whether to go to a licensed or non-licensed doctor.

Having discussed some of the questions and negative aspects of industry registration, the positive benefits for engineer registration for the company should also be emphasized.

1. Registered engineers assure the company of the services of those who have met the prescribed statutory requirements of the law enacted to protect the public health, safety, and welfare.

2. An engineering staff composed of registered professional engineers enhances the prestige and public relations potential of the firm.

3. Engineering registration improves morale of the engineer by attesting to his qualifications, competence, and professional attitude. It also encourages the engineer to take full responsibility for his work.

4. Engineering registration improves company-client relations by attesting to engineering staff competence and satisfies the legal requirements of many states and municipalities requiring project control under a registered engineer.

5. Engineer registration promotes high standards of professional conduct, ethical practice, integrity, and top-quality job performance.

NSPE has recently reaffirmed its policy on the industry exemption subject. It is stated as follows:

> NSPE believes that the state engineering registration laws should apply to all engineers responsible for engineering design of products, machines, buildings, structures, processes, or systems that will be used by the public; and urges management to engage only registered professional engineers for responsible engineering positions. Management is also urged to promote registration among all of its engineering staff.

> The Society is opposed to proposals to exempt engineers in industry from the state registration laws and recommends the phasing out of existing state exemptions in state registration laws.

Elimination of the industry exemption would serve to benefit industry by putting the emphasis on responsible design rather than on belated repairs and insurance claims. The consumer movement and the heightened realization of the engineering profession to its responsibilities in public service and protection has made it obvious that for the industry exemption its time has passed.

8.6

Getting Your P.E.
Why? How?

JOHN D. CONSTANCE

Engineers are pioneers in almost everything they do, with one exception: the securing of professional recognition. The medical profession began to seek a legal basis for its professional standing more than 200 years ago; lawyers began the same process about 80 years ago. Yet engineering has lagged. Although the first engineering licensing law was passed in 1907 (in Wyoming), only in recent years has the profession begun to direct much attention to its legal standing in the community.

Many engineers seem to have little concern for the standing and the status of engineering as a profession. The majority do not belong to technical societies, and nearly 60 percent have not obtained a P.E. license. They apparently do not realize the importance of society activities and registration in shaping professional attitudes.

But without our current engineering registration laws, your right to practice engineering would be restricted by some of the other professions. Some years ago, legislation sponsored by physicians would have reserved the sanitary field for themselves; accountants sought to exclude others from the making of financial reports. Pioneers in the engineering registration movement fought off these attempts, as well as legislation proposed by lawyers that would have deprived engineers of the right to prepare contracts and to engage in arbitration. Real estate brokers even sought to prevent engineers from making appraisals of land and buildings.

Many engineers do not bother to register because they hold positions that do not at present require a P.E. license. But the situation may be changing. Montana early this year [1976] passed a law requiring that the design of all products affecting the public be approved by a registered engineer. It thus became the first state to essentially eliminate the so-called industry exemption, which says that engineers employed by a company engaged in making products for sale are exempt from mandatory registration. Of course, Montana is not an industrial state, but a similar bill is being considered by the Ohio legislature. It would require professional

535

registration for engineers engaged in the design of all products that might affect the life, health, or property, of the user. The Ohio Society of Professional Engineers, author of the bill, is confident of eventual passage. At the National Council of Engineering Examiners meeting in Boston last August, the American Society of Manufacturing Engineers went on record in favor of elimination of the industry exemption, too.

Several other states are moving in the same direction; Illinois, New York, California, and Georgia among them. In North Carolina, the situation is confusing; engineers managed to insert a clause similar to Montana's into the state's new licensing law, but the law still specifically includes the industry exemption as well. However, many observers feel that, if current trends hold, it won't be long before most states have eliminated it.

Elimination of the industry exemption will not mean that every engineer engaged in designing a product that could adversely affect users will have to be registered. In the words of the Ohio bill, the engineer need not be registered if his work "does not include final engineering designs or decisions and is done under the direct supervision of and verified by a person holding a certificate of registration." But the practical application is clear: if you hope to rise to a responsible engineering position, you will eventually have to be registered.

Even now there are a number of good reasons for obtaining your P.E. license. In many states you cannot legally affix the title "Engineer" to your name on a calling card unless you are a registered professional engineer. Or suppose that you become a specialist in some area and plan to do consulting work on a fee basis, either part time or when you retire. Legally you cannot accept this type of work unless you are registered. A contract to do engineering work is not binding unless the engineer is registered. In addition, only registered engineers can testify in court as engineering expert witnesses.

In difficult economic times, a P.E. license might mean the difference between landing a job or not. Engineering registration also tends to unify the profession, helps set minimum standards, and increases public recognition of the professional standing of engineering. All in all, a license is an inexpensive insurance policy for protecting your investment in your professional career.

Engineering registration examinations are currently administered individually by the various states. A man or woman licensed in New York, for example, cannot legally practice in California unless he or she also becomes registered in that state. Although a national engineering exam does not seem imminent, most states subscribe to the standard format devised by the National Council of Engineering Examiners. (Colorado, Illinois, New Jersey, and Pennsylvania were the only states that did not use the NCEE format for the November 1974 examination.)

Under the NCEE format, the test is divided into two sections: the Fundamentals of Engineering and the Principles and Practice of Engineering. Each section consists of two four-hour test sessions, for a total of 16 hours in two full days of written examinations.

The section on fundamentals usually covers the basic subjects of mathematics and physical and applied sciences (dynamics, fluid mechanics, thermodynamics, electrical principles, and mechanical design). The principles and practice section is designed to test judgment in practical situations; the problems are worded so as

Classification of Applicants

Applicants Who Are	Years of Approved Qualifying Experience in Engineering after Graduation	May Qualify For
1. Graduates of ECPD-accredited courses in engineering schools, U.S.	Less than four years	The Fundamentals of Engineering examination and subsequent EIT certification
	Four years or more	The complete written examination. Fundamentals of Engineering and Principles and Practice, and subsequent full P.E. registration or licensure
2. Graduates of nonaccredited courses in engineering schools, U.S. and foreign	Same as above in both cases, plus additional experience time in engineering or ECPD engineering school master's, as determined solely by board of examiners	The Fundamentals of Engineering and Principles and Practice parts of the written examination
3. Graduates of accredited high schools or the approved equivalent	Twelve years minimum or as determined solely by board of examiners	Same as above
4. Graduates of accredited high schools or the equivalent with partial educational credit from engineering school, full-time day or part-time evening	Same as above, less approved time equivalent for educational credits, as determined solely by board of examiners	Same as above

Note: Because boards retain the right to change conditions and make interpretations of the registration laws and since the above table contains space-saving simplifications, engineers are cautioned to verify requirements with their boards. Up-to-date addresses of state boards may be obtained free of charge by writing: Natioanl Council of Engineering Examiners (NCEE), P.O. Box 5000, Seneca, South Carolina 29678.

to require the candidate to call upon his practical experience to develop the solution.

Sample copies of NCEE examinations and the address of your state board can be obtained by writing the National Council of Engineering Examiners, P.O. Box 5000, Seneca, South Carolina 29678. Application forms and other materials may be obtained from your state board of examiners.

All states use open-book examinations, since it is felt that this approximates everyday engineering office practice. The average question takes from 30 to 40 minutes to complete; extremely long answers are discouraged. The fundamentals section is machine-graded, and there is some talk that NCEE is considering machine-grading the principles and practice section also. However, at the present it is felt that machine-grading this section would be inappropriate for the essay-type answers and would ignore the element of creativity in the solutions.

Because an engineer frequently works in many different states, it would seem desirable to develop one standard examination for use nationwide. However, for now, the practical problems in devising a standard examination seem to make such a test unlikely.

In addition to passing the licensing examination, every candidate for registration must fulfill other requirements relating to his education and experience. The general requirements for registration are displayed in the accompanying table. However, these requirements vary slightly from state to state (most states, for example, do not require U.S. citizenship, but a few do).

The list of engineering schools accepted by the state boards also varies from state to state. Some boards accept those schools accredited by the Engineers' Council for Professional Development. Others have compiled their own lists of approved schools. Even if a school is on the approved list, it does not necessarily follow that all its engineering courses are similarly approved. If the applicant's school does not appear on the approved list, he should write to his board for confirmation. He may be required to write to his school, asking it to take steps toward becoming accredited with the board.

The experience requirement can also be subject to a great deal of varying interpretation among the states. It is not only the number of calendar years of experience but also the quality of the experience that counts.

The NCEE defines qualifying experience as "the legal number of minimum years of creative engineering work requiring the application of engineering sciences to the investigation, planning, design, and construction of engineering works. It is not merely the laying out of details of design, nor the mere performance of engineering calculations, writing specifications, or making tests. It is rather a combination of these things plus the exercise of good judgment, taking into account economic and social factors, in arriving at decisions and giving advice to the client or employer, the soundness of which has been demonstrated in actual practice."

Acceptable engineering experience defies exact definition. However, it is predicated on a knowledge of engineering mathematics, physical and applied sciences, properties of materials, and the fundamental principles of engineering design. For the newly graduated engineer, some routine work is unavoidable and may even be desirable. But to obtain the proper kind of experience he should seek the widest possible responsibility. Boards value experience of a broadly diversified character higher than more specialized activities. Boards license engineers, not specialists. They do not allow experience credit for such activities as:

- Teaching of nonengineering subjects in an engineering college or elsewhere.
- Work projects prior to the completion of high school.
- Assignments in branches of the armed forces that do not involve engineering.
- The work periods of a cooperative college curriculum.
- Vacation periods between college semesters.
- Sales work that does not involve the use of engineering knowledge.
- Employment of a nonengineering nature, even though it may be in connection with engineering work.
- The duties of a contractor, superintendent, or foreman on construction work, unless such work involves engineering practices.
- Correspondence school courses in engineering.

A candidate may acquire his experience in his home state or elsewhere. A year of graduate study in engineering leading to a master's degree may be accepted as a year of experience; higher degrees are also seen as comparable to years of experience. In all cases, the candidate must submit a written record detailing his experience and education so that the board can judge its value.

Candidates who have less than the required number of years of experience for licensure may qualify for the Engineer-in-Training examination. The EIT candidate takes the first half of the licensing examination (the section on fundamentals), and, upon its successful completion, is granted an EIT certificate. This does not, however, allow the candidate to practice as a P.E.

A few engineers are able to obtain licensure by endorsement—that is, without taking the written examination. However, in most states, this special consideration befalls only those who can meet rigorous requirements set down by the state boards; it is usually given only to those engineers with long-established practice and recognized standing. Of course, such criteria as "recognized standing" are difficult to define. Most states require at least 15 years of approved engineering experience after graduation from an accredited school of engineering. The mere showing of a large number of years of experience, however, is insufficient to qualify an applicant for exemption. Other criteria are weighed, such as advanced studies, technical society activities, contributions to engineering knowledge and progress, inventions and patents, technical articles published, reputation among fellow engineers, and other important achievements.

Even if you don't qualify for an exemption from the written exam, if you have a great deal of experience you may qualify for a waiver of part of it. The exam usually seems to give more trouble to older than to younger candidates. The state boards appreciate this problem, and are looking into the possibility of giving a different type of examination to older engineers—one that would give them more of an opportunity to demonstrate the benefits of many years of experience.

Younger engineers as well as older ones often complain that the tests are not work-related and fail to take into account the wide diversity of engineering backgrounds. Realistically, no test can cover every detail of a candidate's engineering knowledge. But there are likely to be enough questions in each specialty to enable the candidate to show his or her prowess. And even though college engineering courses differ in some details, there is a core of common material that can be tested.

For more information on the licensing test, contact the secretary of your state board. He or she can furnish you with a list of references and study aids, as well as information about refresher courses. The National Society of Professional Engineers and its state and local chapters can also be helpful in obtaining information. A number of industrial companies provide their qualified engineers with refresher courses. They can also provide more experienced candidates with guidance for exploring the possibility of licensure by endorsement.

8.7

Qualifications for Continued Practice

ALFRED C. INGERSOLL,[1] F. ASCE

ABSTRACT

Progress toward establishing standards for qualifications for continued practice is discussed, with emphasis on the question of mandatory continuing education or professional development for renewal of professional license. Impact of the "consumer society," as well as ASCE policy positions and member attitudes are analyzed. An experimental program in Wisconsin is described, as well as surveys among professional engineers in California and members of the Professional Activities Committee of ASCE. Short- and long-term proposals are offered for assuring the public of the licensed engineer's current competence and assisting the engineer to maintain lifelong competence in an organized way.

It will be the purpose of this paper to provide background for several of the workshops and panels that meet later in this conference, covering subject headings ranging from "Voluntary Continuing Education," to "Mandatory Requirements for Renewal," and including the title of this paper, "Qualifications for Continued Practice."

Since the subject of ethics rates top billing in the title of this ASCE Specialty Conference, it is not of minor importance for me to declare at the outset that I have a potential conflict of interest in speaking about continuing education as a possible mandatory requirement for renewal of the Civil Engineer registration or license. Since I am in the business of offering continuing education programs in engineering (and mathematics)—if not for profit, for survival—what credibility

Grateful acknowledgement for information used in the preparation of this paper is made to C. Allen Wortley of University of Wisconsin Extension, Edwin R. Breihan, F. ASCE, of Horner & Shifrin, Inc., St. Louis, Mo., D. R. Wright of the California Board of Registration for Professional Engineers, Lowell Norseth of the Minnesota Board of Registration for Professional Engineers, Architects and Landscape Architects and J. T. Cobb of the University of Pittsburgh.

[1]Associate Dean for Continuing Education, School of Engineering and Applied Science, and Director, Continuing Education in Engineering and Mathematics, University Extension, University of California, Los Angeles.

can be given any words from me, words encouraging a policy of continuing education (mandatory or voluntary) as a means of maintaining competence? I can tell you that I am not alone in this conflict of interest problem. Let us not forget that our Society, the ASCE, sponsoring this Specialty Conference, is also in the business of offering continuing education. Any pious pronouncements on this subject that come out of this conference, under the imprimatur of ASCE, will be subject to the same credibility problem, from our membership as well as from those on the outside. It has been suggested, on this account, by leading voices in the continuing education community (1), that all of the professional societies should divest themselves of their continuing education business efforts if they are going to enter into any policy recommendations on this subject, but I shall not pursue that point here.

In any event we could observe that ASCE is as pure as the driven snow on this count, not by consciously eschewing a conflict of interest, I should guess, but rather because of the overwhelming aversion of ASCE members to the "forced bussing" notion of mandatory requirements for renewal of professional license.

THE ISSUE OF PROFESSIONAL OBSOLESCENCE

It is important to understand that when we are talking about continuing education or professional development as possible requirements for renewal of registration or license, we are not talking about protecting the public from incompetent professionals who should never have been granted their licenses in the first place, who are practicing outside of their field of expertise, or who are doing shoddy work because they are lazy or burdened with personal problems, etc. These circumstances exist, of course, in any profession and our existing laws are presumed to be adequate to deal with such transgressions.

We are speaking, rather, of dedicated professionals, fully competent at the time of their initial licensure, practicing diligently at the best of their abilities through two or three decades of their careers, but who have simply been overtaken by developments within their profession. It is often set forth that professional obsolescence has been with us for generations, and the marketplace accommodates to the condition, continuously weeding out those who have not kept current in their fields. I can clearly recall, as a small boy in a midwestern city, a value decision taken by our family that I would go to a new dentist in town, rather than the one who had served our family for many years, simply because he was a more recent graduate of dental school.

The argument of the marketplace as the great leveler is especially strongly held by consulting engineers. "Show me a consulting engineer who has not maintained competence while practicing in a fully competitive community," the argument goes, "and I'll show you an engineer who is ready to go out of business."

Because health-care professionals deal so directly with the public—and one incompetent professional can have a life-or-death impact on the patient—there is a larger measure of public acceptance for mandatory requirements for renewal in the health-care fields than in any of the others. Registered nurses have been in the vanguard of the movement for some time, as have pharmacists, nursing-home administrators, chiropractors, dentists, and, most recently, vocational nurses.

Physicians have been establishing their own standards of currency through the Physicians Recognition Award of the American Medical Association.

Accountancy is a field which deals with the public, requires licensure, and where an incompetent professional can have a dollars-and-cents impact on the client. In California (and many other states as well), the Board of Accountancy (2) has required continuing education for renewal of certificate since 1972, with very little objection from the profession, so far as I know.

In another professional field where the public is served, and changes are taking place continuously, the Supreme Courts of Wisconsin, Minnesota and Iowa are now requiring attorneys to take 15 hours of continuing education each year until age 70. If this movement spreads to other states—and with lawyers comprising the majority of all state legislatures—we may well find such requirements imposed by statute on licensed engineers.

Minnesota appears to be the first state in the union to pass enabling legislation that leaves the question of qualifications for renewal of license up to the individual professional boards. The law was passed in 1976, in response to the Nursing Board's request for legislation on relicensure. At first proposed as mandatory, then changed to permissive, the Minnesota law covers all licensed professionals. The Engineers' Board maintains that it has no plans for implementing relicensure qualifications for engineers.

CAN ENGINEERING OBSOLESCENCE HURT THE PUBLIC?

One of the most controversial questions among engineers is whether or not obsolescence, in and of itself, on the part of a licensed professional engineer, can really harm the public. More accurately, the argument should be stated as whether or not the public is significantly *more harmed* by obsolescence of licensed professional engineers than it is by incompetence of engineers generally. Even more controversial is the question of whether a program of mandatory requirements for renewal of license would significantly decrease public exposure to such risk.

Arguments against mandatory requirements for relicensure—so far subscribed to by the vast majority of engineers—are founded on several premises:

a. In most circumstances, the design output of a consulting engineer is checked thoroughly by one government agency or another, so that the client is assured of an independent review of the work for which he has contracted. This is in contradistinction, for example, to the case of the physician who commonly makes a unilateral decision as to the treatment of a patient.

b. The great majority of engineering work in the United States is performed by engineers who are not licensed, either because they work in subprofessional classifications not requiring licensure, or they work in units of industry, government or education which are exempt by law from registration (in all states but Montana). While approximately 65 per cent of the civil engineering population of the U. S. is licensed, only about 28 per cent of the remaining engineering population is licensed.

   c. Since most engineering registration is by Title only (as opposed to Civil Engineering which is usually also covered by a Practice Act), there is no legal requirement of registration for work performed in the noncivil engineering fields in most states.

While I do not contest the validity of these arguments, I see in them a unique opportunity for the civil engineering profession to strike a leadership position on the matter of qualifications for continued practice. I say this for at least two reasons. First, because our profession *is* the most carefully controlled by law. We *are* covered by a Practice Act in most states. Even in exempt industry it is frequently stated that civil engineering work, especially that related to safety and health, shall be performed by registered civil engineers. Thus I believe we are in the best position to take a meaningful stand on the matter.

My second reason is purely chauvinistic. I think civil engineers are the greatest, and should be the natural leaders of the engineering profession. Our history certainly bears me out. ASCE was the first of the engineering societies to get under way in the United States. Civil Engineering has been the first of the engineering fields to achieve Practice Act protection, at least in most states. As may be noted elsewhere in this conference, ASCE was the first society to adopt a Code of Ethics with some real teeth in it. Countless other examples could be cited where civil engineers—and especially ASCE—have exerted real leadership.

But I have not responded to the question, can engineering obsolescence hurt the public? While it is difficult to cite dramatic engineering failures—whether of the Quebec Bridge that we immortalize in our Order of the Engineer, or of the more recent Teton Dam—that could be attributed to engineering obsolescence, one example might be the failure of freeway overpass columns in the San Fernando earthquake of 1971. These columns were designed by the California Division of Highways, following design procedures used successfully, no doubt, for many years. Unfortunately these procedures did not incorporate all that was then known about earthquake-resistant design of reinforced concrete structures, and the plans of one government agency are not reviewed by another, lower-level agency. Thus it was that the circumferential reinforcing was inadequately tied, the columns failed and the public was hurt. There are probably scores of less dramatic instances where the procedures and principles of yesterday, applied to today's design problem, may produce a result that is perfectly acceptable—from the standpoint of the plan checkers and safety requirements—but one that may not be the *best* solution to the problem. Specifically, it may be more expensive than necessary. It is not the job of the Department of Building and Safety to suggest ways in which a particular engineering design could have yielded a lower cost. If the design meets all the safety standards, in my understanding, it gets approved. Additionally, there are always circumstances in which the advance columns of the engineering profession are ahead of the plan checkers. Fire protection in high-rise buildings is one example. Earthquake-resistant design, as noted above, would be another. Just now in Los Angeles we are wrestling over a proposed city ordinance that would mandate strengthening or removal of all buildings constructed before earthquake engineering was applied to the design of new buildings. Engineers are again in a conflict of interest position. We know the right things to do but we are reluctant to speak out for fear of being accused of feathering our own nests.

It is my conviction, then, that the public can be hurt by engineering obsolescence (and presumably, therefore, could be advantaged by a program of mandatory requirements for renewal). The extent to which the public *is* being hurt by engineering obsolescence is still controversial. Even when it is ascertained that the public is being hurt, it is extraordinarily difficult to establish to a certainty that the situation would be improved by the imposition of mandatory requirements for renewal. Our profession considers this issue to be, if not resolved in the negative, at least largely unresolved.

THE ASCE POSITION ON MANDATORY PROFESSIONAL DEVELOPMENT

Within the past year the Board of Direction has adopted a comprehensive policy statement, (3) "Enhancing the Benefits of Engineering Registration," specifically addressing the question of a possible mandatory requirement to demonstrate continuing professional development and competency as prerequisite to relicensure. The policy states that "There is no record proving that the present licensing procedures are not providing adequate protection for the public." It goes on to assert that the main difficulty, as far as the public is concerned, is the extent to which licensing laws are not sufficiently comprehensive or actively enforced. It cites seven specific ways in which ASCE supports licensing of engineers and insuring their competence, ranging from license requirements for various grades of membership to the continuing education and publication programs of the Society.

Significantly, the policy states, "Re-examining the engineer in fundamentals will not establish his competence. Also, keeping a record of the number of hours he devotes to professional development will not establish his competence." I can agree with that all right but the question is a different one. We are continually facing the same problem in admitting students from questionable backgrounds to graduate standing. Entrance examinations do not establish that the student is qualified, but they can often establish that he is *not* qualified. I would ask ASCE policymakers if they cannot conceive of an examination which would, if altogether failed by a practicing engineer, establish to their satisfaction his *in*competence? We know, however, that if there is anything more abhorrent to the practicing engineer than citing his hours of professional development, it is the concept of taking the profesional engineer exam again. They cite with relish the anecdote attributed to the late Supreme Court Justice Hugo Black, who is purported to have said, "I have this terrible nightmare, it is always the same. I am required to sit for the bar exam again."

With such a clear policy statement by ASCE, and good evidence that it is supported by the majority of ASCE membership, I frankly commend the managers of this conference for including consideration of this topic on the program. I can only imagine that they plan to see the main emphasis develop on voluntary programs of professional renewal, which seem to be espoused by one and all as a cardinal virtue of professionalism.

VOLUNTARY CONTINUING EDUCATION AND PROFESSIONAL DEVELOPMENT

As suggested, this subject seems to be relatively noncontroversial. Surely there is little opposition to the general notion that continuing education and, more broadly, professional development are essential to a healthy and happy engineering practice, just as is membership in such organizations as ASCE. Indeed our consumer-oriented society is actively pursuing the argument that most elements of support to the practice of a profession (such as education, licensure by examination, etc.) should really be voluntary rather than mandated by law, unless it can be proved beyond question that certain of these "essentials" are in fact really required in the conduct of that profession. (I note with some amusement that no voice within ASCE suggests that our profession should be *that* free of mandatory requirements.)

If we can agree that fully voluntary continuing education and professional development, pursued individually without penalty for noncompliance or reward for compliance, is a good thing and beyond controversy, let us give our attention to programs, sponsored by organizations, which provide some degree of recognition to the participant. One such program, modeled largely on the aforementioned Physicians Recognition Award, is the Professional Advancement Recognition (PAR) Program, proposed by C. Allen Wortley (4) at the ASCE Annual Convention in 1974. The PAR program proposed a three-year renewal cycle in which the candidate would certify that he has completed a minimum of 100 points of educational activities (roughly equivalent to a two-semester-hour course in a university), plus a minimum of 100 points of professional experience and society activities (25 months of on-the-job engineering experience or society memberships at 10 points per society). Although the PAR program was proposed as an ASCE national effort, it has yet to be adopted by ASCE, even on an experimental basis. On the other hand, the American Consulting Engineers Council has adopted the plan, at least on a provisional basis.

More recently, the Wisconsin Society of Professional Engineers has undertaken a voluntary program, The "Continuing Professional Engineering Development Award," announced only once, in the Janaury 1976 issue of the state society's magazine. The candidate seeking this award was to submit his own record of a minimum of 100 hours in the past two yers of "Type A" educational activities, such as university courses, short courses and institutes, or preparation of technical papers, and another 100 hours of "Type B" professional society activities and independent study. The counting system, it will be noted, is quite similar to that of the PAR program, except for the important difference that the WSPE program does not allow any points for work experience. Similar voluntary programs are being undertaken on a trial basis by the New Jersey and Florida Societies of Professional Engineers.

The Wisconsin Society of Professional Engineers has 1700 members, and the response to the single announcement in the January issue of the magazine elicited 43 responses, submitting an average of 270 total hours. Half were from the field of private practice, the next largest group were from industry, followed by engineers in government, with the engineering educator bringing up the rear. I should surmise that the majority of respondents were civil engineers. The excess hours reported suggest that the only persons submitting records for award consideration were the "front-runners" who would be quite certain of winning the award. This response of almost three per cent from a single journal announcement is a strong indicator that engineers are eager for the opportunity to demonstrate on a voluntary basis what they are doing to combat obsolescence.

One of the most aggressive programs of "voluntary" continuing professional development is that adopted by the Society of Manufacturing Engineers. (2) Until recently (in California) manufacturing engineering has not been a licensed specialty, so the SME certification program is designed mainly to provide SME members peer recognition and to stimulate continuing education. Certification is awarded initially after 10 years of responsible charge and successful completion of a written examination. In order for a member of SME to hold his certificate he must be "recertified" every three years. This is accomplished by re-examination or by proof of having taken certain specific courses or equivalent seminars, clinics, conferences, etc., within the field of certification under which the individual is qualified.

Another of the "new" (i.e., other than "founder") societies, the American Society for Quality Control, has adopted a trial program which involves complex computer systems for recording real-time involvement in continuing education by its members. This is no doubt expensive but is currently being pursued vigorously.

It is a matter of more than passing interest to this observer that these most actively pursued programs of voluntary certification for continuing professional development have been devised and implemented by engineering societies that are still feeling their way in the hierarchy of engineering societies, representing professional fields which have not until very recently been granted the status of licensed specialties. Perhaps it will always be thus: the challenger comes on strong for reform and innovation, while the defender points with pride to the advances of a century and more, maintaining stoutly that no substantive changes are needed.

THE CALIFORNIA SURVEY OF REGISTERED ENGINEERS

In responding to a mandate from the state legislature to study the engineering profession's reaction to a possible requirement of continuing education for relicensure, the California Board of Registration for Professional Engineers appointed in late 1974 a committee of 10 professional engineers to study the matter and submit recommendations (2). Early in its deliberations, the committee learned that statistical data regarding California registered engineers was practically nonexistent. A questionnaire comprised of 23 pertinent questions was mailed to 1500 registered engineers (a 1/30 sampling of all registrants). More than 1250 full replies were received, indicating the depth of interest in this subject. The survey results may be summarized under four headings as follows:

1. *Describe the California registered engineer.* Median age, 52, with 17% over 65 and mostly retired. Educational level, 88% have B. S., additional 27% have higher degrees. Experience, average 24 years, with 79% currently pursuing careers directly related to engineering. Median income, $24,500 per year.

2. *Impact of registered engineers on public health, safety, or welfare; what are*

levels of responsibility?   Only 32.5% claim their firms offer services directly to the public for a fee. Fully 37% (including most of those offering services to the public) work for consulting or construction firms or are self-employed. These are the engineers most concerned with the Engineers' Act. Industries exempt from the Act employ 33% of the registrants, and government agencies employ the remaining 30 per cent. As to job function, 27% term themselves design or development engineers, 21% are consulting engineers, 27% are engineering managers, with 25% in miscellaneous engineering or nonengineering work. A small number (6%) are not working in engineering, and 15% are retired.

3.   *How do registered engineers keep abreast of developments and maintain technical competence?*   Nearly half (48%) of registrants claim two or more hours per week, while 75% spend one or more hours per week in some activity to keep abreast of their field. The most prevalent activity is personal study (88%), followed by all-day seminars (60%), on-the-job study (57%), technical society programs (56%), company training programs (46%), and college extension courses (41%). On the average, registrants devote 13 hours per week on the bona fide study and practice of engineering, the remaining time being consumed in administration, supervision, etc. Most registrants (70%) are dues-paying members of one or more technical societies and spend an average of 1.5 hours per month attending meetings of these societies.

4.   *Would the registered engineers support a mandated program of professional development? As a condition for licence renewal?*   Only 35% feel the State should establish requirements for license renewal that mandate a program of professional development; 48% oppose such requirements and 17% are undecided. If the State did mandate such requirements for renewal, 83% of registrants say they will maintain registration. Those who would drop out are those who now make least use of their license for earning a living.

This is the most extensive survey of engineer attitudes on this subject known to me. It demonstrates that, while only a minority of engineers favor mandatory qualifications for renewal, the vast majority would go along with it if it were made a part of registration law.

THE ASCE SURVEY BY THE TASK COMMITTEE ON MAINTENANCE OF QUALIFICATIONS FOR CONTINUED ENGINEERING PRACTICE

In early 1975, the TCMQCEP surveyed 45 members within the Professional Activity Committee in an attempt to sample attitudes of ASCE people who are greatly concerned with professional matters. (5) With almost no indecision, the members responded: 55% believe there is a need for a Professional Engineer renewal process on a continuing basis after initial registration; 96% oppose a written re-examination on a two or three year cycle; 54% favor a Professional Advancement Recognition for renewal of registration, in which applicant would accumulate (on a 2 or 3 year cycle) points for educational courses, teaching activity, publication of articles, learning experience, society activities and on-the-job experience; 76% oppose the administration of such a program by ASCE

Headquarters; 93% oppose administration of such a program by ASCE Local Sections; 95% oppose administration by ASCE Districts, 97% oppose administration by ASCE Zones; 55% favor assigning the PAR program to special committees of the designated jurisdictional group of State Registration Boards; 58% oppose any ASCE committee member legal responsibility for disallowing registration, even if ASCE should administer the PAR program; 76% favor ASCE involvement in attempting to set up legislative rules; 71% favor ASCE general involvement in PAR; 69% believe there is not sufficient interest in PAR to set up a two-day Specialty Conference to discuss the matter; 59% would not attend such a conference if it were held.

This questionnnaire response seem to say that those *most* concerned with Professional Advancement Recognition in ASCE feel that it is, at best, a good idea for someone else to do, preferably the duly authorized arm of government, through Registration Boards. ASCE would like to have a hand in setting standards, but not in administering the program if it is made mandatory for renewal of license. It should be noted here that the Los Angeles Section, ASCE, proposed in 1973 a pilot plan of professional development required for license renewal, and volunteered to monitor a trial run of the plan for two years. (6).

IMPACT OF THE CONSUMER SOCIETY

Our consumer-oriented society is clearly on the horns of a dilemma, with regard to a position on continuing professional development as a requirement for relicensure. On the one hand consumer advocates must support any move that enhances the quality of professional service rendered to the public (and this could be construed as favoring the continuing professional development requirement), while on the other hand they want to keep the profession as open as possible to free competition.

One can only surmise the direction that consumer interests will take us in the future. With growing acceptance of mandatory qualifications for continued practice in the other professions, however, it seems clear that engineers will be pushed in this direction, and it goes without saying that we are in a stronger position to specify our own rules if we take a leadership position early in the game.

One thing is certain: In California (and a growing number of states) the consumer-oriented public is becoming a larger and larger contender for the engineer's attention. Legislation enacted this year [1977] in California calls for a majority of public members on the Board of Registration for Professional Engineers. Although this majority will not be fully implemented for a few years, the path is clear for all to see: Our profession is going to be increasingly controlled by the public. My personal assessment of consumer/public priorities is that they shall first concern themselves with seeing that the public safety, health and welfare are protected, and then give attention to policies and procedures that unduly limit the number of engineers competing with one another in the marketplace. We should be prepared, in my view, to see within a few years a well-organized attempt to impose mandated qualifications for relicensure in engineering.

A SHORT-TERM PROPOSAL

The program I would propose as a starter would be the simplest imaginable concept to monitor: Each applicant for license renewal would demonstrate 30 contact hours of continuing education in each two-year period, and in any four-year period he would have 30 contact hours minimum in *technical updating,* the remainder being distributed in management, communication, or technical fields at his option. Some system of approval or accreditation of programs qualifying for license renewal credit would have to be adopted. The Continuing Education Unit (7) (CEU) would be one possible approach, already defined and organized through the National University Extension Association. Regular university courses for credit would qualify for twice the number of actual class contact hours.

Another approach would be to consider qualified the credit classes or extension programs of those engineering schools having curricula accredited by the Engineer' Council for Professional Development.

It is seen from the foregoing that I favor a clear requirement *of continuing (or credit) education* rather than the more inclusive professional development concept espoused by so many. I do this because it is inherently the best means of acquiring new knowledge in an organized fashion. If the engineering profession—and especially civil engineers—simply will not buy this idea, then I would ask if it could not at least be required of all applicants for renewal of registration who cannot certify that they have worked an average of at least 20 hours per week (excluding vacation) at engineering per se (as distinct from administration and supervision) since their last registration.

As suggested before, I hold a mandatory program to be desirable. Despite the unsavory aspects of a "controlled society," there is no denying that compliance with a law provides us with the necessary compulsion to make us do the things we know we should be doing anyway. Countless examples of social legislation have demonstrated this point. Of course there would have to be allowances and accommodations for engineers living in remote areas, as well as for those who are retired but wish to retain their license, but the main body of the engineering profession would, I believe, fit easily into the suggested pattern.

A LONG-TERM PROPOSAL

Supplementing such measures as mandatory qualifications for relicensure I would see, on a long-term and continuing basis, a nationwide distribution of "Engineering Professional Development Centers" serving the engineering profession as a whole. Qualifications for continued practice—the title of this paper—goes beyond the question of relicensure. It will be a long time before even half of the engineering work done in this country requires the services of licensed professionals. We must think about upgrading the continuing competence of all practicing engineers, whether they are employed in industry, government, private practice or education.

To this end I have proposed the concept of the Engineering Professional Development Center, (8) probably linked to the extension program of major universities, which would make possible at least one EPDC in every state and probably one in each major city. The Engineering Professional Development Center would serve as an information center on continuing education and professional development opportunities, as a career counseling center, as a library for audio and video cassettes (as well as printed matter) of professional development material, and as a locale for certain continuing education programs, both for after-hours scheduling and full-time intensive institutes and short courses. There could also be opportunities for full-time reschooling for engineers taking a sabbatical leave from work assignments. There might or might not be residential facilities attached to the Center.

Just as we now go to our physician for a periodic physical checkup, I envision the practicing engineer to check into his Professional Development Center, on a purely voluntary basis, perhaps every five or seven years. His permanent professional development record would be on file there, and for a fee he would have the services of a professional development counselor. Here he would learn, perhaps with the aid of confidentially administered tests, the "awful truth" about the condition of his own current competence in his field as well as associated fields with which he should be familiar. Further, he would get counseling as to the economic outlook for his field of employment or practice, and perhaps even for his particular employer. The EPDC would be a kind of "total service station" for the engineer.

Of course the Professional Development Center would have to be supported by the services it renders, just as automobile service stations and medical centers are supported. The sharpest criticism I have found of this concept is not in its cost or who will pay for it, but in the pervasive image of a "big brother" looking deeply into our lives, our employment and our employers, just too much of a welfare state for engineers. To this I can only respond that the program would be strictly voluntary—our libertarian society would not stand for anything else—and it would thrive or fail ultimately on its merits. All of the services I suggest are available now to varying degrees, but they have not been organized in a way that meets the specific needs of practicing engineers. I do not know of another profession that has established such professional development centers, but I find it a stimulating challenge to think of engineering as the natural profession to take the lead in implementing the concept.

I have endeavored to provide a kind of update on qualifications for continued practice, as a definite movement in the engineering community of this country, and to propose a specific short-range plan as well as a longer range concept for "engineer renewal." Whether these plans, or some others—or none at all—are ultimately adopted, I believe it is safe to say that the subject of qualifications for continued practice will be with us for a long time, and I congratulate the organizers of this conference for giving it an important place on the program.

Appendix—References

1.   Frandson, P. E., "Continuing Education of the Professions—Issues, Ethics and Conflicts," National University of Extension Association *Spectator*, pp. 5–10 (Sept., 1975).

2.   "Report on the Professional Development of the California Registered Engineers," Ad Hoc Committee on Professional Development, Donald H. Nack, Chairman, Board of Registration for Professional Engineers, Sacramento, California (Sept. 30, 1975)

3.   Policy Statement 2.4.3, "Enhancing the Benefits of Engineering Registration," American Society of Civil Engineers, New York City (Nov. 2, 1975)

4.   Wortley, C. A., "Proposed Professional Advancement Recognition (PAR) Program," preprint for Oct. 21–25, 1974 ASCE Annual and Environmental Engineering Convention, Kansas City, Mo. (1974)

5.   "Informal Questionnaire for Task Committee on Maintenance of Qualifications for Continued Engineering Practice," E. R. Breihan, Chairman, Horner & Shifrin, Inc., St. Louis, Mo. (Feb. 15, 1975)

6.   "Proposed Professional Development Requirements for License Renewal," report of Task Force Committee on Professional Development, L. L. Lewis, Chairman, Los Angeles Section, ASCE, (Oct. 9, 1973)

7.   "The Continuing Education Unit Criteria· and Guidelines," National University Extension Association, Washington, D. C. (May 1975)

8.   Ingersoll, A. C., 'The Professional Engineer Half a Century Hence," *Civil Engineering*, V. 45, No. 3, pp. 74–76 (March, 1975)

8.8

Continued Professional Development in Florida: PE's Set Up Criteria for Certificate

Engineers as professionals provide special skills to the public. These skills, which have been acquired through education, training and experience, are exercised primarily in the interest of others.

Continuing efforts in maintaining professionalism are an obligation of a professional engineer. Engineers continue to develop professionally and technically by refreshing their basic skills, updating their technical knowledge, and generally broadening themselves by keeping active and abreast of their profession. They continue to develop as professionals through on-the-job experience, either independently or in cooperation with other professionals; through involvement in the activities of their professional/technical organizations; through educational/developmental activities.

The Florida Engineering Society (FES) Certificate of Continued Professional Development has been designed so that professional engineers in any field can qualify. Thus, engineers engaged in academics or engineering management and administration, as well as those engaged in professional practice, can qualify.

PURPOSE OF THE CERTIFICATE

FES's objective is to recognize, encourage, and support those individual engineers who participate regularly in continued professional development. FES strongly believes that all professional engineers must continue to maintain and update their professional as well as technical skills. The certificate is not intended as a means of honoring engineers for acts of charity, long and faithful service, or outstanding accomplishments in the field of engineering. It will be presented to those engineers who earn it by meeting minimum standards.

ELIGIBILITY

Application for the certificate is voluntary. All engineers registered either as professional engineers, engineering interns, or engineers-in-training of the State of Florida, without regard to citizenship or Florida Engineering Society membership become eligible on the effective date of this program, providing they have been so registered for a total of three years. Registration in another state may be substituted for any portion of the aforementioned three years, providing the applicant holds a valid Florida registration at the end of the period for which the application is being submitted.

MINIMUM STANDARDS

Engineers may earn the certificate every three years. Credits for the certificate are earned through professional experience, professional/technical organization activities, and education/developmental activities. Accumulation of 150 points is required during the three-year period for which a certificate is being sought.

A maximum of 75 and a minimum of 25 points are to be acquired during the period being applied for through professional experience and professional/technical organization activities. Each full year of professional-level engineering experience during the three-year period for which the certificate is being sought earns twelve points. Each full year of membership in a professional or technical organization earns five points; each position held by the applicant in the activities of these organizations earn points in accordance with the provisions of the following table.

For an individual to be allowed credit for membership in a professional/technical organization under Category II A, the individual must hold an individual membership in his or her name in the organization being cited. However, individual membership is not required when applying for credit under Category II B. Professional/technical organizations will be honored under this portion of the certificate upon approval of the committee. However, when membership in an organization cannot be attained without membership in a second organization, points will only be allowed for membership in one of these organizations, irrespective of separate dues payment.

A minimum of 75 points are required in educational/developmental activities. A degree of flexibility has been designed into this section of the standards to allow for individualistic approaches to educational/developmental activities. Educational programs taken for academic credit, as well as the non-traditional study programs (i.e., conferences, short courses, seminars, workshops and special training programs), which bear accreditation in the form of continuing educational units (CEU's) earn points on the basis of the point per contact hour of participation by the applicant.

Other educational programs such as academic programs not taken for credit, non-traditional study programs not covered above, and approved in-house training programs earn points on the basis of one-half point per contact hour of participation by the applicant. Course outlines should be submitted for consideration by the committee for non-traditional and in-house programs in this category. Com-

Table of Minimum Standards

	Category	3 Year		Notes
		Minimum Required	Maximum Allowed	
1	Experience	0	36 Points	12 points/year for engineering, engineering management experience during application period
11	Professional/Technical Organization Activities			
	A. Membership in Professional/Technical Organizations	0	0	5 points/year for membership in an approved professional or technical organization
	B. Activities			
	1. Local Activities			
	a. Officer or Director	0	0	2 points/year of service
	b. Committee Member	0	0	1 point/year of service
	2. State or Regional			
	a. Officer or Director	0	0	4 points/year of service
	b. Committee Member	0	0	2 points/year of service
	3. National Activity			
	a. Officer or Director	0	0	5 points/year of service
	b. Committee Member	0	0	3 points/year of service
111	Educational/Developmental Activities			
	A. Educational Programs taken for credit and CEU accredited programs attended	0	0	1 point/contact hour of participation by the applicant.
	B Other Educational Programs (all approved educational programs not covered in A. above)	0	0	0.5 point/contact hour of participation by the applicant.
	C. Self-Study	0	0	2 points/unit of self-study as defined in the minimum standards.
	D. Teaching Educational Programs	0	50 points	1 point/hour of time actually spent in the classroom teaching. Teaching as a part of regular job duties is not eligible.

Table of Minimum Standards (cont.)

| | 3 Year | | |
Category	Minimum Required	Maximum Allowed	Notes
E. Papers for presentation or publication, Publications and Books	0	50 points	10 points/paper, publication or book chapter. Writing as a part of regular job duties is not eligible.
F. Other Meritorious Learning Experience	0	50 points	4 points/calendar day of attendance at annual meetings of state or national professional or technical organizations, not otherwise falling into any of the above categories. Other points may be approved under this category for activities not fitting into other items of this category.

mittee approval of such programs should be obtained prior to submission for credit.

In addition to the above, two points may be earned as a unit of self-study by submission of the following as evidence of a diligent effort. (1) Submission of a one to two page summary of each 16 pages of a technical engineering text giving important points, theory, and related significant equations of formula. (2) Submission of a list of errors (all errors technical or typographical) found on the above mentioned pages and suggested corrections of said errors. (3) Submission of four solved problems from the text's list of problems applicable to the pages of (1) above. Solutions must include checks on the answers and the work should be neatly done.

The documentation listed above is to be attached to the standard application form and the applicant must identify the text and pages being cited for consideration. It is also intended that points be allowed only for completion of the above when the applicant conducts this self-study on his or her own, not as a part of the normal work effort.

A wide variety of topical subjects may be applied to the aforementioned educational programs. However, said subjects should generally enhance the applicant's professional/technical ability to function as a professional engineer and/or serve to achieve the individual's vocational objectives in the general fields of engineering or engineering management. Applicants should keep the basic concepts of the certificate in mind when applying for educational credit. Other points, under the general heading of Educational/Developmental Activities, are earned in accordance with the provisions of the table.

It is recommended that not more than 50 percent of the required points be acquired in any one-year period. A single function or activity may be applied for credit in only one category. Under no circumstances will duplicate credit be allowed.

PROCEDURE

The Professional Development Committee will study and make recommendations to the Florida Engineering Society Board of Directors (or the Executive Committee) with respect to the issuance of Certificates of Continued Professional Dvelopment. The committee will maintain all records, review all applications and make recommendations to the board (or the Executive Committee) as to the acceptability of information submitted by applicants for consideration. Upon approval by the board (or the Executive Committee), a suitable citation will be prepared and presented to the applicant as expeditiously as practicable.

8.9

Continuing Professional Competence—Summary of a Position Paper by the National Council of Engineering Examiners

The issue of Continuing Professional Competence has been under study by a special committee of NCEE for several years. At the Annual Meeting in Norfolk on August 7, 1979, the Council adopted the "Position Paper" submitted by the committee and directed the President to make its position a matter of public record.

A summary of this position is as follows:

NCEE commends and endorses the efforts of the professional/technical societies, engineering schools, and industry in the areas of continuing education and competence for engineers and land surveyors. This voluntary approach is flexible and, while maintaining a cost-effective balance, precludes the need for mandated programs.

NCEE is concerned about the inequity of mandated continuing education for registered engineers, since only about 30 percent of practicing engineers choose to become registered.

NCEE endorses comity between states and encourages each state to evaluate carefully any additional requirements for registration that would tend to interfere with reciprocal licensing between states.

Therefore, NCEE can find no compelling justification for any state to adopt a continuing professional competency requirement as a prerequisite to renewal of a registered professional engineer's or land surveyor's license.

The entire position paper as presented by the committee has been published in the 1979 NCEE Proceedings of the 58th Annual Meeting. The NCEE Continuing Professional Competence Committee will maintain surveillance of the voluntary programs with the objective of updating its position at the next Annual Meeting (1980).

Selected Additional Readings

Brown, R. D., "Continuing Education: What the New Law Is All About," *The Exponent*, August 1977, pp 6–8.

Constance, J. D., "Glossary for Professional Engineer Candidates," *Chemical Engineering*, July 1, 1968, pp. 94–98.

Ellis, P. G., "A Case for Professional Registration," *Engineering Education*, December 1969, pp. 329–330.

Fine, M. S., "Qualifications for Continued Practice: Panacea or Pandora's Box," *Ethics, Professionalism and Maintaining Competence*, American Society of Civil Engineers, March 1977, pp. 248–252.

Fine, M. S., "Qualifications for Continued Engineering Practice: What Are the State Legislatures Asking?" *Professional Engineer*, June 1977, pp. 38–40.

French, D. E., "Iowa P. E.'s Respond to First State Law on Continuing Education for License Renewal," *Professional Engineer*, December 1978, pp. 28–29.

Furman, T. deS., "Specialty Registration for Engineers," *Civil Engineering*, December 1972, pp. 32–34.

Kimmell, P. M., "The Case for National Registration," *Professional Engineer*, March 1972, pp. 45–47.

Kopelman, R., "Career Hurdles and How to Clear Them," *IEEE Spectrum*, February 1977, pp. 66–69.

Lunch, M. F., "Continuing Education and the Law," *Consulting Engineer*, December 1979, pp. 52–55.

Lunch, M. F., "The Industry Exemption and the Drive for Universal Registration," *Professional Engineer*, November 1974, pp. 35–40.

Middleton, W. W., "Mandatory Registration for Industry Engineers is Unreal, Improbable," *Professional Engineer*, March 1976, pp 43–46.

Newman, P. E., "Reciprocal Registration," *Consulting Engineer*, September 1975, pp. 49–51.

Patton, Carole, "The Challenge of Keeping Current," *IEEE Spectrum*, August 1979, pp. 53–59.

Shimberg, B. J., "Should Engineers Climb Aboard the Mandatory C. E. Bandwagon?" *Professional Engineer*, December 1978, pp. 26–27.

Stivers, T. E., "Professional Licensing: The Ancient Rite of Protecting the Public," *Consulting Engineer*, September 1975, pp. 41–44.

Tewkesbury, D. L., "Periodic Registration Review—Impractical and Unjust," *Ethics, Professionalism and Maintaining Competence*, American Society of Civil Engineers, March 1977, pp. 290–303.

Tribus, M., "The Challenge of Continuing Education," *Engineering Education*, May 1977, pp. 744–750.

Tynes, R. A., "Engineering Registration—Today and Tomorrow," *Engineering Issues*, October 1979, pp. 457–463.

About the editors

JAMES H. SCHAUB is Professor and Chairman of the Department of Civil Engineering and Affiliate Professor of History at the University of Florida in Gainesville. A leading member of several professional societies, he has received ASCE's Engineer of the Year Award (Florida section) and an Out-Standing Service Award from the Florida Engineering Society. The author of numerous articles and co-editor of *Engineering and Humanities* (Wiley, 1982), Dr. Schaub received his Ph.D. in civil engineering from Purdue University and is a licensed professional engineer.

KARL PAVLOVIC is Visiting Assistant Professor at the Center for Applied Philosophy and Ethics in the Professions at the University of Florida. A member of the American Philosophical Association and the American Association for the Advancement of Science, he received a Fulbright-Hayes Fellowship in 1976-78 where he studied in Germany and Lilly Post-Doctoral Award in 1978-79. Dr. Pavlovic earned his Ph.D. in philosophy from Purdue University.